Frontiers in Particle Physics
Cargèse 1994

NATO ASI Series

Advanced Science Institutes Series

A series presenting the results of activities sponsored by the NATO Science Committee, which aims at the dissemination of advanced scientific and technological knowledge, with a view to strengthening links between scientific communities.

The series is published by an international board of publishers in conjunction with the NATO Scientific Affairs Division

A	**Life Sciences**	Plenum Publishing Corporation
B	**Physics**	New York and London
C	**Mathematical**	Kluwer Academic Publishers
	and Physical Sciences	Dordrecht, Boston, and London
D	**Behavioral and Social Sciences**	
E	**Applied Sciences**	
F	**Computer and Systems Sciences**	Springer-Verlag
G	**Ecological Sciences**	Berlin, Heidelberg, New York, London,
H	**Cell Biology**	Paris, Tokyo, Hong Kong, and Barcelona
I	**Global Environmental Change**	

PARTNERSHIP SUB-SERIES

1. Disarmament Technologies	Kluwer Academic Publishers
2. Environment	Springer-Verlag
3. High Technology	Kluwer Academic Publishers
4. Science and Technology Policy	Kluwer Academic Publishers
5. Computer Networking	Kluwer Academic Publishers

The Partnership Sub-Series incorporates activities undertaken in collaboration with NATO's Cooperation Partners, the countries of the CIS and Central and Eastern Europe, in Priority Areas of concern to those countries.

Recent Volumes in this Series:

Volume 348—Physics with Multiply Charged Ions
edited by Dieter Liesen

Volume 349—Formation and Interactions of Topological Defects
edited by Anne-Christine Davis and Robert Brandenberger

Volume 350—Frontiers in Particle Physics: *Cargèse 1994*
edited by Maurice Lévy, Jean Iliopoulos, Raymond Gastmans, and
Jean-Marc Gérard

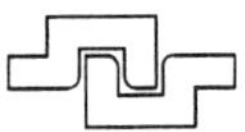

Series B: Physics

Frontiers in Particle Physics

Cargèse 1994

Edited by

Maurice Lévy

Université Pierre et Marie Curie
Paris, France

Jean Iliopoulos

Ecole Normale Supérieure
Paris, France

Raymond Gastmans

Katholieke Universiteit Leuven
Leuven, België

and

Jean-Marc Gérard

Université Catholique de Louvain
Louvain-la-Neuve, Belgique

Plenum Press
New York and London
Published in cooperation with NATO Scientific Affairs Division

Proceedings of a NATO Advanced Study Institute on
Frontiers in Particle Physics,
held August 1–13, 1994,
in Cargèse, France

NATO-PCO-DATA BASE

The electronic index to the NATO ASI Series provides full bibliographical references (with keywords and/or abstracts) to about 50,000 contributions from international scientists published in all sections of the NATO ASI Series. Access to the NATO-PCO-DATA BASE is possible in two ways:

—via online FILE 128 (NATO-PCO-DATA BASE) hosted by ESRIN, Via Galileo Galilei, I-00044 Frascati, Italy

—via CD-ROM "NATO Science and Technology Disk" with user-friendly retrieval software in English, French, and German (©WTV GmbH and DATAWARE Technologies, Inc. 1989). The CD-ROM contains the AGARD Aerospace Database.

The CD-ROM can be ordered through any member of the Board of Publishers or through NATO-PCO, Overijse, Belgium.

```
              Library of Congress Cataloging-in-Publication Data


Frontiers in particle physics : Cargèse 1994 / edited by Maurice Lévy
    ... [et al.].
        p.   cm. -- (NATO ASI series. Series B, Physics ; v. 350)
      "Published in cooperation with NATO Scientific Affairs Division".
      "Proceedings of a NATO Advanced Study Institute on Frontiers in
    Particle Physics, held August 1-13. 1994, in Cargèse, France"--T.p.
    verso.
      Includes bibliographical references and index.
      ISBN 0-306-45129-8
      1. Particles (Nuclear physics)--Congresses.  2. Astrophysics-
    -Congresses.  3. Cosmology--Congresses.   I. Lévy, Maurice, 1922-
    . II. North Atlantic Treaty Organization.  Scientific Affairs
    Division.  III. NATO Advanced Study Institute on Frontiers in
    Particle Physics (1994 : Cargèse, France)  IV. Series.
    QC793.F77  1995
    539.7'2--dc20                                          95-45656
                                                               CIP
```

ISBN 0-306-45129-8

© 1995 Plenum Press, New York
A Division of Plenum Publishing Corporation
233 Spring Street, New York, N. Y. 10013

10 9 8 7 6 5 4 3 2 1

Printed in the United States of America

PREVIOUS CARGÈSE SYMPOSIA PUBLISHED IN THE
NATO ASI SERIES B: SERIES

PREFACE

The 1994 Cargèse Summer Institute on Frontiers in Particle Physics was organized by the Université Pierre et Marie Curie, Paris (M. Lévy), the Ecole Normale Supérieure, Paris (J. Iliopoulos), the Katholieke Universiteit Leuven (R. Gastmans), and the Université Catholique de Louvain (J.-M. Gérard), which, since 1975, have joined their efforts and worked in common. It was the eleventh Summer Institute on High Energy Physics organized jointly at Cargèse by three of these universities.

Several new frontiers in particle physics were thoroughly discussed at this school. In particular, the new energy range in deep-inelastic electron-proton scattering is being explored by HERA (DESY, Hamburg), and Professor A. De Roeck described the first results from the H1 and Zeus experiments, while Professors A.H. Mueller and Z. Kunszt discussed their relevance from the theoretical point of view. Also, the satellite experiments offer new possibilities for exploring the links between astrophysics, cosmology, and particle physics. A critical analysis of these experiments was performed by Professor B. Sadoulet, and Professor M. Spiro made the connection with the results from earth-based neutrino experiments. Finally, much attention was given to the latest results from the TEVATRON (Fermilab, USA), showing further evidence for the long awaited top quark. Professor A. Tollestrup gave a detailed presentation of these results and discussed their importance for the Standard Model.

Also, the ever increasing precision reached by the electro-weak LEP experiments (CERN, Geneva) required a new update on the status of the Standard Model, a task which Professor M. Martinez took upon himself.

On the more theoretical side, it was felt that a series of pedagogical lectures on conformal field theories was required, because of the many important developments in that domain over the last couple of years. They were given by Professors Vl. Dotsenko and L. Baulieu.

Finally, Professor P. Darriulat gave an overview of the accelerator experiments for the next century, the experimental frontiers for the future.

We owe many thanks to all those who have made this Summer Institute possible!

Special thanks are due to the Scientific Committee of NATO and its President for a generous grant. We are also very grateful for the financial contribution given by the C.N.R.S. and the Institut National de Physique Nucléaire et de Physique des Particules (IN2P3).

We also want to thank Ms. M.-F. Hanseler and Ms. S. Poilbois for their efficient organizational help, Mr. and Ms. Ariano and Ms. Cassegrain for their kind assistance in all material matters of the school, and, last but not least, the people from Cargèse for their hospitality.

Mostly, however, we would like to thank all the lecturers and participants: their commitment to the school was the real basis for its success.

M. Lévy   R. Gastmans

J. Iliopoulos   J.-M. Gérard

CONTENTS

Frontiers in Particle Physics
Cargèse 1994

PHYSICS RESULTS FROM THE FIRST ELECTRON-PROTON COLLIDER HERA

Albert De Roeck

Deutsches Elektronen-Synchrotron DESY, Hamburg

1 Introduction

On the 31st of May 1992 the first electron-proton (ep) collisions were observed in the H1 and ZEUS experiments at the newly commissioned high energy collider HERA, in Hamburg, Germany. HERA is the first electron-proton collider in the world: 26.7 GeV electrons collide on 820 GeV protons, yielding an ep centre of mass system (CMS) energy of 296 GeV. Already the results from the first data collected by the experiments have given important new information on the structure of the proton, on interactions of a high energetic photons with matter and on searches for exotic particles. These lectures give a summary of the physics results obtained by the H1 and ZEUS experiments using the data collected in 1992 and 1993.

Electron-proton, or more general lepton-hadron experiments, have been playing a major rôle in our understanding of the structure of matter for the last 30 years. At the end of the sixties experiments with electron beams on proton targets performed at the Stanford Linear Accelerator, revealed that the proton had an internal structure.[1] It was suggested that the proton consists of pointlike objects, called partons.[2] These partons were subsequently identified with quarks which until then were only mathematical objects for the fundamental representation of the SU(3) symmetry group, used to explain the observed multiplets in hadron spectroscopy.[3] This process of probing the internal structure of the proton with lepton beams, termed deep inelastic scattering (DIS), has made a substantial contribution to the development of modern high energy physics over the last three decades. In particular, in addition to the discovery of the partonic content of hadrons, it was established that the quarks carry only about 1/2 of the momentum of the proton (the other half later assumed to be carried by gluons), have a spin of 1/2 and carry fractional electric charge. It was also found that the evolution of the parton momentum distributions in nucleons could be described by perturbative QCD. Furthermore weak neutral currents were discovered. In all, lepton-hadron interactions have proven to be an important testing ground for QCD and the electroweak theory.

Frontiers in Particle Physics: Cargèse 1994
Edited by M. Lévy *et al.*, Plenum Press, New York, 1995

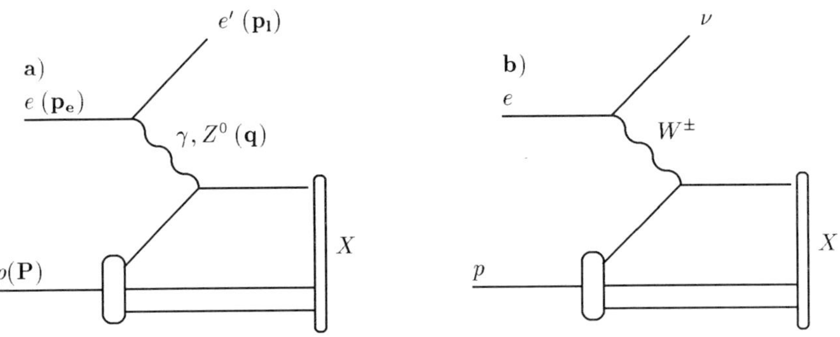

Figure 1: Deep inelastic scattering at HERA for (**a**) a neutral current process, (**b**) a charged current process.

The basic processes for deep inelastic scattering at HERA are depicted in Fig. 1, using the language of Feynman diagrams and the Quark Parton Model (QPM). The incoming electron an proton interact via the exchange of a γ, Z^0 or W boson with a quark in the proton. This quark – often referred to as *current quark* – is being kicked out of the proton, leaving behind a *proton remnant*. Both the current quark and the proton remnant hadronize into a hadronic final state, X in Fig. 1. The proton remnant essentially continues into the proton direction, and will to some extent remain undetected due to the beampipe holes which are inevitable for detectors at a collider. Since both the struck quark and remnant are coloured, one expects a colour connection between these objects. We expect therefore that the region between struck quark and remnant will be filled with particles.

Processes where a photon or Z^0 boson is exchanged (Fig. 1a are termed *neutral current processes*, while processes where a W boson is exchanged (Fig. 1b are termed *charged current processes*. For the latter the outgoing lepton is a neutrino ν. At fixed CMS energy, $\sqrt{s}$, the kinematics of the inclusive ep scattering process, $ep \to lX; l = e, \nu$, is determined by two independent variables, conventionally chosen to be two of x, y and Q^2. These kinematical variables are defined as follows:

$$Q^2 = -q^2 = -(p_e - p_l)^2, x = \frac{Q^2}{2P.q}, y = \frac{P.q}{P.p_e}, W^2 = (q + P)^2 = Q^2 \frac{1-x}{x} + m_p{}^2, \quad (1)$$

where Q^2 is the (minus) four-momentum transfer squared, x the Bjorken-x, y the fraction of the electron energy transfered to the proton in the proton rest system, and W^2 the hadronic invariant mass squared of the system X. In the naive quark parton model, i.e. the parton model with no QCD effects, the x variable measures the fraction of the proton momentum carried by the struck quark. In these definitions p_e, p_l and P denote the four-momenta of the incoming and scattered lepton and the incoming proton respectively, as indicated in Fig. 1. At HERA the centre of mass energy squared $s = 4E_e E_p = 87\,600$ GeV^2, with E_e and E_p the energy of the incoming electron and proton respectively. The deep inelastic regime is generally taken to be that part of the phase space where Q^2 and W^2 are larger than a few GeV^2. As Q approaches the mass of the heavy W and Z^0 bosons, the cross section for W and Z^0 exchange become competitive with the γ-exchange. For small Q, well below these masses, the photon exchange diagram dominates the cross section by orders of magnitude. As an illustration, the analyses of the H1 and ZEUS experiments contain at present a few ten thousand neutral current events and only a few tens of charged current events.

The resolution to resolve the internal structure in the proton is determined by the four-momentum transfer $Q = \sqrt{-q^2}$ between the lepton and the hadron, where q is the four-momentum vector of the exchanged boson in Fig. 1. The smallest distance a virtual photon can resolve is proportional to $1/Q$. The maximum value for the four-momentum transfer Q_{max} is given by the centre of mass energy of the collision $\sqrt{s}$. For example for the "first DIS experiments" at SLAC, the incident electron energy was about 20 GeV, which yields a centre of mass energy (and Q_{max}) of about 6 GeV. Thus distances of the order of 10^{-14} cm, i.e. ten times smaller than the radius of the proton, could be resolved and revealed a new partonic substructure of matter. The obvious question rises: what will happen if still smaller distances are probed? Will a new underlying structure in the partons appear? In order to increase the resolving power experiments with muon and neutrino beams of several hundreds of GeV were used to probe matter, resolving distances of 10^{-15} cm. So far no new substructure has been detected and the quarks and leptons are still considered to be pointlike particles, but deep inelastic scattering has contributed to answering many important questions on the fundamental nature of matter as mentioned above. HERA is the new frontier for DIS, with a gain in Q^2 of two orders of magnitude, such that a spatial resolution of 10^{-16} cm can be reached. As will be explained in chapter 5, another important gain is to reach very small values in x. Indeed since $Q^2 = xys$, x values down to a few times 10^{-5} in the deep inelastic regime can be accessed at HERA, two orders of magnitude smaller than previously achieved.

The kinematic variables in neutral current ep scattering are traditionally determined from the angle θ_e, and the energy, E'_e, of the scattered lepton through the relations:

$$Q^2 = 4E_e E'_e \cos^2(\frac{\theta_e}{2}), \quad y = 1 - \frac{E'_e}{E_e} \sin^2(\frac{\theta_e}{2}) \tag{2}$$

and x can then be determined as $x = Q^2/(sy)$. At HERA we adopt the convention that all polar angles are measured relative to the proton beam direction, termed forward direction in the following. The H1 and ZEUS experiments at HERA are designed to measure both the scattered electron and the hadrons produced from the struck quark and the proton remnant, thus the collision kinematics can be determined from scattered electron, the hadrons or a mixture of both. A variable which has turned out to be particularly useful is y calculated by the Jacquet Blondel[4] method y_{JB}:

$$y_{JB} = \frac{\Sigma_h (E - P_z)_h}{2E_e} \tag{3}$$

where the sum includes all detected hadrons h, which have an energy E and longitudinal momentum component P_z. More methods to calculate the event kinematics will be discussed in chapter 5 on the measurement of the proton structure.

So far we discussed the region where Q^2 and W^2 are larger than a few GeV^2, i.e. the region where the exchanged photon is highly virtual and the hadronic invariant mass is significantly larger than the proton mass. We will start the physics discussion in these lectures however with interactions for which $Q^2 \simeq 0$, i.e. where the exchanged photon is almost on mass shell. These processes are usually termed (almost) real photoproduction processes. For photoproduction (γp) interactions, HERA allows to study collisions with a centre of mass energy approximately one order of magnitude larger than presently achieved in fixed target γp experiments. Due to this increase in CMS energy hard scattering in photon-proton collisions is expected to become clearly visible. This will lead to the production of jets, which can be used for detailed QCD tests and to derive information on the partonic structure of the photon. Additionally, heavy quark flavours (charm and bottom) are expected to be copiously produced in photoproduction interactions, leading to an additional field of interest at HERA.

Next, deep inelastic scattering interactions will be discussed. These will be used to explore the proton structure for the first time at x values down to $\sim 10^{-4}$, and large Q^2 values up to

6000 GeV^2. HERA enters at low x a new kinematical region where it has been speculated that new physics may be observed.[5] Further, in the study of the hadronic final state in DIS, a class of events was observed which have a large rapidity gap between the current jet and the proton remnant, and possibly result of a diffractive-like mechanism. The first electroweak results from HERA have emerged with the measurement of the charged and neutral current cross sections in the high Q^2 range. Finally, a search for exotic particles was performed. HERA is particularly suited for production of s-channel resonances from the fusion of two incoming partons (electron with a quark or gluon from the proton for leptoquarks and leptogluons respectively) or from the electron with an exchanged boson (e.g. to produce excited leptons), since the full centre of mass energy of the collision can be used to produce these states.

Before discussing the physics results, we will first briefly introduce the HERA collider and the experiments. Both experiments have shown similar results on many of the physics topics discussed, but the data are mostly shown only once.

2   The HERA Collider

About 15 years ago several proposals were discussed for electron-proton colliders at DESY, CERN and Fermilab. This appeared to be the obvious possibility for extending the centre of mass energy, and hence the maximum momentum transfer squared Q^2, from $s = 2 \cdot M_p \cdot E_l \leq 1000$ GeV^2 for the ongoing fixed target lepton-proton experiments to $s = 4 \cdot E_e \cdot E_p \leq 10^5$ GeV^2 at HERA. In July 1981 the construction of the Hadron-Elektron-Ring-Anlage (HERA) was proposed to collide 10-30 GeV electrons or positrons off 300-820 GeV protons with a luminosity above 10^{31} cm^{-2}s^{-1}. Ten years after, in October 1991, the first interactions of 12 GeV electrons and 480 GeV protons were observed at DESY. In spring 1992 the first ep collisions at 26.7 x 820 GeV were registered by the detectors. That year both experiments accumulated about 25 nb^{-1} integrated luminosity. For neutral current interactions with $Q^2 > 10$ GeV^2 the cross section is about 100 nb, thus about 2500 events are expected to be produced for the accumulated integrated luminosity. In 1993 the experiments accumulated a total of 500 nb^{-1}. For H1 about 150 nb^{-1} of the 1993 data are of limited use due to a failure of the main magnet of the detector. This year the experiments accumulated as much as 4 pb^{-1}. We still expect HERA to increase the luminosity in the next years, leading to data samples of 20-30 pb^{-1} in 1995 and 1996. Hence, HERA's physics potential is still growing each year.

HERA is an accelerator with both warm and superconducting magnets and cavities. The *proton ring* consists of 104 cells of superconducting magnets (4 dipoles, 2 quadrupoles and correction magnets, 47 m long each cell) for bend and focus. One of the straight sections contains warm cavities, a 52 MHz system accelerating bunches of 1.15 m length at an RF voltage of 280 kV and a 208 MHz cavity system leading to a bunch length of 0.27 m at 820 GeV proton energy. The small extension of the interaction region over a few tens of centimeters has important implications for triggering and analyzing ep interactions. The *electron ring* consists of 416 warm magnet modules (1 dipole, 1 quadrupole, 2 sextupoles and correction dipoles, 12 m long each module). Acceleration is achieved by 82 warm cavities (from PETRA) which are designed to run at about 27 GeV with 60 mA current. The HERA design energy can be reached utilizing the 16 superconducting cavities providing a gradient of about 5 MV/m which gives about 3 GeV more electron energy at the same currents. Synchrotron radiation leads to transverse electron polarization via the Sokolov-Ternov effect.[6] This is a very important feature for the HERMES experiment (see below) and for electroweak studies by the H1 and ZEUS experiments. The achieved level of polarization with 26.7 GeV electron beams at HERA is as large as 60%. Longitudinal polarization of the electrons provides more interesting physics. Spin rotators have been installed and successfully used to convert the transverse polarization to longitudinal polarization at the interaction region. During the luminosity phase operation of the machine now routinely longitudinal polarizations larger than 50% are reached.

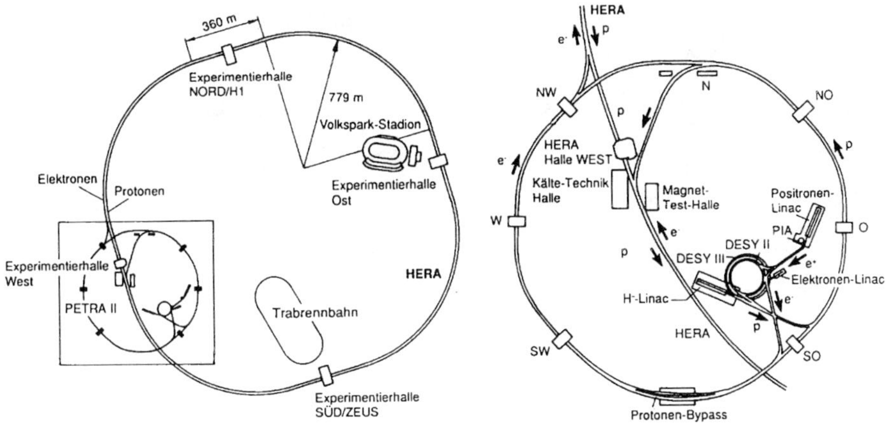

Figure 2: A schematic view of the HERA accelerator complex. The figure on the left shows the HERA ring and location of experimental halls. The figure on the right shows the pre-accelerators for protons and electrons, before injection into HERA.

In order to inject electrons and protons into the HERA ring, various pre-accelerators had to be built or/and reconstructed: three LINACs of 20, 70 and 32 m length for the acceleration of e^- (I), e^+ (II) and negative hydrogen ions (III) to energies of 220, 450 and 50 MeV, respectively; in 1986 the electron synchrotron DESY II replaced the old DESY I acting as the injector of 7 GeV electrons into the PETRA ring where the electrons are accelerated to maximum 14 GeV. A new proton synchrotron (DESY III) of 317 m diameter was constructed to reach 7.5 GeV energy prior to injection into PETRA which in turn provides 40 GeV proton injection energy. An overview of the accelerator system is given in Fig. 2.

HERA is designed to contain 210 e and p bunches. Some of those are $e(p)$ pilot bunches which pass through the detectors without being collided against $p(e)$ bunches coming the other way. These are of particular use for background and timing studies. For data taking and analysis, HERA is a very complicated environment as it combines the disadvantages of e^+e^- machines (large synchrotron radiation and electroweak cross sections) and proton rings (large backgrounds due to protons scattering from residual gas and beampipe wall). For the 1992 analysis the deep inelastic signal to beam background ratio is of the order of 10^{-4}. The bunch crossing frequency is 10.4 MHz, i.e. bunches cross every 96 ns. Table 1 compares some of the relevant characteristics for the years 1992, 1993 and 1994 with the design values.

During 1994 HERA it was realized that longer beam lifetime, and hence better effective luminosity could be reached by changing from e^-p to e^+p collisions. The electron current was found to be limited, probably because of disturbing interference due to too many remaining positive ions in the ring. Since positrons are much less sensitive to positive ions, larger currents, of about 55% of the design value could be reached. The luminosity reached accordingly about $5 \cdot 10^{30}$ cm^{-2} s^{-1}. The result of the change over is shown in Fig. 3, where around day 200 the polarity of the magnets in the electron ring and pre-accelerators were changed. After this short period one sees that the slope of the produced luminosity is much larger than for the first part of the year.

3   The H1 and ZEUS Detectors

Two experiments, H1[7] and ZEUS,[8] were ready to record ep collisions at HERA in summer 1992. Both experiments have tracker, calorimeter and muon detectors and a small angle

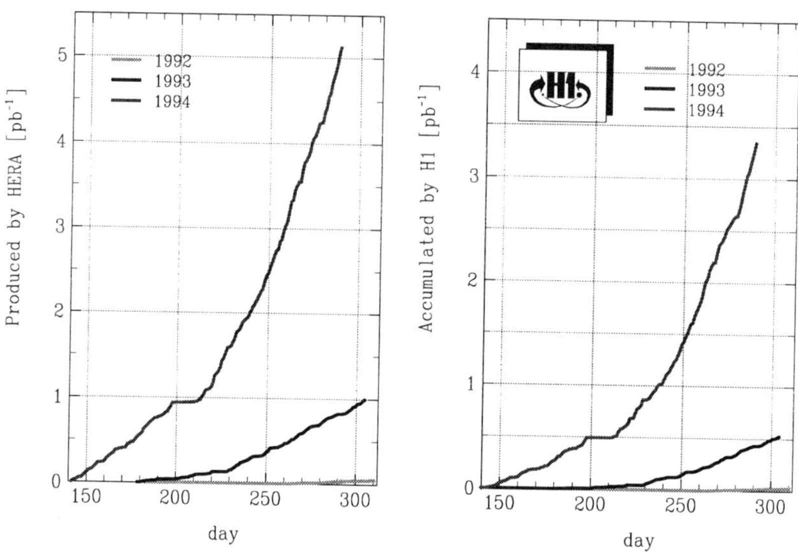

Figure 3: The integrated luminosity produced by HERA (left) and used by the experiments for physics (right) for the years 1992, 1993 and 1994. This result is for the H1 experiment; the ZEUS result is very similar.

parameter	1992	1993	1994	design value
E_p	820 GeV	820 GeV	820 GeV	820 GeV
E_e	26.7 GeV	26.7 GeV	27.5 GeV	30 GeV
nr of bunches	9	84	153	210
p current	2.0 mA	14 mA	54 mA	163 mA
e current	2.5 mA	16 mA	32 mA	58 mA
$\int_{year}$ lumi	50 nb^{-1}	1 pb^{-1}	6 pb^{-1}	50 pb^{-1}

Table 1: Comparison of some HERA parameters reached over the last 3 years of running in collider mode, with their design values. The number of bunches only counts the ep colliding bunches, the integrated luminosity is the one delivered by HERA.

electron tagger system. The experiments are large 4π solid angle detectors, apart from losses in the beampipe. The calorimeter is an important component in the design of both detectors. The H1 collaboration has opted for liquid argon calorimetry, which is well tailored to identify and measure electrons. The large granularity of this calorimeter is exploited to compensate the intrinsically different e/π response by software weighting algorithms. In the backward region H1 has an electromagnetic calorimeter which is made of lead/scintillator stacks. ZEUS has chosen to emphasize on the quality of the hadron measurement by constructing an inherently

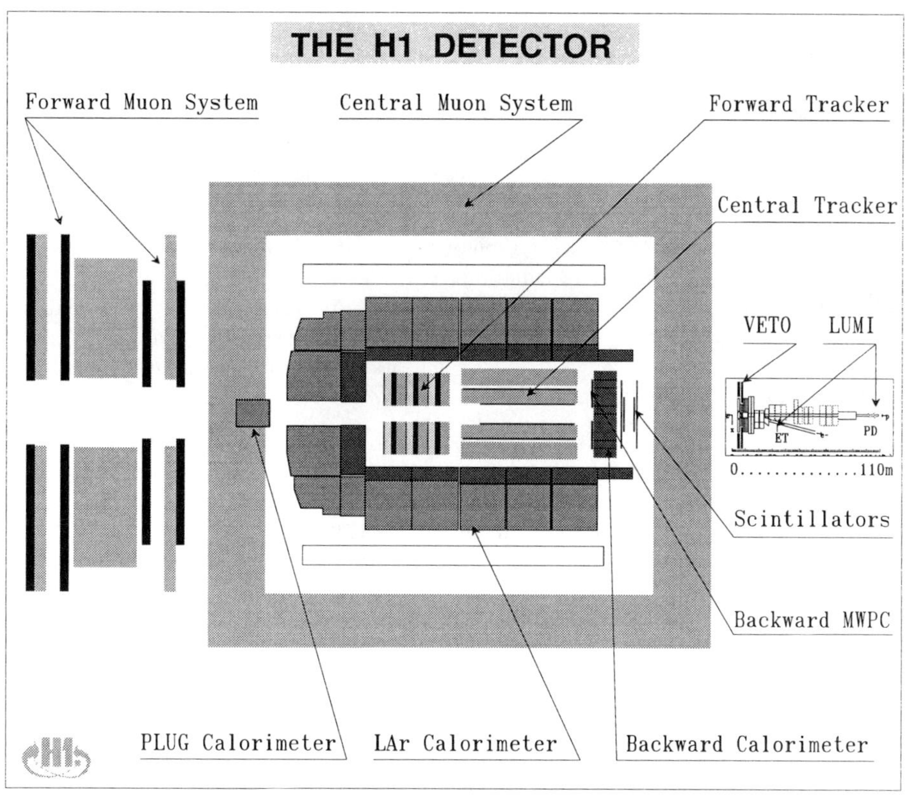

Figure 4: Schematic view of the H1 detector. Note that the luminosity detector, downstream in the electron beam is not on scale.

compensating uranium/scintillator calorimeter.

The H1 detector is schematically shown in Fig. 4. A detailed description of the detector and its performance can be found in.[7] Charged particle tracks are measured in a central tracker (CT), forward tracker (FT) and a backward proportional chamber (BPC). The central tracker consists of two large jet drift chamber modules, two z drift chambers and two multiwire proportional chambers for triggering. Its angular acceptance is $15^0 - 165^0$. The forward tracking detector accepts tracks between 7^0 and 25^0. It consists of three modules of drift and multiwire proportional chambers. The BPC has 4 wire planes and an angular acceptance of $155^0 - 175^0$. A superconducting coil provides a uniform magnetic field of 1.15 T in the tracking region which allows the determination of charged particle momenta. The vertex position of an interaction is determined on an event by event basis from tracks reconstructed in the CT and FT, originating from the interaction region. The presently achieved resolutions for charged track parameters are $\sigma_{r\varphi} = 170$ μm and $\sigma_z = 2$ mm for the CT and $\sigma_{r\varphi} = 170$ μm and $\sigma_{xy} = 210$ μm for the FT.

The tracking detectors are surrounded by calorimeters. The Liquid Argon (LAr) calorimeter[9] consists of an electromagnetic section with lead absorber and a hadronic section with stainless steel absorber. The total depth of the electromagnetic part varies between 20 and 30 radiation lengths whereas the total depth of both calorimeters varies between 4.5 and 8 interaction lengths. The LAr calorimeter covers the angular range between 4° and 153°.

Test beam measurements of LAr calorimeter modules have demonstrated energy resolutions of about $0.12/\sqrt{E/\ \text{GeV}} \oplus 0.01$ for electrons and about $0.5/\sqrt{E/\ \text{GeV}} \oplus 0.02$ for charged pions.[7,9,10] The electromagnetic energy scale is verified to a 3% accuracy in the H1 detector by comparing the measured track momentum of electrons and positrons with the corresponding energy deposition in the calorimetric cells. The absolute scale of the hadronic energy is presently known to 6% as determined from studies of the p_T balance for deep inelastic scattering events.

The Backward Electromagnetic calorimeter (BEMC) is made of 88 lead-scintillator sandwich stacks, each with a depth of 22 radiation lengths, corresponding to about one interaction length. It covers the angular range of $155^0 < \theta_e < 176^0$. A 1.5 cm space resolution for the reconstructed centre of gravity of an electromagnetic cluster has been achieved. The energy scale of the BEMC is known to 1.7% and its resolution is described by $\sigma(E)/E = 0.1/\sqrt{E} \oplus 0.42/E \oplus 0.03$, where E is in GeV.

The calorimeters are surrounded by an iron structure, which is instrumented with streamer tubes to act as a backing calorimeter and muon filter. The forward region is equipped with a forward muon spectrometer, consisting of a toroidal magnet and drift chambers. The total angular coverage of the H1 detector for muons is $4^o - 171^o$.

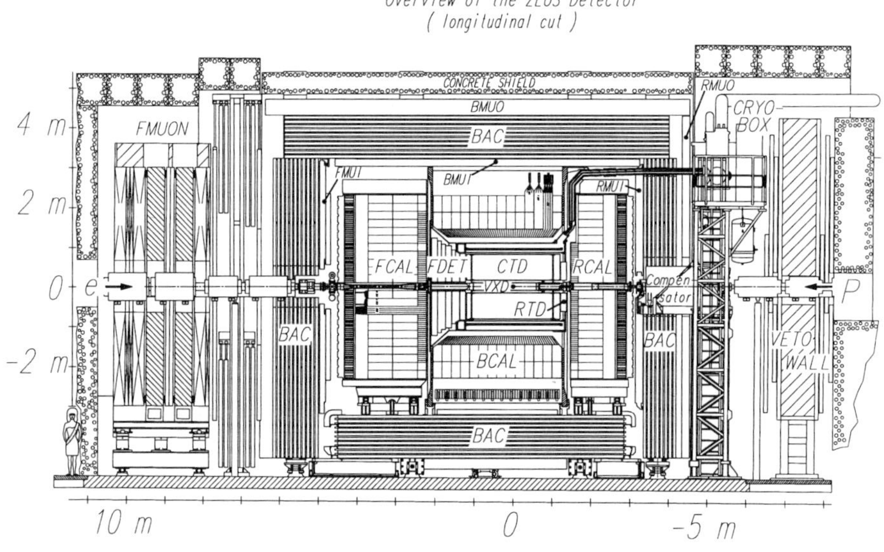

Figure 5: Schematic transversal view of the ZEUS detector. The electron and photon tagger detectors, downstream of the electron beam, are not shown.

A view of the ZEUS detector is shown in Fig. 5, together with a typical collaborator in the left down corner. Charged particles are measured by the inner tracking system consisting of a vertex detector (VXD), a central tracking detector (CTD), and a forward tracking detector (FTD). The VXD consists of 12 layers of axial sense wires. The CTD has 9 superlayers (5 axial and 4 small angle stereo), each with 8 layers of sense wires. A superconducting solenoid surrounds the inner tracking system, and produces a magnetic field of 1.43 T. Similar resolutions for charged tracks as for H1 have been achieved for the CTD.

An uranium-scintillator calorimeter (CAL) surrounds the solenoid. It is divided into a forward (FCAL), barrel (BCAL) and rear (RCAL) calorimeter. In total the calorimeter covers the full azimuthal angle and polar angle range from 2.6^0 to 176^0. The calorimeter is subdivided longitudinally into an electromagnetic part and two (one) hadronic parts in the FCAL, BCAL (RCAL), representing a total depth of 7 to 4 absorption lengths. The scintillator plates form 5 x 20 cm^2 (10 x 20 cm^2) cells in the electromagnetic section and 20 x 20 cm^2 cells in the hadronic sections of FCAL, BCAL (RCAL). From test beam results the electromagnetic energy resolution is $\sigma(E)/E = 0.18/\sqrt{E}$ and the hadronic energy resolution is $\sigma(E)/E = 0.35/\sqrt{E}$. Compensation has been checked up to 3%. A small tungsten-silicon calorimeter (BPC) positioned at the beampipe behind the RCAL tags electrons scattered with Q^2 down to 0.5 GeV^2

The iron yoke is instrumented with proportional tube chambers and LST chambers, and serves as a backing calorimeter and muon filter. For the identification and momentum measurement of muons, the yoke is magnetized to 1.6 T with copper coils. In the forward direction a spectrometer of two iron toroids and drift- and LST-chambers (FMUON) identifies muons and measures their momenta up to 100 - 150 GeV/c.

At HERA the luminosity is measured with the elastic bremsstrahlung reaction $ep \rightarrow ep\gamma$ which, according to the Bethe-Heitler[11] cross section formula, depends on the secondary energies E_e' and E_γ only. The experiments have installed luminosity monitor systems to measure both energies with an electron tagger for very small angle scattering at about -30 m downstream the electron beam and a photon detector at about -100 m. These detectors are electromagnetic calorimeters using TlCl/TlBr crystals for H1[12] and a Pb/SCSN38 scintillator sandwich for ZEUS.[13] The integrated luminosity measurement for the 1993 data was quoted to be accurate to 4.5 (3.5) % for H1 (ZEUS).[14,15]

Apart from H1 and ZEUS, two more experiments HERMES[16] and HERA-B,[17] have been approved for the HERA physics program (conditionally for HERA-B). The HERMES experiment is designed to make use of the electron beam polarization at HERA. It is planed to install a polarized target (protons, deuterons and ^{3}He) to measure the proton and neutron spin dependent structure functions. Besides the scattered electron also the final state will be detected in HERMES allowing for semi-inclusive charged hadron cross section asymmetry measurements. Data taking could start as early as 1995. HERA-B is designed to make use of the high intensity of the HERA proton beam to study b-quark production and decay in fixed target pp collisions. These decays will be used to study CP violation (i.e. the subtle disregard of physics for invariance under simultaneous particle-antiparticle and left right reversal, observed so far only in the decays of neutral kaons). The pp collisions are produced by exposing thin wires in the beampipe to the halo of the proton beam. The experiment is optimized to study the channel $B^0 \rightarrow J/\psi$, with the J/ψ meson decaying into a lepton pair, which has the advantage that the theoretical predictions are particularly clean and model independent. Other channels will be studied as well. The schedule is to have a full detector available for data taking in 1998.

4 Photoproduction

In this chapter we consider interactions for which the four-momentum transfer Q^2 is small, in all cases less than a few GeV^2, and in most cases even less than 0.01 GeV^2, depending on the detection method used. Thus the virtuality of the exchanged photon is small and these interactions are called real photon collisions or *photoproduction* interactions. The scattered electron is not detected in the central detector, but for a fraction of the events the electron is detected in the small angle electron taggers of the experiment.

4.1 Introduction: Photoproduction Processes

Real photons can interact with matter *directly* through the pointlike coupling of this gauge particle with partons of the hadrons, or via the so called *hadronic* component of the photon. Additionally the photon can split up into a quark-antiquark pair before the interaction, which does not form a bound hadronic state; this will be referred to as the *anomalous* component (see below). The hadronic component is expected to dominate the total photoproduction cross section[18] and is phenomenologically described by the Vector Dominance Model (VDM). Here the photon is pictured to couple to and fluctuate into a vector meson which has the same quantum numbers as the photon: e.g. $\rho(770), \omega(782), \phi(1020), J/\psi(3097)$. This vector meson interacts subsequently with the proton. Hence photon-hadron collisions are expected to follow largely the same phenomenology as hadron-hadron collisions. Consequently the majority of the γp are expected to be of rather soft nature, as for hadronic collisions.

In hadron-hadron scattering some fraction of the interactions exhibit hard scattering features, leading to jets with large transverse energy in the final state. These jets originate from the occasional hard scattering between constituents –quarks or gluons– from both hadrons, and are well described by QCD theory. In this hadronic picture of the γp interaction, we expect similar hard scatters to occur in photoproduction. The study of these processes is an important aspect of the HERA physics program.

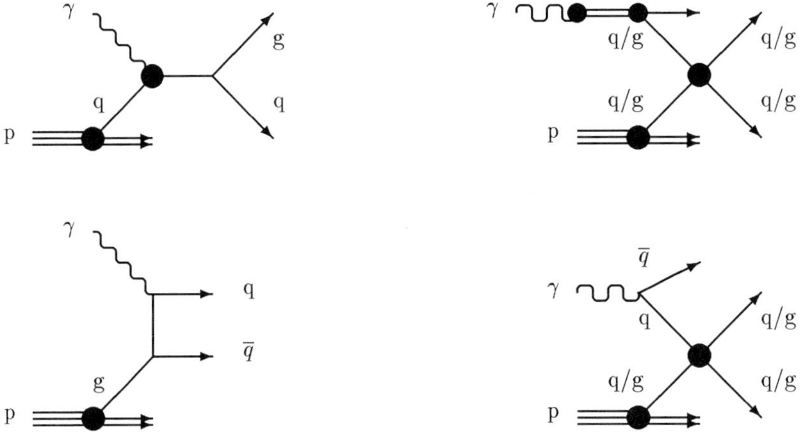

Figure 6: Hard γp processes: left part with the direct processes QCD Compton (top) and photon-gluon fusion (bottom); right part with resolved processes from the hadronic VDM component (top) and anomalous or pointlike component (bottom).

The leading order (LO) QCD diagrams leading to hard γp process are pictured in Fig. 6. The diagrams on the left result from the direct interaction of a photon with a quark from the proton and are called a *direct photon processes*. The most important direct processes are the photon-gluon fusion (PGF) and the QCD Compton (QCDC) process. The hadronic VDM component of the γp interactions contributes to hard scattering with similar diagrams as the ones in hadron-hadron interactions. Such diagrams are termed *resolved photon processes*. In fact the hadronic VDM component constitutes only part of the resolved processes. Additionally resolved processes are expected to have a contribution of the so called *anomalous component* or pointlike component, mentioned above, which results from the direct splitting of the photon in a $q\bar{q}$ pair that does not form a bound state such as a vector meson. Such diagrams are not present in hadronic collisions and are a special feature of the photon, together with the direct interactions. In case of resolved processes one can define the photon as having a structure, described by a structure function or parton densities in the photon, similar to the structure function of a hadron. These parton densities are a measure of the probability to find in the photon a parton with a certain fraction of the original photon momentum. The analysis of hard processes in γp interactions at HERA will contribute to the exploration of the structure of the photon, as will be shown below.

The interactions of real photons with matter have been studied in fixed target experiments with photon beams, and in $\gamma\gamma$ interactions at e^+e^- colliders. However, the different components of the γp interaction (direct/resolved) have not yet been unambiguously isolated by these experiments. Due to the colliding beam environment of HERA, centre of mass energies of 200 GeV can be reached for almost real γp collisions, roughly one order of magnitude larger than what has been reached so far in fixed target experiments. The large centre of mass energy should allow a clean separation of the resolved and direct components and, similar to high energy hadronic interactions, clear jet production and jet structures should become visible.

At HERA two methods are used for isolating photoproduction interactions.

- Tagged events. For this sample the small angle electron tagger is used, located at about 30 m downstream of the interaction point, to detect the scattered electron. This limits the acceptance for the virtuality of the incident photons to the range 3×10^{-8} GeV^2 $< Q^2 < 10^{-2}$ GeV^2 (4×10^{-8} GeV^2 $< Q^2 < 2 \times 10^{-2}$ GeV^2) for H1 (ZEUS). Since for this method the energy of scattered electron, E_e', is measured, the energy of the interacting photon is simply $E_\gamma = E_e - E_e'$. The fractional energy of the photon $y \simeq E_\gamma / E_e$ (see eqn. 2) as measured by the small angle electron detector is required to be in the interval $0.25 < y < 0.7$, where the acceptance can be well controlled. This range in y corresponds to the CMS energy interval of the γp system ($W_{\gamma p}$) from 150 GeV to 250 GeV, with an average of about 200 GeV. The tagging efficiency for events in this Q^2, y region amounts to about 50%.

- Untagged events. For this sample there is no requirement on the scattered electron in the tagger. The main requirement is that no electron should be detected in the main detector, which means that Q^2 is smaller than about 4 GeV^2. Generally, there is no restriction on y required. The y of the photon is not measured directly but can be deduced from y_{JB} (eqn. 3) calculated from the hadrons. Untagged event samples are roughly a factor 5 to 10 larger in statistics compared to the tagged samples

An example of a photoproduction event, observed in the H1 detector, is shown in Fig. 7. The upper left corner shows the energy deposit of the electron in the small angle electron tagger. The central detectors show the activity of the γp hadronic final state.

In the following, we will first discuss some aspects from soft processes in γp interactions. These constitute the bulk of the collisions. Then hard scattering processes are discussed and the hadronic structure of the photon is explored. A special class of events, so called diffractive events, will be studied in terms of hard scattering in the subsequent section. Finally, the production of heavy quark flavours are considered.

4.2 Soft Processes in Photoproduction

Following the classification proposed by Storrow,[19] soft or "low-p_T" collisions are those interactions where the differential p_T cross section is well described by an exponential fall off. The dynamics of soft physics is generally regarded to be of a non-perturbative nature and not to be calculable in QCD. In hadron-hadron collisions this has been an active field of research since

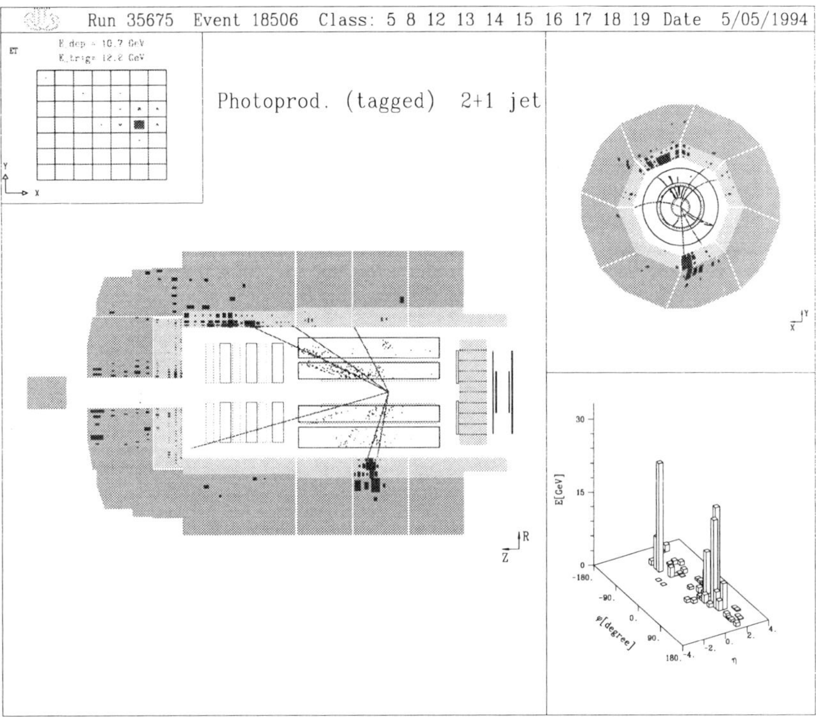

Figure 7: Photoproduction event with the electron tagged in the small angle electron tagger (upper left corner), observed in the H1 detector.

many years. Although progress has been made, details on the dynamics and rôle –if any– of partons in soft hadronic collisions have not yet been fully understood. It remains however one of the challenges in high energy physics to get a better insight in these processes.[20] In this section we will discuss general aspects of multi-particle production in dominantly soft collisions in γp interactions.

Fig. 8 shows the differential p_T cross section for charged particles in γp interactions at HERA, compared with measurements from proton-antiproton collisions, at a CMS energy of 200 GeV. The dominant part of the cross section shows indeed an exponential fall off, up to p_T values of 1-2 GeV/c. Thus most γp interactions are soft interactions. At larger p_T values the data behave more like a power law. This is the region we identify with hard scattering, where we can hope to use perturbative QCD to describe the scattering process, and will be studied in detail in the next section.

Soft hadron-hadron collisions are traditionally subdivided into elastic and inelastic diffractive, and inelastic non-diffractive processes. As a result of the similarity with hadron-hadron collisions one expects a diffractive scattering component in the γp cross section. Diffractive scattering involves the exchange of energy-momentum between the incident hadrons, but no exchange of quantum numbers. Due to the interaction both or either one of the incident particles can dissociate into a multi-particle cluster. Fig. 9 shows the elastic and the inelastic diffractive process. For the latter the proton dissociates but the vector meson keeps its original identity, called single proton diffractive dissociation. Further processes are single vector meson diffractive dissociation and double diffractive dissociation. Diffraction is phenomenologically described by the exchange of an object called the Pomeron,[21] postulated by I. Pomeranchuk. The exact nature and the very question whether this object is a particle state or has any

12

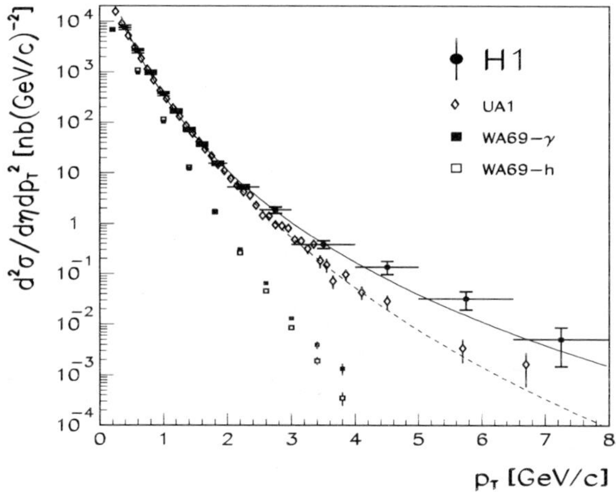

Figure 8: The inclusive ep cross section for charged particles in photoproduction (full circles) measured by the H1 experiment in the kinematical region $| \eta | < 1.5$, $Q^2 < 10^{-2}$ GeV^2 and $0.3 < y < 0.7$, at an average $W_{\gamma p} \approx 200$ GeV. Also shown are cross sections for $p\bar{p}$ collisions measured by the UA1 collaboration (open diamonds) at $W_{\gamma p} \approx 200$ GeV for $| \eta | < 2.5$, normalized to the H1 data at $p_T = 1.5$ GeV/c. The rectangles show the shape of the cross section measurements by the WA69-collaboration at $W_{\gamma p} \approx 18$ GeV, for γp (filled rectangles) and for hadron-proton data (open rectangles).

particle like properties, is far from being resolved. As it turns out HERA will be perhaps the ideal machine to study this bizarre object, which nevertheless represents a remarkably successful[22] phenomenology. In section 4.4 we will show how one can learn more about the Pomeron in γp collisions.

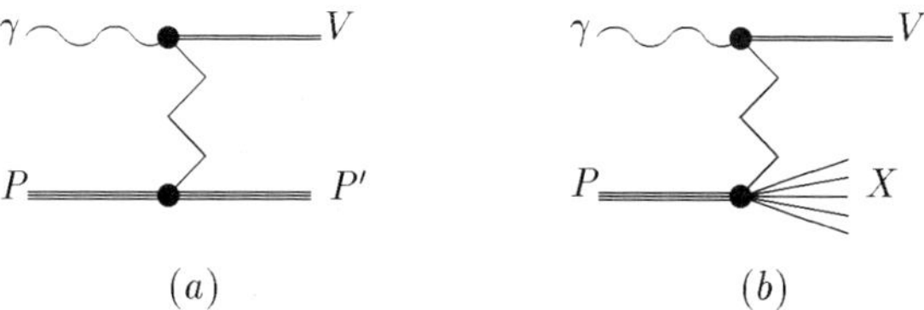

Figure 9: Examples of diffractive processes in γp: (a) elastic scattering, (b) inelastic single proton diffraction dissociation.

The total photoproduction cross section in the HERA energy region is an important measurement, due to speculations based on data from cosmic air showers.[23] These measurements have suggested, albeit with limited statistical significance, an anomalously high muon component in photon induced air showers in the PeV energy range in the laboratory frame. This has lead to predictions for the total photoproduction cross section in the HERA energy region which ranged from 100 to 700 μb.[24]

The total photoproduction cross section has been derived by ZEUS and H1 at the centre of mass energy of $\simeq 200$ GeV, from the measured ep cross section, using the Weizsäcker-Williams approximation[26] for the photon flux. In these analyses tagged photoproduction samples have been used. The result of the total cross section measurement is shown in Fig. 10

13

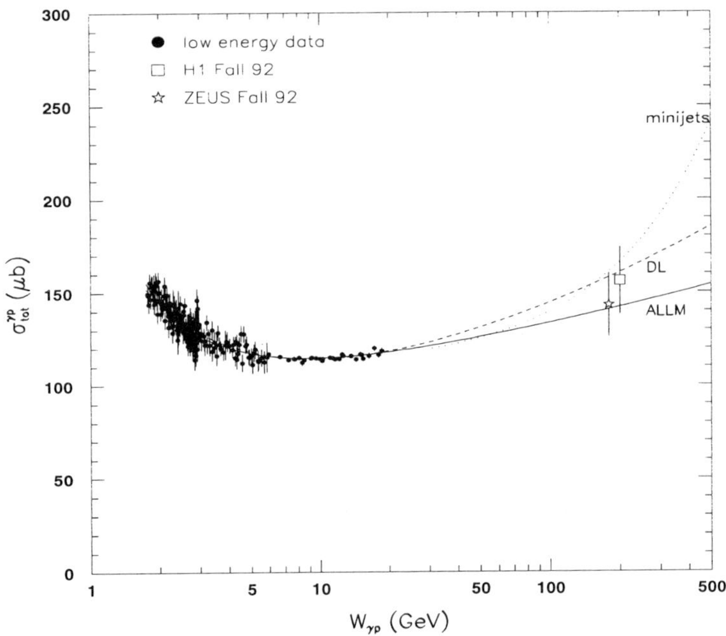

Figure 10: Energy dependence of the total γp cross section. Data from H1 (square) and ZEUS (star) are compared to model predictions from[22] (solid line) and[25] (dashed line). The dotted line is obtained with the PYTHIA Monte Carlo program using the Ansatz $\sigma = \sigma^{soft} + \sigma^{jet}(s)$ for a minimum $p_T = 2$ GeV/c for the partonic collision.

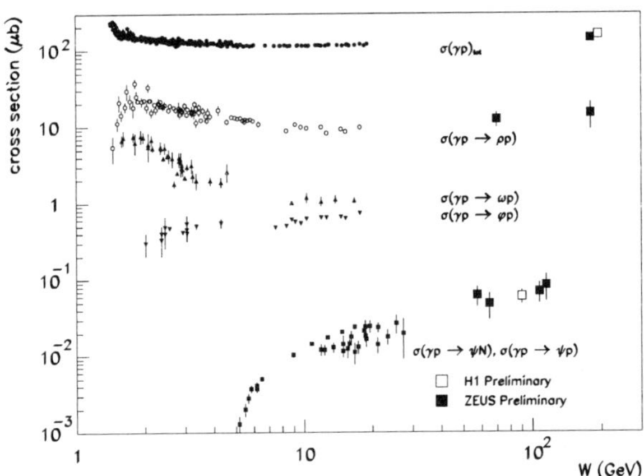

Figure 11: Photoproduction cross section measurements from HERA and lower energy experiments. The J/ψ production is discussed in section 4.5.

together with the lower energy data. The measured value is $156 \pm 2(\text{stat}) \pm 18(\text{syst})$ μb at $\langle W_{\gamma p} \rangle = 200$ GeV for H1,[12] and $143 \pm 4(\text{stat}) \pm 17(\text{syst})$ μb at $\langle W_{\gamma p} \rangle = 180$ GeV for ZEUS,[27]

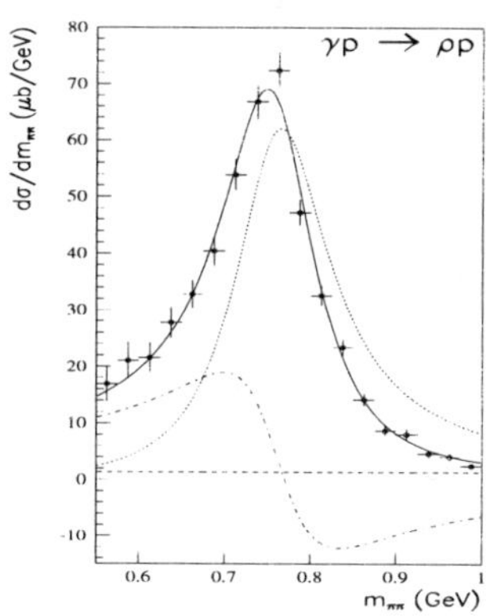

Figure 12: Distribution of the $\pi^+\pi^-$ invariant mass for elastic $\rho^0(770)$ candidates, measured with the ZEUS detector, with the curves showing the resonant (dotted line) and non-resonant (dashed line) production, and interference (dashed dotted line) between the two.

where $W_{\gamma p}$ denotes the CMS energy of the γp collision. The systematic error is dominated by the acceptance corrections for photoproduction events in the central detector. The total photoproduction cross section is found to rise only weakly with the increasing centre of mass energy, as predicted by Regge inspired models.[22,25] No spectacular rise, as suggested by the photon induced air shower data, is observed!

Using tagged events global event characteristics have been used by the ZEUS collaboration to estimate the fraction of non-diffractive, inelastic and elastic diffractive components to be 64.0%, 23.3% and 12.7% respectively.[27] Assuming that 82% of the elastic cross section is due to $\rho^0(770)$ production leads to an indirect measurement of the cross section $\sigma(\gamma p \rightarrow \rho^0 p)$ of 14.8 ± 5.7 μb, and is shown in Fig. 11.

A direct measurement of the $\rho^0(770)$ elastic cross section has been made by ZEUS from the untagged γp event sample at $W_{\gamma p} \sim 50$ GeV.[28] The $\pi^+\pi^-$ mass spectrum for events with two oppositely charged tracks detected in the central tracker is shown in Fig. 12. The deviation from a Breit-Wigner shape of the $\rho^0(770)$ mass spectrum is well known and caused by the interference (dashed-dotted) of resonant (dotted) and non-resonant (dashed) $\pi^+\pi^-$ production.[29] The preliminary result for the cross section $\sigma(\gamma p \rightarrow \rho^0 p)$ is 12.5 ± 2.8 μb, and is shown in Fig. 11. Also the elastic γp cross sections show a small rise, if any, with increasing energy $W_{\gamma p}$.

H1 has further studied inclusive properties of soft hadronic collisions. Fig. 13 shows the corrected multiplicity distribution for a sample of tagged photoproduction events plotted in the KNO[30] variable $z = n/\langle n \rangle$, with n the multiplicity of the event. The average $W_{\gamma p}$ is about 200 GeV. The data are compared with results from $p\bar{p}$ collisions at 540 GeV in the same pseudo-rapidity region, from the UA1 experiment. The distributions clearly look quite similar.

Particle correlations have already shown to be a useful tool to explore the dynamics of soft interactions for multi-particle data. One of the traditional investigations are correlations between identical bosons, so called Bose-Einstein (BE) correlations. The production of two identical bosons from two particle sources is governed by an amplitude which is symmetrized with respect to interchange of the bosons, resulting in an enhanced probability of emission if the bosons have similar momenta. As such, BE correlations were thought to provide information on the space-time structure of the region from which the particles originate i.e. the size

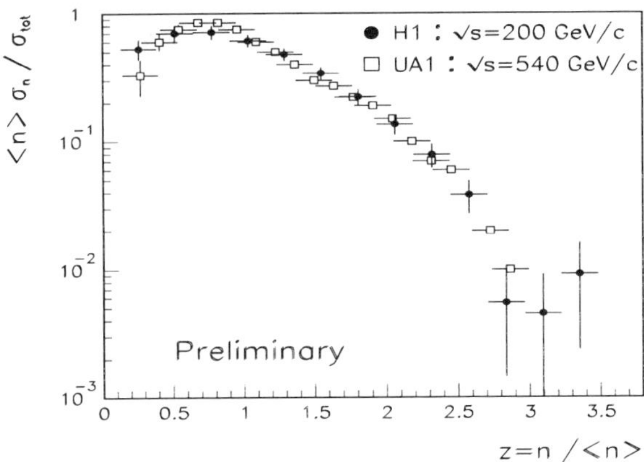

Figure 13: The γp multiplicity distribution for charged particles in the kinematical region $|\eta| < 1.5$, $Q^2 < 10^{-2}$ GeV^2 and $0.3 < y < 0.7$, at an average $W_{\gamma p} \simeq 200$ GeV (full circles), measured by the H1 experiment. The data are compared with results from $p\bar{p}$ interactions (open rectangles), and presented in KNO form.

of the particle emitting source, and on "freeze-out" properties of hadronization. Recent ideas however[31] tend to relate the strength of the BE effect observed in data to effects such as the string tension, rather than the particle emission volume.

H1 has presented preliminary results on BE correlations, based on tracks reconstructed in the central tracker, with a $p_T > 250$ MeV/c and $|\eta| < 1.5$ GeV, for tagged γp interactions. The charged particles were identified as pions, using dE/dx information. Figure 14 shows the ratio $R(Q^2_{BE}) = N(Q^2_{BE}(\pi^-\pi^-))/N(Q^2_{BE}(\pi^+\pi^-))$ of the Q^2_{BE} distributions for like-sign and unlike-sign pion pairs, where here Q^2_{BE} is defined as the (minus) square of the four-momentum difference between the two bosons: $Q^2_{BE} = -(q_1 - q_2)^2$. The unlike-sign pion pair sample does not exhibit a BE effect and is used as a reference. This sample however contains correlations resulting from particle decays ($K^0, \rho^0(770)$), in regions indicated in the figure. The $R(Q^2_{BE})$ distribution is shown for both data and Monte Carlo (which does not include the BE effect). A clear BE enhancement in like-sign pairs is observed for $Q^2_{BE} < 0.1$ GeV^2 in the data. A fit of the form $R(Q^2_{BE}) = 1 + \lambda \exp(-\beta Q^2_{BE})$ yields a radius $r_{BE} = 0.197\sqrt{\beta} = 1.04 \pm 0.04 \pm 0.1$ fm, $\lambda = 0.54 \pm 0.04 \pm 0.07$. The $\rho^0(770)$ and K^0 regions have been excluded for the fit. This parameterization corresponds to the assumption of a Gaussian shape of the source in the centre of mass of the pion pair. Here λ is the correlation strength and the radius r_{BE} a measure for the spatial dimension of the pion source. A comparison for different processes is shown in Fig. 14. The meson-proton data can be considered as reference for this γp measurement. Comparing with the (highest available energy) meson-proton data, recorded at ~ 10 times smaller CMS energy, does not reveal any significant energy dependence of r_{BE} or λ.

In all, soft γp interactions show a multi-particle production and cross section behaviour similar to hadron-hadron collisions. HERA can in this sense be considered as a "meson-proton" collider, and the γp data can be used to study soft hadronic collisions at a center of mass energy of 200 GeV.

4.3 Hard Processes in Photoproduction

Hard scattering in γp interactions is expected from partonic collisions between quarks and gluons of the incident proton and the resolved photon, and from the direct production diagrams

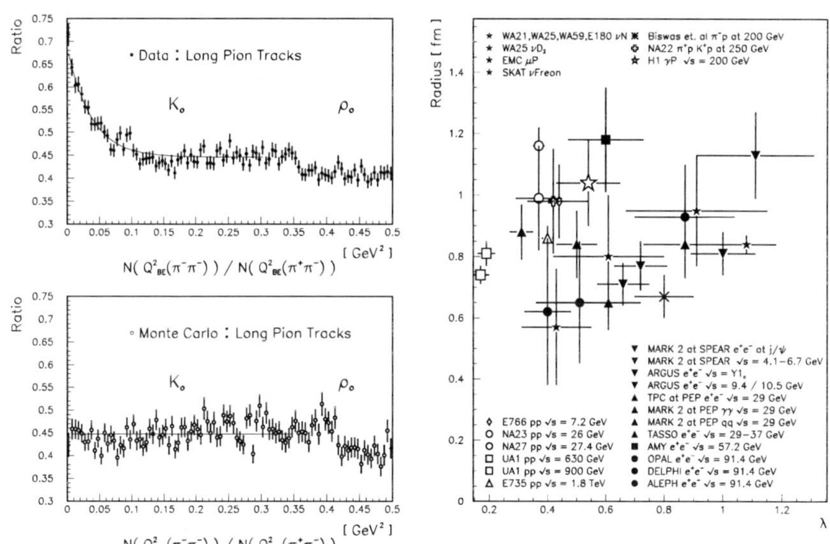

Figure 14: Preliminary distributions of Bose-Einstein correlations from H1: (left) the ratio $R(Q^2_{BE}) = N(Q^2_{BE}(\pi^-\pi^-))/N(Q^2_{BE}(\pi^+\pi^-))$ for data (top) and Monte Carlo (bottom); (right) a compilation of the radius and correlation strength (λ) values for different experiments, including the H1 result.

(photon-gluon fusion and QCD Compton scattering) as depicted in Fig. 6. In this section we will address the questions: do we have evidence for hard scattering in γp interactions, and, if yes, do we find any evidence for both the resolved and direct γp production? For resolved collisions this picture leads to the introduction of a "photon structure function", describing the probability for finding partons in the photon which carry a momentum fraction x_γ of the photon. One of the challenges at HERA is to measure the x_γ distribution in the photon: to measure the photon structure.

A high p_T tail, characteristic for a hard scattering process, is expected to be observed in the inclusive p_T spectrum of charged particles. Indeed, from Fig. 8 the presence of such a large p_T tail is evident. If hard parton scattering dynamics is the cause of this high p_T part of the data, we can compare it with QCD calculations. This is shown in Fig. 15, where the high p_T part is compared with a next to leading order (NLO) calculation[32] including resolved and direct processes. The agreement between the data and the calculation is very good. It also shows that indeed the resolved processes dominate at the lower p_T end of the distribution, but constitute only about 70% of the cross section at large p_T values, the remaining part are direct processes. Fig. 8 also displays the $p\bar{p}$ data at $\sqrt{s} = 200$ GeV. The high p_T tail in γp interactions is clearly larger. The effect due to the different structure function for the proton and photon (when taken to have the same parton distributions as a meson) cannot explain fully the discrepancy observed for the high p_T part of the differential cross section.[33] Hence extra non-VDM contributions are needed to explain the γp cross section, such as the direct and the anomalous component. Note that the later is often taken to be part of the photon structure function.

The next step towards establishing hard scattering in γp interactions is the observation of jets. In fact, the event shown in Fig. 7 shows clear jet structures, visible in the different detector views and in the energy flow in the $\eta - \varphi$ plot. Events such as this one were found

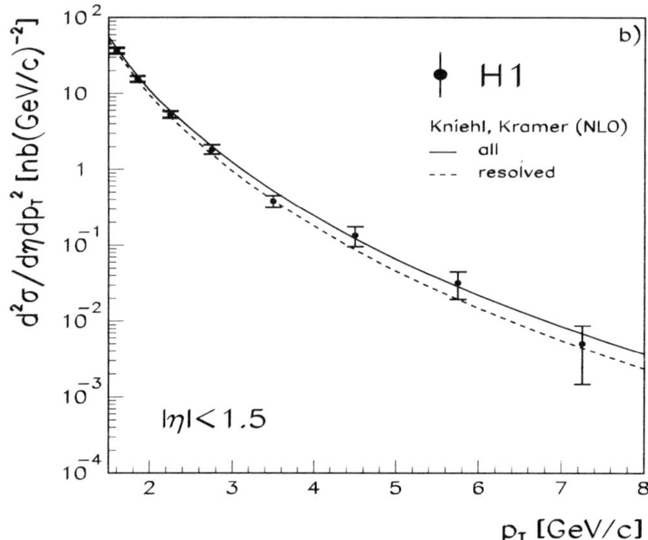

Figure 15: The measured single particle cross section of H1 from Fig. 8 (full circles), compared in the $p_T > 1.5$ GeV/c region with an analytical NLO QCD calculation.[32] The solid line represents the sum of the resolved (dashed line) and direct photon contributions.

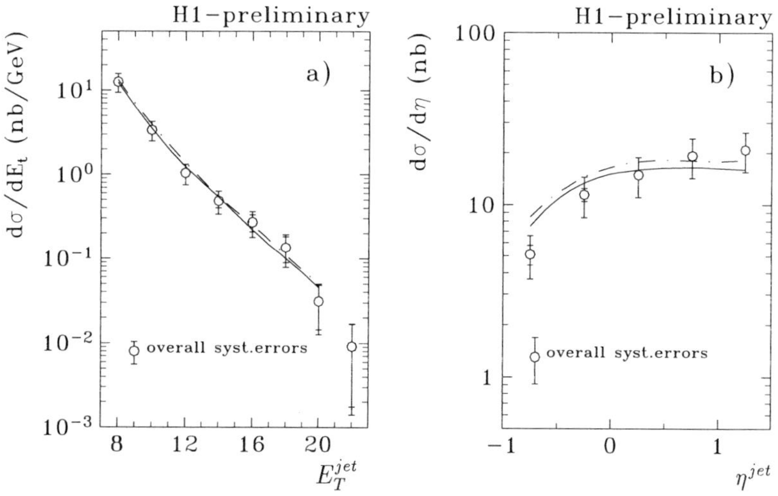

Figure 16: Preliminary H1 data showing: (a) the inclusive differential ep cross section $d\sigma/dE_T^{jet}$ integrated in the pseudo-rapidity interval $-1.0 < \eta^{jet} < 1.5$; (b) inclusive ep cross section $d\sigma/d\eta^{jet}$ for jets with $E_T^{jet} > 7$ GeV. The inner error bars represent statistical errors, the outer error bars the statistical and systematic errors added in quadrature. The overall systematic uncertainty is shown separately. The measurement is compared to PYTHIA predictions using the GRV-LO[34] (full line) and LAC2[35] (dashed-dotted line) parton distributions for the photon.

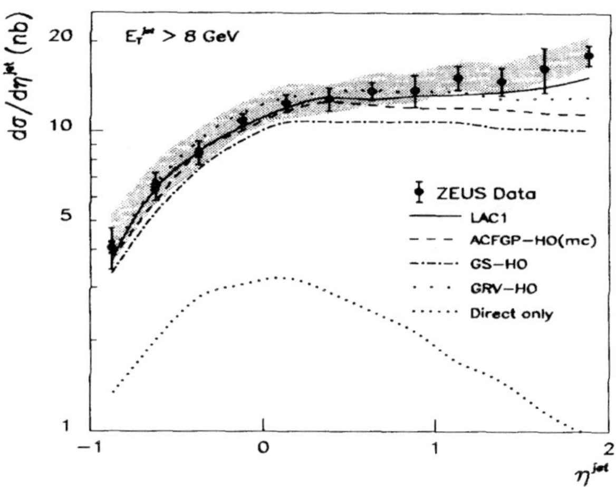

Figure 17: Measured differential ep cross section $d\sigma/d\eta^{jet}$ of the ZEUS experiment for inclusive jet production integrated over $E_T^{jet} > 8$ GeV, for $Q^2 < 4$ GeV^2 and $0.2 < y < 0.85$. The shaded band displays the uncertainty due to the energy scale of the jets. The data are compared with LO QCD calculations using PYTHIA, for different parton distributions in the photon.

soon after the startup of HERA, in the late spring of 1992. To be more quantitative a jet algorithm is used. For the H1 analysis presented in Fig. 16 jets within the range $-1 < \eta^{jet} < 1.5$ a selected using a cone algorithm,[36] requiring $E_T^{jet} > 7$ GeV in a cone with radius $R = \sqrt{\Delta\eta^2 + \Delta\varphi^2} = 1.0$ in the space of pseudo-rapidity η and azimuthal angle φ (in radians). In Fig. 16a the ep jet cross section, corrected for detector smearing to the cross section at the level of the final state hadrons (the hadron level), is shown as function of E_T^{jet}. A sample of tagged events is used for this analysis. The data follow an $E_T^{-5.6}$ dependence. The η dependence of the jet cross section is shown in Fig. 16b. The figures also show a LO QCD predictions calculated using the PYTHIA[37] Monte Carlo program for hard photon-hadron processes, using different assumptions for the parton density distributions – or structure – of the photon. These parton density distributions describe the results from $\gamma\gamma$ interactions from e^+e^- experiments. PYTHIA includes both direct and resolved processes. The QCD predictions describe the data quite well for the selected parton density parameterizations. In Fig. 17 the differential jet cross section from ZEUS is shown for jets with $E_T^{jet} > 8$ GeV.[38] The data show the same level of agreement with the LO QCD calculations.

A quantity of particular interest is the momentum fraction, x_γ, of the parton in the photon involved in the hard scattering. For direct processes $x_\gamma = 1$, since the full momentum of the photon enters the hard scattering, while for resolved processes $x_\gamma < 1$. The measurement of the distribution of x_γ is analogous to the measurement of Bjorken-x in deep inelastic scattering (see next chapter) and is a direct measure of the hadronic structure of the photon. The jet kinematics can be used to determine x_γ of the parton involved in the hard scattering collision. Indeed, for a LO QCD $2 \rightarrow 2$ scattering process x_γ can be approximately reconstructed as follows:

$$x_\gamma = \frac{E_T^{jet1}e^{-\eta^{jet1}} + E_T^{jet2}e^{-\eta^{jet2}}}{2E_\gamma}, \tag{4}$$

where the indices refer to the two jets resulting from the two partons involved in the hard scattering (see Fig. 6).

First, we will establish the presence of a direct component in the γp cross section from 2-jet event studies. In an analysis performed by the ZEUS collaboration[39] events with at least two jets have been selected using a cone algorithm on a grid in pseudo-rapidity η and azimuthal

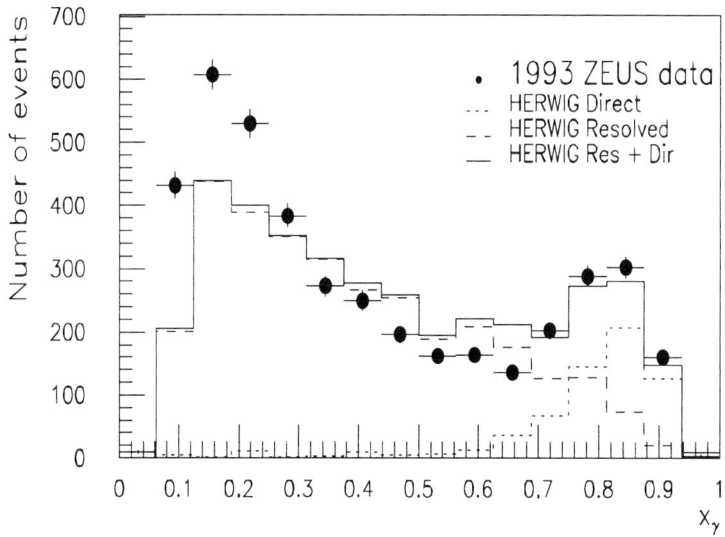

Figure 18: Reconstructed x_γ distribution for ZEUS (full circles). The histograms represent the prediction of the LO QCD calculation from HERWIG for the direct component (dotted line), the resolved component (dashed line) and the sum of these two contributions (full curve).

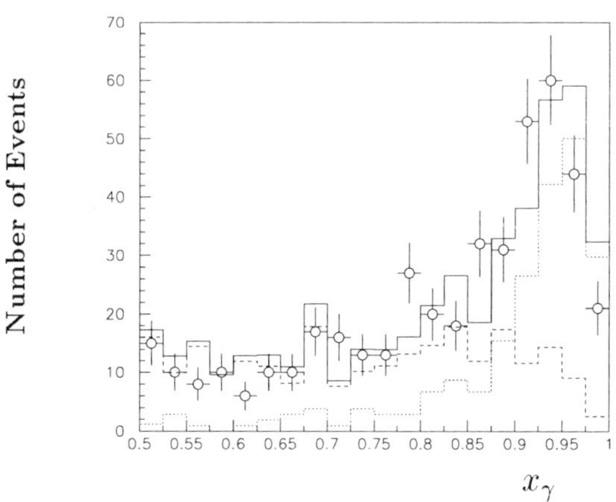

Figure 19: Preliminary reconstructed x_γ distribution for H1 (open circles). The histograms represent the prediction of the LO QCD calculation from PYTHIA, with full detector simulation, for the direct component (dotted line), the resolved component (dashed line) and the sum of these two contributions (full curve).

angle φ with cone radius $R = 1$. Jets are required to have $E_T^{jet} > 6$ GeV and to be in the pseudo-rapidity interval $-1.125 < \eta^{jet} < 1.875$. For this analysis untagged γp interactions were used. The photon energy E_γ is given by $E_\gamma = yE_e$ and the y is reconstructed from the hadronic energy flow measured with the calorimeter using eqn. 3. The reconstructed

x_γ spectrum is shown in Fig. 18. The curves in Fig. 18 are leading order QCD calculations, this time using the HERWIG[40] Monte Carlo program for hard γp processes. The calculations include a full detector simulation and assume the LAC1 parameterization[35] for the parton distributions of the photon. The Monte Carlo predictions are also shown separately for the direct and the resolved component. In the high x_γ region ($x_\gamma > 0.6$) the data are well described by the sum of the two contributions: the resolved component alone is not able to reproduce the "peak" at $x_\gamma \approx 0.8$. Note that the measurement of x_γ in direct process extends to values smaller than 1, due to the experimental resolution. In the region of $x_\gamma < 0.5$ the direct contribution is negligible and the events observed in the data sample can only be explained by resolved photon processes. The prediction for this region depends critically on the assumed parton distributions for the photon, and will be explored below. Fig. 19 shows the x_γ distribution at high x_γ for untagged events, as measured by the H1 collaboration. Here jets with an $E_T^{jet} > 8$ GeV and $-1 < \eta^{jet} < 3$ have been used. The distance in η between the jets was required to be less than 1.5. The results are compared with predictions from the PYTHIA Monte Carlo program using the GRV parton distributions for the photon and proton. Again, the data can only be explained if a direct component is included in the data. In all, the x_γ distribution shows that both classes of processes, direct and resolved, shown in Fig. 6, are present in the data.

From Figs.15 and 18 it is clear that the bulk of hard scattering photoproduction events at HERA is due to resolved processes. In a resolved process only a part of the original photon momentum enters the hard subprocess and the rest is carried by other –spectator– partons, as is shown in Fig. 6. These spectator partons fragment into a photon remnant, similar to the proton remnant introduced earlier, and which is expected to appear in the detector close to the original photon direction. A study of this photon remnant was performed by the ZEUS collaboration. A clustering algorithm, called k_T algorithm[41] was used for the jet search. With this algorithm all calorimeter cells are grouped in three clusters, excluding the proton remnant (most of which disappears in the beampipe anyway). The clusters are ordered according to their p_T and the following cuts are applied for event selection: $p_T^{cluster1,2} > 5$ GeV, $\eta^{cluster1,2} < 1.6$ and $\eta^{cluster3} < -1$. Monte Carlo studies show that the selected sample is dominated by resolved photoproduction. The third cluster is associated with the photon remnant. Fig. 20 shows the average total transverse and longitudinal energy of this third cluster with respect to the cluster axis, as a function of the cluster energy. The data are compared with predictions from a Monte Carlo calculation (based on the PYTHIA[37] generator and including a full detector simulation) in which the fragmentation of the remnant is treated the same way as the hard jets. The good agreement between data and the Monte Carlo predictions demonstrates the jet-like properties of the photon remnant.

Finally, hard scattering events in photoproduction data will be used to retrieve information on the "partonic structure" of the photon. The quark content of the photon has been measured in $\gamma\gamma$ interactions in e^+e^- experiments, down to $x_\gamma \geq 0.007$.[42] For these measurements a highly virtual photon is used to probe a real photon. As for an ordinary hadron the real photon is expected to have a gluon content as well but, since a virtual photon does not couple directly to gluons, the gluon content is not directly accessible in these measurements. Hence to date only poor constraints on the gluon density in the photon[43] exist. Furthermore there is no momentum sum rule for the photon and therefore the present predictions for the gluon distribution from different parton parameterizations of the photon differ wildly.

At HERA a parton from the proton rather than a photon is used to probe the photon structure. These partons evidently interact with both the quarks and the gluons in the photon, giving for the first time direct access to the gluon content of the photon. The price one has to pay is that at HERA one measures in this way always the sum of the quark and gluon component of the photon. Therefore, to isolate the gluon part one has to subtract the part induced by the quarks, using e.g. the measurements from e^+e^- experiments.

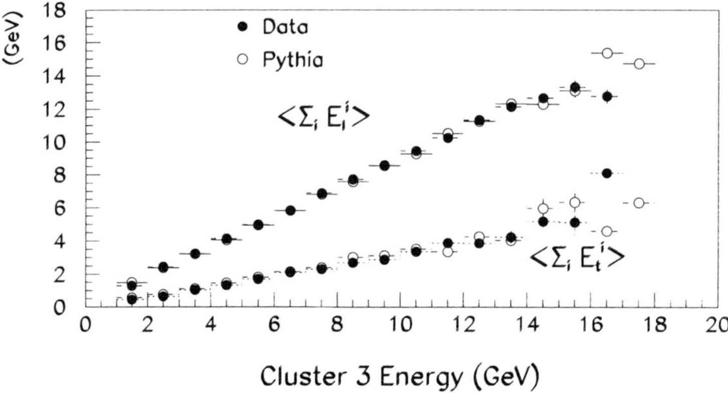

Figure 20: Average transverse $\langle \sum_i E_t^i \rangle$ and longitudinal $\langle \sum_i E_l^i \rangle$ energy as a function of the third cluster energy. The sum runs over all hadrons belonging to the cluster. The preliminary ZEUS data (full circles) are compared with Monte Carlo prediction (open circles).

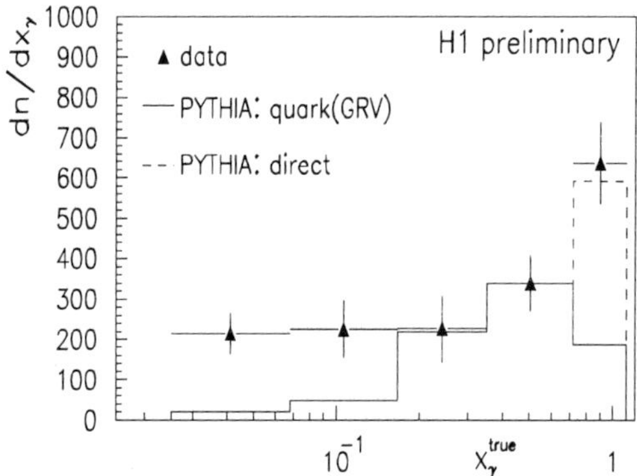

Figure 21: The distribution of the corrected parton momentum x_γ. The preliminary H1 data (full triangles) are compared to the PYTHIA prediction of the direct component (dashed line) and the quark part of the resolved contribution (full line) using the GRV-LO parameterization. Only statistical errors are shown.

A first attempt to constrain the gluon in the photon was made by H1, using jets with a minimum transverse energy $E_T^{jet} > 7$ GeV in the pseudo-rapidity interval $-0.2 < \eta^{jet} < 2.5$, which give access to the partons in the range $0.03 < x_\gamma < 1$. A sample of tagged events is used so that the photon energy E_γ was given by the difference between the beam and the tagged electron energies. A problem encountered during this analysis was that the hard scattering Monte Carlo programs, like PYTHIA, give a poor description of the energy flow at large pseudo-rapidities. The data show a much higher pedestal energy between the jets compared to the predictions of these Monte Carlo generators. Allowing for events with multiple scattering, i.e. apart from the principally interacting partons, additionally partons from the proton and photon remnant can interact, these energy flows can be substantially improved. An unfolding

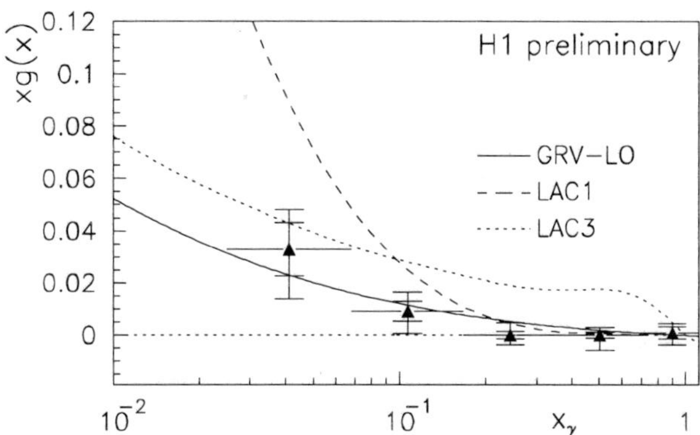

Figure 22: The gluon density in the photon (triangles), as measured by the H1 experiment compared with the GRV-LO (full) and the LAC1 (dotted) distributions at the average scale of 60 GeV^2. The data are preliminary.

procedure[44] is used to convert the observed x_γ distribution into the "true" x_γ distribution, shown in Fig. 21. The correlations between x_γ and x_γ are taken from the PYTHIA Monte Carlo program, introducing unavoidably a certain model dependence in the result. The data are compared to the LO QCD prediction, calculated with PYTHIA and normalized to the integrated luminosity, for the direct component and the quark part of the resolved photon contribution. These calculations use the GRV-LO parameterization[34] of the photon parton densities. The sum of the two components gives a good description of the data for $x_\gamma > 0.2$. The excess of events over the quark part of the resolved contribution in the region $x_\gamma < 0.2$ can be attributed to the gluon content of the photon.

After subtraction of the predicted direct contribution and the quark part of the resolved component predicted from e^+e^- data, the gluon density in the photon can be extracted. The result is shown in Fig. 22 where the gluon density $x_\gamma g(x_\gamma)$ at an effective scale ≈ 60 GeV^2 is given. It is important to note that this result was achieved in the framework of a leading order interpretation of the data. The inner error bars in Fig. 22 represent the statistical and the outer error bars the statistical and systematic errors added in quadrature. The dominating systematic errors are the uncertainty on the hadronic energy scale and the correction for the imperfect description of the energy flow by the Monte Carlo generator. The gluon content in the photon is restricted to small x_γ values as expected. Despite the large error bars the data already constrain the parton distributions in the photon and discriminate between different parameterizations. The measurement presented in Fig. 22 is compared to the LAC1, LAC3 and GRV-LO parameterizations of the photon parton distributions. The LAC3 distribution assumes a large gluon component at high x_γ. This scenario is clearly disfavoured by the data. The dashed curve shows the prediction for LAC1 assuming a very large gluon component at small x_γ. The GRV parton density parameterization gives the best description of the data

In summary, hard scattering and jets have been unambiguously observed in γp interactions at HERA. We have established the presence of both direct and resolved processes. The jet inclusive cross sections behave as expected from (LO) QCD. The 2-jet events give access to the parton distributions in the photon. For the first time a (LO) gluon extraction was performed relying on the validity of the PYTHIA model for the description of γp hard scattering interactions.

4.4 Hard Scattering in Diffractive Processes

In section 4.2 it was discussed that between 30% and 40% of the γp interactions are diffractive events. From the experimental and phenomenological point of view these events exhibit gaps in rapidity which are not exponentially suppressed.[45] Such a gap results from the absence of colour flow between the systems connected by the exchanged Pomeron as is shown in the examples given in Fig. 9.

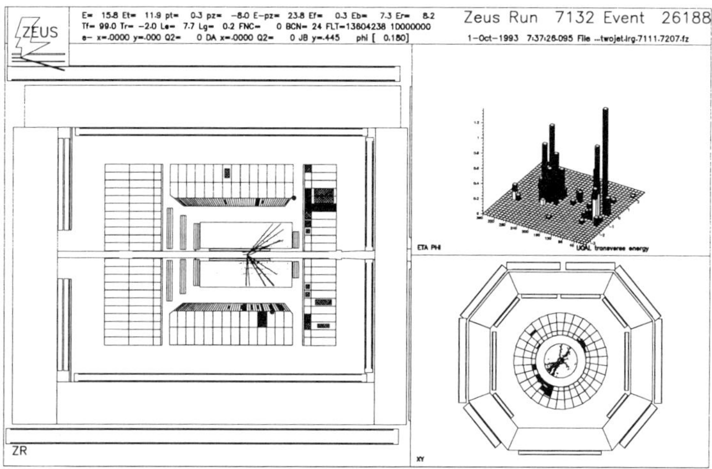

Figure 23: A display of a ZEUS photoproduction event with a large rapidity gap in the forward direction.

The technique used so far by both experiments to isolate events with rapidity gaps at HERA is based on the variable η_{max}. The η_{max} of an event is defined as the largest measured value of pseudo-rapidity for which activity is detected. The maximum value which can be reached is determined by the acceptance of the detector. Both experiments have used the main calorimeter for their initial studies, leading to maximum reachable values in the laboratory system of η_{max} of 4.3 for ZEUS and 3.7 for H1. Here η_{max} of an event is defined to be the largest η for which a cluster or condensate with energy larger than 400 MeV is found. Small values of η_{max} indicate that there is a large region between η_{max} and the detector edge with no activity in the detector. A measured η_{max} value equal to 0 means for the ZEUS detector that there is an empty gap of 4.4 units in rapidity in the forward direction. Such an event (with $\eta_{max} \approx 0$) is shown in Fig. 23.

The distribution of η_{max} for a sample of tagged photoproduction events is shown in Fig. 24 for H1 data. For small η_{max} values, i.e. for large gaps, the data clearly do not show an exponential decrease, and hint towards diffraction as the underlying dynamic process. The data are compared with predictions of a diffractive model (**sd**) and a model which does not contain diffractive events (**nd**). Clearly, the non-diffractive model describes well the values at large η_{max}, but fails completely in the region where the gap gets large. The diffractive model on the other hand gives a rather good description of the region $\eta_{max} < 2$. Hence diffractive production is a plausible interpretation for the events with a large rapidity gap in γp interactions.

Phenomenologically, the observed properties of the diffractive cross section in the framework of Pomeron exchange are described by triple-Regge theory.[46] This interpretation however gives no information on the details of the hadronic final states produced in diffractive

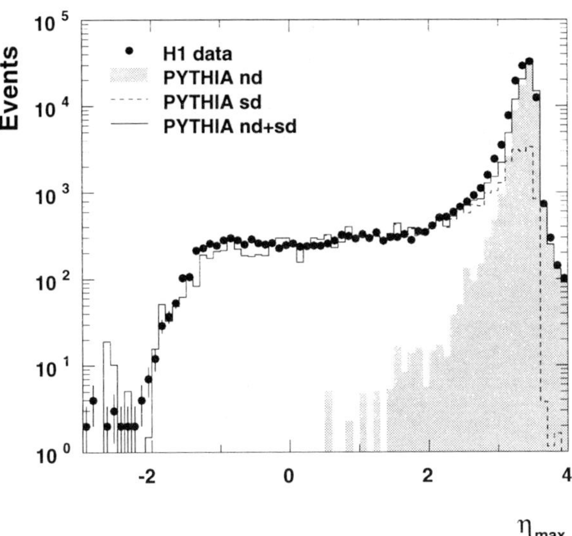

Figure 24: Maximum peudo-rapidity η_{max} distribution in γp events observed in the H1 detector, compared to a diffractive (dashed line) and a non-diffractive (shaded area) Monte Carlo model, and their sum (full line).

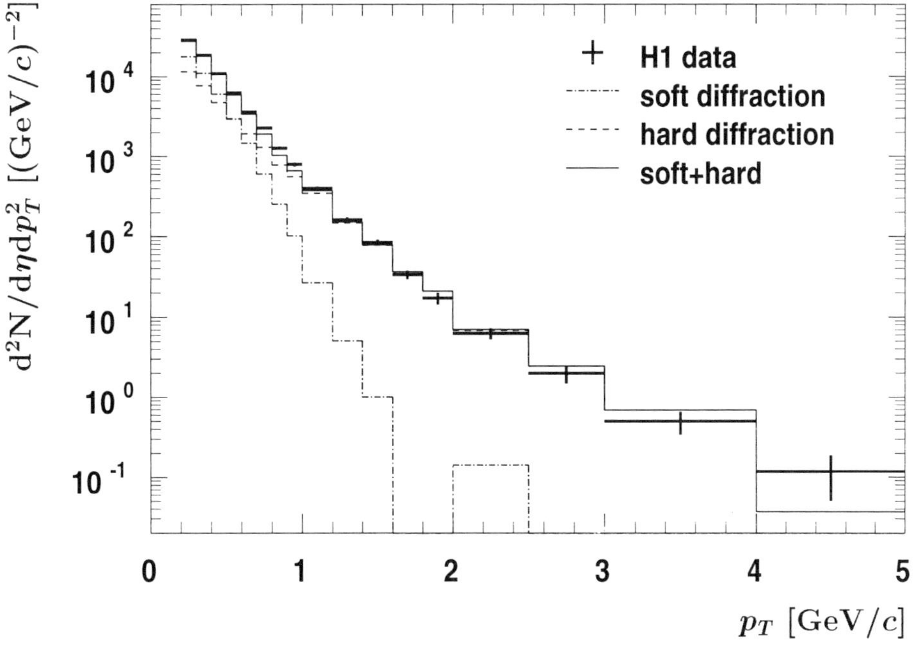

Figure 25: Transverse momentum distribution of charged particles for events with a large pseudo-rapidity gap ($\eta_{max} < 1.5$), measured by the H1 experiment, compared to Monte Carlo predictions explained in the text.

events. Traditionally the final state in diffractive dissociation is assumed to be described by a multiperipheral[47] type of model in which particles are distributed throughout the final state

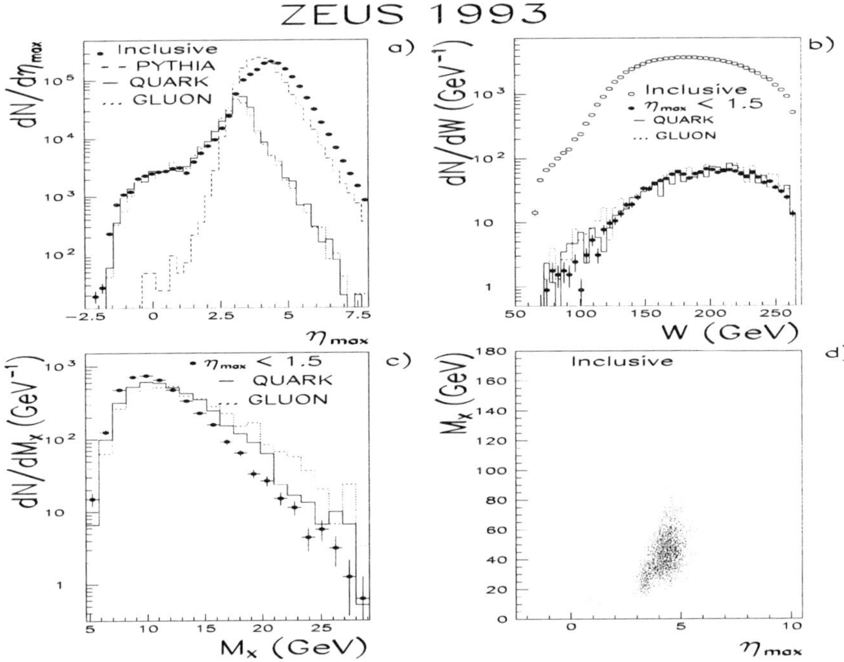

Figure 26: (**a**) The distribution of η_{max}, the pseudo-rapidity of the most forward calorimeter condensate with an energy above 400 MeV for the ZEUS photoproduction sample with $E_T > 5$ GeV along with the predictions from a non-diffractive model (dashed line) and model for diffractive hard scattering, with a quarkonic (solid line) or gluonic (dotted line) Pomeron of type $G0$. (**b**) The distribution in W for all events and for those with $\eta_{max} \le 1.5$. (**c**) The mass of the hadronic system M_X for events with a large rapidity gap as defined by $\eta_{max} \le 1.5$ along with model predictions. (**d**) A scatter plot of the mass of the hadronic system, M_X, versus η_{max}.

phase space with limited transverse momentum. This approach has been used successfully so far for comparisons with the available measurements of multiplicity and rapidity distributions of charged particles from the diffractive system. On the other hand, in modern QCD language it is tempting to consider the Pomeron as a partonic system[48] which can be probed in a hard scattering process. Models based on this idea assume that the Pomeron behaves as a hadron and the concept of a Pomeron structure function is introduced.[49-51] In contrast to the approach of assuming limited p_T phase space, these models predict that, similar to high energy hadron-hadron scattering, high mass diffractive dissociation exhibits the production of jets and a large p_T tail in the differential transverse momentum distribution. Thus hard hadron-Pomeron scattering events should be observed in diffractive hadronic collisions at high energies. The UA8 collaboration has shown evidence for jet production in diffractive $p\bar{p}$ events,[52] interpreted as resulting from collisions of partons from the proton with partons from the Pomeron. Furthermore, within this partonic picture, these data have shown sensitivity to the parton distribution in the Pomeron. On the latter relatively little information is known. Scenarios exist in which the Pomeron is pictured to consist either dominantly of quarks or gluons. For the distribution functions of the partons one assumes either a "hard" distribution, $\beta g(\beta) \sim \beta(1 - \beta)$, (hereafter labeled "$G0$"), or a "soft" distribution, $\beta g(\beta) \sim (1 - \beta)^5$, (hereafter labeled "$G5$"). The variable $\beta = x_{i/\mathbb{P}}$ is the fraction of the Pomeron momentum carried by the struck parton i involved in the interaction. The results from high p_T jet production in

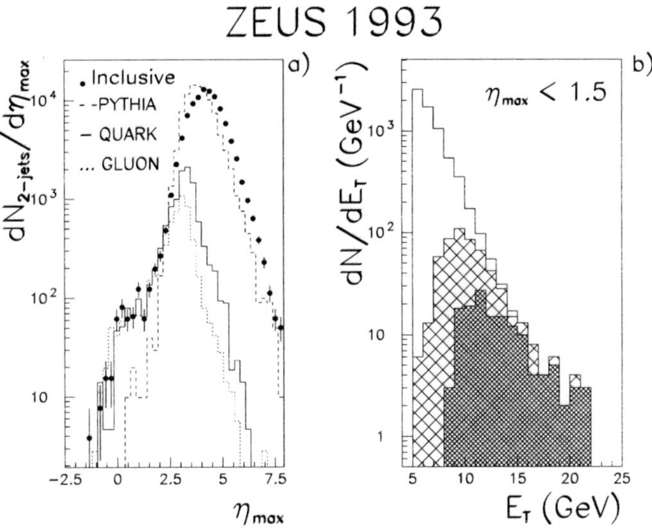

Figure 27: (**a**) The distribution of η_{max} for the ZEUS photoproduction sample with $E_T > 5$ GeV and two or more jets along with the prediction from a non-diffractive model (dashed line) and a diffractive model, with a quarkonic (solid line) and gluonic (dotted line) Pomeron of type $G0$. (**b**) The distribution of the total observed transverse energy E_T for the photoproduction event sample with a large rapidity gap and, in addition, for the subsample of those events with at least one (cross hatched area) and at least two (shaded area) jets in the final state.

diffractive proton-antiproton interactions mentioned above favour the hard $G0$ distribution.

In the present studies, the agreement with the partonic collision picture is tested with the aid of Monte Carlo programs. For this study we use a model which explicitly includes diffractive hard scattering: POMPYT1.0.[53] This model assumes the emission of a Pomeron at the proton vertex. The resulting photon-Pomeron interaction is simulated as the hard scattering of the photon (direct process) or partons in the photon (resolved process) with partons in the Pomeron according to LO QCD calculation for the hard scattering processes.

We show results from H1[54] and ZEUS[55] on the observation of hard processes in γp diffractive events. In Fig. 25 the transverse momentum distribution is shown for charged particles from events with a large rapidity gap ($\eta_{max} < 1.5$). The presence of a clear large p_T tail – similar to the one for all γp events as shown in Fig. 15 – is clearly visible. The data are compared to predictions of models with (hard diffraction; POMPYT) and without (soft diffraction) hard partonic scattering. It shows that the predicted shape of the first model is consistent with the data at large p_T values, while the model without diffractive hard scattering does not describe the data.

Next events are preselected which have a total E_T larger than 5 GeV. ZEUS data[55] for η_{max}, W and M_X are shown in Fig. 26. Here W is the total hadronic invariant mass of the event, M_X is the visible hadronic invariant mass of the system for all hadrons with $\eta < \eta_{max}$, i.e. the diffractive dissociated system. The POMPYT model predictions were calculated with the $G0$ parton distributions for the Pomeron, assuming either the quark and gluon hypothesis for the Pomeron structure. In Fig. 26a a clear shoulder is seen at small η_{max} indicating the presence of rapidity gap events in this large E_T sample. Values of $\eta_{max} > 4.3$, which are outside the acceptance of the calorimeter, occur when energy is deposited in many contiguous cells around the beampipe in the forward direction. It is interesting to note that at large

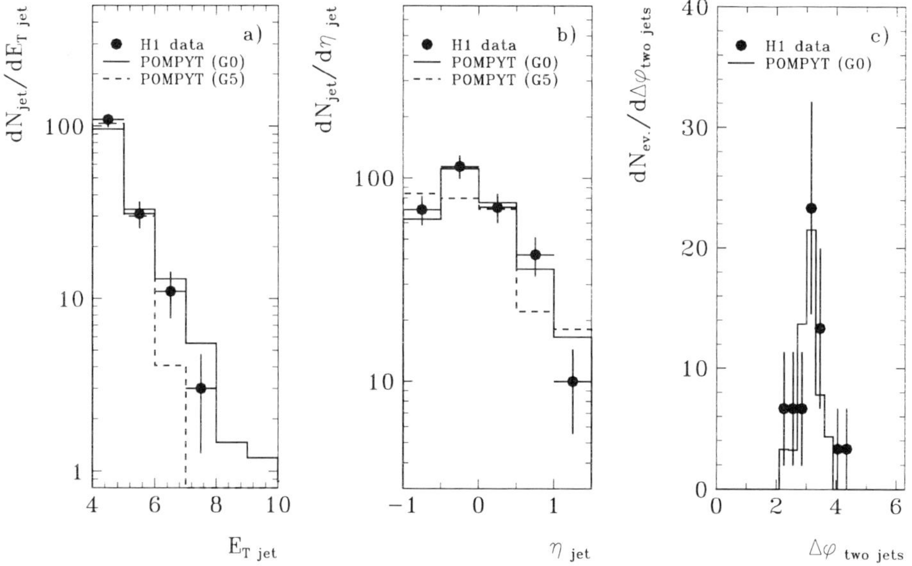

Figure 28: (**a,b**) Inclusive jet distributions for large pseudo-rapidity gap events ($\eta_{max} < 1.5$) measured with the H1 detector: transverse energy E_T^{jet} and pseudo-rapidity η^{jet}. (**c**) Distribution of the azimuthal angle $\Delta\varphi$ between the jets for 2 jet events. The data are compared with Monte Carlo predictions assuming a Pomeron with hard ($G0$; full line) and soft ($G5$; dashed line) gluon momentum distribution.

W the data show a rather constant or slowly rising behaviour (Fig. 26b) consistent with the interpretation that the rapidity gap data are predominantly resulting from Pomeron exchange. The deviations close to the maximum W and at W below 150 GeV result from acceptance effects. Fig. 26c shows that the distribution of the visible hadronic mass dN/dM_X falls steeply with increasing M_X as expected for diffractive phenomena. In this high E_T sample jets we search for jets which have an $E_T^{jet} > 4$ GeV within a cone of radius equal 1. The sample is found to consist of 91.4% zero-, 6.5% one-, 2.0% two and 0.1% three or more-jet events. The η_{max} distribution for events with at least 2 jets and event E_T distribution are shown in Fig. 27. At large E_T the sample predominately consist of 2-jet events. The 2-jet event sample will allow us to study the Pomeron structure with future high statistics data. The parton densities in the photon will be measured at HERA and then used in jet analyses of diffractive events to unfold the parton densities of the Pomeron using the same technique as discussed in section 4.3. Characteristics of the jets are shown in Fig. 28. The model for diffractive hard scattering describes the data well. In particular the back-to-back behaviour of the jets for events where two jets were detected is clearly seen.

To restrict to a region where the data show dominantly hard scattering features, a comparison is made of the H1 data with the POMPYT model by increasing the minimum E_T requirement for the events to 9 GeV. The 1 and 2-jet event fractions are then 38.7% and 13.4% respectively, and the ratio (2 jets)/(1 jets) is 0.35 ± 0.09. These results are compared with POMPYT predictions in Table 2, assuming the Pomeron consists predominately of gluons. The (2 jets)/(1 jets) ratio, which is only weakly sensitive to the remaining soft diffractive contribution, compares favourably with the prediction of a hard Pomeron parton distribution. It depends however somewhat on the divergence limit ($\hat{p}_T^{min}$) used in the LO QCD matrix element calculation for the hard partonic scattering process.

Sample	1 jet events(%)	2 jet (%)	2-jet/1-jet
Data (142 events)	38.7	13.4	0.35 ± 0.09
POMPYT $G0$ ($\hat{p}_T^{min} = 2$ GeV)	46.4	10.1	0.22 ± 0.05
POMPYT $G5$ ($\hat{p}_T^{min} = 2$ GeV)	27.3	-	< 0.1

Table 2: Jet rates: data compared to POMPYT Monte Carlo calculations for γp events with $E_T > 9$ GeV and $\eta_{max} < 1.5$, and for jets with $E_T^{jet} > 4$ GeV and $-1 < \eta^{jet} < 1.5$

In summary, we have observed hard processes in photoproduction with a large rapidity gap at HERA. Interpreted in terms of diffractive scattering, the results are sensitive to the parton distributions in the Pomeron (hard/soft; quark/gluon), where in general a harder type of parton distribution is prefered. Future precise data will allow to extract the parton distributions of the Pomeron.

4.5 Heavy Flavour Production

An important aspect of photoproduction studies at HERA is the production of heavy flavours, in particular charm quarks. Apart from properties of charmed mesons and baryons, heavy flavour studies in photoproduction are expected to give information on the gluon density in the proton and on the dynamics of strong interaction physics in kinematical regions in which the physics descriptions range from non-perturbative phenomenological approaches to perturbative QCD.

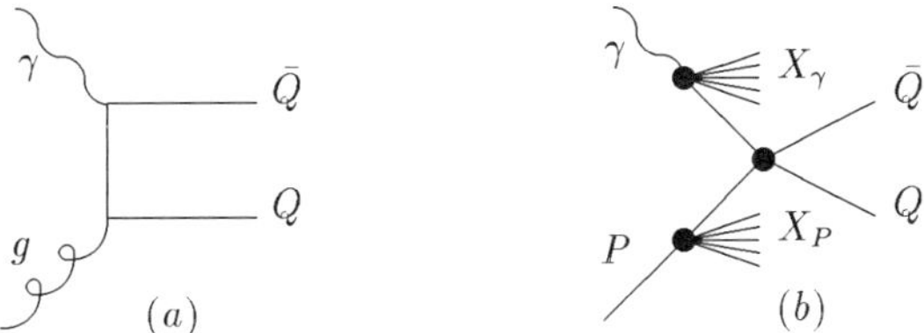

Figure 29: Mechanisms for heavy quark photoproduction: (**a**) direct photoproduction process, (**b**) resolved photoproduction process

The luminosity collected by the experiments so far allows us to study (a) J/ψ production, particularly in the elastic channel, (b) the total charm production cross section from semi-leptonic decays and (c) the production of D^{*0} mesons. Photoproduction LO QCD processes leading to inelastic production of heavy flavours are shown in Fig. 29a for direct and Fig. 29b for resolved production. The production of J/ψ mesons is in this context described by the colour singlet model,[58,57] where the $Q\bar{Q}$ pair can end up in a bound state via the emission of a gluon, which connects with the proton remnant. The photon-gluon fusion process (Fig. 29a) shows that these events give direct information on the gluon density in the proton, since the gluon enters at the Born level for this diagram. Measuring the cross section of these events will be one of the experimental handles at HERA to extract the gluon content of the proton, which will be elaborated on in the chapter on deep inelastic scattering. Elastic production of J/ψ mesons can in this QCD prescribtion be described by the exchange of 2 gluon.[59] On the other hand, for J/ψ production vector meson dominance contributions are expected for interactions with a small p_T. The photon can fluctuate into e.g. a J/ψ particle which is observed in the final state if the meson-proton interaction is elastic (9a)or single proton diffractive dissociation (9b).

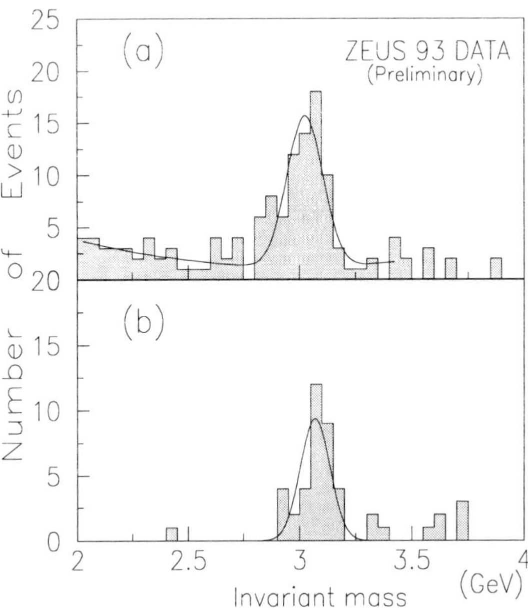

Figure 30: Invariant mass of lepton pairs in two-prong photoproduction events from ZEUS; (a) e^+e^-, (b) $\mu^+\mu^-$. The curves are fits to the preliminary data.

Both H1 and ZEUS have studied the production of "elastic" J/ψ mesons.[60,28] The J/ψ mesons are identified by their decay into leptons (e or μ) which are required to be the only particles visible in the detector. In fact the data selected this way contain a mixture of true elastic events and those where the proton breaks up into fragments (like in a diffractive dissociation process) which remain in the beampipe. In total 48 (40) muon (electron) pair events with 22 (10) in a region of ± 225 MeV around the nominal J/ψ mass, enter the H1 analysis, and 148 events the ZEUS analysis. The invariant mass of the lepton pairs is shown in Fig. 30. A clear signal is seen around the J/ψ mass of 3.097 GeV.

The cross section of $\sigma(\gamma p \to J/\psi + X)$ is $(56 \pm 13 \pm 14)$ nb at a mean $W_{\gamma p}$ of about 90 GeV for H1. In Fig. 31 the measured cross section for J/ψ production is shown as function of $W_{\gamma p}$, together with the preliminary ZEUS data. For comparison a diffractive and a QCD based model prediction are shown. Both models have free parameters which can be adjusted to the data. In the QCD model the "K-factor" was adjusted to describe the data at low $W_{\gamma p}$ and the MRSD–' or MRSD0' parton distributions[62] were chosen for the proton. The MRSD–' has a steeply rising gluon distribution for decreasing x while the MRSD0' has a rather flat gluon distribution (see chapter 5). Hence Fig. 31 shows the sensitivity of the measurement to the gluon distribution. Clearly MRSD–' describes the data better. In the diffractive model a substantial amount of proton dissociation has to be added to come close to the data.

A global way to search for heavy flavour production is the analysis of semi-leptonic decays into muons: $c\bar{c} \to \mu + X$. The H1 collaboration performed an analysis where the production of c and b quarks is tagged via high p_T muons. Events with a reconstructed muon with transverse momentum $p_T > 1.5$ GeV/c in the polar region $30° \leq \theta \leq 130°$ are selected. A severe problem for this measurement is the background which is dominated by muons from $\pi^\pm$, $K^\pm$ decays and fake muons: these are estimated to be responsible for half of the observed

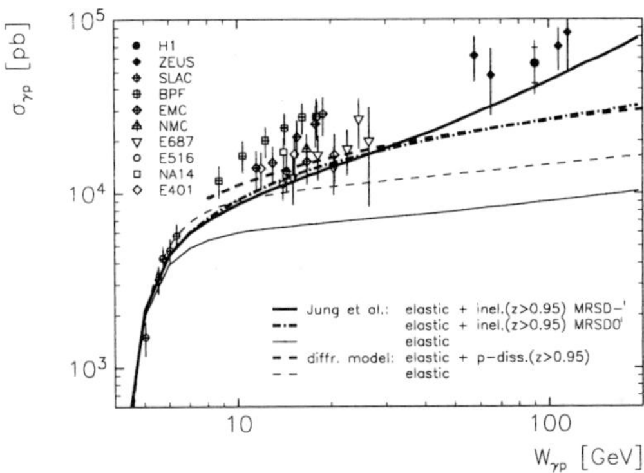

Figure 31: Total cross section for $\gamma p \to J/\psi + X$ from H1 and ZEUS. The data at lower CMS energies are from previous experiments; they were corrected with the new J/ψ decay branching ratio[56] and include systematic errors (added in quadrature). The dashed curves show the predictions from the VDM model in PYTHIA,[37] the thin dashed line is the elastic contribution only, the thick dashed line includes proton dissociation. The thick full line shows the QCD model by Jung et al.[57] with the MRSD–' parton density functions, the dash-dotted line with MRSD0'. The thin full line represents the purely elastic contribution in the QCD model.

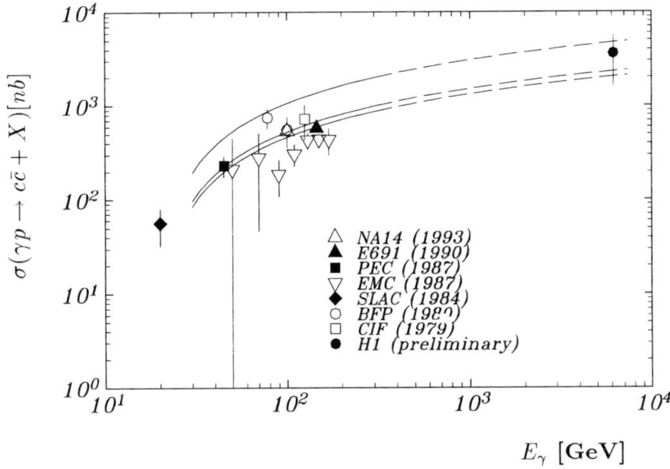

Figure 32: Total cross-section for $\gamma p \to c\bar{c} + X$, including the new preliminary data from H1. The solid curves show QCD calculations[61] for $m_c = 1.5\,\mathrm{GeV}$ and the dashed curves extrapolations to HERA energies assuming a logarithmic rise.

muons. After background subtraction the preliminary total photoproduction cross section derived is $\sigma(\gamma p \to c\bar{c} + X) = (3.6 \pm 0.8 \pm 1.8)\,\mu b$ at a mean $W_{\gamma p}$ of about 114 GeV. The result is shown in Fig. 32 together with earlier measurements at lower values of $W_{\gamma p}$ and a QCD calculation by Ellis and Nason.[61]

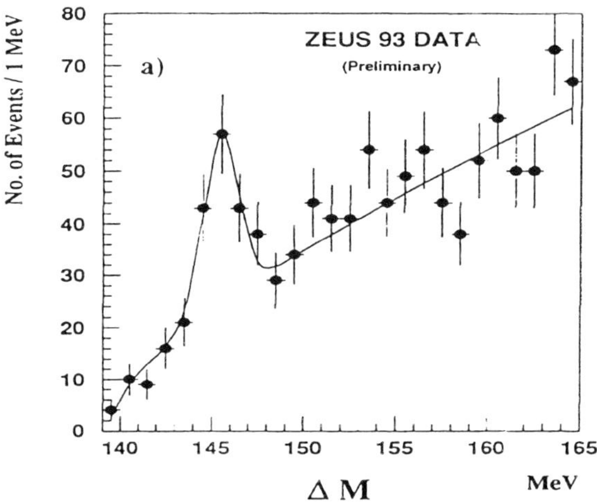

Figure 33: Distribution of the mass difference $m(K\pi\pi_s) - m(K\pi)$ from ZEUS. The enhancement around 145 MeV indicates the production of the D^* meson

The production of c quarks is further studied by tagging $D^{*\pm}(2010)$ mesons in the decay chain $D^{*\pm} \to D^0\pi_s^\pm$, $D^0 \to K^\pm\pi^\mp$. The mass difference distribution $m(K\pi) - m(D^0)$ shows a clear D^* peak (Fig. 33. ZEUS quotes a preliminary cross section of $\sigma(ep \to D^{*+} + X) = (1.5 \pm 0.3 \pm 0.3)$ nb[28] in the kinematic region of $p_T(D^*) > 1.7$ GeV and $\eta(D^*) < 1.5$. From this measurement a total charm cross section of $\sigma(ep \to c\bar{c} + X)$ between 1 and 1.7 μb is extrapolated, depending on the parton densities used for proton and photon.

5 Deep Inelastic Scattering

The centre of mass energy squared $s = 87600$ GeV^2 at HERA opens a completely new kinematical domain to study deep inelastic scattering. Four-momentum transfers Q^2 of up to a few times 10^4 GeV^2 and x values down to about 10^{-4} can be reached. Compared to fixed target experiments, this is an extension of the kinematical domain by almost two orders of magnitude in both x and Q^2. Another advantage of the experiments at HERA is their ability to detect the full hadronic final state, apart from losses in the beampipe. This allows one to determine the kinematical variables from the scattered electron, the hadronic final state, or a mixture of both. Further it enables one to study properties of hadron production in the final state.

For values of Q^2 well below the mass squared of the W and Z^0 gauge bosons, the dominating process is the photon exchange process, thus most DIS events produced at HERA are neutral current events: the scattered lepton is an electron. A typical neutral current deep inelastic event is shown in Fig. 34 in the H1 detector. The scattered electron, detected in the BEMC calorimeter, is well isolated from the hadronic final state, which generally balances the electron in p_T. Note that around the beampipe, in the proton direction, some activity is seen which can be attributed to the fragmentation of the proton remnant or to the colour flow between current and remnant.

In this chapter we will discuss new results on the structure of the proton: the F_2 structure function measurements and attempts to extract the gluon distribution from these data.

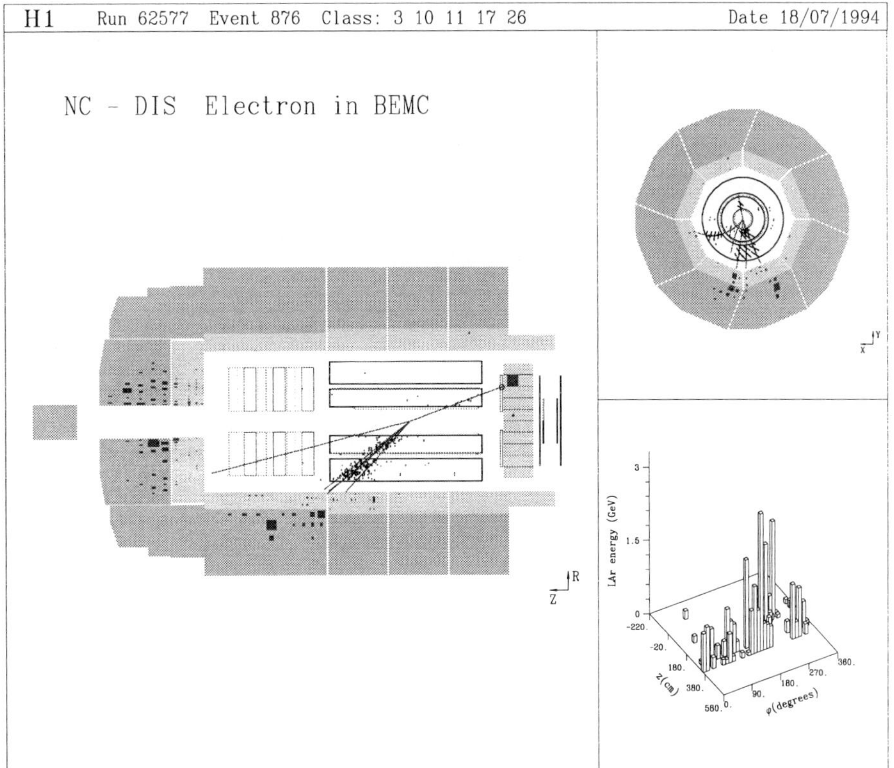

Figure 34: Example of a neutral current deep inelastic scattering event in the H1 detector. The electron is detected in the BEMC calorimeter, the hadronic final state in the main liquid argon calorimeter and central tracker.

Further we will discuss the hadronic final state in DIS events, showing that with the present available data, the fragmentation of the current quark in lepton-hadron scattering and quarks produced in e^+e^- annihilation interactions is quite similar. However, the region between the current quark and the proton remnant turns out to be less understood, leaving room for the onset of new QCD effects. Finally, a quite different type of events has been found at HERA, in which no colour flow is seen between the system including the current quark and the proton remnant. These so called "rapidity gap" events are not yet fully understood, but turn out to be compatible with diffractive processes, similar to the ones discussed in the chapter on photoproduction. Hence HERA is likely to shed light on the dynamics of diffractive processes, known for about 30 years in hadronic physics, but not yet unambiguously explained within QCD.

5.1 Structure Functions

An introduction to the subject of structure functions can be found in the lectures of A. Mueller, in these proceedings. Basically, the structure function F_2 of the proton is derived from the

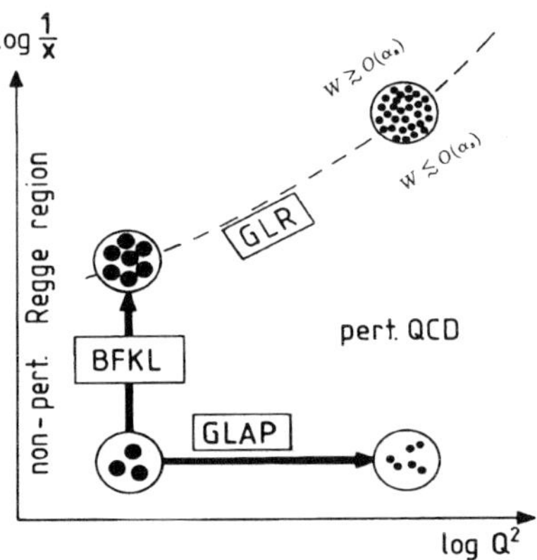

Figure 35: Schematic evolution of the quark densities in various (x, Q^2) regions according to the dominant dynamical effects. The dashed line is the theoretical limit of validity of perturbative QCD.

one-photon exchange cross section:

$$\frac{d^2\sigma}{dxdQ^2} = \frac{2\pi\alpha^2}{xQ^4}(2 - 2y + \frac{y^2}{1+R})F_2(x, Q^2) \tag{5}$$

where $d^2\sigma/dxdQ^2$ is the Born cross section, i.e. corrected for radiative events. Radiative events are events where a photon is emitted, mostly from the incoming or scattered electron. Consequently the x, Q^2 values measured from the scattered electron do not correspond to to the true "x_t, Q_t^2" values of the ep genuine interaction. At higher $Q^2 > 1000$ GeV^2, effects due to Z^0 exchange have to be taken into account, but these remain small ($\sim 5\%$) for the data presented here. R is the ratio of the longitudinal to transverse photon absorption cross section and relates the structure functions F_1 and F_2 by $R = F_2/xF_1 - 1$. This quantity has not yet been measured at HERA, but calculations using the NLO QCD relations[63] lead us to expect that the effect on the F_2 measurements presented here is at most 5%.

Structure function measurements are a key ingredient for the determination of parton distributions in nucleons. These parton distributions $f_i(x)$ give the probability to find a parton i with a momentum fraction x from the original proton momentum. Precise knowledge of parton distributions is important, for example for the calculation of production rates of hadronic processes which can be described by perturbative QCD. Since future high energy colliders (e.g. the LHC) will be sensitive to x values of $O(10^{-3} - 10^{-4})$, measurements at HERA will be a key input for understanding the production rates for QCD processes. Moreover, the dynamics that generates parton distributions at low x is a field of strong theoretical interest. In particular, novel perturbative QCD effects could show up at x values below 10^{-3}, as we will discuss below.

The behaviour of the parton distributions at small values of x was, prior to HERA data, dictated by data from fixed target muon-proton experiments, which include measurements down to $x = 8.10^{-3}$.[80] For smaller x values there was no direct guide from data, leaving room for theoretical speculation on the behavior of the parton distributions in this region.[65] In fact, extrapolations of parton distributions varied wildly in the low x region. These extrapolations were often motivated by either studying the asymptotic behaviour of perturbative QCD evolution equations, or by expectations based on Regge theory. Until now mostly the Dokshitzer-Gribov-Lipatov-Altarelli-Parisi (DGLAP) QCD evolution equations[66] have been used to study and successfully interpret the Q^2 evolution of the available deep inelastic data from fixed target experiments, which are essentially limited to the region $x > 10^{-2}$.

However, the linear evolution equation particularly adapted to study the small x region is the Balitskii-Kuraev-Fadin-Lipatov (BFKL) equation.[67] This is an evolution equation in the variable x, i.e. it relates the parton densities at a value x_0 to any value x. This equation can so far only be used to predict the evolution of the gluon density and is calculated up to LO in QCD. The BFKL equation predicts a characteristic $x^{-\lambda}$ behaviour of the gluon density at small x, with $\lambda \sim 0.5$. At low x the sea quark distribution is expected to be driven by the gluon distribution, thus the $F_2(x, Q^2)$ evolution at small x is expected to reflect the behaviour of $xg(x)$. A $x^{-0.5}$ behaviour of the gluon density will result in a rapid growth of $F_2(x, Q^2)$ with decreasing x.

This perturbative QCD result has to be contrasted with the expectations of the Regge limit where it is expected that $xg(x) \sim x^{1-\alpha_{\mathbb{P}(0)}}$, with the soft Pomeron intercept $\alpha_{\mathbb{P}(0)} \simeq 1$, hence $xg(x) \sim$ constant. This could be a valid scenario for not too large Q^2, but for $Q^2 > 10$ GeV^2 eventually perturbative effects as prescribed by the DGLAP equations, have to become more and more visible.

The DGLAP evolution equations can in fact also cause F_2 to rise at low x. These equations, contrary to the BFKL equation, prescribe the evolution in Q^2, i.e. one can calculate parton densities for $Q_0^2 \to Q^2$, but one needs an explicit non-perturbative input distribution of the x behaviour at a starting Q_0^2. The different approach in evolution for BFKL and DGLAP is pictured in Fig. 35. In principle using the DGLAP equations at low x is questionable since these account only partially for the $\ln 1/x$ terms which become large at low x. The choices for the non-perturbative input at the scale Q_0^2 can range from a flat – Regge inspired soft Pomeron – behaviour, to an already steeply rising – BFKL inspired – behaviour. It will be demonstrated in the next section that the rise of F_2 at small x is either a result of the choice of the non-perturbative input, or a result from the lever arm in evolution in Q^2.

Since $F_2(x, Q^2) \sim \sigma_{tot}^{\gamma^*}$ a continuing increase of F_2 can lead to an unphysical blowup of the cross section. Therefore, it is expected that at very small x the rise should be damped by a new mechanism. A proposed scenario is that at small x the parton densities become so large that annihilation and recombination of parton pairs will start to compete with the parton decay processes included in the standard evolution equations. These "screening" or "shadowing" effects damp the fast increase of the parton density. Such processes have been included in the Gribov-Levin-Ryskin (GLR) equation,[68] and the qualitative results are shown in Fig. 35. In the $x - Q^2$ plane a region will be reached where strong non-linear effects due to parton recombination become important. The border line of this region is often termed the "critical" line. It is however not clear if HERA data will finally probe this new region, since the position of the critical line depends strongly on the strength of the rise of the gluon density at small x. However, an observed strong rise of $F_2(x, Q^2)$ at HERA considerably enhances the probability of observing these novel effects at small x values.

Both the H1 and ZEUS experiments have released new data on structure function measurements[69,15] at small x. We have noted in chapter 1 that, to determine the kinematical variables x and Q^2, we can use two out of four experimentally accessible quantities: energy E_e' and angle θ_e of the scattered electron, and energy E_h and average angle θ_h of the hadron flow. The ultimate method is a global fit of all observed quantities, which requires a level of understanding of the detector response and of the error correlations that the experiments have not yet achieved. In total four methods are currently used in the analyses to reconstruct the event kinematics. The electron method (1), as given in eqn. 2 is the method used so far in all fixed target experiments. It remains at HERA the most precise way to reconstruct Q^2 in the whole kinematic range. However at low y ($y < 0.1$) the measurement of x becomes poor[70] and at large y ($y > 0.8$) the radiative corrections to apply to the observed cross section to extract the Born cross section are very large.[71] The mixed method (2) used by the H1 collaboration in 1992 takes Q^2 from the electron according to eqn. 2 and y from the hadronic variables (y_{JB}) according to eqn. 3. The resolution of y_{JB} is better than the resolution of y_e for low y values but becomes inferior at large y values. For the double angle method (3)[72] only the angles of the scattered electron and the hadronic system are used. The method is almost independent of energy scales in the calorimeters but, at very low y, the method is very sensitive to noise in the calorimeters. The variables y and Q^2 are reconstructed from

$$y_{DA} = \frac{\sin\theta_e(1 - \cos\theta_h)}{\sin\theta_h + \sin\theta_e - \sin(\theta_e + \theta_h)} \tag{6}$$

$$Q_{DA}^2 = 4E_e^2\frac{\sin\theta_\gamma(1 + \cos\theta_e)}{\sin\theta_h + \sin\theta_e - \sin(\theta_e + \theta_h)} \tag{7}$$

$$\tan\frac{\theta_h}{2} = \frac{\Sigma_h(E - P_z)_h}{P_{T,h}^2} \tag{8}$$

A new method used by the H1 collaboration,[69,73] called the Σ method (4), combines y from the following expressions: :

$$y_\Sigma = \frac{\Sigma_h(E - P_z)_h}{(E - P_z)_e + \Sigma_h(E - P_z)_h} \tag{9}$$

where the sum runs over all hadrons in the nominator and over all hadrons plus the scattered electron in the denominator, and Q^2

$$Q_\Sigma^2 = \frac{E_e'^2\sin^2\theta_e}{1 - y_\Sigma} \tag{10}$$

In this method the energy of the incident electron at the interaction is reconstructed, which reduces drastically the sensitivity to the main radiative process. The resolution in x at low y is good enough to allow the H1 collaboration to reach $y = 0.01$. The resolution at large y is worse but less sensitive to radiative corrections than when using only the measurement of the scattered electron. For precision measurements of the structure function all of the different methods are used to control the systematics of event smearing and radiative corrections.

For the final presentation of the results, ZEUS uses the double angle method, while H1 uses the electron method at high y (roughly $y > 0.15$) and the Σ method at low y. The distribution of the events in the $x - Q^2$ plane is shown in Fig. 36 for ZEUS, together with the regions covered by data from fixed target experiments.

Already the analysis of the 1992 data revealed the interesting and perhaps somewhat unexpected result that the proton structure function F_2 rises strongly towards low x.[74,75] However, the significance was limited due to the statistics. The high statistics of the 1993 data enables us to make a more precise measurement and extend the analysis to higher values of $Q^2 (\leq 2000 \text{ GeV}^2)$. Also very low values of $Q^2 (4-8 \text{ GeV}^2)$ could be studied, due to a few hours of data taking when the interaction point was shifted by 80 cm towards the proton direction in order to increase the detector acceptance for electrons scattered under very small angles. Fig. 37 shows the result for F_2, obtained by H1 and ZEUS from the '93 data. The strong rise observed in the '92 data is definitely confirmed with much higher statistical significance. The data of both experiments are found to agree nicely. In Fig. 39 a summary plot is shown with the new data from HERA and the preliminary data from E665[76] as a function of Q^2 at fixed x, compared to published fixed target data and to the GRV parameterization[77]

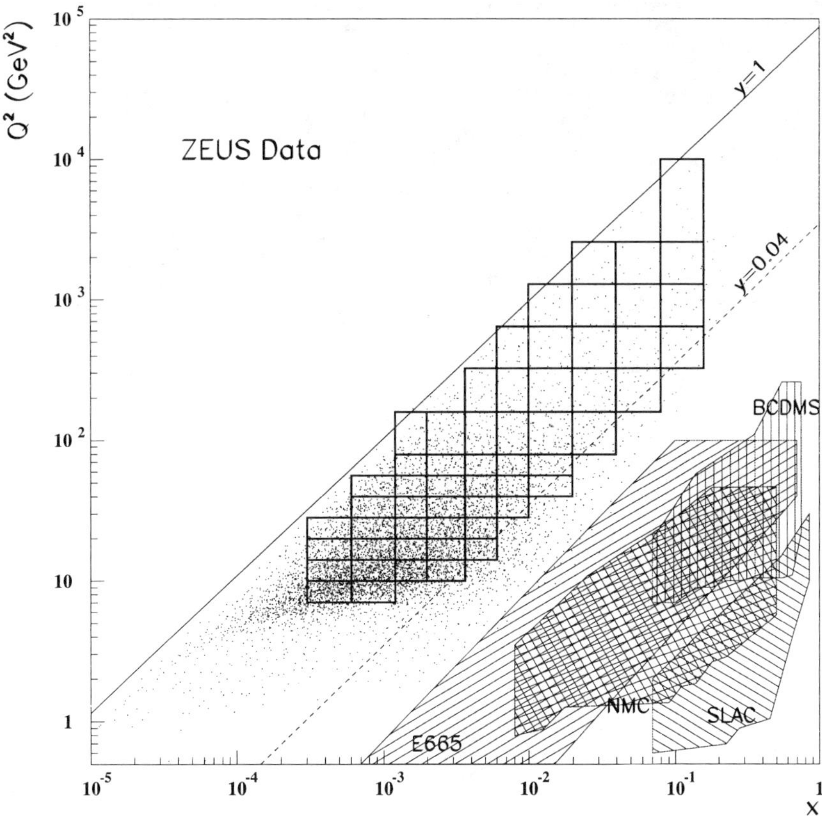

Figure 36: Distribution of the neutral current event sample in the $x - Q^2$ plane. The regions covered by fixed target experiments are shown, together with $x - Q^2$ bins used by ZEUS for the F_2 measurement.

for parton densities (see below). The HERA data agree with a smooth extrapolation from SLAC,[78] BCDMS,[79] NMC[80] and E665 data as well as with the GRV parameterization (see next section). Positive scaling violations are clearly visible at low x and are more and more pronounced as x decreases.

5.2 Comparison of F_2 with Model Predictions

In Fig. 37 predictions for F_2 are shown, calculated from parton density parameterizations which were available prior to the data from HERA. All these calculations assume a certain shape of the x behaviour at small x –where no measurements existed– at a Q_0^2 value and use DGLAP equations to get predictions at other values of Q^2. For the MRS[81] distributions two different scenarios were proposed for the behaviour for $x \to 0$ at a starting $Q_0^2 = 4$ GeV^2: a flat, Regge inspired behaviour (MRSD0'; $\sim x^0$) and a singular, Lipatov inspired behaviour (MRSD–'; $\sim x^{-0.5}$). These parton distributions are evolved in Q^2 with the DGLAP evolution equations, and show that a flat input becomes indeed rather singular for $Q^2 >> Q_0^2$. However the MRSD–' remains steeper than the MRSD0' for all Q^2 values. The CTEQ1MS approach

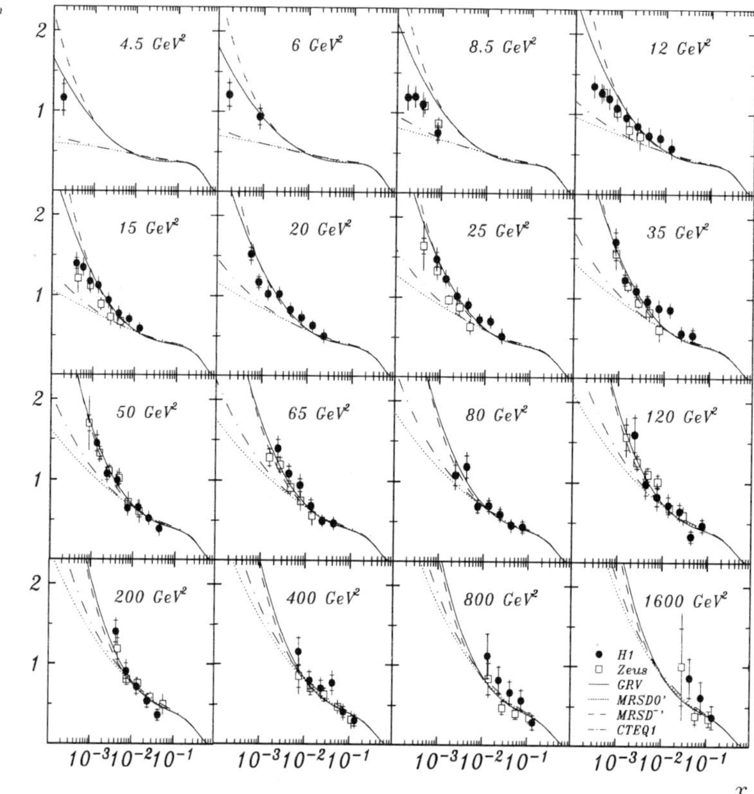

Figure 37: The proton structure function from H1 (full points[69]) and ZEUS (open points[15]) as a function of x for different values of Q^2. The inner error bars show the statistical error, the outer error bars include the systematical error added in quadrature. An overall normalization uncertainty of 4.5% for H1 and 3.5% for ZEUS is not shown. The curves represent pre-HERA fits to previous data.

is similar to the MRSD−' approach, but here the sea-quark distributions are not forced to be strongly coupled to the gluon distribution. The parameterization shown contains a singular gluon distribution, but the F_2 extrapolated in the small x region turns out to be rather flat in x. This results in an F_2 prediction from CTEQ1MS which is close to the MRSD0' calculation. The data clearly exclude the MRSD0'/CTEQ1MS scenarios, and favour more a scenario such as MRSD−'. For $Q^2 < 15$ GeV² however, the MRSD−' prediction tends to rise too fast compared to the data.

The GRV calculation assumes that parton distributions at a very low Q^2, namely $Q_0^2 = 0.3$ GeV², have a valence quark behaviour, i.e. they are expected to vanish for $x \to 0$. The functional form used is $x^\alpha (1 - x)^\beta (\alpha > 0)$ for the parton distributions to fit to fixed target data. These data fix the prediction and there is little or no freedom left for further adjustments at HERA. Despite the valence behaviour Ansatz, which results in a dramatic decrease of the parton distributions at small x for small Q^2, the predictions show a strong rise over the measured Q^2 range. This is a result of the long lever arm used in the DGLAP evolution from

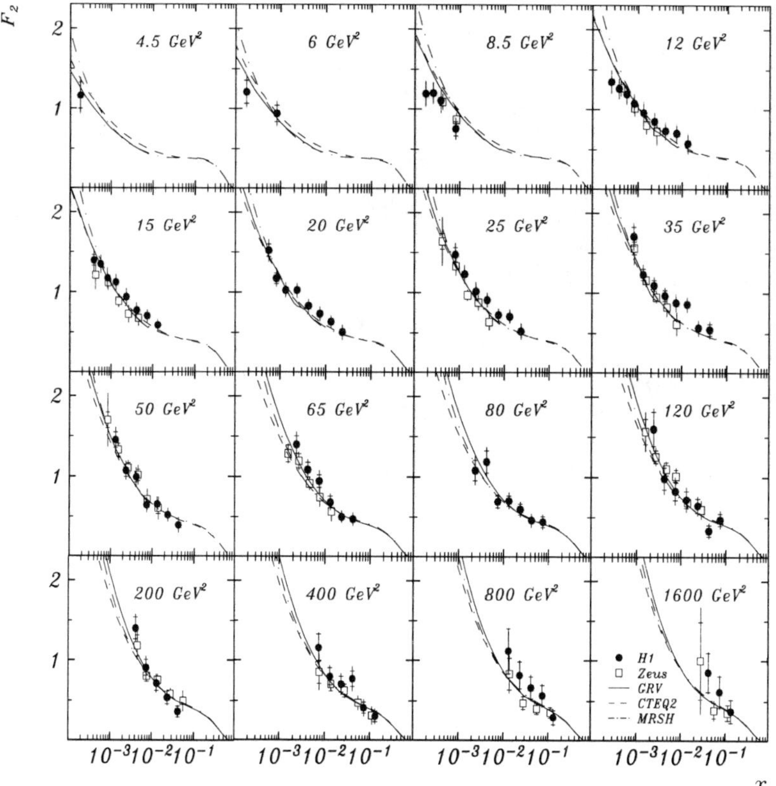

Figure 38: The proton structure function from H1 and ZEUS (same data as Fig. 37). The curves show fits including 1992 HERA data.

the starting scale Q_0^2 to measured Q^2 values. Note that differences are apparent between the MRSD–' and GRV at low x and low Q^2, slightly favouring the GRV distributions.

The 1992 HERA data clearly gave a major clue on the behaviour of F_2 in the new small x domain, for the region $Q^2 \geq 8.5$ GeV². These data were subsequently used in fits by the MRS and CTEQ group to produce new parameterizations. The results are shown in Fig. 38. The new CTEQ and MRS distributions now evidently show better agreement with the data. Note that also the new, lower Q^2 region is rather well described. The GRV distributions were updated w.r.t. to the treatment of the charm quark threshold in the evolution, which affects mainly the lower Q^2 region. These distributions also show a good agreement with the data, which is less trivial, since these do not include the HERA measurements in the fit. In general one can say that parameterizations using the DGLAP equations are able to describe our data, provided a suitable non-perturbative input is chosen (e.g. $\sim x^{-0.3}$ for MRSH[82]), or a large lever arm is taken for the DGLAP evolution (GRV).

In Fig. 40 we have a closer look at the low Q^2 region, comparing the data with parameterizations based on the BFKL – instead of the DGLAP – evolution equations, and with some recent prediction of Regge theory inspired models.

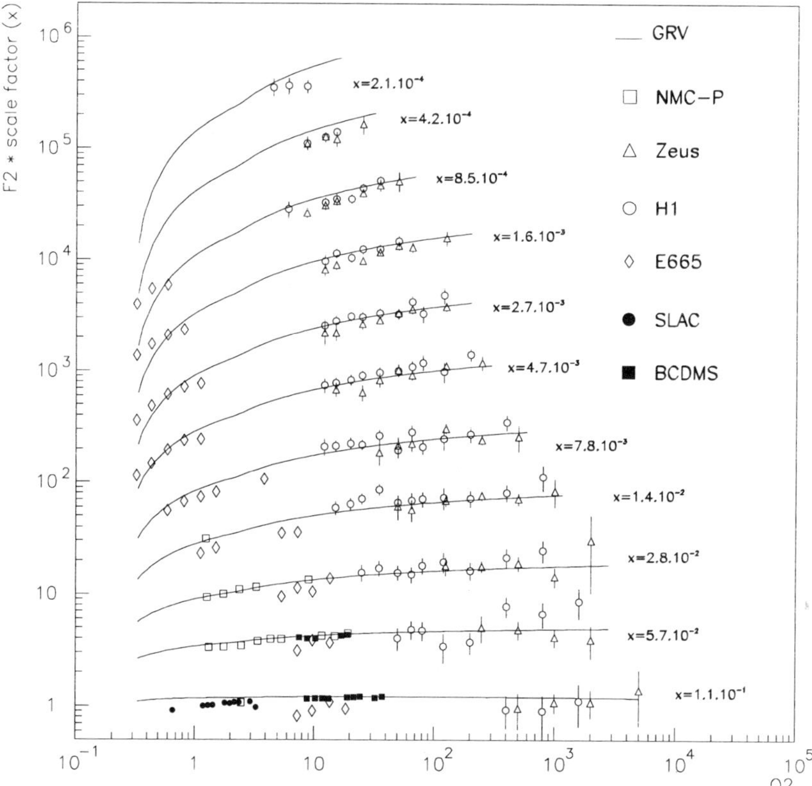

Figure 39: $F_2(x, Q^2)$ at fixed x values as a function of Q^2 from H1 and ZEUS together with data points from E665 (preliminary), SLAC, NMC and BCDMS in the same x bins as the HERA experiments. The data points of fixed target experiments have been slightly rebinned in x to match the HERA values. The error bars show the total errors, except those of E665 which are only statistical. For clarity of the picture, common factors which are different for the different x values have been applied to all data sets.

The Regge theory motivated parameterization relates the structure function to Reggeon exchange phenomena which successfully describe e.g. the rise of the total cross section in hadron-hadron collisions and γp interactions. Using the "bare" instead of the "effective" Pomeron intercept to guide the calculations, the new CKMT predictions[83] rise faster with x compared to former DOLA calculation.[50] The latter ones were already shown to be significantly below our '92 data. The CKMT curves were calculated using the Pomeron intercept $1 - \Delta$, with $\Delta = 0.25$ and without the QCD evolution term, hence they are compared the data in the lowest Q^2 bins only. The newly measured F_2 at small Q^2 values, down to 4.5 GeV^2 opens a new region for testing this Regge assumption. Fig. 40 shows that the parameterization undershoots the data at low Q^2.

The Durham group has used the BFKL evolution equations to predict the x dependence of F_2 at low Q^2.[84] Here we show the sets AKMS1 and AKMS2. The difference between these two sets is the effect of gluon shadowing at very small x. AKMS1 does not include shadowing, while AKMS2 represents the scenario of "hot spots", i.e. it is assumed there are small regions in the proton where shadowing has set in. From Fig. 40 it is shown that these parameterizations describe the data with a similar quality to the ones shown in Fig. 38, based on the DGLAP evolution equations. A similar calculation was presented in.[85]

Shadowing was recently studied in[86] with the GLR equation, and found that it will be necessary to detect electrons at smaller scattering angles at HERA to have a chance to identify these effects. Furthermore, restoring the momentum sum rule in the GLR equation introduces additional "anti-shadowing" terms, which will reduce the shadowing effect.[87]

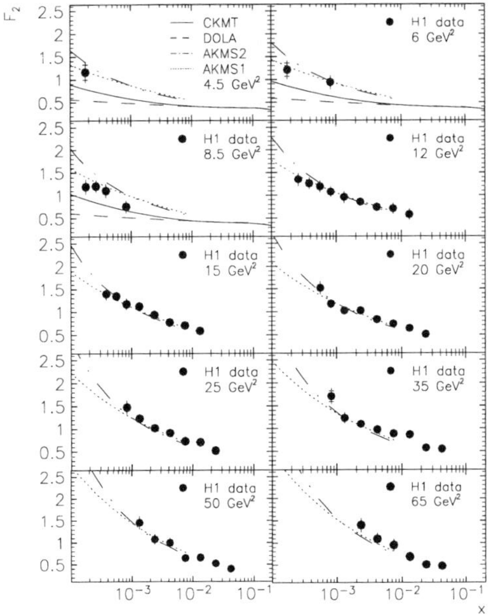

Figure 40: The proton structure function from H1 as a function of x for low Q^2 values, compared with model predictions described in the text: DOLA (dashed lines), CKMT (solid line), AKMS without shadowing (dash-dotted line), AKMS with shadowing (dotted line)

In summary, it turns out that the present data can be described by both approaches: assuming the parton evolution to be dictated by DGLAP or BFKL evolution equations. Presently proposed Regge inspired models are somewhat disfavoured by the data but can most likely be rescued by lowering the Q_0^2 value of the initial non-perturbative part and allow for more QCD evolution lever arm. In the currently covered x, Q^2 range the effect of shadowing is probably small. More precise future data and, in particular, data at lower x values could shed important light on these assumptions and have the potential to discriminate between scenarios. Such data will become available at HERA in the near future from special runs and by the improved coverage of the detectors for DIS events with the electron scattered under small angles.

5.3 QCD Interpretation of F_2 and Determination of the Gluon Density

In this section the data will be analysed in the framework of perturbative QCD. LO and NLO QCD fits to F_2 will be shown, as well as approximate methods to extract the gluon.

In section 5.2 it was shown that the GRV distributions are in accord with the data. These distributions are essentially generated by the DGLAP equations and probed in a region ($Q^2 > 5\,\mathrm{GeV}^2$) far away from the scale of the starting distributions ($Q_0^2 = 0.3\,\mathrm{GeV}^2$). A similar analysis[89] showed that evolving a flat input distribution with the DGLAP equations at a scale of $Q_0^2 = 1\,\mathrm{GeV}^2$ leads to a strong rise of F_2 at low x in the region measured at HERA. An interesting feature is that if pure DGLAP evolution is the underlying dynamics of the rise, the structure function should exhibit scaling in the two variables $\sqrt{(\ln 1/x)(\ln \ln Q^2)}$ and $\sqrt{(\ln 1/x)/(\ln \ln Q^2)}$ at sufficiently large Q^2 and low x values.[88,89] This confirms a prediction from 1974[90] where the asymptotic form of $F_2(x, Q^2)$ at small x had been calculated based on the operator product expansion and renormalization group at leading perturbative order.

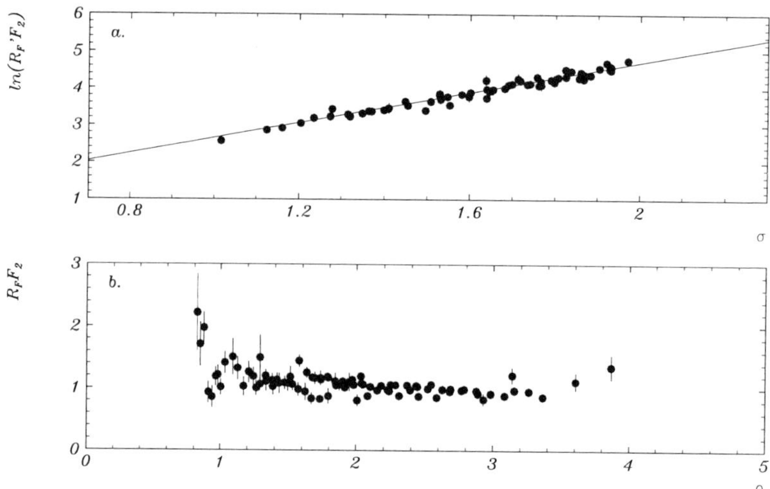

Figure 41: The rescaled structure functions $R'_F F_2^p$ and $R_F F_2^p$ plotted against (a) σ and (b) ρ where $\sigma = \sqrt{\ln \frac{x_0}{x} \ln \frac{t}{t_0}}$, $\rho = \sqrt{\ln \frac{x_0}{x} / \ln \frac{t}{t_0}}$ and $t_0 = \ln Q_0^2/\Lambda^2$. The starting values are $x_0 = 0.1$ and $Q_0^2 = 1\,\mathrm{GeV}^2$. R_F and R'_F are simple rescaling factors to remove the trivial model-independent part of the prediction, given in the text.

In order to test the prediction of double asymptotic scaling we present the F_2 data in a different way.[89] The variables ρ and σ are defined as $\sigma \equiv \sqrt{\ln \frac{x_0}{x} \ln \frac{t}{t_0}}$ and $\rho \equiv \sqrt{\ln \frac{x_0}{x} / \ln \frac{t}{t_0}}$ with $t \equiv \ln(Q^2/\Lambda^2)$. The starting values of the evolution x_0 and Q_0^2 are chosen to be $x_0 = 0.1$ and $Q_0^2 = 1\,\mathrm{GeV}^2$. To present the data as a linear dependence on σ in the region of scaling, the F_2 is rescaled by a factor $R'_F \equiv N\sqrt{\sigma}\rho e^{\delta\sigma/\rho}$, with $\delta = 61/45$ for four flavours and three colours. Fig. 41 shows clearly a linear rise of $\ln(R'_F F_2)$ with σ, confirming scaling in this variable in the range of the data. For this figure only data with $\rho^2 > 2$ are included, which means that points with $x > 0.02$ are excluded, and Λ was taken to be 240 MeV. The LO prediction for the slope of σ with $\ln R'_F F_2$ is 2.4, but higher order corrections are expected to reduce the slope[91] somewhat. A linear fit $\ln R'_F F_2 = a\sigma + c$ to our data gives a (preliminary) slope $a = 2.07 \pm 0.03(\mathrm{stat.})$. Changing Λ by 40 MeV changes the value of the slope by 5%.

Scaling in ρ can be shown by multiplying F_2 with the factor $R_F \equiv R'_F e^{-2\gamma\sigma}$. We observe scaling for $\rho \geq 1.5$ in Fig. 41. This figure is interesting also because the presence of a 'hard Pomeron' behaviour, as given by the BFKL dynamics, is expected to violate the scaling by producing a rise at high ρ. With the available data a moderate increase at high ρ is not excluded. However, the inclusion of higher loop corrections is expected to give a rise at high ρ.[91] In all, double asymptotic scaling seems to work quite well in the region of our data.

Does this approach work at all x and Q^2? Fig. 39 shows that the GRV parameterization is above the preliminary E665 points at small x. Clearly higher twists and other low Q^2 effects can invalidate to use these predictions in that region, but it will be definitely interesting to see whether these discrepancies persist in the final data of E665 after evaluation of the systematic errors.

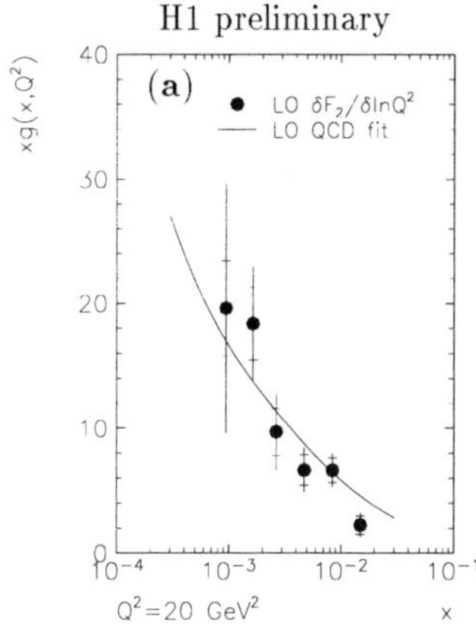

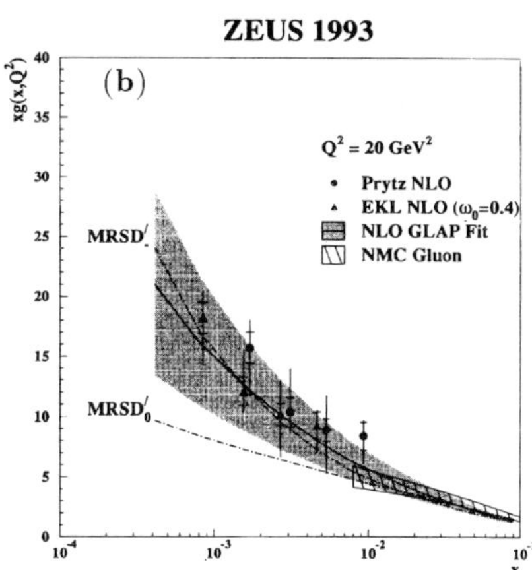

Figure 42: (**a**) The preliminary gluon density $xg(x, Q^2)$ as a function of x at $Q^2 = 20$ GeV^2 as obtained from a LO QCD fit (full line), and from an analysis following the LO Prytz prescription. (**b**) Preliminary gluon density $xg(x, Q^2)$ from the ZEUS collaboration using a NLO QCD fit, the NLO Prytz method and the NLO EKL method. In the EKL method it is assumed that F_2 behaves in $x^{-\omega_0}$ with $\omega_0 = 0.4$. Also shown are the parameterizations MRSD–' and MRSD0'.

Next, we turn to the extraction of the gluon distribution $xg(x)$ in the proton. A first glimpse of the gluon was already presented in section 4.5, in the discussion of the production of heavy quark flavours in γp interactions. Here we will extract the gluon distribution from the structure function measurements. Assuming the DGLAP equations to be valid, the behaviour of $dF_2/d\ln Q^2$ can be used to extract the gluon density in the proton. H1 performed a leading order QCD fit to the F_2 data using only its own data.[39] The result is shown in Fig. 42. Parameters relevant to the high x region have been fixed to values known from fixed target experiments; Λ_{QCD} was set to 240 MeV and the momentum sum rule was imposed. Free parameters are the exponent λ in the gluon distribution $xg(x) \sim x^{-\lambda}$ and the exponent and normalization of the quark-singlet distribution. The χ^2 of the fit is 65 for 86 degrees of freedom, which shows again that the data with the current precision can be described by LO QCD and DGLAP evolution. We obtain $\lambda = 0.38 \pm 0.08$ at $Q^2 = 20$ GeV^2. The fit is shown in Fig. 42a for a Q^2 of 20 GeV^2 as a function of x. ZEUS performed a next to leading order fit, including the NMC and BCDMS data.[28] The NLO fit takes the functional form for the singlet, valence, non-singlet and gluon distribution form the MRS parameterizations. From the fit a value of $\lambda = 0.35$ for a $Q^2 = 7$ GeV^2 is obtained. The result is shown in Fig. 42b.

Several approximative methods have been used to deconvolute the gluon density. The method proposed by J. Prytz[92] consists of neglecting the quark contribution and doing a Taylor expansion of the splitting function around $x = \frac{1}{2}$, leading to a very simple LO expression of the gluon density :

$$xg(x, Q^2) \approx \frac{27\pi}{20\alpha_s(Q^2)} \frac{\partial F_2(\frac{x}{2}, Q^2)}{\partial \ln Q^2} \qquad (11)$$

It is a crude approximation which holds to within 20% at $x = 10^{-3}$.[93] Approximate NLO corrections have been calculated.[94,92] The method of Ellis, Kunszt and Levin (EKL) consists in solving the DGLAP evolution equation in momentum space.[95] This leads to the following relation,

$$xg(x, Q^2) = f_1 \otimes \frac{\partial F_2(x, Q^2)}{\partial(\ln Q^2)} + f_2 \otimes F_2(x, Q^2) \qquad (12)$$

where f_1 and f_2 are known functions to fourth order in α_s and depend on the slope of F_2 in x. The relation is only valid when F_2 has a steep rise at low x. The results for both approximations are shown in Fig. 42.

The errors on the gluon distribution are still large but the message is clear, at $Q^2 = 20$ GeV^2 $xg(x)$ rises by about a factor 5 to 10 as x decreases from 10^{-1} to 10^{-3}. It exhibits a $x^{-\lambda}$ behaviour with $0.2 < \lambda < 0.5$. The NLO gluon data have also been compared to the MRSD0' and MRSD-' parameterizations. The gluon data disfavour the MRSD0' parameterization, in accord with the F_2 measurements and the results of heavy quark production in γp.

In summary, it is now unambiguously established that the structure function rises at small x for Q^2 values down to 4.5 GeV^2. Models which do not predict such strong rise are disfavoured by the data. From the scaling violations of F_2 it follows that also the gluon distribution rises strongly with decreasing x. The inclusive F_2 measurement has turned out not to be conclusive (yet) on the question of whether HERA data at low x are in a new region where conventional DGLAP fails and BFKL evolution has to be used instead. Therefore, it was suggested that additionally exclusive final states should be studied. These are expected to show sensitivity to QCD evolution in the initial state, a topic which will be discussed in the next section.

5.4 The Hadronic Final State: Spectra and Multiplicities

In the naive quark-parton model (QPM) the transverse momentum of the scattered electron is balanced by a single jet resulting from the hadronization of the struck quark, usually called the current jet. Higher order QCD processes modify this picture. Examples of first order processes are shown in Fig. 43, namely photon-gluon fusion (PGF) and QCD Compton (QCDC) processes. These processes can lead to multi-jet final states and can be used to determine e.g. the strong coupling constant α_s or to make a direct measurement of the gluon distribution $xg(x)$. An important question is: what is the nature of the quark kicked out by the exchanged boson? Is it the same object as a quark created in e.g. e^+e^- annihilation? The observed jet universality observed in available e^+e^- and lepton nucleon and data at lower energies clearly hint in that direction, but new tests at higher energies are essential. In this section we will compare the spectra and event multiplicities with those from e^+e^- and low energy lepton-hadron experiments to further check this hypothesis in the HERA kinematical domain.

Predictions for the properties of hadronic final states are available in the form of analytical calculations and Monte Carlo models, which are in general based on standard QCD evolution. In this report we will refer only to two of the currently available Monte Carlo programs: the MEPS and CDM models. The MEPS model is an option of the LEPTO generator[96] based on DGLAP dynamics. MEPS incorporates the QCD matrix elements up to first order, with additional soft emissions generated by adding leading log parton showers. The CDM model[97] provides an implementation of the colour dipole model of a chain of independently radiating dipoles formed by emitted gluons. Since all radiation is assumed to come from the dipole formed by the struck quark and the remnant, photon-gluon fusion events have to be added and are taken from the QCD matrix elements. It is claimed that CDM should approach more the BFKL type of evolution,[98] although it does not explicitly include the BFKL evolution equation.

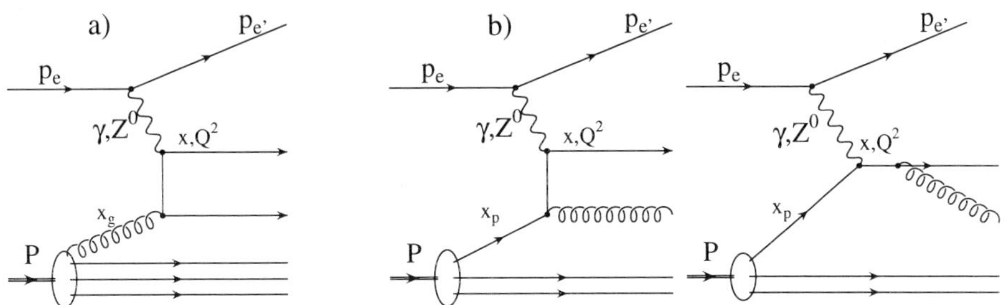

Figure 43: Feynman diagrams for (**a**) the photon-gluon fusion process and, (**b**) the QCD-Compton process

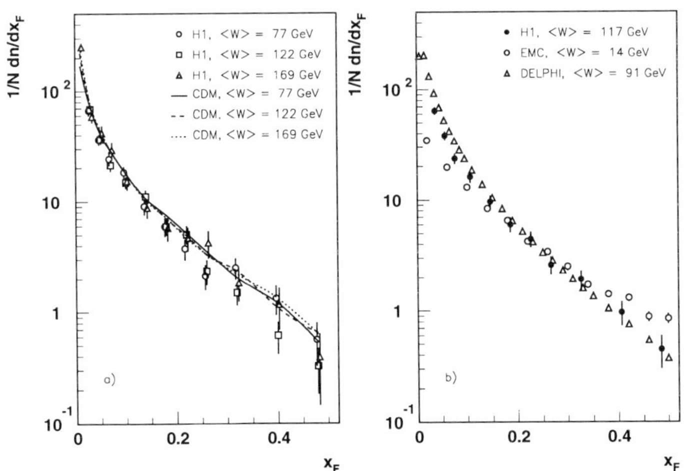

Figure 44: Scaled charged particle spectrum x_F in the hadronic CMS frame. The distributions are normalized to the number of events, and n refers to the number of charged particles in a given bin. In (**a**) the H1 data for three different W bins are shown, together with the CDM Monte Carlo prediction. In (**b**) the H1 data are compared with data from EMC and DELPHI. The DELPHI data are divided by two to account for the two jets in e^+e^- annihilation.

Charged particle production has been studied by both collaborations. Results are presented in the hadronic CMS and in the Breit frame. The hadronic CMS frame is defined as the centre of mass system of the incoming proton and the virtual photon, i.e. the centre of mass system of the hadronic final state with invariant mass W. The z-axis is defined by the direction of the virtual photon. In the quark parton model, the scattered current quark and the proton remnant are back to back along the z-axis. Traditionally the current quark region is defined by all particles with longitudinal momentum $p_z > 0$. A further linear boost along the z-axis from the hadronic centre of mass frame can give a system in which the exchanged current is entirely space-like, having just a z-component of momentum $-Q$. This is called the Breit frame and has been claimed[99] to be the preferred system to study current quark properties because the separation of remnant from current region is theoretically easier to handle. In the simple QPM picture the convention is used that the incident parton approaches with momentum $+Q/2$, absorbs the photon, and leaves with momentum $-Q/2$, in what is called the current hemisphere.

In Fig. 44 the x_F distribution of charged particles is shown for DIS events in the CMS system, compared to results from e^+e^- interactions. The data are corrected for detector effects.[100] The variable x_F is defined as $2p_z/W$, hence Fig. 44 shows essentially the result of the current quark fragmentation. Compared to lepton-hadron collisions at lower energy (EMC, $\langle W \rangle$=14GeV), the HERA data (H1, $\langle W \rangle$=117 GeV) are clearly falling steeper, an indication for QCD induced scaling violations, which have been observed in e^+e^- interactions as well.[101] The HERA data compare quite well with the e^+e^- data (DELPHI, $\langle W \rangle$ = 91 GeV) at a similar hadronic centre of mass energy for the region of $x_F > 0.15$.

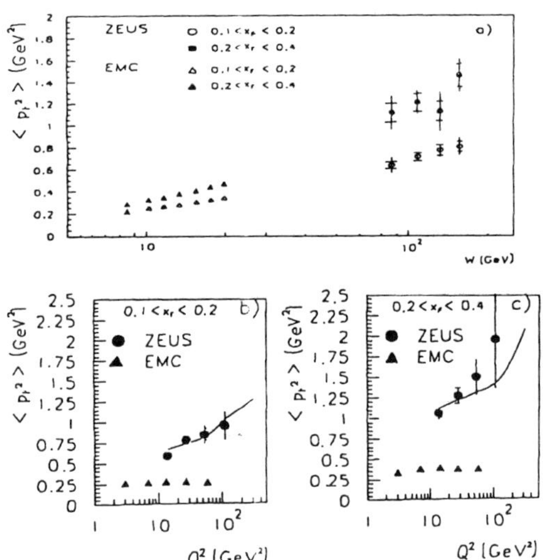

Figure 45: Preliminary ZEUS data showing the rise in track $\langle p_T^2 \rangle$ in the hadronic centre of mass frame as function of W and Q^2 for different x_F intervals.

Next we investigate the $\langle p_t^2 \rangle$ in the CMS frame, in the current quark region. In the simple QPM diagram the quark and proton remnant are back to back, along the z-axis such that the p_T comes essentially from the fragmentation. The QCD diagrams shown in Fig. 43 and similar higher order diagrams are expected to give a significant rise of the p_T with increasing CMS energy. This is shown in Fig. 45 for the lower energy (EMC) and the HERA (ZEUS) data. A clear rise is observed with increasing W. The $\langle p_T^2 \rangle$ and the rise with $\langle W \rangle$ is larger for high x_F values compared to lower ones. Fig. 45 also shows a comparison made as function of Q^2 instead of W. It is tempting to conclude that W and not Q^2 is the variable controlling the increase of $\langle p_t^2 \rangle$, however one has to note that x_F is not a Lorentz invariant variable.

In Fig. 46a the average charged multiplicity is shown for the current quark region as

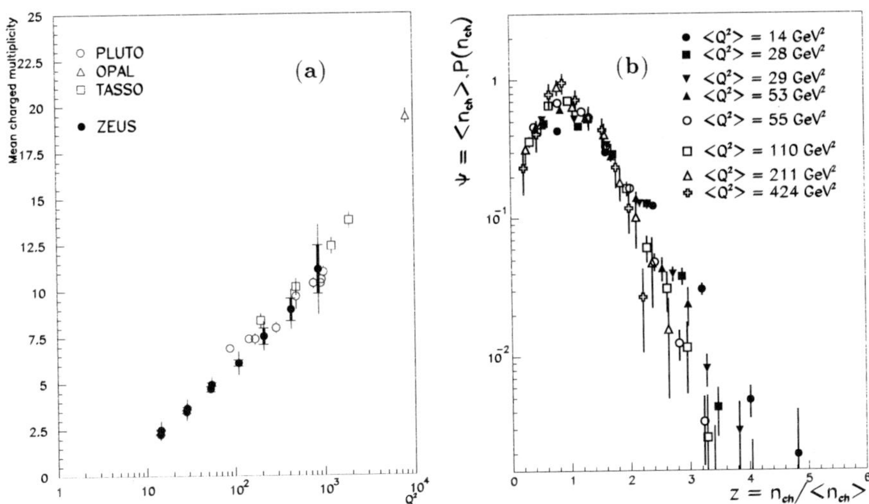

Figure 46: (a) Preliminary ZEUS data on charged multiplicity in the Breit current region as a function of Q^2. The ZEUS data are scaled by a factor 2. The fat error bars are statistical errors, the thin ones statistical and systematic errors added in quadrature. Also shown are e^+e^- data from PETRA and LEP. (b) The KNO plot for ZEUS data (preliminary) at different values of Q^2. Only statistical errors are shown.

function of Q^2, in the Breit frame, for e^+e^- data and preliminary HERA data. Since the current quark has the momentum $Q/2$, Q is the natural scaling variable in the Breit system. The HERA data are scaled up by a factor 2 to account for the two jets in the electron positron annihilation data. The data show good agreement with the e^+e^- data. In Fig. 46b, the shape of the multiplicity is presented in KNO form: the distribution of events multiplied by $\langle n \rangle$ is plotted in the variable $z = n/\langle n \rangle$, allowing a comparison of distributions from different kinematical regions. The KNO spectra are found to be approximately independent of Q^2. Finally we show in Fig. 47 the fragmentation function for HERA and e^+e^- data. The hadronic fragmentation variable x_p is defined as the fraction of the QPM quark momentum carried by the hadron. In the Breit frame we have $x_p = 2p_h/Q$. It is common to show these particle spectra as function of $\ln(1/x_p)$, where Modified Leading Log QCD Approximations (MLLA) coupled with the assumption of Local Parton Hadron Duality (LPHD) predict a Gaussian shape for the data, confirmed by the data in Fig. 47 for seven Q^2 intervals. The area is a measure of the average multiplicity. The peak of the distribution is moving to larger $\ln(1/x_p)$ values with increasing Q^2. This is more clearly demonstrated in Fig. 47, where the ep data are also compared with the e^+e^- data, showing again excellent agreement. The data show a slope different from the expectation of a phase space model. This deviation is sometimes claimed to be due to colour coherence effects, resulting from interference in gluon emission.

In summary distributions concerning quark fragmentation in ep and e^+e^- data show a very good agreement. Within the sensitivity of the present data, the performed studies show no evidence that quarks kicked out of the proton and quarks created in e^+e^- annihilation are different objects, or behave differently.

5.5 The Hadronic Final State: Jets

Already with the data collected in 1992 multi-jet events have been observed and measured[102,103] in DIS events. With the 1993 data we are able to use the jet rates for quantitative measurements. Here we will show:

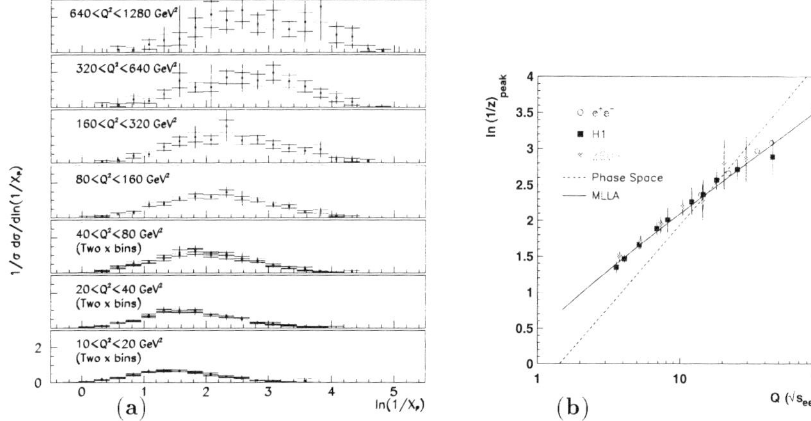

Figure 47: (**a**) Charged particle spectra $\ln(1/x_p)$ with $x_p = 2p/Q$ in the Breit current region from ZEUS (preliminary), forming the "hump backed" plateau. The inner error bars are statistical only. (**b**) Peak position of the $\ln 1/x_p$ distributions as a function of Q. The ZEUS and H1 data (preliminary; statistical errors only) are compared with data from e^+e^- annihilation

- the jet rates at the parton level

- extraction of the gluon distribution in the proton from jets

- extraction of the strong coupling constant α_s

- demonstration of the angular asymmetry.

Jets are experimentally defined using a jet algorithm. So far mainly the *JADE* algorithm[104] and the *cone* algorithm[36] have been used for jet studies at HERA. For the JADE algorithm resolution (y_{cut}) dependent jet multiplicities are determined by calculating scaled invariant masses y_{ij} defined as

$$y_{ij} = \frac{m_{ij}^2}{W^2}, \qquad \text{with} \qquad m_{ij}^2 = 2E_i E_j (1 - \cos \theta_{ij}),$$

that is, neglecting the masses of clusters i and j. The invariant mass of the hadronic system W is chosen as the scale. Clustering is repeated until y_{ij} is above the jet resolution parameter y_{cut} for all clusters. The jet resolution parameter is necessary both for the assignment of final states with soft and nearly collinear partons to a given cross section class and for the regularization of infinities in the theoretical expressions. The remaining clusters are counted as jets. The loss of a large fraction of the proton remnant jet in the beampipe is compensated in the jet algorithm by introducing a pseudoparticle carrying the missing longitudinal momentum of the event. The cone algorithm on the other hand searches for cones with an $E_T > E_T^{min}$ in the azimuthal φ angle and pseudo-rapidity η space, within an area with fixed radius $R = \sqrt{\Delta\varphi^2 + \Delta\eta^2}$. The cone radius is chosen to be $R = 1$, the same as for the photoproduction analyses discussed in section 4.

At HERA we use the following terminology. In the quark parton model (QPM), one jet arises from the struck quark scattering into the detector, while the proton remnant leads to another jet. This is called a "1+1" jet event configuration. The jet of the proton remnant is generally lost in the beampipe. Due to QCD processes to $O(\alpha_s)$, such as gluon radiation in

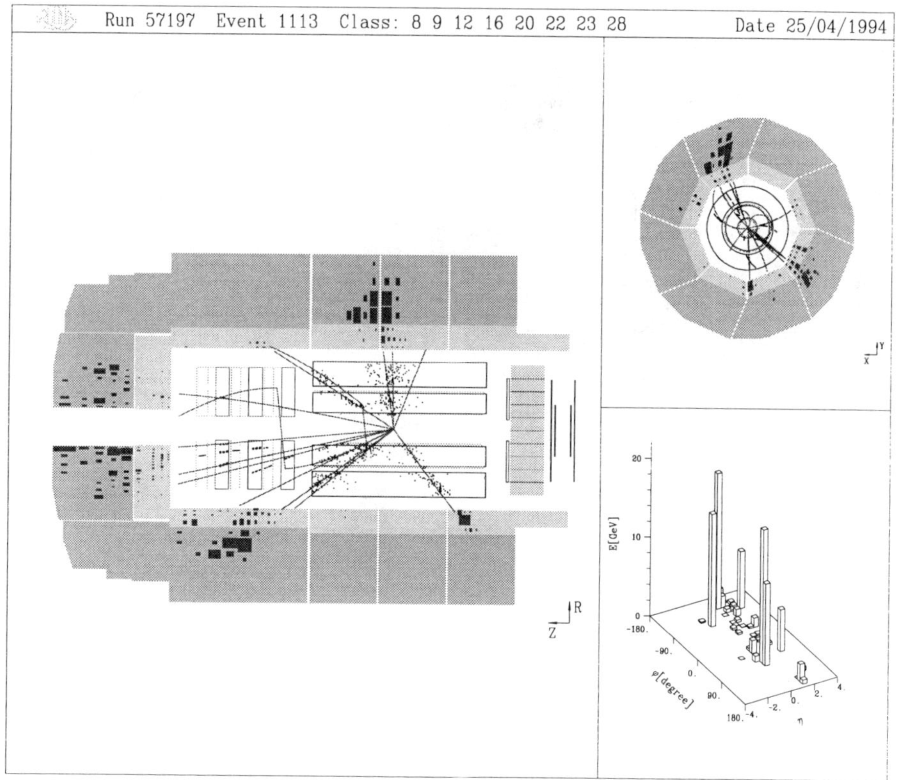

Figure 48: Example of a (2+1) jet deep inelastic scattering event in the H1 detector.

the initial or final state or photon-gluon fusion, a further jet can appear. These events are termed 2+1 jet events. Fig. 48 shows a 2+1 jet event in the H1 detector. Two well separated jets are visible in the detector and in the energy flow plot, with a possible third jet close to the proton remnant direction.

Fig. 49 shows jet profiles and jet rates for 1+1 and 2+1 measured using the JADE algorithm as function of the cut-off parameter y_{cut} by the ZEUS experiment, corrected from observed jets in the detector to jets at the parton level.[28] These corrections were made with the MEPS model. The results are compared with NLO calculations using the programs PROJET[105] and DISJET.[106] The calculations agree rather well with the measurements.

H1 used the measured 1+1 and 2+1 jet rates to extract values of α_s as function of Q^2, identified with the scale in the renormalization group equations (RGE). This technique has been used at e^+e^- colliders before and uses the fact that the hard emission of a gluon in e.g. the QCDC diagram is suppressed by a factor α_s w.r.t. the QPM diagram. It is however not a priori clear that the same technique can be used in ep collisions due to the presence of strongly interacting partons in the initial state. Problems arising in this respect are multiple gluon emission, particularly in the initial state, and the limited precision on the knowledge of parton densities in the proton. The measurement is performed as follows.[107] In NLO QCD the cross sections for 1+1 and 2+1 jet events are given by

$$\sigma_{1+1}(Q^2, y_{cut}) = A_{1+1,0}(Q^2) + \alpha_s(Q^2)A_{1+1,1}(Q^2, y_{cut}) \tag{13}$$

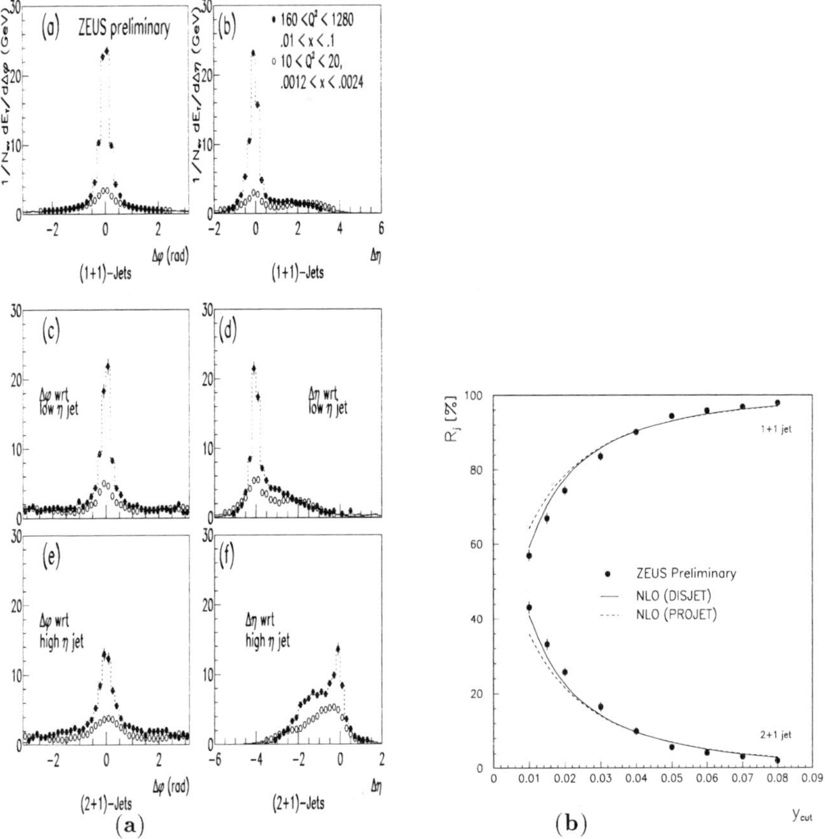

Figure 49: (a) Jet profiles from ZEUS for 1+1 and 2+1 jet events. Plotted is the energy flow transverse to the beam axis around the jet axis in azimuth φ and pseudo-rapidity $\eta = -\ln\tan\theta/2$, where θ is the angle of the energy deposition w.r.t to the proton beam axis. The distance in pseudo-rapidity $\Delta\eta$ is measured such that the proton direction is towards the right. The data from a high and a low Q^2 sample are shown. (b) The preliminary 1+1 and 2+1 jet rates R_j from ZEUS as a function of y_c. Calculations up to next to leading order from the programs DISJET and PROJET for fixed $\lambda = 0.312\,\mathrm{GeV}$ are compared to the measurements.

and

$$\sigma_{2+1}(Q^2, y_c) = \alpha_s(Q^2)A_{2+1,1}(Q^2, y_c) + \alpha_s^2(Q^2)A_{2+1,2}(Q^2, y_c). \tag{14}$$

The terms $A_{i,j}$ contain the hard scattering matrix elements (without the strong coupling constant) and the parton densities of the incoming proton. The first index stands for the jet multiplicity as defined above. The second index indicates the order α_s^j to which the process is calculated. The parameter y_{cut} is the jet resolution parameter, chosen to be 0.02 for this analysis. Using eqns. (13) and (14) the ratio of the 2+1 jets to the full cross section (which consists almost exclusively of 2+1 and 1+1 jet events) can also be expressed as a power series in α_s which is correct to $\mathcal{O}(\alpha_s^2)$. This ratio, corrected to the parton level, can be measured at different values of the scale Q^2, as shown in Fig. 50a. In Fig. 50b the measured ratios are converted to values of α_s. The results give a good description of the jet rates from a QCD calculation in NLO and the running of α_s with Q^2. It has however turned out that the α_s values calculated this way at low Q^2 are rather sensitive to the QCD model used to correct the

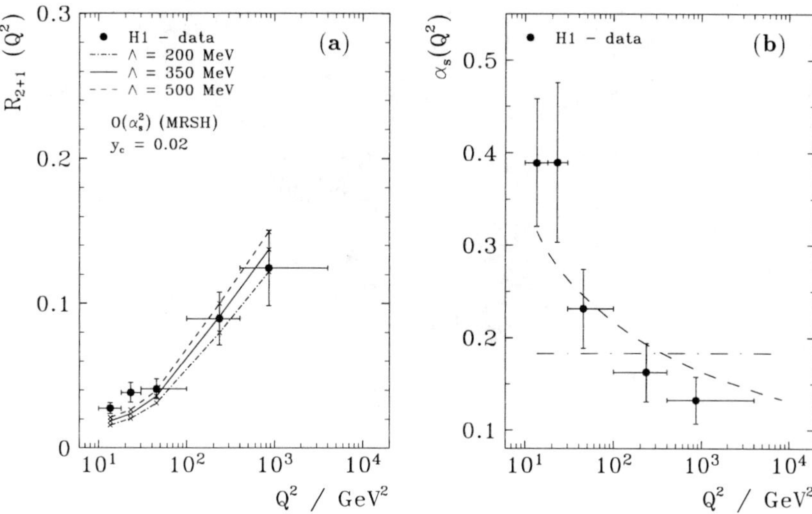

Figure 50: (a) Comparison of jet rates at the parton level from H1 data to QCD in next to leading order using PROJET, for various Λ values and the MRSH parton density. The vertical error bars correspond to the statistical error of the data and the correction factors. (b) The measured value of α_s as a function of Q^2. The fit to the RGE prediction (falling dashed curve) is shown. For comparison the fit to the Ansatz of constant α_s is also included.

observed jets at the detector level, towards the parton level and to a lesser extend both to the cuts used to suppress higher order contributions (beyond NLO) and the limited knowledge of parton distributions in the proton. Therefore for a quantitative determination of α_s only the two highest Q^2 points were taken. A fit of those points to the RGE with the MRSH parton distributions leads to the value

$$\alpha_s(M_z^2) \;=\; 0.123 \pm 0.018 \,.$$

The statistical and systematical errors are added in quadrature. The largest contributions to the systematic errors are given by the dependence of the correction factors on the QCD model to correct to the parton level, and the current 5% uncertainty of the hadronic energy scale of the LAr calorimeter in H1. The value can be compared with $\alpha_s = 0.119 \pm 0.010$ obtained from the LEP experiments using the same observable in NLO,[108] and with $\alpha_s(M_Z^2) = 0.117 \pm 0.005$ from the world average.[56] The agreement between the α_s values determined from the same observable in deep inelastic ep scattering and e^+e^- annihilation again demonstrates the coherence and consistency of the underlying QCD picture.

Inspecting Fig. 43, shows that the gluon distribution of the proton enters the interaction at the Born level for the PGF diagram. Just as for photoproduction events, one can attempt to measure the gluon distribution in LO by trying to isolate this PGF diagram contribution. Heavy flavour tagging would be a natural choice, but the present statistics does not allow such measurement at this stage. Instead a region was selected were the 2+1 jet events are dominantly produced by the PGF mechanism, and the remaining background was estimated by QCD jet cross section programs and subtracted. Jets with $E_T > 3.5$ GeV are selected

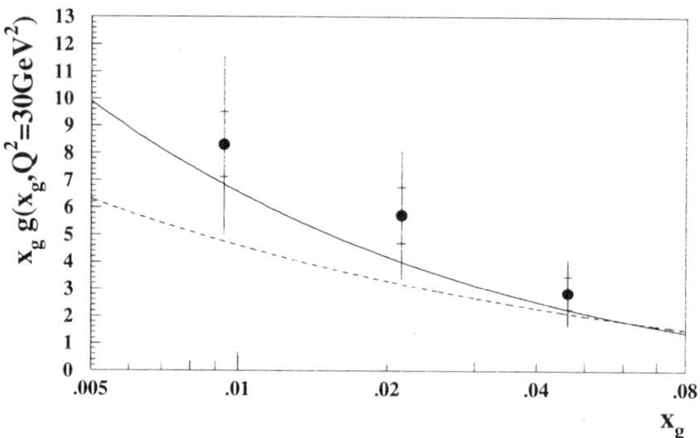

Figure 51: The preliminary gluon distribution as a function of x at $Q^2 = 30$ GeV^2 as obtained from 2 jet events.

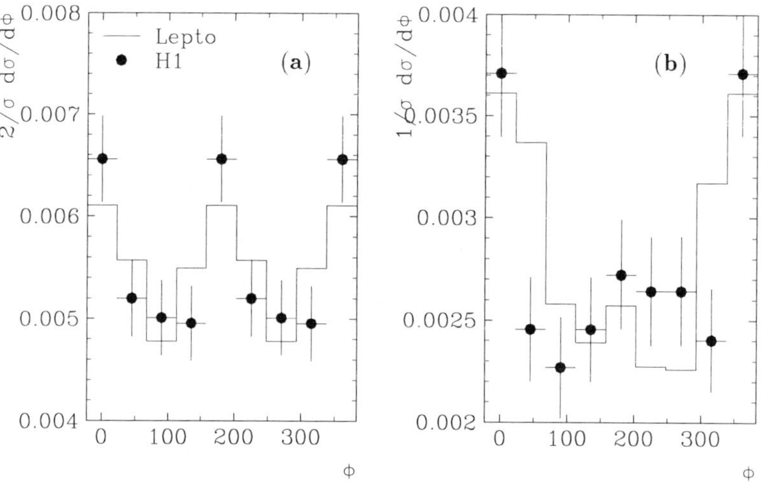

Figure 52: Preliminary azimuthal distribution[110] of the 2-jet plane with respect to the electron scattering plane in the hadronic centre of mass frame, using the H1 data; (a) including both jets of each event; (b) selecting only the most energetic jet. Only statistical errors are given.

in the angular range $10° < \Theta_{jet} < 150°$, using the cone algorithm. The invariant mass of the 2-jet system has to be larger than 10 GeV, and the pseudo-rapidity difference between the two jets is required to be less than two units. The QCDC background prediction of the program PROJET was used, folded with the experimental acceptance calculated with Monte Carlo studies using the MEPS model. The resulting 2+1 event rates were corrected to cross sections. The gluon distribution is extracted at a $Q^2 = 30$ GeV^2, by a bin by bin reweighting of the gluon distribution used in PROJET, using the measured and predicted

PGF cross section. The scale of the gluon extraction was taken to be the p_T of the hard scattering process. The preliminary result is shown in Fig. 51 and is compatible with the gluon distribution determined from scaling violations of the structure function data, discussed in section 5.3. A considerable rise is seen for decreasing x. The large systematical errors result mainly from present differences found when different jet algorithms are used, and are expected to improved in the near future.

A further study examines the azimuthal asymmetry of the 2-jet plane with respect to the electron scattering plane in the centre of mass frame. QCD predicts that QCDC and PGF diagrams exhibit an asymmetry in the distribution of the azimuthal angle between these planes.[109] Preliminary results are shown in Fig. 52. for jets with an $E_T > 4$ GeV (in hadronic CMS).[110] The MEPS Monte Carlo follows the general trend of the data. Future studies will increase the precision of the current methods and exploit new techniques for the extraction of the PGF component from the data.

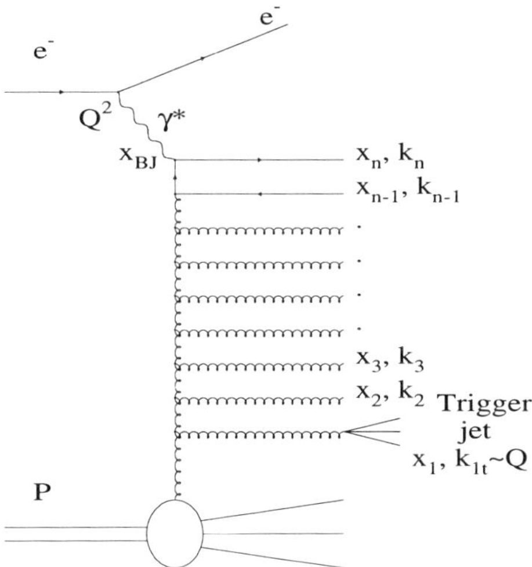

Figure 53: Parton showers in the ladder approximation. The selection of forward jets in DIS events is illustrated.

5.6 BFKL and DGLAP Evolution Revisited

We turn back to the question of DGLAP and BFKL QCD dynamics at low x, introduced in the discussion on the structure function F_2. For events at low x, hadron production in the region between the expected current jet and the proton remnant is of particular interest, since it is expected to be sensitive to effects of the BFKL dynamics. Indeed the initial state QCD radiation is a testing ground for the BFKL and DGLAP hypothesis. This is depicted in Fig. 53, showing that before the struck quark is hit by the virtual photon, it may emit a number of gluons. The figure indicates the proton momentum fractions x_i and transverse momenta k_i (virtualities) of the quarks and gluons which are emitted. In the DGLAP scheme the cascade follows a strong ordering in transverse momentum $k_n^2 >> k_{n-1}^2 >> ... >> k_1^2$, while there is only a soft (kinematical) ordering for the fractional momentum $x_n < x_{n-1} < ... < x_1$. For the BFKL scheme the cascade follows a strong ordering in fractional momentum $x_n <<$

$x_{n-1} << ... << x_1$, while there is no ordering in transverse momentum. In fact the transverse momentum undergoes a random walk type of diffusion: the k_i value is not to far from the k_{i-1} value, but it can be both larger or smaller.[111] Therefore BFKL evolution is expected to produce more E_T in the region between the current and remnant for low x events, compared to DGLAP evolution. Hence, the E_T flow measurement probes the evolution dynamics for small x processes.

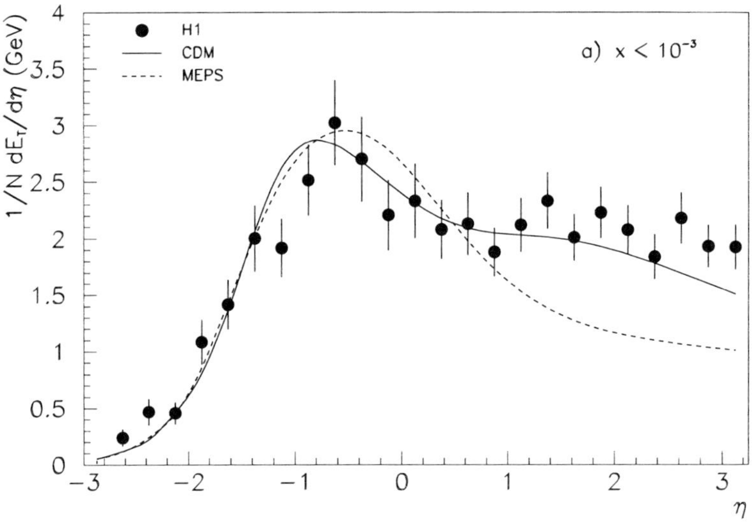

Figure 54: Transverse energy flow E_T in the laboratory system as a function of the pseudo-rapidity η with $x < 10^{-3}$. The proton direction is to the right. The error bars contain the statistical and systematic errors added in quadrature, except for an overall 6% energy scale uncertainty.

The transverse energy flow has been studied by H1.[100] Fig. 54 shows the transverse energy flow corrected for detector effects as a function of the pseudo-rapidity η, in the laboratory system for values of $x < 10^{-3}$. Away from the current quark the data show a plateau of $E_T \approx 2$ GeV per unit of rapidity. The CDM and LEPTO model predictions are compared to the data. While the CDM model describes the data reasonably well, the DGLAP based MEPS model fails to describe the plateau away from the current quark and clearly undershoots the data in this region. Recently analytical calculations predicting the transverse energy flow at the parton level have been performed[112] both for DGLAP and BFKL scenarios. The result for the BFKL at the parton level is shown in Fig. 55, for $\langle x \rangle = 5.7 \cdot 10^{-4}$ and $\langle Q^2 \rangle = 15$ GeV^2. The BFKL calculation predicts a fairly flat plateau at low x with $E_T \approx 2$ GeV per unit of rapidity. The E_T is considerably lower for predictions based on DGLAP parton showering dynamics. The analytic DGLAP calculation yields about 0.6 GeV E_T at the parton level. The effect of the additional E_T contribution due to fragmentation effects is shown in Fig. 55, by the the histogram, using the LEPTO model. The discrepancy with the data remains large.

one notices that BFKL dynamics predicts a fairly flat plateau at low x with $E_T \approx 2$ GeV per unit of rapidity. The E_T is considerably lower for predictions based on DGLAP parton showering dynamics.

Another possible footprint of the BFKL dynamics is the rate of jets produced in a DIS event with the following characteristics.[113,114] The transverse size $1/k_j^2$ of the selected jet should be close to $1/Q^2$ and the momentum fraction x_j of the jet should be as large as possible, whereas

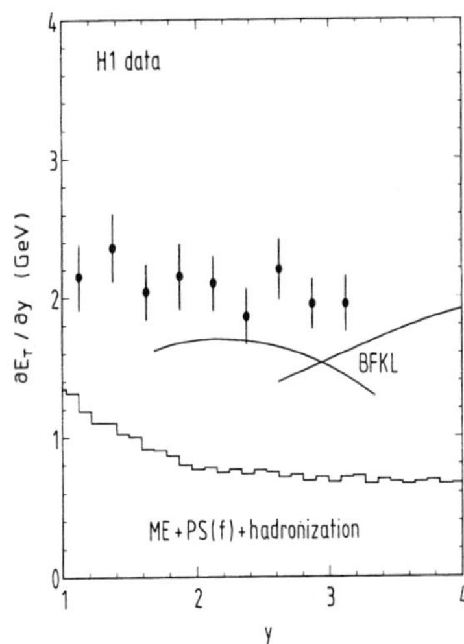

Figure 55: The same data as Fig. 54 (η is termed y here). The continuous curve show the BFKL predictions at the parton level of $x = 5.7 \cdot 10^{-4}$ and $Q^2 = 15$ GeV^2, which correspond to the average values of the data sample. The histogram is the MEPS Monte Carlo estimate, including hadronization.

x range	data	MEPS MRSD0(−)'	CDM
$2 \cdot 10^{-4} - 2 \cdot 10^{-3}$	$128 \pm 12 \pm 26$	69 (53)	32
$2 \cdot 10^{-4} - 1 \cdot 10^{-3}$	$85 \pm 9 \pm 17$	37 (27)	21
$1 \cdot 10^{-3} - 2 \cdot 10^{-3}$	$43 \pm 7 \pm 9$	32 (26)	11

Table 3: Number of DIS events with a selected forward jet compared to Monte Carlo predictions. (Preliminary.)

the momentum fraction x_{Bj} of the quark struck by the virtual photon should be as small as possible. The process is shown in Fig. 53. The rate of those jets is sensitive to the type of evolution dynamics since for the DGLAP case, due to the strong ordering of k, there is little room for the evolution in Q^2 if $k_j^2 \approx Q^2$ while for the BFKL Ansatz the gluon radiation is governed by the ratio x_j/x. Hence for a low x event the phase space for emission of a high x_j jet is large. Therefore we expected the jet rate to be higher for the BFKL than for the DGLAP scenario. In a sample of DIS events with $Q^2 \approx 20$ GeV^2 and $2 \cdot 10^{-4} < x < 2 \cdot 10^{-3}$ we have counted the jets with $x_j > 0.05$ and $0.5 < k_j^2/Q^2 < 6$. The resulting number of events, corrected for background contribution, is given in table 3 and compared to expectations of the MEPS and CDM models simulated in our detector. These predictions were found not to depend significantly on the parameterization of the structure function and generally tend to be below the observations in the data. The size of the errors do not allow yet a firm conclusion. We can however notice that the rate of jets rises with decreasing x. This is expected from BFKL dynamics as an analytical calculation[114] has demonstrated. At the parton level, with the same cuts as for this analysis, the BFKL evolution yields 75 and 36 events for the low and high x bin respectively. Without BFKL evolution, i.e. only taking into account the box diagram, the calculation expects 25 and 20 events respectively. So the tendency of the data is there, but a correction from the measured jet rates to the parton level and more statistics will be necessary before we can make quantitive comparisons with the analytical calculations.

5.7 DIS Events with Rapidity Gaps

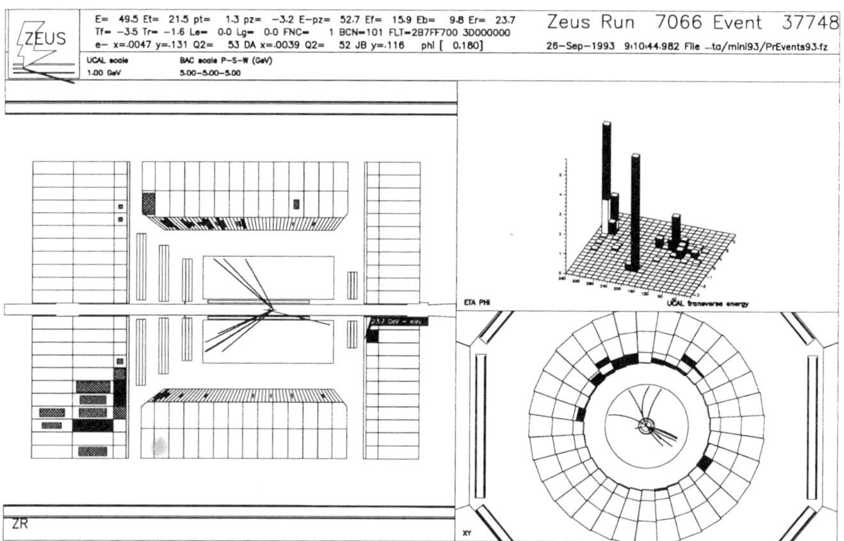

Figure 56: Display of a deep inelastic event with a large rapidity gap observed in the ZEUS detector. There is no energy deposited in a large region of rapidity in the proton direction, to the left in the figure.

Analysing the data taken in 1992, the ZEUS and H1 collaboration observed a peculiar class of DIS events[116,117] which, unlike the majority of the events, had no energy flow in a region around the beampipe in the forward proton direction. Some activity is expected in the forward direction for "conventional" DIS processes (Fig. 1), due to the colour connection between the struck quark and the proton remnant. However, these events show a region with no activity – a gap – in the forward detector region. An example of such an event is shown in Fig. 56, which can be compared to a conventional DIS event shown in Fig. 34. In chapter 4 a class of events with similar characteristics was found in photoproduction interactions and was interpreted as diffractive scattering. An indicative variable to tag diffractive events is the η_{max} of the event, as introduced in section 4.4. The η_{max} distribution for DIS events is shown in Fig. 57 for the ZEUS experiment and can be compared with a corresponding distribution in γp interactions in Fig. 26. A similar behaviour is seen in both distributions. The rate of DIS events with a small η_{max} (i.e. large gap) is substantially above the expectations of standard DIS models,[97] showing these events are not included in the conventional DIS Monte Carlo programs. It may be tentatively assumed that the DIS events which exhibit a rapidity gap are connected with diffractive scattering, in a way similar as for photoproduction.

In fact, it was anticipated that the HERA collider should provide a rather unique possibility to study diffractive dissociation at short distances[118] and that the rapidity gap would be a powerful criterion to eliminate conventional deep inelastic background.[119] An example of a model to explain these events in terms of diffraction is shown Fig. 58b, compared to the conventional DIS process in Fig. 58a. The model assumes that the photon scatters off the partonic content of an object in, or emitted by, the proton, called the Pomeron (introduced in sections 4.2 and 4.4). The Pomeron is not colour connected with the proton and hence there is a region between the proton remnant and the hadronic system probed by the virtual photon

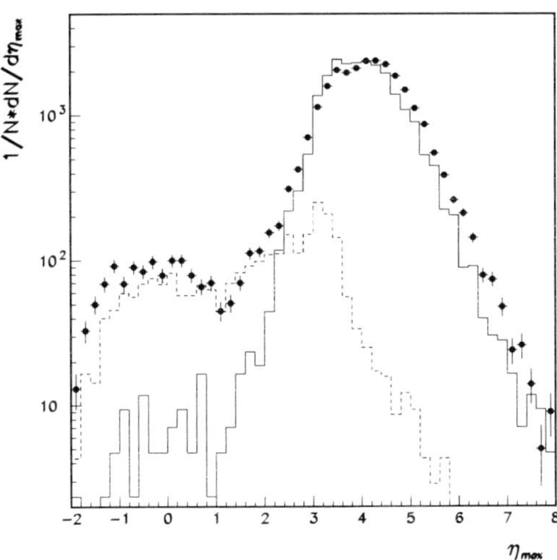

Figure 57: The distribution of the variable η_{max}, the rapidity of the most forward energy deposit above 400 MeV in the calorimeter. The solid circles are ZEUS data points. The full histogram is the CDM Monte Carlo, and the dashed histogram is the POMPYT Monte Carlo.

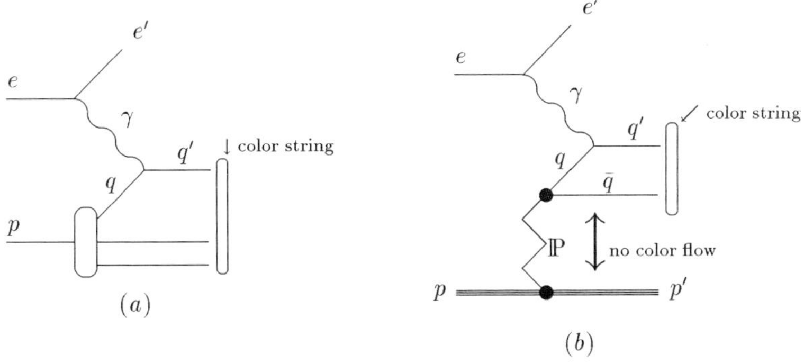

Figure 58: Model for the hadronic final state in usual deep inelastic scattering (**a**) and diffractive deep inelastic scattering (**b**).

without colour flow. This can lead to an observable gap in the detector, and such models can account for the observed η_{max} distribution, as is shown in Fig. 57. These processes are implemented in the Monte Carlo programs POMPYT[53] and RAPGAP.[115] Note however that the mere observation of events with gaps does not unambiguously prove that these events are indeed of a diffractive nature, including Pomeron exchange. In fact other colourless exchange, such as meson exchange, also leads to events with gaps.

In the case that diffraction represents the underlying dynamics of the rapidity gap events then these events can be used to probe the partonic content of the Pomeron. These events were most likely also present in the data of fixed target DIS experiments (in fact exclusive $\rho^0(770)$ production has been reported; see below), but the large centre of mass energy of HERA and the capability to detect the hadronic final state in H1 and ZEUS enables these events to be

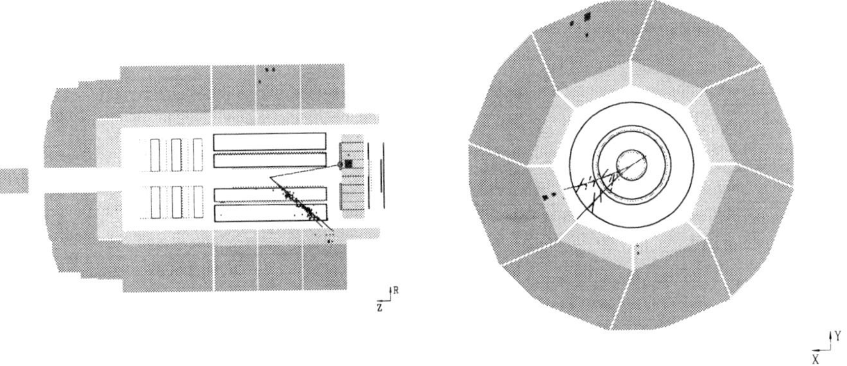

Figure 59: Example of an exclusive $\gamma^* \to \rho^0 p$ event candidate. Apart from the scattered electron only two charged particles are detected.

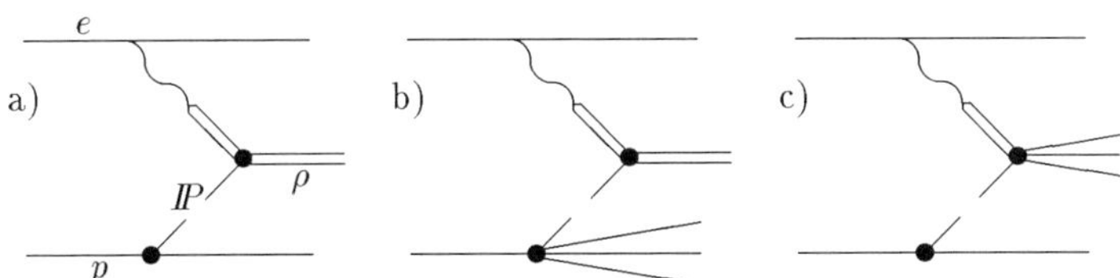

Figure 60: Diagrams which illustrate the VDM description of the rapidity gap events: (**a**) $\rho^0(770)$ production via elastic VDM; (**b**) VDM $\rho^0(770)$ production with soft dissociation of the proton; (**c**) VDM photon interaction followed by soft dissociation of the vector meson. The double dissociation diagram is not shown.

isolated with a topological selection, such as the η_{max} selection. Hence the structure of the Pomeron can be measured and questions on the shape of the parton distributions and on whether quarks or gluons dominate the structure of the Pomeron (see section 4.4) can be studied. So far our knowledge of the structure of the Pomeron is essentially based on the results of the UA8 $p\bar{p}$ experiment.[52]

We will argue below that such a mechanism of deep inelastic scattering on a Pomeron is indeed compatible with many of our observations, but other processes are likely to be needed to give a full description of these events. This results from the observation of events which have, apart from the scattered electron, only two charged particles in the detector (Fig. 59). The invariant mass of these particle pairs, shown in Fig. 61, reveals a clear $\rho^0(770)$ vector meson peak. Thus, there is an exclusive vector meson component in the data which amounts to about 10% of all diffractive events. A possible interaction mechanism for this process is a vector meson dominance contribution (VDM), where the photon fluctuates into a vector meson similar to photoproduction processes. Hence exclusive leptoproduction of vector mesons can be elastic (Fig. 60a) or followed by soft dissociation of the proton (Fig. 60b). H1 has shown that the shape of the low η_{max} ($\eta_{max} < 1.5$) distribution can be also reproduced by a VDM motivated simulation.[120] The ZEUS collaboration has studied the exclusive production of vector mesons in the range $7 < Q^2 < 25$ GeV^2 and $0.01 < y < 0.25$. The preliminary

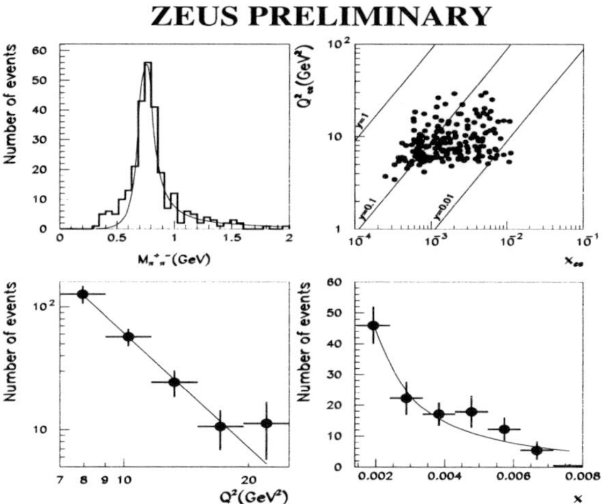

Figure 61: (**a**) Mass ($\pi^+\pi^-$) spectrum of particle pairs with a Breit-Wigner fit to the data; (**b**) Q^2 versus x for ρ^0 events; (**c**) corrected ep Q^2 distribution for ρ^0 events. The line is an exponential fit to the data; (**d**) corrected ep x distributions for ρ^0 events. The curve is a $x^{-1.6}$ functional form.

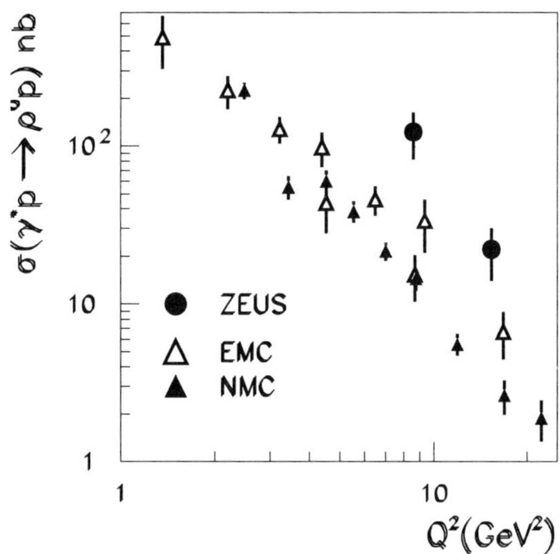

Figure 62: ρ^0 cross section for virtual photoproduction as a function of Q^2. The mean γ^*p centre of mass energy W is about 10 GeV in the EMC/NMC data points and 100 GeV in the ZEUS data points.

data on the x and Q^2 dependence is given in Fig. 61 corrected for detector acceptance and resolution. After fitting the Q^2 distribution to an exponential form, the power of the Q dependence obtained was $-8.2 \pm (\text{stat})^{+1.4}_{-0.3}(\text{syst})$. The x distribution falls off with increasing x and is reasonably described by the form $x^{-1.6}$. The cross section in the range $7 < Q^2 < 10$ GeV^2 for $\gamma^* \to \rho^0 p$ is $123 \pm 15(\text{stat}) \pm 39(\text{syst})$ nb, to be compared with a prediction of 165

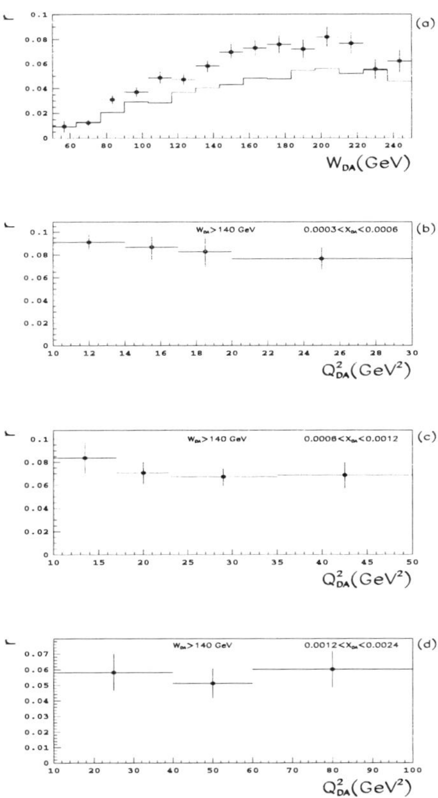

Figure 63: The ratio of DIS events with a large forward rapidity gap ($\eta_{max} < 1.5$) to all DIS events measured by ZEUS, uncorrected for detector effects, as function of (**a**) the total hadronic energy W and (**b-d**) the Q^2 of the interaction, for different x intervals. The variables W and Q^2 are calculated with the double angle method. The histogram in (a) shows the detector acceptance as function of W. Figs. (**b-d**) are shown for $W > 140$ GeV.

nb.[121] The production of vector mesons at high Q^2 has already been observed in fixed target experiments.[122] In Fig. 62 the ZEUS results are compared with these data. At $Q^2 = 8.6$ GeV^2, the ZEUS collaboration measures a $\gamma^* p \to \rho^0 p$ cross section which is about 3 times larger than those of EMC[123] and NMC[124] at the same Q^2 value. Note that both HERA experiments cannot observe the hadronic system on the proton side, since it disappears in the beampipe. Hence this measurement involves the subtraction of events where the vector meson production is associated with proton dissociation which is not observed. This part has been estimated to be 10% in Monte Carlo studies by the ZEUS collaboration. Proton tagging devices are being commissioned by both HERA experiments, which will allow in future to select samples of elastic events. The elastic production of $\rho^0(770)$ will remain a hot topic at HERA, since it has been shown that it can be used to probe BFKL dynamics.[125]

A sample of rapidity gap events is defined by the cut $\eta_{max} < 1.8 (< 1.5)$ for H1 (ZEUS). The total observed fraction of rapidity gap events in the DIS sample amounts to about 6(5)%. However, the cut on η_{max} has selected only part of the diffractive events, namely those events were the rapidity gap is visible in the detector. After an acceptance correction which depend on the models used to describe the events, the H1 and ZEUS experiments get an estimate

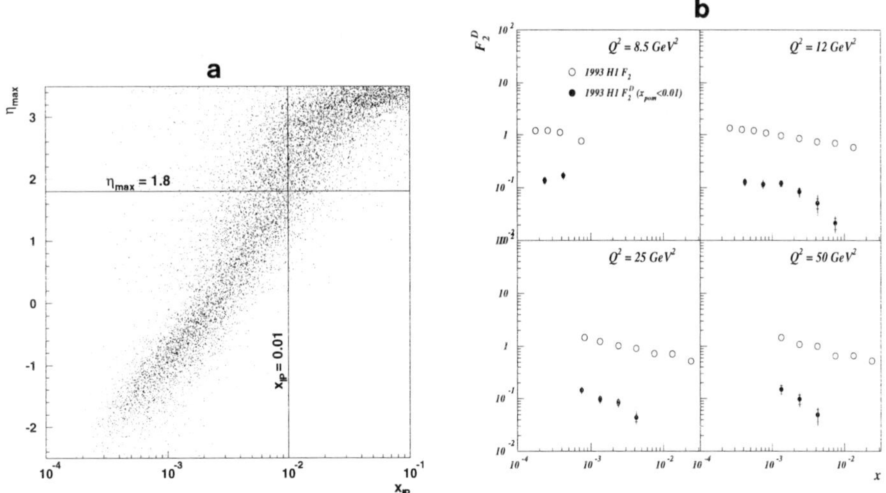

Figure 64: (a) Correlation between η_{max} and $x_{\mathbb{P}}$ from Monte Carlo studies by the H1 collaboration. b) The H1 results on the diffractive contribution F_2^D to F_2 for $x_{\mathbb{P}} < 0.01$.

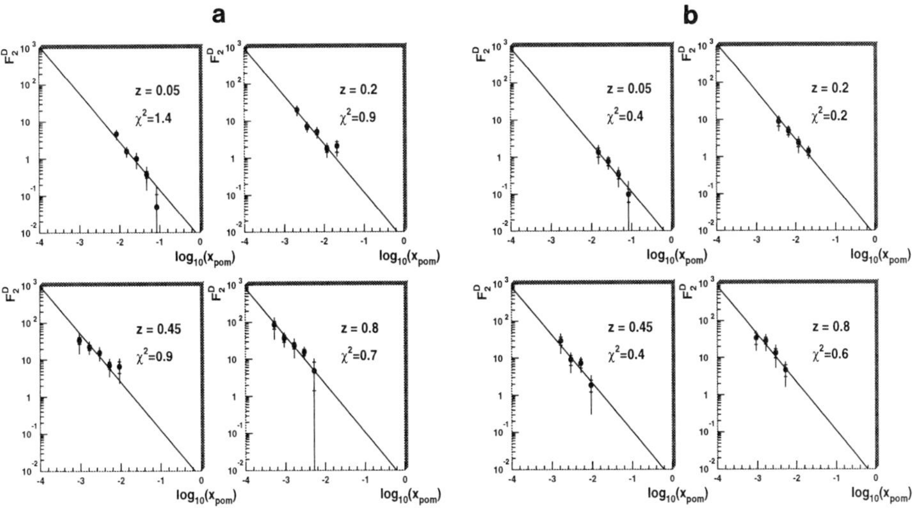

Figure 65: The diffractive part of the structure function $F_2^{D(3)}(x, Q^2, x_{\mathbb{P}}, t)$ as a function of $x_{\mathbb{P}}$ for (a) $Q^2 = 15$ GeV^2 and (b) $Q^2 = 30$ GeV^2. The data points are preliminary results from the H1 experiment. The straight line is a fit of the $x_{\mathbb{P}}^\alpha$ behaviour with $\alpha = -1.3 \pm 0.1$.

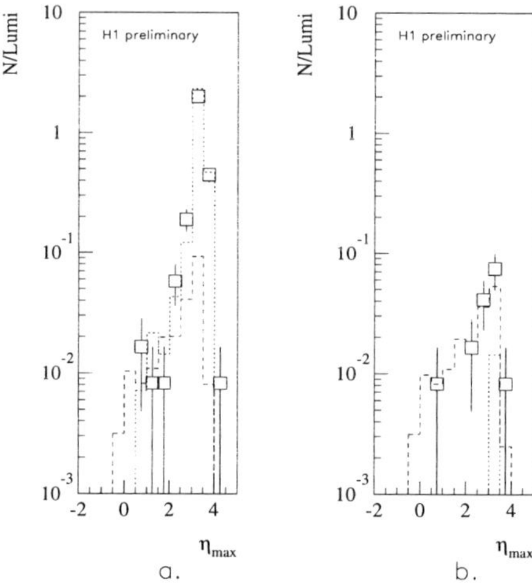

Figure 66: η_{max} distribution of (**a**) all (2+1) jet events and (**b**) those satisfying the forward detector selection, described in the text obtained by the H1 collaboration (preliminary). The prediction from the deep inelastic scattering Monte Carlo[96] is shown with the dotted line and the dashed line shows the prediction from hard diffractive scattering.[115]

of the corrected fraction of diffractive events in the DIS sample which varies between 10 and 15 %.[120,28]

The rapidity gap events at HERA selected with the η_{max} cut are distributed over the entire Q^2 and x range covered by the sample of detected DIS events. Fig. 63 shows the ratio of rapidity gap events to all DIS events as a function of Q^2 in narrow x regions as measured by the ZEUS collaboration. The data are restricted to values of $W > 140$ GeV where Monte Carlo calculations show that the acceptance is flat. It is striking that this ratio has no significant dependence on Q^2, which is consistent with a leading twist QCD production mechanism. The similarity in the Q^2 dependence of rapidity gap events with normal DIS events is as expected in models based on partonic structure of the pomeron. However, given the uncertainties in the Q^2 dependence of VDM models for highly virtual photons it is possible to reproduce the ratio observed in Fig. 63 in a VDM-like picture[120] as well.

Next we investigate the contribution of diffractive events to the structure function F_2. A variable $x_{\mathbb{P}/p}$ is defined which is the momentum fraction of the proton carried by the Pomeron. In the H1 analysis the cut on $\eta_{max} < 1.8$ selects mainly events with $x_{\mathbb{P}/p} < 0.01$, as shown in Fig. 64a and hence with $x < 0.01$ because the variables x and $x_{\mathbb{P}/p}$ are related by :

$$x = x_{\mathbb{P}/p} \frac{Q^2}{Q^2 + M_X^2}. \tag{15}$$

Here M_X is the invariant mass of the final state hadronic system observed in the detector. In the following the data are corrected for acceptance to the region $x_{\mathbb{P}/p} < 0.01$ using various models, the difference of which is taken into account for the systematic errors. Thus we define diffraction as the sample of events for which $x_{\mathbb{P}/p} < 0.01$. Note that Regge analyses performed on hadron-hadron collisions at low energies suggest that for $x_{\mathbb{P}/p} < 0.1$ Pomeron exchange is dominant whereas for larger values the contribution from other Reggeons is not negligible.[126] Hence we have some justification to call these events diffractive production, but there is no proof at this stage.

Following[53] one can define $F_2^{D(4)}$:

$$\frac{d\sigma(ep \to epX)}{dx_{\mathbb{P}/p} dt dx dQ^2} = \frac{2\pi\alpha^2}{Q^4 x} \left(2\left(1-y\right) + y^2\right) F_2^{D(4)}(x, Q^2, x_{\mathbb{P}/p}, t) \tag{16}$$

as the unintegrated diffractive contribution to the structure function. The contribution of the longitudinal structure function has been neglected in eqn. 16. Here t is the square momentum transfer between the incident and the outgoing proton or proton dissociative system, a quantity we cannot measure with sufficient resolution with the present detector setup. The integral of $F_2^{D(4)}$ over $x_{\mathbb{P}/p}$ and t gives the contribution from the diffractive events to the structure function $F_2(x, Q^2)$, which we call F_2^D :

$$F_2^D(x, Q^2) = \int_{10^{-4}}^{10^{-2}} \int_{t_{min}}^{\infty} F_2^{D(4)}(x, Q^2, x_{\mathbb{P}/p}, t) dx_{\mathbb{P}/p} dt. \tag{17}$$

The resulting diffractive part of the structure function $F_2^D(x, Q^2)$, defined for $x_{\mathbb{P}/p} < 0.01$ is shown together with the total inclusive structure function $F_2(x, Q^2)$ in Fig. 64b. For $x < 10^{-3}$, $F_2^D(x, Q^2)$ contributes about 10% to $F_2(x, Q^2)$. Clearly the diffractive events cannot explain the rise of F_2 at low x. The x dependence of F_2^D as x approaches 10^{-2} has to be taken with caution, since the cut on $x_{\mathbb{P}/p} < 0.01$ forces F_2^D to zero at $x = 10^{-2}$. Analysing F_2^D as a function of Q^2 reveals that there are no significant scaling violations observable, within the present experimental errors.[69]

An important characteristic of many models for hard diffraction is the factorization of $F_2^{D(4)}$ into a Pomeron flux term and a Pomeron structure function :

$$F_2^{D(4)}(x, Q^2, x_{\mathbb{P}/p}, t) = f(x_{\mathbb{P}/p}, t) F_2^{\mathbb{P}/p}(\beta, Q^2) \tag{18}$$

where $\beta = \frac{x}{x_{\mathbb{P}/p}}$ is the fraction $x_{q/\mathbb{P}}$ of the $\mathbb{P}$ momentum carried by the quark interacting with the virtual boson, and $f(x_{\mathbb{P}/p}, t)$ is the pomeron flux factor. If the factorization is true, then

$$F_2^{D(3)} = \int_{t_{min}}^{\infty} F_2^{D(4)} dt \tag{19}$$

should have the same $x_{\mathbb{P}/p}$ behaviour independent of β and Q^2.

In a dedicated analysis, the H1 collaboration has replaced the η_{max} selection by a set of cuts based on forward detectors. These detectors are mainly sensitive to secondaries produced by forward going hadrons interacting in collimators close to the proton beam axis.[127] The detectors are sensitive to particles produced in the pseudo-rapidity region $3.6 < \eta < 6.6$. The advantage is that this selection gives access to higher values of $x_{\mathbb{P}/p} < 0.1$. The resulting diffractive sample has been divided into four bins of Q^2 ($Q^2 = 8.5, 15, 30, 60$ GeV^2) times four bins of β ($\beta = 0.05, 0.2, 0.45, 0.8$). The $Q^2 = 15$ GeV^2 and $Q^2 = 30$ GeV^2 bins are shown as an illustration in Fig. 65. In all the bins the dependence of $F_2^{D(3)}(x, Q^2, x_{\mathbb{P}/p})$ on $x_{\mathbb{P}/p}$ can be fitted by a simple expression :

$$F_2^{D(3)}(x, Q^2, x_{\mathbb{P}/p}) = x_{\mathbb{P}/p}^{\alpha} F_2^{\mathbb{P}/p}(\beta, Q^2) \tag{20}$$

with (preliminary)

$$\alpha = -1.3 \pm 0.1. \tag{21}$$

Within the present errors the diffractive cross section is compatible with factorization. Furthermore, in Regge theory for hadronic exchange, the $x_{\mathbb{P}}$ dependence of the flux factor is related to the leading Regge trajectory $\alpha(t)$ via $x_{\mathbb{P}}^{-[2\alpha(t)-1]}$. Hence the leading trajectory has $\alpha(t) \sim 1.15$. Neglecting the presumably small t dependence, we find that the leading trajectory is close to the trajectory of the soft Pomeron, for which $\alpha_0 = 1.08$. More details are given in.[127] This is the first evidence that the rapidity gap events can be interpreted as diffraction!

Finally, we study jet production in diffractive events. High E_T jets are expected if the underlying process of these events is hard scattering of the virtual photon with constituent partons of the Pomeron. ZEUS has analysed the data for jets, using a cone algorithm, requiring a jet to have an E_T larger than 4 GeV. The analysis shows that in the laboratory frame 15%

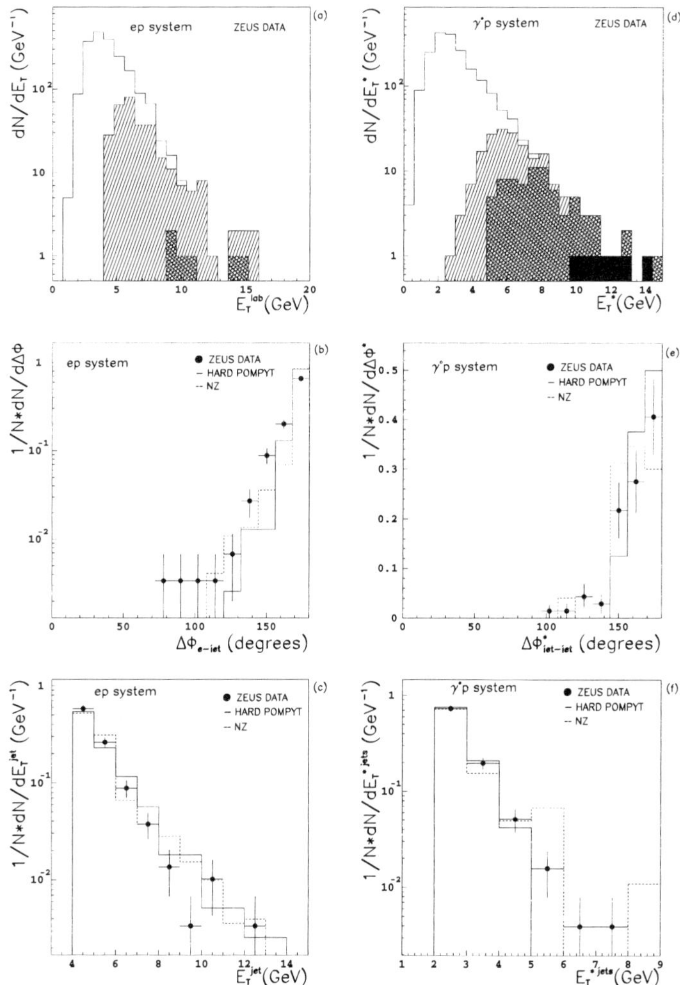

Figure 67: (**a**) The distribution of the total hadronic transverse energy seen in the calorimeter, E_T, for DIS events with a large rapidity gap and those with, in addition, ≥ 1 (hashed) and ≥ 2 jets (cross-hashed). A jet is required to have at least 4 GeV transverse energy with respect to the beam direction. (**b**) The difference in azimuthal angle between the scattered electron and the jet. (**c**) The jet transverse energy in the laboratory for events in the DIS sample with a large rapidity gap. (**d**) The total hadronic energy transverse to the virtual photon direction, E_T^*, for DIS events with a large rapidity gap and those with, in addition ≥ 1 (hashed), ≥ 2 (cross-hashed) or 3 jets (solid) in the final state. Here a jet is required to have at least 2 GeV with respect to the virtual photon direction. (**e**) The difference in azimuthal angle between the the two jets in the $\gamma^* p$ centre-of-mass system (2-jet sample). (**f**) The distribution of the jet energy transverse to the virtual photon direction for the 1– and 2–jet samples. In figures (**b**), (**c**), (**e**) and (**f**) the data are shown as black dots with errors and the results from the POMPYT and NZ[51] models as full and dashed histograms respectively.

of the rapidity gap events are of the 1-jet type with a negligible 2-jet production rate, Fig. 67. With a lower jet transverse energy cut of 2 GeV, a small 2-jet production rate is observed in the $\gamma^* p$ centre of mass frame. A similar analysis is shown in Fig. 66 for the H1 collaboration. Events with 2 jets with $E_T > 3.5$ GeV and an invariant mass of the two jets of $m_{ij} > 10$ GeV have been observed. In the figure the η_{max} distribution is shown for the events with two jets, with and without a rapidity gap requirement, as given by the forward selection. The data without rapidity gap requirement are well described by the standard DIS Monte Carlo calculation, while the Monte Carlo calculation based on RAPGAP accounts well for the data with a rapidity gap. Since the forward selection gives access to larger $x_{\mathbb{P}/p}$ values, larger M_X values can be reached compared to the η_{max} cut analysis. Thus jet production is less suppressed by the available phase space. The observation of two jet events in rapidity gap events is consistent with the assumption that these events are produced in the interaction of the virtual photon with partons in the Pomeron.

The properties of the rapidity gap events in deep inelastic scattering at HERA can be summarized as follows.

- After acceptance correction, the diffractive events represent about 10% of the DIS sample.

- The Q^2 dependence is similar to all DIS events.

- The rapidity gap events cannot explain the rise of F_2.

- About 10% of the observed rapidity gap events are exclusive vector mesons with or without proton dissociation.

- In the laboratory frame 15% of the rapidity gap events are of the 1-jet type with $E_T^{jet} \geq 4$ GeV

- The diffractive cross section can be factorized in a Pomeron flux term and a Pomeron structure function.

The interpretation of the events is still subject to discussion and further studies, but the physics potential is clearly very large.

6 Electroweak Measurements

One of the major physics topics conceived at HERA are studies of the electroweak theory. Indeed, the high Q^2 range accessible at HERA enables studies in the region of $Q^2 \simeq M_{W,Z^0}^2$, where W and Z^0 exchange is no longer mass suppressed w.r.t. photon exchange and becomes of competitive size. The exchange of the charged W results in events with a spectacular signature due to the escaping neutrino (see Fig. 1) with generally large p_T. Hence, these charged current events are characterized by a large missing transverse momentum, p_T^{miss}. This characteristic is exploited to isolate charged current events from other processes. A typical charged current event is shown in Fig. 68.

In the past decades the weak charged current has been extensively studied in νN scattering experiments.[128] The total cross section was found to rise linearly with the neutrino beam energy. The cross section has the form

$$\sigma \sim s \frac{1}{(1 + \frac{Q^2}{M_W^2})^2}. \tag{22}$$

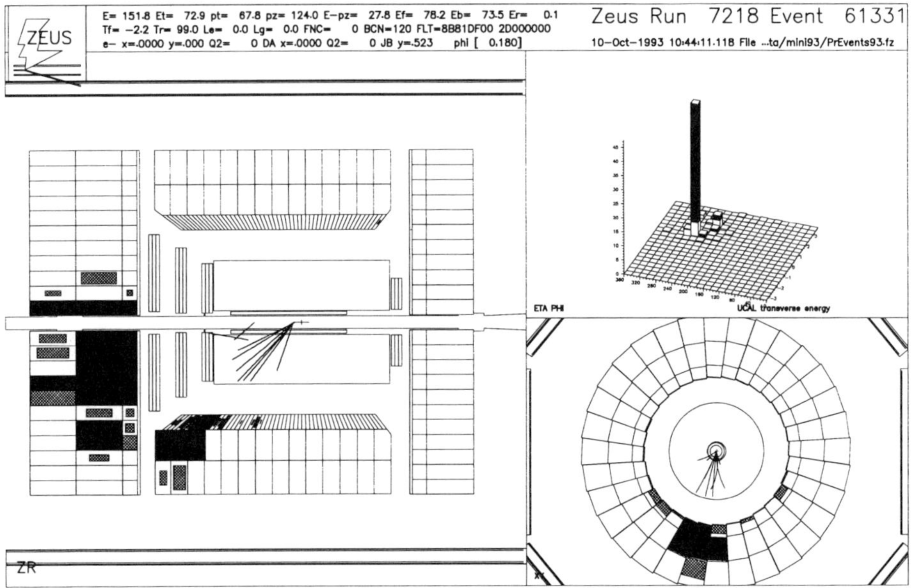

Figure 68: A charged current event in the ZEUS detector

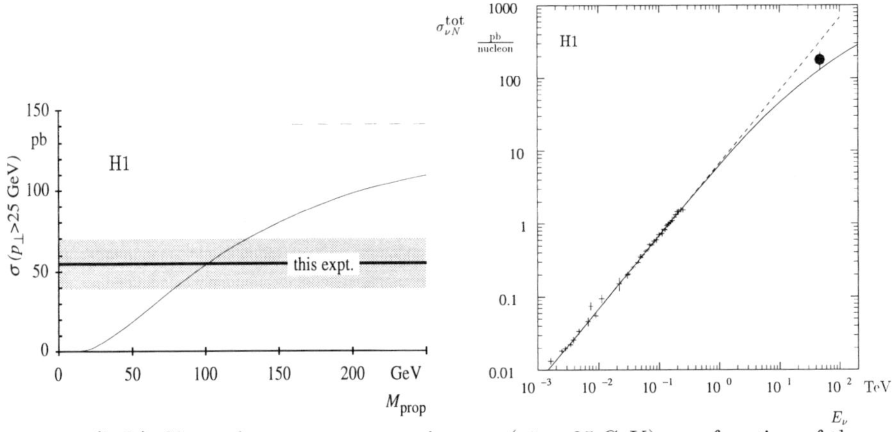

Figure 69: **(left)** Charged current cross-section $\sigma_{CC}(p_t^\nu > 25\,GeV)$ as a function of the propagator mass. The shaded band shows the H1 measurement ($\pm 1\sigma$) and the thin curve the theoretical expectation. The dashed line indicates the asymptotic case $M_{prop} = \infty$.
(right) The energy dependence of the νN cross section. The crosses represent the low energy neutrino data while the full point has been derived from the H1 measurement at HERA. The straight dashed line is the extrapolation from low energies assuming $M_W = \infty$ while the curve represents the predicted cross section including the W propagator with $M_W = 80.22\,\mathrm{GeV}$.

The squared CMS energy, s, is proportional to the incoming lepton energy in a fixed target experiment. Clearly, if $Q^2 \sim M_W^2$ the effect of the W propagator becomes visible and the cross section is expected to deviate from linearity. However the beam energies in fixed target experiments were too low to observe the effect of the propagator mass which we know since some year to expect to be $M_W = 80$. At HERA the equivalent fixed target energy is 50 TeV, so the effect of the W propagator should become visible for the first time in charged current

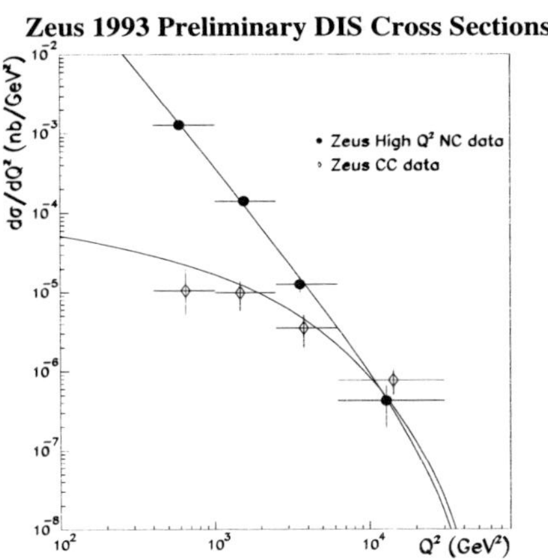

Figure 70: Preliminary cross section for neutral current (full circles) and charged current (open circles) scattering as a function of Q^2 together with predictions from simulation.[96]

interactions.

The H1 collaboration has measured the total charged current cross section for events with $p_T^{miss} > 25$ GeV.[129] After all selection criteria and background filters 14 events remain in the sample. The prediction from the electroweak theory for the cross section for $p_T^{miss} > 25$ GeV is 40.9 pb. The measurement gives $55 \pm 15 \pm 6$ pb, in good agreement with theory. The result is shown in Fig. 69a as function of the propagator mass. The result is consistent with a propagator mass of the known W resonance, $M_W = 80$ GeV. An infinite propagator mass is excluded by five standard deviations. The measured ep cross section can be converted to an equivalent νN cross section, by extrapolating to $p_T^{miss} = 0$ and taking into account the relevant flavour contributions. The result is shown in Fig. 69b. It shows clearly that the cross section at HERA deviates from a linear dependence on the neutrino energy.

The ZEUS collaboration has measured the Q^2 dependence[130] of the cross sections $ep \to \nu X$ and $ep \to eX$. As expected the two cross sections seem to become equal at a Q^2 of about 10^4 GeV^2, the scale where the electromagnetic and weak forces are unified. This expectation is confirmed by the Born level calculations, obtained with the program LEPTO, shown in the figure. Clearly these are just the first appetizers of electroweak physics at HERA. Much more is expected, when much larger statistics data samples will become available.

7 Searches Beyond the Standard Model

For every new high energy collider the search for new particles and phenomena is a "must". Due to its large centre of mass energy and the presence of an electron in the initial state, HERA is particularly suited to look for leptoquarks (gluons) and excited leptons, which can be produced either by fusion of the incoming lepton with a quark (gluon) of the proton, or fusion of the lepton with an exchanged boson.

The 1993 data have been analysed but so far no clear signal has been found for any of these channels. Limits have been deduced[131,132] which depend on the coupling of these new particles with the lepton and quark or exchanged boson. For leptoquarks the results are shown in Fig. 71 and for leptogluons in Fig. 72 for the H1 experiment. For couplings as large as the electromagnetic coupling, the excluded mass range is typically below 230 GeV for various types of scalar and vector leptoquarks. New limits for excited electrons and neutrinos are

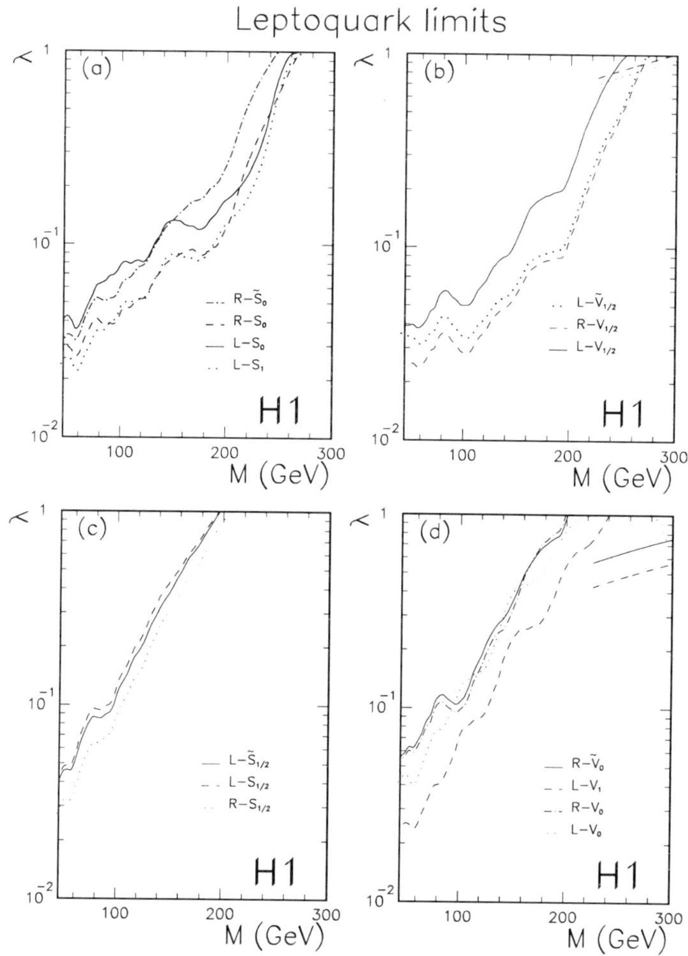

Figure 71: Upper limits at 95% C.L. as a function of mass on the couplings $\lambda_{L,R}$ for scalar and vector leptoquarks decaying into (**a,b**) lepton+q and (**c,d**) lepton+$\bar{q}$. The regions above the curves are excluded. The limits on λ_L for S_0, S_1, V_0 and V_1 combine $e + X$ and $\nu + X$ decays. The additional lines at high masses in (**b**) and (**d**) represent the result of the indirect search via the contact term analysis.

shown in Fig. 73 from ZEUS. Both experiments have produced similar limits on Leptoquarks and excited leptons.

ZEUS has presented results on the first search for excited quark (q^*) production through electroweak coupling. They are is complementary to searches at $p\bar{p}$ colliders – which reached a mass limit of 540 GeV[133] – and hold for excited quark production via the gluon coupling. The limits are shown in Fig. 74.

The H1 Collaboration made a search for R-parity violating supersymmetric squarks. In the accessible range of couplings, the squarks have mainly leptoquark-like signals. The rejection limits obtained for leptoquarks can be re-interpreted as a function of the squark masses. The results are shown in Fig. 75. Assuming couplings of electromagnetic strength masses of squarks in an R-parity violating susy environment below about 239 GeV can be excluded, depending on the photino mass

The search for new bosons or eq compositness can be considerably extended beyond the

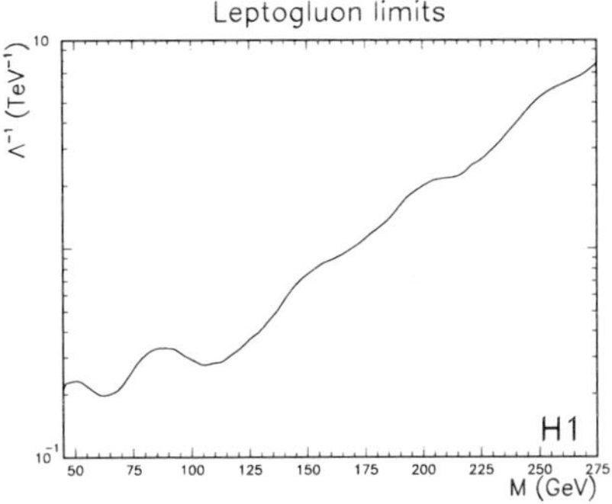

Figure 72: Upper limit at 95% C.L. for the inverse of the scale parameter Λ versus the mass M for leptogluons. The region above the curve is excluded

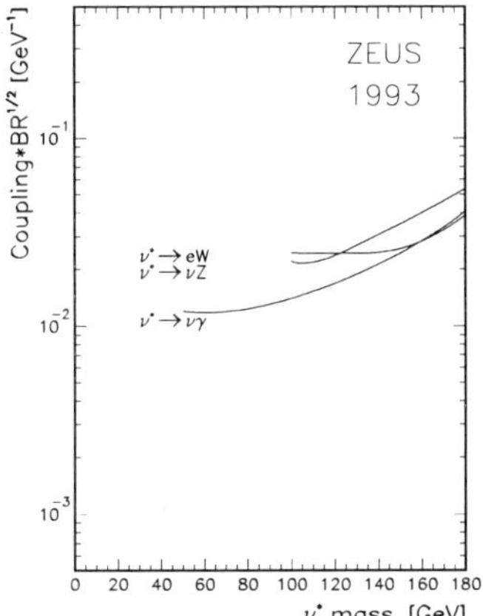

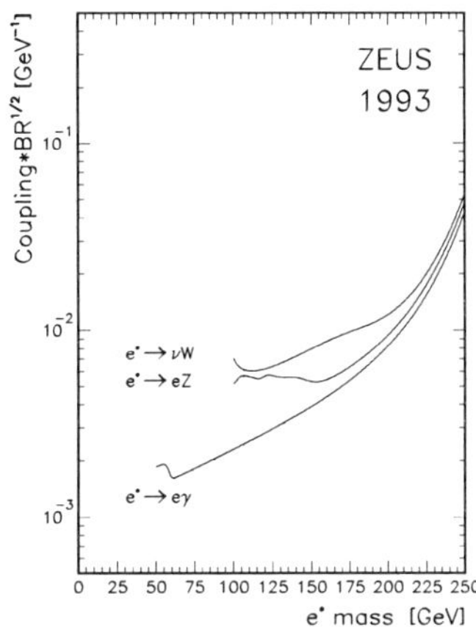

Figure 73: Rejection limits at the 95% C.L. for (**a**) e^* and (**b**) ν^* for different decay modes. Regions above the curves are excluded. Decay modes of the W and Z bosons are combined.

kinematic production limit of HERA through the study of indirect effects from virtual particle exchange. Such virtual effects are conveniently described by contact interactions and show up as deviations from the Standard Model at high Q^2 values. H1 has made a contact interaction analysis for leptoquarks and compositness. The leptoquark limits are shown in Fig. 71. The result on the compositness scale Λ is with the 1993 statistics $\Lambda > O(1\text{TeV})$. Similar limits from e^+e^- and $p\bar{p}$ colliders yield values roughly a factor 2 to 3 higher.

In the 1994 data one event was found with an unusual topology. The event is shown in Fig. 76, where one sees a large hadronic activity o one side of the detector and an isolated penetrating track on the other side of the detector. This penetrating particle is found to be compatible with a high energy muon emerging from the interaction vertex. No scattered electron is seen the detector. The hadronic system and the muon candidate have a large transverse momentum, $p_T = 41.1 \pm 4.2$ GeV and for $p_T = 23.4 \pm 2.4^{+7}_{-5}$ GeV, for the hadronic

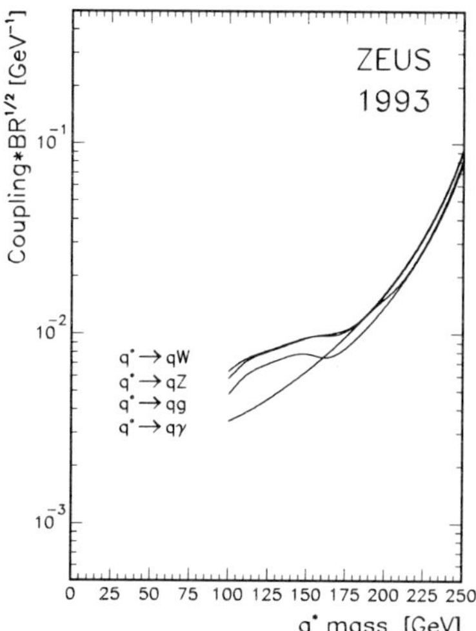

Figure 74: Rejection limits at the 95% C.L. for q^* for different decay modes. Regions above the curves are excluded. Note that these are limits on q^* production through electroweak coupling

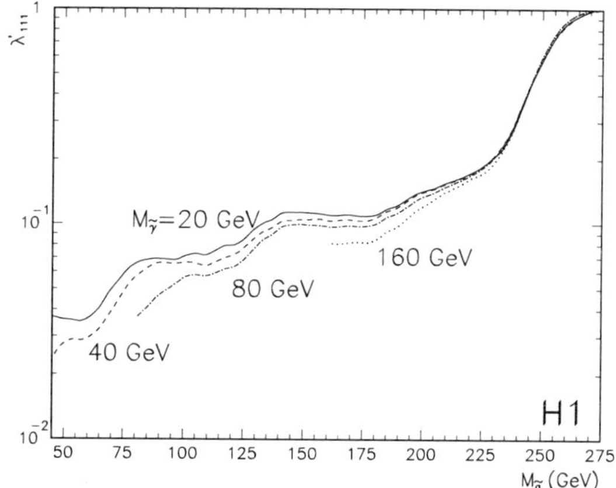

Figure 75: Rejection limits at the 95% C.L. for the couplings λ'_{111} as a function of the squark mass for various values of the photino mass. Regions above the curves are excluded. Note that these limits combine all charged and neutral decays of the $\bar{d}$ and $\bar{u}$.

system and muon respectively. The azimuthal angle between the muon and hadronic system is 183^0, i.e. they are essentially back to back. There is room for missing longitudinal momentum in the electron beam direction of more than 20 GeV and $p_T^{miss} = 18.7 \pm 4.8^{+5}_{-7}$ GeV. The muon candidate has the same charge as the incident lepton (which is a positron, since this event was found in the second part of the 1994 luminosity period). This event has been analysed[134] and the most probable Standard Model interpretation is the production and leptonic decay of a W boson, for which 0.03 events are expected given the total collected luminosity. Evidently this leaves room for speculation on more exotic interpretations. To settle this matter the approximately fivefold larger luminosity for the 1995 data taking period is eagerly awaited for...

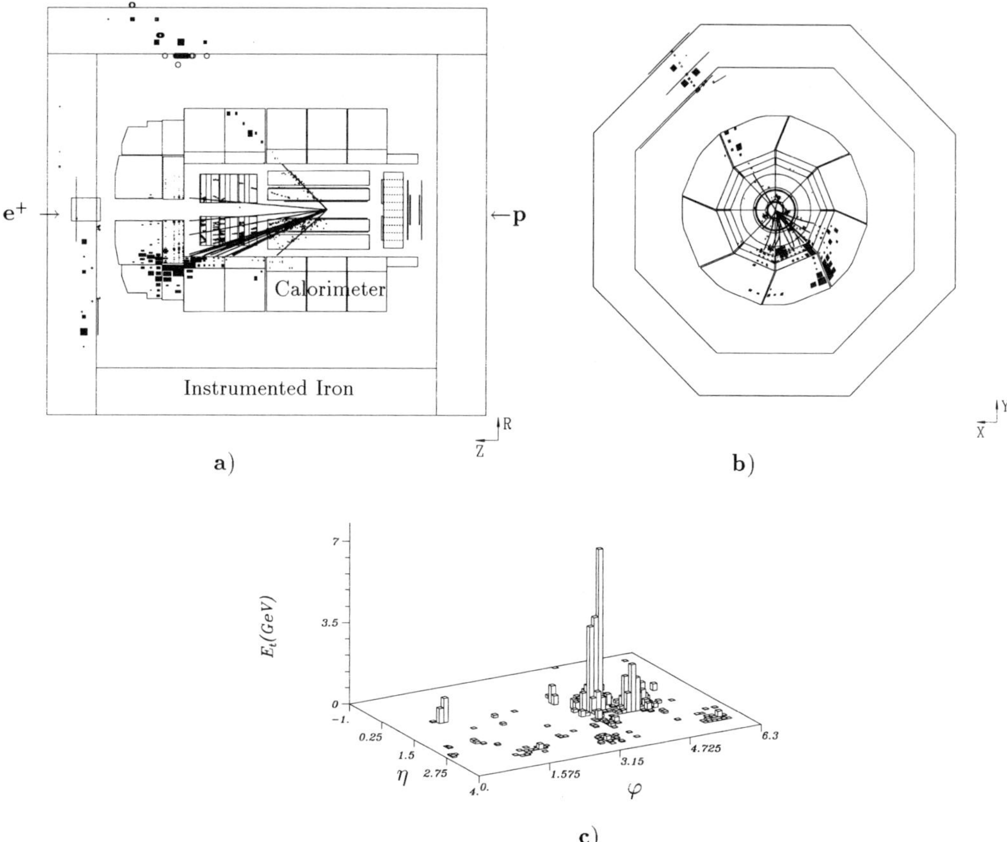

Figure 76: Event display : (**a**) $R - z$ view (**b**) $R - \varphi$ view and (**c**) transverse calorimetric energy

8   Conclusions

After two years of experimenting at the new ep collider HERA many new results have been obtained. In this report we have presented results on interactions of high energy photons with matter, and showed that similar to hadronic interactions, hard scattering is observed in these collisions. The different photoproduction processes have been isolated, and a first attempt was made to measure the structure of the photon at HERA. A new region has been explored for deep inelastic scattering interactions. The proton structure is probed to very small values of Bjorken-x, showing a large increase of F_2 with decreasing x. Events with large rapidity gaps have been observed and are identified as diffractive scattering. These events eventually will allow us to study structure of the Pomeron, an object from used to describe diffractive phenomena, but so far rather poorly known. The first electroweak results became available by studying the production of charged current events. searches for new, exotic phenomena were made, but no evidence for the breakdown of the standard model has been found. Many of the topics in this report will strongly benefit from the increase in luminosity of HERA, expected for the next year(s). Clearly, HERA and its physics program still have a bright future ahead.

Acknowledgment I would like to thank my colleagues from the H1 and ZEUS collaborations for their efforts in accumulating this large amount of nice results. In particular I wish to thank J. Bartels, D. Cussans, M. Erdmann, J. Feltesse, T. Haas, M. Kuhlen, R. Martin, J. Phillips, G. Rädel, J. Riedlberger, H. Rick, U. Stösslein and J. Withmore for discussion, critical reading of the text and valuable help with some of the figures.

References

[1] M. Breidenbach et al., Phys. Rev. Lett. **23** (1969) 935.

[2] R.P. Feynmann, *Photon-Hadron Interactions*, W.A. Benjamin Co., New York, (1972).

[3] M. Gell-Mann and Y. Ne'eman, *The Eightfold Way*, W.A. Benjamin Co., New York, (1964).

[4] F. Jacquet and A. Blondel, Proceedings of *the Study for an ep Facility for Europe*, ed. U.Amaldi, DESY 79/48 (1979) 391.

[5] J.Bartels and J.Feltesse, Proceedings of the Workshop *Physics at HERA*, ed. W. Buchmüller and G. Ingelman, Hamburg (1992) 133;
Workshop on *Deep Inelastic Scattering*, April 1992, Teupitz, Germany, J. Blümlein and T. Riemann editors;
E.M. Levin, Proceedings of the Internatinal Conference on *QCD, 20 Years Later*, ed. P.M. Zerwas and H.A. Kastrup, Aachen (1992).

[6] A.A. Sokolov and M. Ternov, Dokl. Sov. Akad. Nauk. **8** (1964) 1203.

[7] H1 Collaboration, *The H1 Detector at HERA*, DESY preprint DESY 93-103 (1993).

[8] ZEUS Collaboration, *The ZEUS Detector*, Status Report (1993).

[9] H1 Calorimeter Group, B. Andrieu et al., NIM **A336** (1993) 460.

[10] H1 Calorimeter Group, B. Andrieu et al., NIM **A336** (1993) 499.

[11] H. Bethe and W. Heitler, Proc. Roy. Soc. **A146** (1934) 83.

[12] H1 Collaboration, T. Ahmed et al., Phys. Lett. **B299** (1993) 374.

[13] ZEUS Collaboration, M. Derrick et al., Phys. Lett. **B293** (1992) 465.

[14] H1 Collaboration, T. Ahmed et al., *Experimental Study of Hard Photon Radiation Processes at HERA*, in litt.

[15] ZEUS Collaboration, M. Derrick et al., *Measurement of the Proton Structure Function F_2 from the 1993 HERA Data*, DESY preprint DESY 94-143 (1994).

[16] HERMES Collaboration, *A Proposal to Measure the Spin Dependent Structure Functions of the Neutron and the Proton at HERA*, DESY PRC 93/06 (1993).

[17] HERA-B Collaboration, *HERA-B: An Experiment to Study CP Violation in the B System Using an Internal Target at the HERA Proton Ring*, DESY-PRC 94/02 (1994).

[18] G. Schuler and T. Sjöstrand, Phys. Rev. **D49** (1994) 2257.

[19] J.K. Storrow, J. Phys. **G19** (1993) 1641.

[20] J.D. Bjorken, *Geometry of Multihadron Production*, summary talk of the 24th International Symposium on Multiparticle Dynamics, Vietri Sul Mare, September 1994.

[21] I. Pomeranchuck, Ya. JETP, 7 (1958), 499.

[22] A. Donnachie and P.V. Landshoff, Phys. Lett. **B296** (1992) 227.

[23] G. Yodh, Nucl. Phys. **B** (Proc. Suppl.) 12 (1990) 277.

[24] A. Levy, Proceedings of the Workshop *Physics at HERA*, ed. W. Buchmüller and G. Ingelman, Hamburg (1992) 481.

[25] H. Abramovicz et al., Phys. Lett. **B269** (1991) 465.

[26] C.F Weizsäcker, Z. Phys. **88** (1934) 612; E.J. Williams, Phys. Rev. **45** (1934) 729.

[27] ZEUS Collaboration, M. Derrick et al., Z. Phys. **C63** (1994) 391.

[28] ZEUS Collaboration, F. Barreiro et al., *New Results from HERA*, DESY preprint, DESY 94-204 (1994).

[29] P. Söding, Phys. Lett. **B19** (1966) 702.

[30] Z. Koba, H.B. Nielsen, P. Olesen, Nucl. Phys. **B40** (1972) 317.

[31] E.A. De Wolf, contributed talk to ICHEP94, 27th Int. Conf. on High Energy Physics, Glasgow, July 1994, to be published.

[32] B.A. Kniehl and G. Kramer, Z. Phys. **C62** (1994) 53.

[33] H1 Collaboration, I. Abt et al., Phys. Lett. **B328** (1994) 176.

[34] M. Glück, E. Reya and A. Vogt, Z. Phys. **C53** (1992) 651.

[35] H. Abramowicz, K. Charchula and A. Levy, Phys. Lett. **B269** (1991) 458.

[36] J. E. Huth et al., Fermilab-Conf-90/249-E (1990).

[37] H. Bengtsson and T. Sjöstrand, Comp. Phys. Comm. **46** (1987) 43; T. Sjöstrand, CERN-TH-6488-92 (1992).

[38] ZEUS Collaboration, M. Derrick et al., *Inclusive Jet Differential Cross Sections in Photoproduction at HERA*, DESY preprint DESY 94-176 (1994).

[39] H1 Collaboration, V. Brisson et al., *New Results from HERA on Deep Inelastic Scattering at low x, the Proton Structure Function, Jets in Photoproduction, Heavy Flavour Production and Searches for New Particles*, DESY preprint DESY 94-187 (1994).

[40] C. Marchesini et al., Comp. Phys. Comm. **67** (1992) 465.

[41] S. Catani, Yu.L. Dokshitzer and B.R. Webber, Phys. Lett. **B285** (1992) 291.

[42] OPAL Collaboration, R. Akers et al., Z. Phys. **C61** (1994) 199.

[43] AMY Collaboration, R. Tanaka et al., Phys. Lett. **B277** (1992) 215.

[44] V. Blobel, DESY preprint DESY 84-118, and Proceedings of the 1984 CERN School of Computing, Aiguablava (Spain), CERN 1985.

[45] J.D. Bjorken, talk given at the 21nd Annual SLAC Summer Intitute Topical Conference, Stanford, August 1993.

[46] K. Goulianos, Phys. Rep. **101** (1983) 169.

[47] D. Amati et al., Nuovo Cimento **26** (1962) 896.

[48] F. E. Low, Phys. Rev. **D12** (1975) 163; S. Nussinov, Phys. Rev. Lett. **34** (1975) 1286.

[49] G. Ingelman and P. Schlein, Phys. Lett. **B152** (1985) 256.

[50] A. Donnachie and P. V. Landshoff, Phys. Lett. **B191** (1987) 309.

[51] N. N. Nikolaev and B. G. Zakharov, Z. Phys. **C53** (1992) 331.

[52] UA8 Collaboration, A. Brandt et al., Phys. Lett. **B297** (1992) 417.
UA8 Collaboration, R. Bonino et al., Phys. Lett. **B211** (1988) 239.

[53] P. Bruni and G. Ingelman, Proc. of the Europhysics Conference, Marseilles, France, July 1993, p. 595.

[54] H1 Collaboration, T. Ahmed et al., *Observation of Hard Processes in Rapidity Gap Events in γp Interactions at HERA*, DESY preprint DESY 94-198 (1994).

[55] ZEUS Collaboration, M. Derrick et al., *Observation of Hard Scattering in Photoproduction Events with a Large Rapidity Gap at HERA*, DESY preprint DESY 94-210 (1994).

[56] M. Aguilar-Benitez et al., PDG, Phys. Rev. **D45** (1992).

[57] H. Jung, D. Krücker, C. Greub and D. Wyler, Z. Phys. **C60** (1993) 721.

[58] E.L. Berger, D. Jones, Phys. Rev. **D23** (1981) 1521.

[59] M.G. Ryskin, Z. Phys. **C57** (1993) 89.

[60] H1 Collaboration, T. Ahmed et al., Phys. Lett. **B338** (1994) 507.

[61] R.K. Ellis, P. Nason, Nucl. Phys. **B312** (1989) 551.

[62] A.D. Martin, W.J. Stirling, R.G. Roberts, Phys. Lett. **B306** (1993) 145, Erratum **B309** (1993) 492.

[63] G. Altarelli, G. Martinelli, Phys. Lett. **B76** (1978) 89.

[64] P. Amaudruz et al., Phys. Lett. **B295** (1992) 159.

[65] J. Bartels and J. Feltesse, Proceedings of the Workshop *Physics at HERA*, ed. W. Buchmüller and G. Ingelman, Hamburg (1992) 133.

[66] V.N. Gribov and L.N. Lipatov, Sov. Journ. Nucl. Phys. **15** (1972) 438 and 675;
G. Altarelli and G. Parisi, Nucl. Phys. **B126** (1977) 298 ;
Yu.L. Dokshitzer, Sov. Phys. JETP **46** (1977) 641.

[67] E.A. Kuraev, L.N. Lipatov and V.S. Fadin, Phys. Lett. **B60** (1975) 50;
Zh.E.T.F **72** (1977) 377.

[68] V.N. Gribov, E.M. Levin and M.G. Ryskin, Phys. Rep. **100** (1983),1.

[69] H1 Collaboration T. Ahmed et al., *A Measurement of the Proton Structure Function F_2*,
DESY preprint DESY 95-006 (1995).

[70] J. Feltesse, in Proceedings of the Workshop Proc. HERA Workshop (DESY, 1987) p.33.

[71] D. Yu. Bardin et al., Z. Phys. **C42** (1989) 679;
M. Böhm and H. Spiesberger, Nucl. Phys. **B294** (1987) 1081;
J. Blümlein, $O(\alpha^2)$ *Radiative Corrections to Deep Inelastic ep scattering for different kinematical variables*, DESY preprint DESY 94-044 ;
A. Akhundov et al., CERN Preprint CERN-TH.7339/94.
H. Spiesberger et al., Proceedings of the Workshop *Physics at HERA*, ed. W. Buchmüller and G. Ingelman, Hamburg (1992) 798.

[72] S. Bentvelsen, P. Kooijman and J. Engelen, Proceedings of the Workshop *Physics at HERA*, ed. W. Buchmüller and G. Ingelman, Hamburg (1992) 23.

[73] U. Bassler and G. Bernardi, DESY preprint DESY 94-231 (1994), submitted to Nucl. Inst. and Meth.

[74] H1 Collaboration, I. Abt et al., Nucl. Phys. **B407** (1993) 515.

[75] ZEUS Collaboration, M. Derrick et al., Phys. Lett **B316** (1993) 412.

[76] E665 Collaboration, H. Melanson et al., contributed talk to ICHEP94, 27th Int. Conf. on High Energy Physics, Glasgow, July 1994, to be published.

[77] M. Glück, E. Reya and A. Vogt, Z. Phys. **C53** (1992) 127 and Phys. Lett. **B306** (1993) 391.

[78] L.W. Whitlow et al., Phys. Lett. **B282** (1992) 475.

[79] BCDMS Collaboration, A.C. Benvenuti et al., Phys. Lett. **B223** (1989) 485.

[80] NMC Collaboration, P. Amaudruz et al., Phys. Lett. **B295** (1992) 159.

[81] A.D. Martin, W.J. Stirling, R.G. Roberts, Phys. Rev. **D47** (1993) 867.

[82] A.D. Martin, W.J. Stirling, R.G. Roberts, Proc. Workshop on *Quantum Field Theoretical Aspects of High Energy Physics*, Kyffhässer, Germany, eds B. Geyer and E.M. Ilgenfritz, Leipzig (1993) p.11.

[83] A. Capella et al., Phys. Lett. **B337** (1994) 358.

[84] A.J. Askew, J. Kwieciński, A.D. Martin, P.J. Sutton, Phys. Rev. **D47** (1993) 3775;
A.J. Askew, K. Golec-Biernat, J. Kwieciński, A.D. Martin, P.J. Sutton, Phys. Lett. **B325** (1994) 212;
A.J. Askew, J. Kwieciński, A.D. Martin, P.J. Sutton, Phys. Rev. **D49** (1994) 4402.

[85] N.N. Nikolaev and B.G. Zakharov, Phys. Lett. **B327** (1994) 149.

[86] K. Golec-Biernat, M.W. Krasny and S. Riess, Phys. Lett. **B337** (1994) 367.

[87] W. Zhu, D. Xue, Kang-Min Chai and Zai-Xin Xu, Phys. Lett. **B317** (1993) 200.

[88] D.W. McKay and J.P. Raltson, Nucl. Phys. **B** (Proc. Suppl.) 18C (1990) 86.

[89] R.D. Ball and S. Forte, Phys. Lett. **B335** (1994) 77;
R.D. Ball and S. Forte, Phys. Lett. **B336** (1994) 77.

[90] A. De Rujula et al., Phys. Rev. **D10** (1974) 1649.

[91] R.D. Ball and S. Forte, CERN preprints CERN-TH.7421/94 and CERN-TH.7422/94.

[92] K. Prytz, Phys. Lett. **B311** (1993) 286.
K. Prytz, Rutherford-Appleton Laboratory preprint : RAL-94-036.

[93] H1 Collaboration, I. Abt et al., Phys. Lett. **B321** (1994) 161.

[94] K. Golec-Biernat, Phys. Lett. **B328** (1994) 495.

[95] R.K. Ellis, Z. Kunszt and E.M. Levin, Fermilab preprint : Fermilab-PUB-93/350-T.

[96] G. Ingelman, LEPTO 6.1, Proceedings of the Workshop *Physics at HERA*, ed. W. Buchmüller and G. Ingelman, Hamburg (1992) 1366.

[97] L. Lönnblad, ARIADNE version 4.03, Comp. Phys. Commun. **71** (1992) 15, and references therein.

[98] L. Lönnblad, talk at the DESY QCD Institute, DESY, September 1994.

[99] B.R. Webber, J. Phys. **G19** (1993) 1567.

[100] H1 Collaboration, I. Abt et al., Z. Phys. **C64**, (1994) 377.

[101] DELPHI Collaboration, P. Abreu et al., Phys. Lett. **B311** (1993) 408.

[102] ZEUS Collaboration, M. Derrick et al., Phys. Lett. **B306**, (1993) 158.

[103] H1 Collaboration, I. Abt et al., Z. Phys. **C61**, (1994) 59.

[104] JADE Collaboration, W. Bartel et al., Z. Phys. **C33** (1986) 23.

[105] D. Graudenz, Phys. Lett. **B256**, (1991) 518;
D. Graudenz, Phys. Rev. **D49**, (1994) 3291;
D. Graudenz, PROJET4.1, CERN-TH.7420/94.

[106] T. Brodkorb, J.G. Körner, Z. Phys. **C54**, (1992) 519;
T. Brodkorb, E. Mirkes, Univ. of Wisconsin, MAD/PH/820 (1994).

[107] H1 Collaboration, T. Ahmed et al., *Determination of the Strong Coupling Constant from Jet Rates in Deep Inelastic Scattering*, DESY preprint DESY 94-220 (1994).

[108] S. Bethke, J.E. Pilcher, Annual Review of Nuclear and Particle Science **42**, 251 (1992).

[109] V. Hedberg, G. Ingelman, C. Jacobsson and L. Jönsson, Proceedings of the Workshop *Physics at HERA*, ed. W. Buchmüller and G. Ingelman, Hamburg (1992) 331.

[110] C. Jacobsson, *Jet Azimuthal Angle Asymmetries in Deep Inelastic Scattering as a Test of QCD*, Phd thesis, University of Lund, (1994), unpublished,

[111] J. Bartels, H. Lotter, Phys. Lett. **B309** (1993) 400.

[112] K. Golec-Biernat, J. Kwieciński, A. D. Martin and P. J. Sutton, Phys. Rev. **D50** (1994) 217;
K. Golec-Biernat, J. Kwieciński, A. D. Martin and P. J. Sutton, Phys. Lett. **B335** (1994) 220.

[113] A.H. Mueller, Nucl. Phys. **B** (Proc. Suppl.) 18C (1990) 125; J. Phys. **G17** (1991) 1443;
J. Bartels, A. De Roeck, M. Loewe, Z. Phys. **C54** (1992) 635;
W.K. Tang, Phys. Lett. **B278** (1992) 363.

[114] J. Kwieciński, A.D. Martin, P.J. Sutton, Phys. Rev. **D46** (1992) 921.

[115] H. Jung, DESY preprint DESY 93-182.

[116] ZEUS Collaboration, M. Derrick et al., Phys. Lett. **B315** (1993) 481.

[117] H1 Collaboration, A. De Roeck, *Results from the H1 Experiment*, to appear in Proc.
of the Europhysics Conf. on HEP, Marseille, France, July 1993. DESY preprint : DESY
94-005 (1994);
J. B. Dainton, *Results from the H1 Experiment at HERA*, in Proc. XVI International
Symposium on Lepton Photon Interactions, Cornell, Ithaca, USA, August 1993.

[118] G. Ingelman and K. Prytz, Z. Phys. **C58** (1993) 285.

[119] M.G.Ryskin and M.Besançon, Proceedings of the Workshop *Physics at HERA*, ed. W.
Buchmüller and G. Ingelman, Hamburg (1992) 215 and references therein.

[120] H1 Collaboration, T. Ahmed et al., Nucl. Phys. **B429** (1994) 477.

[121] S.J. Brodsky et al., Phys. Rev. **D50** (1994) 3134.

[122] NMC Collaboration, M. Arneodo et al., *Exclusive ρ^0 and phi Muoproduction at Large
Q^2*, CERN preprint CERN-PPE-94-146 (1994).

[123] EMC Collaboration, J.J. Aubert et al., Phys. Lett. **B161** (1985) 203 and J. Ashman
et al., Z. Phys. **C39** (1988) 169.

[124] NMC Collaboration, P. Amaudruz et al., Z. Phys. **C54** (1992) 239.

[125] J. Nemchik, N.N. Nikolaev and B.G. Zakharov, Phys. Lett. **B341** (1994) 228.

[126] G. Alberi and G. Goggi, Phys. Rep. **74** (1981) 1.

[127] H1 Collaboration, T. Ahmed et al. *First Measurement of the Deep Inelastic Structure
of Proton Diffraction.*, in litt.

[128] For a review see D. Haidt and H. Pietschmann, Lanolt-Börnstein New Series I/10,
Springer (1988).

[129] H1 Collaboration, T. Ahmed et al., Phys. Lett. **B324** (1994) 241.

[130] T. Haas, *Recent Results from ep Scattering at HERA*, DESY preprint DESY 94-160
(1994).

[131] H1 Collaboration, T. Ahmed et al., Phys. Lett. **B340** (1994).
H1 Collaboration, T. Ahmed et al., Z. Phys. **C64** (1994) 545.

[132] ZEUS Collaboration, M. Derrick et al., *A Search for Excited Fermions in Electron-
Proton Collisions at HERA* DESY preprint DESY 94-175.

[133] CDF Collaboration, F. Abe et al., Phys. Rev. Lett. **72** (1994) 3004.

[134] H1 Collaboration, T. Ahmed et al., *Observation of an $e^+p \rightarrow \mu^+X$ Event with High
Transverse Momenta at HERA* DESY preprint DESY 94-248 (1994).

DIFFRACTIVE PRODUCTION OF DIJETS AT HERA:
A SIMPLE MODEL

M. Diehl

DAMTP
University of Cambridge
Silver Street
Cambridge CB3 9EW
England

In Albert de Roeck's lectures we have heard about an object called pomeron, which appears in certain processes at HERA. This talk is about a simple model of this object and its application to a particular process. I will first give a short introduction to the pomeron and to the reaction I propose to study. Then I will describe the Landshoff-Nachtmann model, which gives a simple description of the pomeron in QCD. In the third part I will present some predictions of this model for diffractive production of dijets at HERA, in photoproduction or in deep inelastic scattering. In the case of photoproduction the results are rather peculiar.

INTRODUCTION

The pomeron was invented to describe the leading energy behaviour of hadronic reactions in the limit where the centre-of-mass energy becomes large, but the momentum transfer remains fixed. More specifically, consider elastic scattering of two hadrons, with the usual Mandelstam variables s and t (fig. 1). Pomeron exchange gives $s^{\alpha(t)}$ for the leading s-dependence of the amplitude. The exponent is called the pomeron trajectory, and has been fitted from experiment to a linear behaviour

$$\alpha(t) = 1 + \epsilon + \alpha' t \tag{1}$$

with $\epsilon \approx 0.08$ [1] and $\alpha' \approx 0.25\,\text{GeV}^{-2}$ [2]. Comparing reactions with different hadrons shows that the pomeron has the quantum numbers of the vacuum.

There are other possible exchanges, with different quantum numbers and different trajectories, which can be related to known mesons (ρ, ω, f, a etc.). All these exchanges are described in the framework of Regge theory [3], which gives a good description

Frontiers in Particle Physics: Cargèse 1994
Edited by M. Lévy *et al.*, Plenum Press, New York, 1995

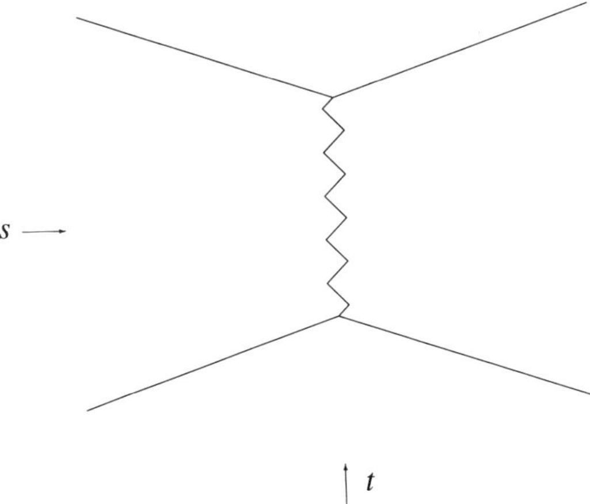

Figure 1: Elastic hadron-hadron scattering. The zigzag line stands for a pomeron, which gives the leading s-dependence of the transition amplitude.

of hadronic reactions at high energies and small momentum transfers. However, the connection to QCD, which we believe to be the theory that underlies strong interactions, is not well known. In particular, we would like to know what the pomeron is in terms of QCD. Since its domain is that of small momentum transfers, i. e. of large distances, perturbation theory cannot be applied, at least not in a straightforward manner. It has long been proposed to describe the pomeron by the exchange of two gluons which couple to a colour singlet. This is one of the ingredients of the model I will describe.

Electron-proton collisions at HERA give us an opportunity to study the pomeron in more detail. In events where the proton is scattered diffractively, i. e. where it remains intact and loses only a tiny fraction of its momentum, one can expect that it has radiated a pomeron, which can interact with a real or a virtual photon emitted by the electron. Because of the very high energy available at HERA, this interaction can be hard enough to break up the pomeron and thus reveal something about its structure.[1] For purely kinematic reasons, the final state of the photon-pomeron collision is well separated from the diffractively scattered proton in rapidity. Events with such a rapidity gap have indeed be seen at HERA [5, 6].

In this talk I will consider a specific final state of the pomeron-photon interaction, namely a quark-antiquark pair, which gives rise to two jets (fig. 2). I will impose a minimum transverse momentum for the jets to allow them to be identified experimentally, and to ensure that there is a hard momentum scale in the reaction, even when the photon is real.

THE LANDSHOFF-NACHTMANN MODEL OF THE POMERON

I will now briefly describe the Landshoff-Nachtmann (LN) model of the pomeron [7, 8]. It follows the idea that pomeron exchange can be described in QCD by the exchange of two gluons forming a colour singlet. These gluons do of course interact, but as a simple approximation we just take two noninteracting gluons. This does not give

---

[1]Similar reactions have already been observed in hadron-hadron collisions, where one of the colliding particles radiates a pomeron which is hit hard by the other [4].

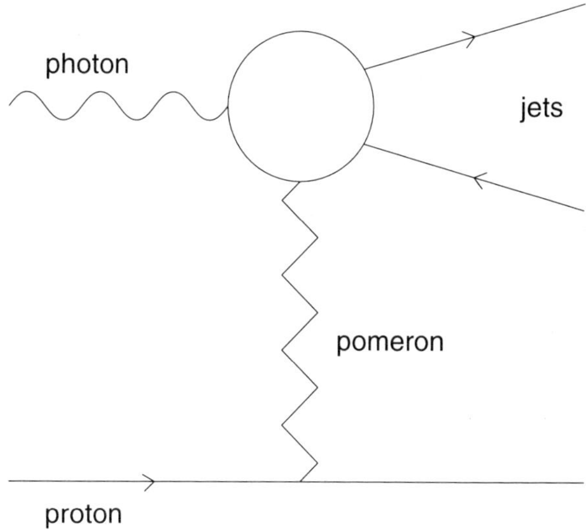

Figure 2: Diffractive production in a photon-proton collision of a quark-antiquark pair forming a dijet.

the energy dependence typical for the pomeron, and for the hadron-hadron scattering in fig. 1 we get e. g. an amplitude proportional to s. We modify this to $s^{\alpha(t)}$, having in mind that the change of the exponent from 1 to $\alpha(t)$ is produced by the interaction between the two gluons.

Since the total momentum they carry is small, their individual momenta can be small, too, and we are in a region of nonperturbative physics. To take this partly into account we do not use the perturbative gluon propagator, $-g_{\mu\nu}/l^2$ in Feynman gauge, but the full two-point function $-g_{\mu\nu}D(l^2)$ which one would obtain if one could solve QCD. At small squared momenta l^2 it will presumably be very different from the perturbative one, whereas for large l^2 one can expect the perturbative form to be valid.[2] Of course we have not solved QCD and do not know $D(l^2)$, but it turns out that in applications one only needs some of its moments, which can be fitted to experiment. In particular, the moment we need for our present investigation has been determined from exclusive ρ-production in deep inelastic scattering [8], i. e. the process shown in fig. 2, but with a ρ instead of $q\bar{q}$.

The next issue we have to address is how the two gluons couple to the proton. If we take the simple picture of a proton consisting of constituent quarks, we have two different types of diagrams to consider: one where both gluons couple to the same quark, and one where they couple to different quarks (fig. 3).

Let me give an argument why under certain assumptions the second type can be neglected. The first point is that in the propagator $D(l^2)$ there is a mass scale μ_0, which gives the momentum range over which $D(l^2)$ decreases (the nonperturbative gluon propagator must contain some scale, because it has the same dimension as $1/l^2$ but is not proportional to it). It can be related to one of the moments of $D(l^2)$, and has been determined to be $\mu_0 \approx 1.1\,\mathrm{GeV}$ [8]. It turns out that in the kinematical regime we are working in, we have $l^2 \approx l_T^2$, where l_T is the component of l which is transverse to the momenta of the proton and of the photon. The main contribution to the loop

---

[2]To be more precise, the propagator $D(l^2)$ used in the LN model has the perturbative part $1/l^2$ subtracted off at large l^2, so that the contribution from the latter has to be added afterwards, but this need not concern us here.

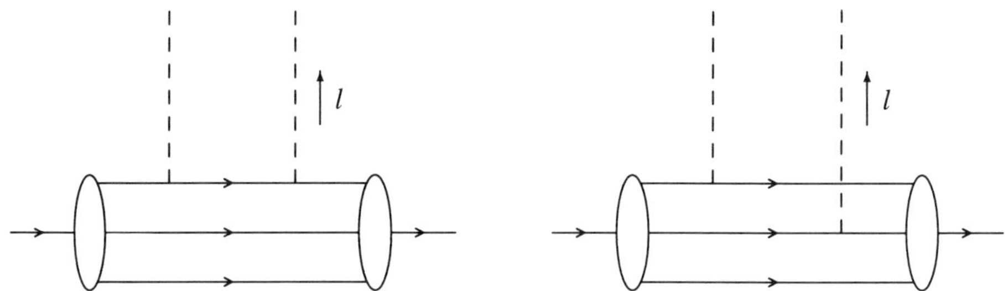

Figure 3: Typical diagrams showing the two possibilities for two gluons to couple to the quarks in a proton. The blobs stand for the proton wave function and the dashed lines for nonperturbative gluons.

integral over l will therefore come from the region

$$|l_T^2| \lesssim \mu_0^2 \approx 1.2\,\mathrm{GeV}^2 \ . \tag{2}$$

Now compare the momenta of the quarks before and after the interaction with the two gluons in fig. 3. In the first diagram the momentum of the upper quark is changed by the net momentum of the two gluons, i. e. the momentum of the pomeron. In the second, however, two quarks suffer an additional momentum transfer of l, in particular their transverse momenta are changed into opposite directions by l_T. If this change is too large it will be difficult for them to "fit" into a proton again, in other words, the proton wave function will suppress the second diagram. It is plausible that the relevant scale for this suppression is given by the inverse proton radius R^{-1}, so that diagrams of the second type will only be relevant if

$$|l_T^2| \lesssim R^{-2} \approx 0.04\,\mathrm{GeV}^2 \ . \tag{3}$$

Comparing with eq. (2) we see that this is only a small part of the important region of integration. If we assume that $D(l^2)$ is finite at $l^2 = 0$ (or has a sufficiently weak singularity), so that the region of very small l_T^2 is not enhanced by the gluon propagator, we can therefore neglect diagrams where the gluons couple to different quarks in the proton.

A more intuitive (but somewhat more handwaving) argument in position space goes by identifying $\mu_0^{-1} \approx 0.2\,\mathrm{fm}$ as the transverse size of the pomeron. This is smaller than the mean distance between two constituent quarks, which is of the order of the proton radius, and therefore the pomeron is "too small" to couple coherently to different quarks.

In the limit where only the diagrams with the gluons coupling to the same quark are important, the amplitude is clearly proportional to the number of constituent quarks in a hadron. This nicely reproduces the additive quark rule, which observes just this proportionality in comparing total cross sections for hadron-hadron and hadron-meson scattering (remember that by the optical theorem total cross sections are proportional to forward scattering *amplitudes*). One also finds that the spin structure of the proton-pomeron coupling involves the proton vector current, which is related to the isoscalar electromagnetic form factor $F_1(t)$ of the nucleon. This gives a good description of the differential cross section at small t in elastic pp and $p\bar{p}$-scattering [2, 9].

An important remark is that considering only the coupling of the gluons that make the pomeron to the constituent quarks of the proton is less a question of neglecting that

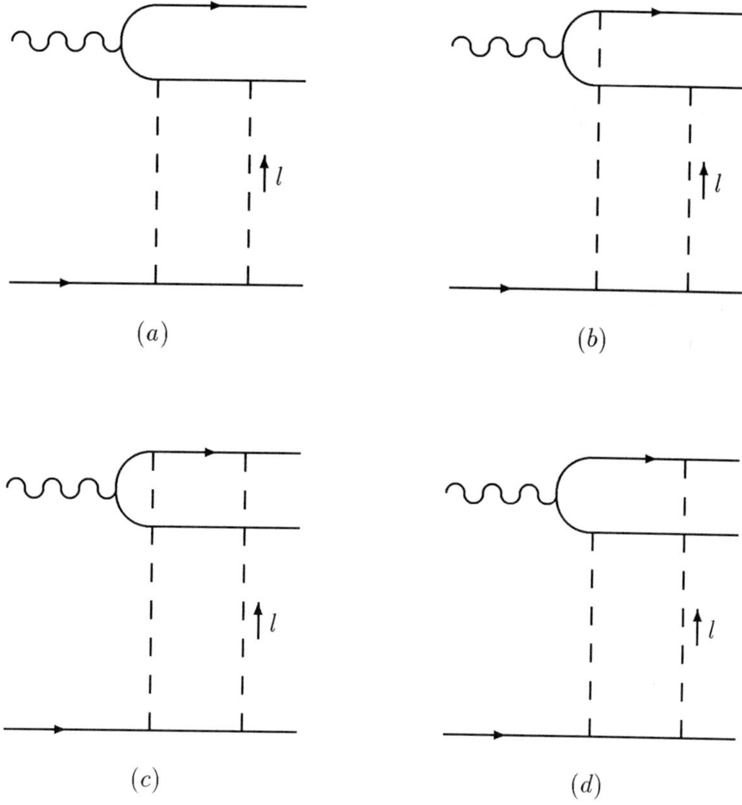

Figure 4: Feynman diagrams contributing to the imaginary part of the amplitude for $p + \gamma \to p + q\bar{q}$, which is leading in the centre-of-mass energy. The pomeron is approximated by two nonperturbative gluons (dashed lines).

the proton structure is much more complicated, but rather of what one considers as part of the pomeron and what as part of the proton, i. e. a question of factorisation. When l^2 is not too large, the pomeron (or rather our approximation of it) consists of two weakly virtual gluons, and one could also interpret them as two gluons within the *proton*.

I have now presented all the ingredients one needs for calculating in the LN model the cross section for diffractive production of a $q\bar{q}$-pair with large transverse momentum p_T. It turns out that the leading behaviour in the photon-proton centre-of-mass energy is given by the imaginary part of the amplitude, which can be obtained from the unitarity of the scattering matrix by putting the intermediate lines in the relevant Feynman diagrams on shell. One then has four diagrams to calculate (fig. 4). It is important to notice that diagrams (b) and (d), where the gluons do not couple to a quark line directly one after another, cannot be dropped as in the case of the coupling to the proton. The argument to discard such diagrams I have just given does not apply here, because the kinematics are quite different. Namely, one of the quark lines coupling to the photon can be highly virtual, which is very good since it justifies the calculation of the upper part of the diagrams in ordinary perturbation theory. In fact, taking only diagrams (a) and (c) gives a result that badly violates electromagnetic gauge invariance, only the sum of all four diagrams is gauge invariant.

Let me now talk about the results one obtains in the LN model for the reaction we are interested in [10]. I should mention that the two-gluon model of the pomeron has also been applied to this reaction in a purely perturbative framework, i. e. using the ordinary perturbative gluon propagator [11, 12], and that some results are quite similar in the two approaches.

As I have announced in the beginning, one finds something rather surprising for photoproduction of light quarks, namely that the cross section is *very* small. In fact, for photons with a small virtuality Q and transverse polarisation the cross section behaves like Q^4 plus something tiny. This means that it decreases even faster as Q^2 goes to zero than the one for longitudinal photons, which vanishes like Q^2 (the longitudinal cross section *has* to vanish because of gauge invariance). To see what is going on one can perform all loop integrations except the one over l_T^2. One then sees that there is complete cancellation unless

$$|l_T^2| > |p_T^2| \; , \tag{4}$$

i. e. the transverse momentum and thus the virtuality of the exchanged gluons must be larger than the transverse momentum of the produced jets. At such large momenta there is of course a strong suppression from the gluon propagators, and this explains why the result we obtain is so small.

We must however be a bit cautious. We have done a calculation with two noninteracting gluons and put in by hand the pomeron trajectory $\alpha(t)$ of eq. (1) to take into account gluon interactions. This trajectory is however taken from processes such as hadron-hadron scattering, where the main contribution to the loop integration comes from *soft* gluons. Another point is that we do not know whether the condition (4) for the loop integration still holds when gluon interactions are included in the calculation.

If we look at the production of heavy quarks, the cancellation at small gluon momenta is no longer present and small l^2 dominate, just as in hadron-hadron scattering. For photoproduction of a $c\bar{c}$-pair we find a rate which is not very large but should be observable at HERA. With a minimum transverse jet momentum of $p_T = 3\,\mathrm{GeV}$ at parton level and a typical photon-proton centre-of-mass energy around $200\,\mathrm{GeV}$ the photon-proton cross section is of the order of $300\,\mathrm{pb}$.

In a similar way the cancellation disappears even for light quarks if Q^2 is large enough, say $Q^2 > 5\,\mathrm{GeV}^2$, and one finds again cross sections that should not be too small to be measured at HERA. Because of the behaviour at small Q^2 mentioned above, longitudinal photons dominate the cross section at lower values of Q^2, but around 10 to $15\,\mathrm{GeV}^2$ transverse photons take over. For charm quarks the cross section decreases slowly from its value at $Q^2 = 0$, in deep inelastic scattering it is comparable in size with the one for production of the three light quark flavours.

To conclude let me say that the LN model predicts that at HERA diffractive events with just a pair of high-p_T jets and the scattered proton and electron in the final state should be seen in deep inelastic scattering, with a significant fraction of charm quark jets. Under the assumption that the simple approximation of noninteracting gluons gives a qualitatively correct picture in photoproduction as well, then charm should be strongly enhanced compared with light flavours if the photon is real. Should experiment find an important rate for high-p_T jets not coming from heavy quarks, then, under the same assumption, their production could not be explained by exchange of the "soft" pomeron we know from phenomenology, because the gluons in the pomeron would have to be hard. This would be rather opposite to conventional wisdom, which expects

that diffractive photoproduction is the domain of the soft pomeron, and that if a hard
pomeron is to be seen it should be in deep inelastic scattering, where Q^2 sets a hard
scale.

ACKNOWLEDGEMENTS

I am grateful to Peter Landshoff for suggesting this work, for many discussions, and
for reading the manuscript. This research is supported in part by the EU Programme
"Human Capital and Mobility", Network "Physics at High Energy Colliders", Contract
CHRX-CT93-0357 (DG 12 COMA), and in part by Contract ERBCHBI-CT94-1342.
It is also supported in part by PPARC.

REFERENCES

[1] A Donnachie and P V Landshoff, Phys. Lett. **B296** (1992) 227

[2] A Donnachie and P V Landshoff, Nucl. Phys. **B231** (1984) 189

[3] P D B Collins, *An Introduction to Regge Theory and High Energy Physics*
(Cambridge University Press, Cambridge, 1977); for a shorter account see also:
A D Martin and T D Spearman, *Elementary Particle Theory* (North Holland,
Amsterdam, 1970)

[4] A Brandt et al., UA8 Collaboration, Phys. Lett. **B297** (1992) 417

[5] ZEUS Collaboration, M. Derrick et al., Phys. Lett. **B315** (1993) 481; preprint
DESY 94-063 (1994)

[6] T. Greenshaw (H1 Collaboration), talk given at the *XXIXth Rencontres de
Moriond*, Méribel, France, March 1994, preprint DESY 94-112 (1994)

[7] P V Landshoff and O Nachtmann, Z. Phys. **C35** (1987) 405;

[8] A Donnachie and P V Landshoff, Nucl. Phys. **B311** (1988/89) 509

[9] A Donnachie and P V Landshoff, Nucl. Phys. **B244** (1984) 322

[10] M Diehl, preprint DAMTP-94-60 (1994)

[11] A H Mueller, Nucl. Phys. **B335** (1990) 115; M G Ryskin, Sov. J. Nucl. Phys. **52**
(1990) 529; E Levin and M Wüsthoff, preprint DESY 92-166 (1992)

[12] N N Nikolaev and B G Zakharov, Z. Phys. **C53** (1992) 331; Phys. Lett. **B332**
(1994) 177

DEEP INELASTIC SCATTERING
AND SMALL-X PHYSICS*

A.H. Mueller[†]

Department of Physics
Columbia University
New York, New York 10027

1. INTRODUCTION

This is a slightly extended version of lectures given in Cargèse in August, 1994. The first part reviews the parton model and operator product expansions of deep inelastic lepton-nucleon scattering. While the discussion is self-contained it may seem, perhaps, a bit rushed. There are many more leisurely expositions of this material available in standard textbooks for the reader who finds the present discussion somewhat brief. The second part gives a general discussion of small-α behavior with an emphasis on a qualitative understanding of small-x behavior. Achieving high densities of spatially overlapping partons, and how this occurs in the BFKL pomeron, is the focal point of the discussion. The third part gives a more technical treatment of the BFKL pomeron and how it may be possible to measure it.

In order to guide the reader toward filling in many of the details which are left out of these lectures, I have included specific problems within the body of the text. The problems range from almost trivial to challenging and are labelled by an E,M, or H signifying that a particular problem is easy, of medium difficulty, or hard.

2. THE PARTON MODEL AND THE
OPERATOR PRODUCT EXPANSION

In this lecture a general description of deep inelastic lepton-nucleon scattering will be given in terms of the parton model and the operator product expansion.

2.1 Cross Sections and Structure Functions

The deep inelastic lepton-nucleon reaction is illustrated in Fig.1. The cross section for scattering of an unpolarized lepton off an unpolarized nucleon is given in terms

*Lecturers given at "Frontiers in Particle Physics," Cargese, Aug. 1-12, 1994.
[†]This work is supported in part by the Department of Energy under grant DE-FG-2-94ER 40819

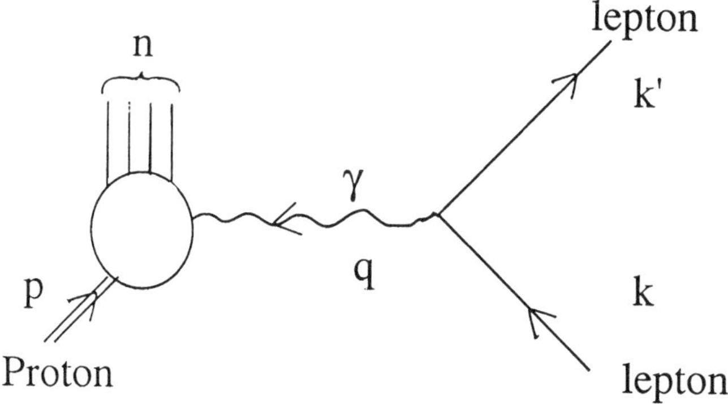

1. Deep Inelastic Lepton-nucleon Scattering.

of structure functions W_1 and W_2 as

$$\frac{d\sigma}{dE'd\Omega'} = \frac{4\alpha_{em}(E')^2}{Q^4}[W_2 cos^2\theta/2 + 2W_1 sin^2\theta/2] \tag{1}$$

in the rest frame of the nucleon. E' is the energy of the outgoing lepton and θ is the lepton's angle of scattering. α_{em} is the usual fine structure constant and $Q^2 = -q_\mu q^\mu$ is the invariant momentum transfer carried by the virtual photon to the nucleon.

W_1 and W_2 are structure functions defined in terms of a structure tensor $W_{\mu\nu}$,

$$W_{\mu\nu}(p,q) = \frac{4\pi^2 E_p}{m}\int d^4x\ e^{iqx}(p|j_\mu(x)j_\nu(0)|p), \tag{2}$$

by

$$W_{\mu\nu} = -(g_{\mu\nu} - \frac{q_\mu q_\nu}{q^2})W_1 + \frac{1}{m^2}[p_\mu p_\nu - \frac{p\cdot q}{q^2}(p_\mu q_\nu + p_\nu q_\mu) - \frac{(p\cdot q)^2}{(q^2)^2}q_\mu q_\nu]W_2. \tag{3}$$

In the above m is the nucleon mass, and a spin average over nucleon spin orientations is assumed but not explicitly indicated.

2.2 The Bjorken Frame

In describing deep inelastic lepton-nucleon scattering in the parton model it is important to refer to a particular frame, the Bjorken or infinite momentum frame. In that frame the proton and virtual photon momenta take the form

$$p = (p_0, p_x, p_y, p_z) \approx (p + \frac{m^2}{2p}, 0, 0, p) \tag{4}$$

and

$$q = (q_0, \underline{q}, q_z = 0) \tag{5}$$

as p becomes arbitrarily large. In terms of the two invariants Q^2 and $\nu = p \cdot q/m$ one finds $q_0 = m\nu/p$ becomes small as p becomes large so that $\underline{q}^2 = Q^2$ as $p \to \infty$. In what follows we shall generally take Q^2 and $x = Q^2/2p \cdot q$ as the two independent invariants on which W_1 and W_2 can depend.

2.3 Physical Basis of the Parton Model

In discussing the physics basis of the parton model it is useful to consider $T_{\mu\nu}$ defined as in (2) but with $j_\mu(x)j_\nu(0)$ replaced by $T(j_\mu(x)j_\nu(0))$ with T the usual time-ordering symbol. Then

$$W_{\mu\nu} = 2Im T_{\mu\nu}. \tag{6}$$

$T_{\mu\nu}$ is the forward elastic scattering amplitude for virtual photons on a nucleon which we now take to be a proton.

It is convenient to imagine the interaction picture time evolution of a proton. The proton consists of three valence quarks along with a quark-antiquark sea and gluons. The sea and gluons are created and reabsorbed with the passage of time. In the proton's rest system the typical time between interactions should be $1/\Lambda$ since $\Lambda \approx 200 MeV$ is the only genuine scale in light quark QCD. In the Bjorken frame, we can expect this typical scale to be time-dilated so that $\frac{p}{m\Lambda}$ becomes the natural scale for virtual fluctuations. Now the lifetime of the virtual photon, the time between its emission by the electron and absorption by a quark, appearing in $T_{\mu\nu}$ is given by

$$\tau_\gamma = \frac{1}{|\vec{q}| - q_0} \approx \frac{1}{Q} \tag{7}$$

in the Bjorken frame. Thus, one may view the photon as being absorbed *instantaneously* by some quark in the proton so long as we use the Bjorken frame.

Suppose the quark which absorbs the photon has longitudinal momentum k. (In a moment, we shall see that $k = xp$ with $x = \frac{Q^2}{2p \cdot q}$ as defined earlier.) Then, upon absorbing the virtual photon, the struck quark becomes highly virtual with a lifetime k/Q^2, and since this time is much shorter than the normal interaction time between quarks in the proton the struck quark must re-emit the photon before any interactions with the other quarks and gluons in the proton take place.

Finally, since the transverse momentum of the absorbed photon is $|\underline{q}| = Q$ the photon must be absorbed, and re-emitted, over a transverse coordinate region having $|\Delta\underline{x}| \approx 1/Q$. That is, the quark which absorbs the virtual photon, the struck quark, is pointlike (bare) down to a transverse size $|\Delta\underline{x}| \approx 1/Q$.

Thus, our picture of $T_{\mu\nu}$, and hence of $W_{\mu\nu}$, is that the scattering by the virtual photon takes place essentially instantaneously and over a very small, almost pointlike, spatial region. Since the photon interacts only with a single quark we expect $T_{\mu\nu}$, and $W_{\mu\nu}$, to be given in terms of the number density of quarks in the proton times the $T_{\mu\nu}$, or $W_{\mu\nu}$, of an individual quark. We stress that this picture of deep inelastic scattering

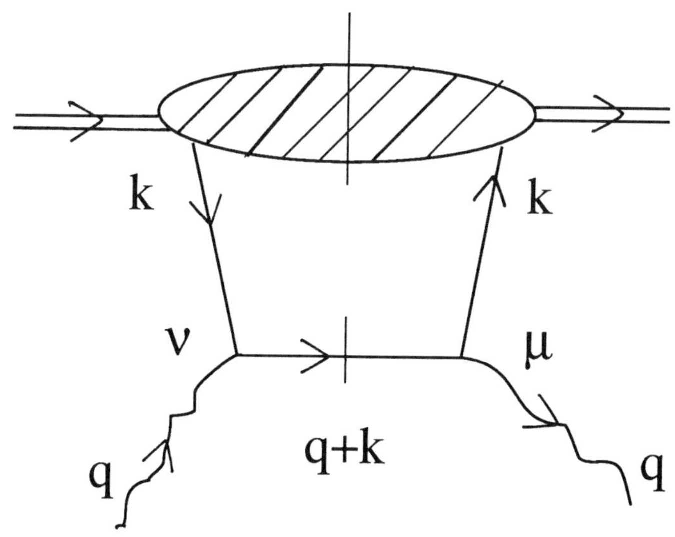

2. Deep Inelastic Lepton-nucleon Scattering in the Parton Model.

as a measurement of the number density of quarks in the wavefunction of the proton only holds in the infinite momentum frame of the proton.

2.4 The "Naive" Quark Parton Model

Now let us put the words of the previous section into formulas. Once the arguments as to what time scales are relevant for the photon absorption and re-emission have been made, and they are more directly made for $T_{\mu\nu}$ than for $W_{\mu\nu}$, it is convenient to deal directly with $W_{\mu\nu}$. Consider the graph in Fig.2 where $W_{\mu\nu}$ is explicitly given in terms of the $W_{\mu\nu}$ of a quark line, q + k, where the vertical line on q + k indicates that one must put the quark on shell as demanded by the ordinary (not time-ordered) product in (2). Let $\Gamma^f_{\mu\nu}$ be the lower part of the graph in Fig.2, which is shown in Fig.3 for explicitness. f is the flavor of the struck quark. Then

$$\Gamma^f_{\mu\nu} = e_f^2 2\pi\delta((k+q)^2)\gamma_\mu\gamma\cdot(k+q)\gamma_\nu \tag{8}$$

where e_f is the electric charge, in units of the proton's charge, of the quark of flavor f, and where we have taken the quark masses to be zero for simplicity.

It is convenient to use light-cone vector notation

$$v_\pm = \frac{1}{\sqrt{2}}(v_0 \pm v_3) \tag{9}$$

for an arbitrary four-vector v_μ. Then $v \cdot m = v_\mu m_\mu = v_+ m_- + v_- m_+ - \underline{v} \cdot \underline{m}$ where $\underline{v}$ and $\underline{m}$ represent the x and y components of v_μ and m_μ. Then

$$\delta((k+q)^2) = \delta(-Q^2 + k^2 + 2k_+ q_- + 2k_- q_+ - 2\underline{k} \cdot \underline{q}). \tag{10}$$

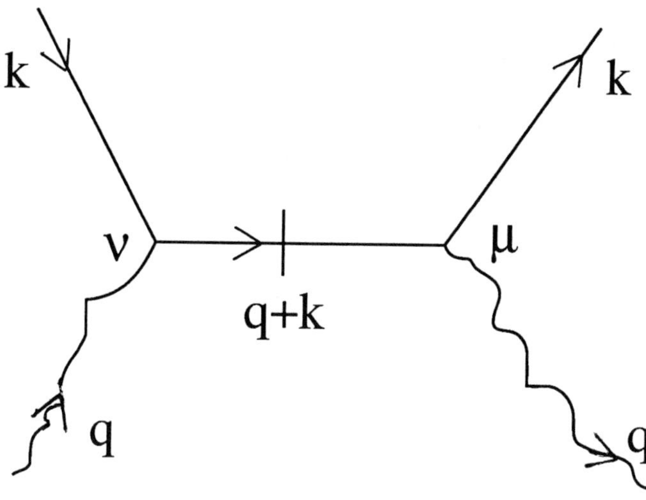

3. Virtual Photon Scattering of a Quark, from the Lower Part of the Graph Shown in Fig.2.

In the naive parton model one supposes that $\underline{k}^2$ and k^2 are of size $\wedge^2$. This is indeed equivalent to our assumption, in the previous section, that the only time scale in the rest frame of the proton is $\tau = 1/\wedge$. Further, in the Bjorken frame $k_+ >> k_-$ while $q_+ = q_-$ so that one may simplify (10) by keeping only the Q^2 and $2k_+q_-$ terms in the δ-function. Thus

$$\delta((k+q)^2) \approx \frac{x}{Q^2}\delta(x - \frac{k_+}{p_+}) \tag{11}$$

with x as defined earlier. Eq.(11) says that the longitudinal momentum fraction of the proton's momentum carried by the struck quark is x.

The $\gamma-$matrix factors in (8) can be simplified when one realizes that it is permissable to replace $\gamma \cdot (k+q)$ by γ_+q_-. This is possible because γ_+ will ultimately turn into a p_+ so that γ_+q_- is of size $p \cdot q$ while no other terms in $\gamma \cdot (k+q)$ can possibly be of the same order. Thus

$$\Gamma^f_{\mu\nu} = e^2_f \frac{\pi}{p_+}\delta(x - \frac{k_+}{p_+})\gamma_\mu\gamma_+\gamma_\nu \tag{12}$$

and using (12) with (2) gives

$$W_{\mu\nu} = \frac{(2\pi)^3 E_p}{2mp_+}\Sigma_f e^2_f \int d^4k\delta(x - \frac{k_+}{p_+})A^f_{ab}(p,k)(\gamma_\mu\gamma_+\gamma_\nu)_{ba} \tag{13}$$

where A represents the upper portion of the graph of Fig.2. a and b are Dirac indices. Still in the Bjorken frame one finds from (3)

$$W_{ij} = -g_{ij}W_1 + \frac{q_iq_j}{q^2}(W_1 + \frac{(p \cdot q)^2}{q^2m^2}W_2) \tag{14}$$

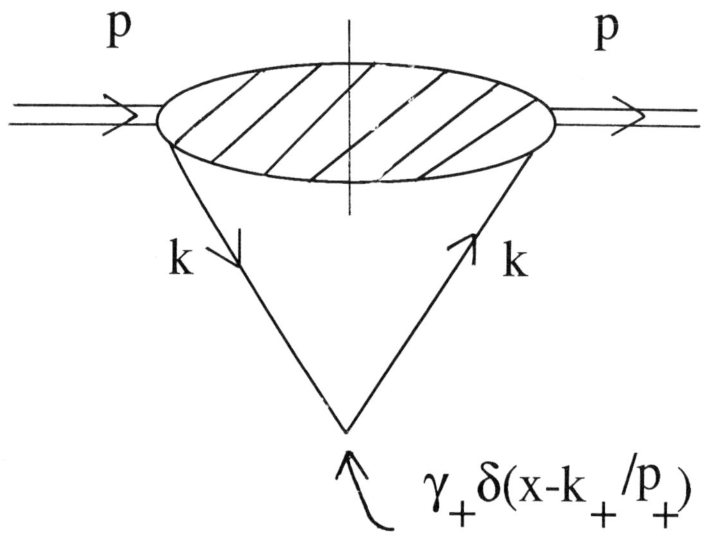

4. The Quark Parton Distribution in a Proton.

with $i, j = 1, 2$. W_{ij} is manifestly symmetric in i, j so in evaluating W_{ij} from (13) we may replace $\gamma_i\gamma_+\gamma_j$ by

$$\frac{1}{2}(\gamma_i\gamma_+\gamma_j + \gamma_j\gamma_+\gamma_i) = -\frac{1}{2}\{\gamma_i, \gamma_j\}\gamma_+ = -g_{ij}\gamma_+. \tag{15}$$

Using (15) in (13) and comparing with (14) one finds

$$W_1 + \frac{(p \cdot q)^2}{q^2 m^2}W_2 = 0$$

or

$$\nu W_2 = 2mxW_1 \tag{16}$$

along with

$$\nu W_2 = \sum_f e_f^2 x P^f(x) \tag{17}$$

where

$$P^f(x) = \frac{(2\pi)^3 2E_p}{2p_+} \int d^4 k \delta(x - k_+/p_+) A_{ba}^f(p, k)(\gamma_+)_{ab}. \tag{18}$$

Eq.(18) is illustrated in Fig.4.

Equation (16) is the Callan-Gross[1] relation which follows from the spin 1/2 nature of the charge carrying constituents of the proton. Eqs. (17) and (18) say that νW_2 depends only on x and not on Q^2. This is known as Bjorken scaling.[2] The essential ingredient in obtaining Bjorken scaling was our assumption that k^2 and $\underline{k}^2$ are of the same order of magnitude as $\wedge^2$. In the next section we shall see that this is not an exact result in QCD and we shall extend our treatment from this naive (scaling) parton model to the more precise QCD improved parton model.

Problem 1(M-H). Show that

$$\int_0^1 dx\ x^{n-1} P^f(x) = \frac{(2\pi)^3 2E_p}{(2p_+)^n} (p|\tilde{q}_f \gamma_{\mu_1} i\overset{\leftrightarrow}{\partial}_{\mu_2} i\overset{\leftrightarrow}{\partial}_{\mu_3} \cdots i\overset{\leftrightarrow}{\partial}_{\mu_n} q_f|p)|_{\mu_i=+}.$$

This result relates moments of the structure function νW_2 to matrix elements of local operators.

2.5 The QCD Improved Parton Model (The DGLAP Equation)

Refer back to problem 1. It is clear that this cannot be a result which is generally true since the left-hand side of the equation is, using (17), gauge invariant while the right-hand side is not gauge invariant. Indeed, the result given in problem 1 can be given in a more generally correct way as

$$\int_0^1 dx\ x^{n-1} P^f(x) = \frac{(2\pi)^3 2E_p}{(2p_+)^n} (p|\tilde{q}_f \gamma_{\mu_1} i\overset{\leftrightarrow}{D}_{\mu_2} i\overset{\leftrightarrow}{D}_{\mu_3} \cdots i\overset{\leftrightarrow}{D}_{\mu_n} q_f|p)|_{\mu_i=+} \tag{19}$$

where $D_\mu = \partial_\mu - igA_\mu$ is the gauge covariant derivative in QCD.

What has gone wrong? The result given in problem 1 seemingly follows straightforwardly from the discussion given in sec.2.3 with the graph shown in Fig.2 being the mathematical representation of the physics of the naive parton model. The point, however, is the following. Although the physics discussion given in sec.2.3 is correct that physics does not necessarily have a manifest realization in terms of Feynman graphs. Indeed, the physics of the parton model is only manifestly realized in a particular gauge, the light-cone gauge with $A_+ = 0$. If $A_+ = 0$ then $D_+ = \partial_+$ and (19) agrees with the result stated in problem 1. This is an important lesson for us. Parton model ideas will only be expressed simply in terms of field theory concepts in light-cone gauge.

Referring back to (19) we can notice another problem. The local operators appearing on the right-hand side of (19) cannot all be expected to have zero anomalous dimensions. In fact, none of them have zero anomalous dimensions. Thus, the right-hand side of (19) must depend on a normalization scale while the left-hand side of (19) would seem to have no room for such a normalization scale dependence. In fact, the only natural scale on the left-hand side of (19) would be Q^2 since νW_2 can depend on Q^2 and x. The naive parton model result given by (17) and (18) is not quite right since the integration over d^4k in (18) is divergent in the large $\underline{k}^2$ region. This would seem to vitiate the discussion just below (10) where we argue that $\underline{k}^2$ and k^2 could be dropped with respect to Q^2. The integration over $d\underline{k}^2$ is logarithmically divergent with Q^2 furnishing the natural cut off for the integration in (18). Since most of the logarithmic contribution does come from $\underline{k}^2/Q^2 << 1$ the discussion leading to (18), as well as the discussion given in sec.2.3 remains valid, and all that is necessary is to cut off the divergent integration in (18) at a scale Q^2 to obtain what has come to be called the QCD improved parton model. Let's see how this happens in a little more detail.

The main dependence on $\underline{k}^2$ in (18) can be obtained by considering the graph shown in Fig.5. Let Γ_+ be the part of the graph including the lines $k, k_1 - k$ and the vertices attached to those lines, that is the part of the graph shown in Fig.6. Then

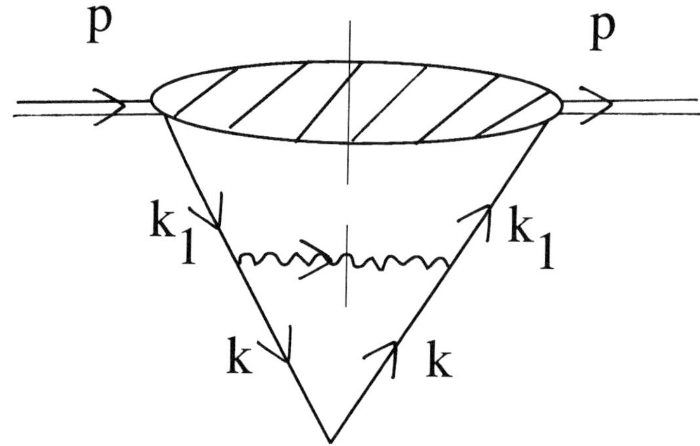

5. The First Nontrivial Correction to the Naive Quark Model Parton Distribution.

$$\Gamma_+ = \frac{g^2 C_F}{(2\pi)^4} \int \frac{d^4 k \Theta(Q^2 - \underline{k}^2)}{[k^2]^2} \delta(x - k_+/p_+)[-2\pi\delta((k_1 - k)^2)]$$

$$\cdot \; \gamma_\alpha \gamma \cdot k \gamma_+ \gamma \cdot k \gamma_\beta \left(g_{\alpha\beta} - \frac{\eta_\alpha (k_1 - k)_\beta + \eta_\beta (k_1 - k)_\alpha}{\eta \cdot (k_1 - k)} \right), \tag{20}$$

where the last factor on the right-hand side of (20) comes from the gluon propagator in light-cone gauge. g is the coupling of the quarks to the gluon, the line $k_1 - k$ in Fig.6, while C_F is the Casimir operator for the fundamental representation of SU(3). $C_F = \frac{N_c^2 - 1}{2N_c} = 4/3$. $\eta \cdot v = v_+$ for any vector v_μ.

Problem 2(M). Consider

$$L_o = -\frac{1}{4}(\partial_\mu A_\nu^i - \partial_\nu A_\mu^i)(\partial_\mu A_\nu^i - \partial_\nu A_\mu^i)$$

with $\eta \cdot A^i = A_+^i = 0$. Show that the gluon propagator is

$$D_{\alpha\beta}^{ij}(k) = \frac{-i\delta_{ij}}{k^2}\left(g_{\alpha\beta} - \frac{\eta_\alpha k_\beta + \eta_\beta k_\alpha}{\eta \cdot k}\right).$$

It is convenient to write $d^4 k = dk_+ dk_- d^2\underline{k}$ in (20). Then

$$\int dk_- \delta((k_1 - k)^2) = \frac{1}{2(k_1 - k)_+} \tag{21}$$

and

$$k_- = k_{1-} - \frac{(\underline{k}_1 - \underline{k})^2}{2(k_1 - k)_+}. \tag{22}$$

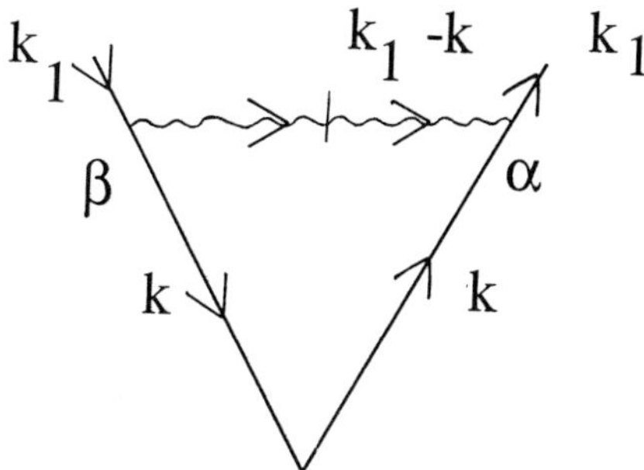

6. The Lower Part of the Graph Shown in Fig.5.

Problem 3(M). Show that

$$k^2 = \frac{-1}{1-z}[(\underline{k} - \frac{1}{z}\underline{k}_1)^2 - z(1-z)k_1^2]$$

where $z = x/x_1 = k_+/k_{1+}$.

Problem 4(M). Show that

$$[g_{\alpha\beta} - \frac{(k_1 - k)_\alpha \eta_\beta + (k_1 - k)_\beta \eta_\alpha}{\eta \cdot (k_1 - k)}]\gamma_\beta\gamma \cdot k\gamma_+\gamma \cdot k\gamma_\alpha = -2\underline{k}^2\gamma_+\frac{1+z^2}{(1-z)^2}$$

as far as terms linear in $\underline{k}^2$ are concerned.

Using (21) and (22) as well as the results of problems 3 and 4 in (20) gives

$$\Gamma_+ = \gamma_+\frac{1+z^2}{1-z}\frac{1}{x_1}\int_{\underline{k}_1^2}^{Q^2}\frac{d\underline{k}^2}{\underline{k}^2}\frac{\alpha(\underline{k}^2)C_F}{2\pi} \tag{23}$$

as far as the logarithmic part of the integration is concerned. We have used the fact that the QCD running coupling enters, when a more complete calculation is done, and depends on the transverse momentum of the gluon. Since the γ-matrix structure of (23), a simple γ_+, is the same as the vertex shown in Fig.4 one can write

$$Q^2\frac{\partial}{\partial Q^2}P^f(x, Q^2) = \frac{\alpha(Q^2)}{2\pi}\int_x^1\frac{dx_1}{x_1}\gamma_{qq}(x/x_1)P^f(x_1, Q^2) \tag{24}$$

where $\gamma_{qq}(z) = C_F\frac{1+z^2}{1-z}$. Including self-energy corrections on the line k in Fig.4 adds a $\delta(z-1)$ term to γ giving the final result for γ

$$\gamma_{qq}(z) = C_F\left(\frac{1+z^2}{1-z}\right)_+ \tag{25}$$

where

$$\int_0^1 dz\,[h(z)]_+ f(z) = \int_0^1 dz\ h(z)[f(z) - f(1)] \tag{26}$$

defines the general function $[h(z)]_+$.

For completeness we include gluonic parton distributions and break quark parton distributions into a flavor singlet and non-singlet case. If the flavor non-singlet distribution $\Delta_{ff'}(x, Q^2)$ and the flavor singlet distribution $\Sigma(x, Q^2)$ are defined by

$$\Delta_{ff'}(x, Q^2) = P^f(x, Q^2) - P^{f'}(x, Q^2)$$

$$\Sigma(x, Q^2) = \sum_f (P^f(x, Q^2) + P^{\bar{f}}(x, Q^2)) \tag{27}$$

then Δ obeys an equation identical to (24) while Σ and the gluon distribution $G(x, Q^2)$ obey

$$Q^2 \frac{\partial}{\partial Q^2} \begin{pmatrix} \Sigma(x, Q^2) \\ G(x, Q^2) \end{pmatrix} = \frac{\alpha(Q^2)}{2\pi} \int_x^1 \frac{dx_1}{x_1} \begin{pmatrix} \gamma_{qq}(x/x_1) & \gamma_{qG}(x/x_1) \\ \gamma_{Gq}(x/x_1) & \gamma_{GG}(x/x_1) \end{pmatrix} \begin{pmatrix} \Sigma(x_1, Q^2) \\ G(x_1, Q^2) \end{pmatrix}, \tag{28}$$

where

$$\gamma_{qG}(z) = N_f[z^2 + (1-z)^2] \tag{29a}$$

$$\gamma_{Gq}(z) = C_F \frac{1 + (1-z)^2}{z} \tag{29b}$$

$$\gamma_{GG}(z) = 2C_A\left[\frac{z}{(1-z)_+} + \frac{1-z}{z} + z(1-z)\right] + \frac{11C_A - 2N_f}{6}\delta(x-1) \tag{29c}$$

with $C_A = N_c = 3$. These equations are the DGLAP equations and they furnish the basis for determining the Q^2-dependence of parton distributions in QCD. A few comments on our procedure are in order.

The exact way in which the $\underline{k}^2$-integral in (20) and in (23) is cut off is unimportant as far as the derivation of (24) and (28) are concerned so long as the scale of the cut off is given by Q^2. This is the same as saying that the renormalization scale dependence of the operators on the right-hand side of (19) is unique at the leading logarithmic level. Indeed, if one renormalizes the operators on the right-hand side of (19) at a scale Q^2, then the Q^2 dependence which that renormalization introduces into the P^f on the left-hand side of (19) is exactly the same as given by (24). The Q^2-dependence of $P^f(x, Q^2)$ is uniquely given at the leading logarithmic level, the level described by (24), but is no longer unique beyond leading logarithms. Beyond leading logarithms, for example, when $\underline{k}^2/Q^2$ is of order 1 in (20), there are color charge density correlations on a spatial scale $|\Delta \underline{x}| \sim 1/Q$ in the proton so that the virtual photon probe no longer acts like a point-like probe of individual quark components of the proton. Beyond the leading logarithmic approximation one still defines quark and gluon distributions in the proton, but these distributions depend on the scheme used in renormalizing the operators appearing in the operator product expansion, which scheme dependence reflects the ambiguity is separating the probe measuring the partons from the measured partons at this level of precision.

3. SMALL-x BEHAVIOR (GENERAL DISCUSSION)

In this section the general properties of small-x behavior of structure functions will be discussed along with the theoretical motivation for studying small-x physics. A very heuristic discussion of the BFKL[6-8] pomeron and the contrasting pictures between BFKL evolution and DGLAP evolution is also given.

3.1 DGLAP Evolution at Small-x

At very small values of x parton distributions are driven by gluonic dynamics. We can see this by referring back to (28) and noticing that $\gamma_{Gq}(z)$ and $\gamma_{GG}(z)$ are singular as $z \to 0$ while $\gamma_{qq}(z)$ and $\gamma_{qG}(z)$ are regular as $z \to 0$. Thus, one may determine small-x behavior of parton distributions by first determining the small-x behavior of $G(x, Q^2)$ and then using (28) to determine Σ in terms of G and some initial distribution for Σ. The equation for $G(x, Q^2)$ is

$$Q^2 \frac{\partial}{\partial Q^2} G(x, Q^2) = \frac{\alpha(Q^2)}{2\pi} \int_x^1 \frac{dx_1}{x_1} \gamma_{GG}(x/x_1) G(x_1, Q^2).$$ (30)

The DGLAP equation always keeps leading logarithms in Q^2. (That is if one imagines fixing the coupling, α, in (30) then there will be one power of $\ell n\, Q^2$ for each factor of α.) As x becomes small one can also take the leading $\ell n\, 1/x$ approximation to (30) by keeping only the singular part of $\gamma_{GG}(z)$ as $z \to 0$. That is we take $\gamma_{GG}(z) = \frac{2C_A}{z}$ as given in (29c). Taking $\alpha(Q^2) = \frac{1}{b\, \ell n\, Q^2/\Lambda^2}$ with $b = \frac{33-2N_f}{12\pi}$ and noting that $\ell n\, Q^2/\Lambda^2 \; Q^2 \frac{\partial}{\partial Q^2} = \frac{\partial}{\partial \ell n\, \ell n\, Q^2/\Lambda^2}$ one can obtain from (30) the equation

$$\frac{\partial}{\partial \ell n\, 1/x} \frac{\partial}{\partial \ell n\, \ell n Q^2/\Lambda^2} x G(x, Q^2) = \frac{C_A}{\pi b} x G(x, Q^2).$$ (31)

Asymptotically, one can write the solution to (31) as

$$x G(x, Q^2) = \int_x^1 \frac{dx_1}{x_1} K(Q^2, x/x_1, Q_0^2) x_1 G(x_1, Q_0^2)$$ (32)

so long as the initial distribution $x G(x, Q_0^2)$, does not grow too rapidly as $x \to 0$. K is given by

$$K(Q^2, x/x_1, Q_0^2) = \frac{1}{2\pi} \{ \frac{C_A}{\pi b} \ell n(\frac{\ell n Q^2/\Lambda^2}{\ell n Q_0^2/\Lambda^2}) \}^{\frac{1}{4}} [\ell n\, x_1/x)]^{-3/4} exp\{2\sqrt{\frac{C_A}{\pi b} \ell n x_1/x \ell n(\frac{\ell n Q^2/\Lambda^2}{\ell n Q_0^2/\Lambda^2})} \}.$$ (33)

Eqs.(32) and (33) lead to growth of xG of the form

$$x G(x, Q^2) \propto exp\, 2\sqrt{\frac{C_A}{\pi b} \ell n\, 1/x \ell n\, \ell n Q^2/\Lambda^2}\, .$$ (34)

Thus, the DGLAP equation leads to parton distributions which grow moderately rapidly at small values of x.

Problem 5(H). Defining

$$G_n(Q^2) = \int_0^1 dx\; x^{n-1} G(x, Q^2),$$

show that (32) can be written as

$$G_n(Q^2) = K_n(Q^2, Q_0^2) G_n(Q_0^2).$$

Using the fact that (30) is equivalent to the equation

$$Q^2 \frac{\partial}{\partial Q^2} K_n(Q^2, Q_0^2) = \gamma_n(\alpha)Q^2)) K_n(Q^2, Q_0^2)$$

with

$$\gamma_n(\alpha(Q^2)) = \frac{\alpha(Q^2)}{2\pi} \int_0^1 dx\; n^{n-1} \gamma_{GG}(x) \approx \frac{\alpha(Q^2) C_A}{\pi(n-1)}$$

show that

$$K(Q^2, x/x_1, Q_0^2) = \int \frac{dn}{2\pi i} (x_1/x)^{n-1} e^{\int_{Q_0^2}^{Q^2} \gamma_n(\alpha(\lambda^2))}$$

and that this equation leads to (33). In the above n-integral the integration goes parallel to the imaginary axis and to the right of the point n=1. The integral is most easily done by a saddle point approximation in the n-plane.

Problem 6 (M). When α is fixed (independent of Q^2) show that the solution to the DGLAP equation in the leading double logarithmic approximation (leading logs in $\ell n\; 1/x$ and in $\ell n\; Q^2$) is

$$xG(x, Q^2) \propto exp\{2\sqrt{\frac{\alpha C_A}{\pi} \ell n\; 1/x\; \ell n\; Q^2/Q_0^2}\},$$

3.2 Leading and Nonleading Logarithms in DGLAP

Generically, the DGLAP equation takes the form

$$Q^2 \frac{\partial}{\partial Q^2} P(x, Q^2) = \int_x^1 \frac{dx_1}{x_1} \gamma(x/x_1, \alpha(Q^2)) P(x_1, Q^2) \tag{35}$$

where we now include the running coupling in the anomalous dimension function, γ. Although γ and the parton distribution P, either a quark or a gluon distribution, are not unique beyond order α, nevertheless, within a given scheme of renormalization one has a definite γ and P. Combined with coefficient functions, E, calculated in the same scheme one can calculate structure functions systematically in terms of $\alpha(Q^2)$ and some initial parton distribution $P(x, Q_0^2)$. Thus,

$$\nu W_2(x, Q^2) = \int_x^1 \frac{dx_1}{x_1} P(x/x_1, Q^2) E(x_1, (Q^2)) \tag{36}$$

gives νW_2 as an, in principle, scheme independent quantity. γ can be expanded in powers of α as

$$\gamma(z, \alpha) = \alpha \gamma^{(1)}(z) + \alpha^2 \gamma^{(2)}(z) + \cdots . \tag{37}$$

Keeping only $\gamma^{(1)}$ in (35) gives the conventional leading logarithmic approximation. Keeping $\gamma^{(1)}$ and $\gamma^{(2)}$ terms in (35) gives the next-to-leading logarithmic term also. This, along with the order $\alpha(Q^2)$ term in E is what is known as the second order formalism in describing structure functions.

Thus the DGLAP equation can always be used to describe structure functions. What is needed is the anomalous dimension $\gamma(z, \alpha)$, the coefficient function $E(\alpha)$, and the initial value $P(x, Q_0^2)$ for solving (35). At moderate values of x this is a very efficient way of describing structure functions. At small values of x it is not clear that the DGLAP equation is very useful. If $\alpha(Q^2)\ell n\, 1/x \geq 1$, one needs to keep terms like $[\alpha(Q^2)]^{n_1}[\ell n Q^2/\Lambda^2]^{n_2}(\ell n 1/x)^{n_3}$ in νW_2. For $n_2 = 0$ such terms are in the coefficient function E and one must evaluate E to all orders in α. For $n_2 \neq 0$ but $n_2 << n_1$ one also needs to keep very high order terms in the anomalous dimesnion function.

It has been suggested that when x is very small it may be more efficient to try and resume all terms involving $(\alpha(Q)\ell n\, 1/x)^n$. (Such terms are governed by another equation, the BFKL equation, which we shall consider shortly.) Whether this is indeed the case or not is unclear at present. Recent data[9-11] from HERA show that νW_2 rises rapidly as x becomes small, however, it is difficult to say, at present, whether that rise is a manifestation of DGLAP evolution, using low order anomalous dimensions and coefficient functions along with an initial distribution which does not rise rapidly with decreasing x. The structure functions determined by GRV[12] do get such an increase from DGLAP evolution, but at the price of taking the initial parton distribution at $Q_0^2 = 0.3 GeV^2$, an uncomfortably low value. On the other hand, MRS[13] take $Q_0^2 = 4 GeV^2$ and cannot obtain a good fit to the data from a second order DGLAP formalism without taking initial parton distributions which increase strongly at small x. This rise of the initial distribution at small x could be a manifestation of BFKL evolution or it might be simply a rewriting of GRV. Good data over a wide range of Q^2, including low values of Q^2 might help to settle this ambiguity.

3.3 A Physical Picture of DGLAP Evolution

The DGLAP equations given by (24) and (28) describe how parton number densities change as Q^2 changes. Recalling that $1/Q$ is the transverse coordinate resolution of the virtual photon probe in deep inelastic scattering one can interpret the DGLAP equation as governing the change in quark and gluon densities as one changes the spatial resolution of the probe that measures those densities. For example, one might measure a quark at a given momentum fraction x_0 of the proton and at a given transverse spatial resolution $1/Q_0$. If that exact same quark is measured at a finer spatial resolution, $1/Q > 1/Q_0$, it may be found to be a quark and a gluon or a quark along with several quark-antiquark pairs and some gluons. The DGLAP equation is the equation that describes the structure of quarks and gluons known to be pointlike at some initial scale $1/Q_0$. This is schematically shown in Fig.7 where the change in partonic structure of the proton is illustrated when one changes the resolution of measurements from $1/Q_0$ to $1/Q$. At very small values of x these number densities grow rapidly with Q^2, as seen

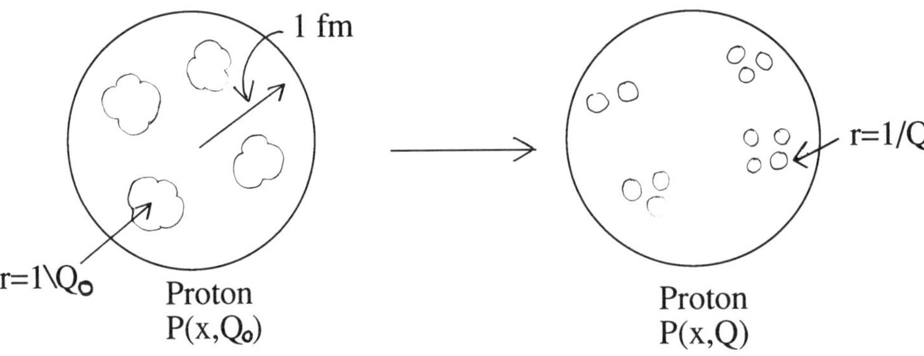

7. The Picture of DGLAP Evolution as One Goes from a Transverse Spatial Resolution of $1/Q_0$ to $1/Q$. The Blobs Inside the Proton Are Partons, Either Quarks or Gluons.

from (34), however, the increasing number of partons tend to be nonoverlapping since they are mainly generated by looking at smaller spatial scales.

3.4 The Physical Picture of BFKL Evolution (Simplified Version)

The BFKL equation is much more difficult to derive than the DGLAP equation and we shall not give a derivation, although we shall state the equation and discuss its solutions later on. In this section, an intuitive picture of BFKL evolution will be given. The essential features in this simple picture are: (i) BFKL evolution is an evolution from high longitudinal momentum partons to low longitudinal momentum partons. (ii) The evolution occurs at a fixed transverse momentum or, equivalently, the evolution occurs over a fixed transverse area $\Delta x_\perp^2 \approx 1/Q^2$ of the proton. As we shall see later, there is some nontrivial structure in transverse momentum but this dependence is much less pronounced than in DGLAP evolution. One can view the x-evolution of the BFKL equation as creating the small-x part of the wavefunction of a proton or simply as the dressing of a high momentum quark or gluon in the proton with low-x gluons. The description given below starts with a bare large-x gluon and describes the evolution which builds up the small-x cloud around the parent gluon.

To that end, consider a high energy gluon having a light-cone momentum p_+ and splitting into two gluons having momentum $p - k_1$ and k_1 respectively with $k_{1+}/p_+ <<$ 1. The element of probability for the soft emission is

$$dP_1 = \frac{\alpha C_A}{\pi} \frac{dk_{1\perp}^2}{k_{1\perp}^2} \frac{dk_{1+}}{k_{1+}}. \tag{38}$$

Eq.(38) is the result obtained from the low-x limit of $\gamma_{GG}(x)$ given in (29c) and, except for the factor of C_A, it is the same formula that describes soft photon emission from an electron. We shall make the approximation that all transverse momenta are fixed

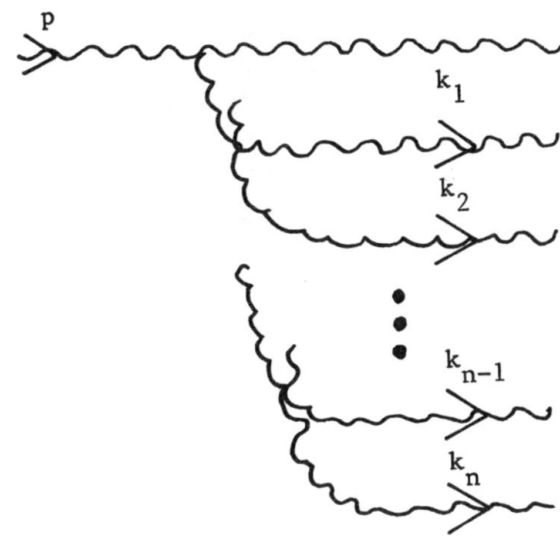

8. The Emission of the Gluon k_N Coherently Off Higher Momentum, Longer Lived, Gluons.

near a given scale Q, and so we replace $dk_{1\perp}^2/k_{1\perp}^2$ by c, a constant, in (38). Define $y_1 = \ell n(k_{1+}/p_+)$, then $dy_1 = dk_{1+}/k_{1+}$ and

$$dP_1 = c\frac{\alpha C_A}{\pi}dy_1. \tag{39}$$

The probability of soft gluon emission is small because of the $\alpha(Q^2)$ in (39), but it is also directly proportional to the longitudinal phase space available. Thus, in order that an emission take place one requires a rapidity interval Δy_1 given by

$$\Delta y_1 = (c\frac{\alpha C_A}{\pi})^{-1}. \tag{40}$$

The lifetime of the fluctuation of the gluon p into the two gluons k_1 and $p - k_1$ is given by

$$\tau_1 = \left[\frac{k_{1\perp}^2}{2k_{1+}} + \frac{k_{1\perp}^2}{2(p - k_1)_+}\right]^{-1} \approx \frac{2k_{1+}}{k_{1\perp}^2} = (\frac{p_+}{Q^2})e^{-y_1} \tag{41}$$

a time which is determined by the softer of the the gluons.

Emission of a gluon having longitudinal momentum $k_{2+} << k_{1+}$ occurs off the two gluons $(p - k_1, k_1) \approx (p, k_1)$ which can be considered as free particles during the time of emission of gluon k_2 since $\tau_2 << \tau_1$.

Continue the process of emmisions to the point where N-1 gluons have been emitted into the wavefunction of the original high momentum gluon p. We now wish to calculate the Nth emission of a gluon k_{N+} where $k_{N+} << k_{i+}$ for $i < N$. The process is illustrated in Fig.8 where the Nth emission occurs coherently off the preexisting gluons which are frozen in longitudinal momentum and transverse coordinate space during the time of emission of $k_N, \tau_N \approx \frac{2k_{N+}}{Q^2}$. The Nth emission occurs over a transverse area proportional to $k_{N\perp}^{-2} \sim 1/Q^2$. We suppose the N preexisting gluons occupy an area comparable to

Q^{-2} and that the Nth gluon is emitted off a finite fraction N′ of the previously emitted gluons $k_{N-1}, k_{N-2} \cdots p$. (The Nth gluon would need to have a transverse momentum much less than Q to "see" the total charge of the preexisting gluons. Since its transverse momentum is Q it "sees" a finite fraction of the charges, added coherently, of the preexisting gluons.) We suppose that the charges of these N′ gluons are randomly distributed in color space so that the effective charge for the Nth emission is

$$g_N = \sqrt{\frac{c_1}{c} N} g \tag{42}$$

with c_1 a constant. Thus,

$$dP_N = c_1 \frac{\alpha C_A}{\pi} N dy_N \tag{43}$$

so that the rapidity internal necessary for the Nth emission is

$$\Delta y_N = \left(c_1 \frac{\alpha C_A}{\pi} N \right)^{-1}. \tag{44}$$

The rapidity interval necessary in order that N gluons be emitted is

$$Y_N = \left(c_1 \frac{\alpha C_A}{\pi} \right)^{-1} \sum_{i=1}^{N} \frac{1}{i} = \left(c_1 \frac{\alpha C_A}{\pi} \right)^{-1} \ell n N. \tag{45}$$

Inverting (45) to give N(Y) one finds

$$N(Y) = e^{c_1 \frac{\alpha C_A}{\pi} Y}. \tag{46}$$

or

$$\frac{dN}{dy} = c_1 \frac{\alpha C_A}{\pi} e^{c_1 \frac{\alpha C_A}{\pi}}. \tag{47}$$

(It is $\frac{dN}{dy}$ which should be compared to the gluon density $xG(x, Q^2)$ with $Y \sim \ell n \, 1/x$.) An exact treatment of BFKL evolution gives $c_1 = 4 \, \ell n \, 2$ and shows that there is a prefactor proportional to $(\alpha C_A Y)^{-\frac{1}{2}}$ on the right-hand side of (47). However, the basic picture presented here is a rough description of how BFKL evolution works. The key ingredient is that in forming the wavefunction soft gluons are emitted off the color charge fluctuations of the previously emitted gluons. The picture is of an unstable evolution of charge fluctuations growing, in Y, as indicated in (46).

From this discussion it should now be clear that the gluons corresponding to BFKL evolution overlap each other much more than in DGLAP evolution. This is illustrated in Fig.9 where the gluons created by BFKL evolution starting from a particular high momentum gluon in the proton are shown as a "hot spot" of many gluons, localized in a small transverse spatial region of the proton. It is precisely the picture of BFKL evolution which makes small-x physics so exciting. When the available range of rapidity ($\ell n \, 1/x$) becomes so large that many small-x gluons begin to live in the same transverse spatial region of the proton one can expect that these gluons will no longer behave as free partons. It is natural that gluons will be absorbed as well as emitted leading to a quasi-equilibrium. This is the "saturation" regime[14] of small-x physics where perturbation theory breaks down because of the high density of spatially overlapping

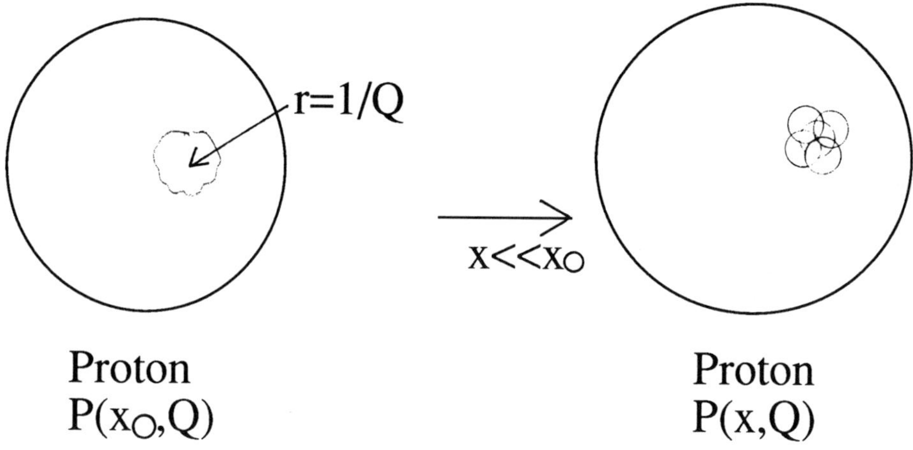

9. The Picture of BFKL evolution as one goes from a measurement of a high momentum fraction, x_0, gluon to measurement of small-x gluons in the proton.

gluons. This occurs when on the order of $1/\alpha$ gluons occupy the same spatial area $1/Q^2$, in which circumstance the gluons form a strongly interacting system even though α is small. The field strength, averaged over an area $1/Q^2$, is

$$1/Q^2 F_{\mu\nu} \sim \sqrt{\text{Number of gluons}} \sim 1/g. \tag{48}$$

When field strengths of size 1/g are created a highly nonlinear circumstance arises where perturbation theory is not reliable. Thus, the small-x problem leads to a new regime of QCD where individual parton-parton interactions are weak, but where the number of partons is so large that the system becomes strongly nonperturbative. Reaching this regime is the ultimate goal of small-x physics.

4. THE BFKL EQUATION AND ITS CONSEQUENCES

In this section, we shall describe the BFKL equation and properties of the solution of that equation at high energy. In the next section, we shall describe some ways in which one might possibly measure the BFKL pomeron, however, from a theoretical point of view I think that high energy heavy onium-heavy scattering is by far the best context in which to describe this physics. And it is to that scattering that we now turn.

4.1 Onium-Onium Scattering at High Energy[15-17]

We imagine an onium ground state for quarks so heavy that $\alpha(R^2) << 1$ with R the onium radius. Then as one scatters one onium on another onium perturbative QCD should be the appropriate tool with which to calculate the cross section. Define the forward onium-onium scattering amplitude A to have normalization such that the total onium-onium cross section is given by

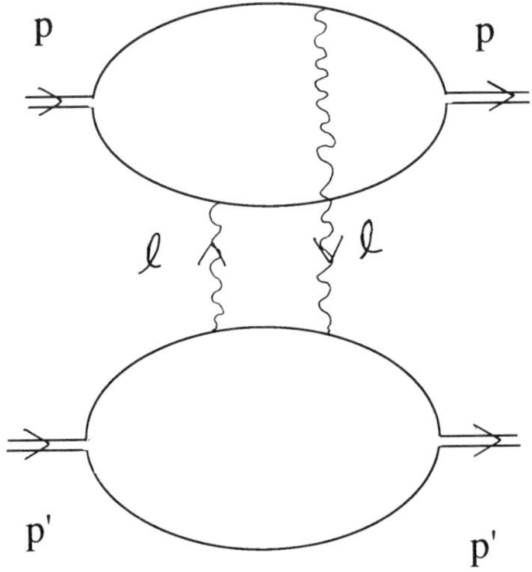

10. Onium-onium Scattering in the Two-Gluon Exchange Approximation. This is one of 20 graphs corresponding to the various ways the gluons can connect to the heavy quark or antiquark in each onium.

$$\sigma = -2\, Im\, A(Y), \tag{49}$$

for center of mass energy $E = 2M cosh Y/2$ where M is the onium mass. Then

$$A = -i \int d^2x \int_0^1 dz_1 \int d^2x' \int_0^1 dz_1' \Phi(\underline{x}',z_1')\Phi(\underline{x},z_1)F \tag{50}$$

where $\Phi(\underline{x},z)$ is the square of the onium light-cone wavefunction with a transverse coordinate separation of the heavy quark and antiquark given by $\underline{x}$ and the longitudinal momentum fraction of the heavy quark given by z. In the two gluon exchange approximation, one of the four graphs of which is illustrated in Fig.10, $F = F^{(0)}$ is given by

$$F^{(0)}(\underline{x},\underline{x}') = -\frac{\alpha^2(N_c^2-1)}{2N_c^2} \int \frac{d^2\ell}{[\underline{\ell}^2]^2}(2 - e^{-i\underline{\ell}\cdot\underline{x}} - e^{i\underline{\ell}\cdot\underline{x}})(2 - e^{-i\underline{\ell}\cdot\underline{x}'} - e^{i\underline{\ell}\cdot\underline{x}'}) \tag{51}$$

which gives

$$F^{(0)}(\underline{x},\underline{x}') = -\frac{\pi\alpha^2(N_c^2-1)}{N_c^2}x_<^2(1 + \ell n\, x_>/x_<) \tag{52}$$

where $x_<(x_>)$ is the lesser (greater) of x, x' with $x = |\underline{x}|, x' = |\underline{x}'|$. Thus, in the two gluon exchange approximation onium-onium scattering has a constant cross section at high energy proportional to α^2. The appropriate α here is $\alpha(R)$ with R the onium radius.

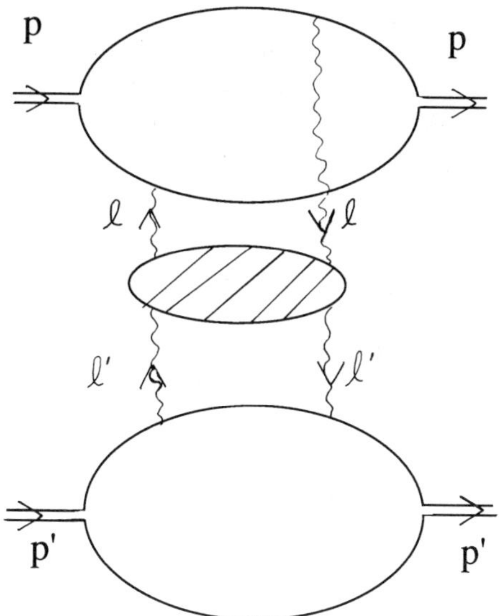

11. Onium-onium Scattering in the Leading Logarithmic Approximation.

Problem 7(H*). Derive (51)

When α is small, but Y is large, it makes sense to keep only those higher order corrections in α that are accompanied by a factor of Y. The problem of calculating all $(\alpha Y)^n$ corrections is exactly the problem the BFKL equation solves. In this leading logarithmic approximation, where all $(\alpha Y)^n$ terms are kept,

$$F(\underline{x},\underline{x},'Y) = -\frac{\alpha^2(N_c^2 - 1)}{N_c^2} \int \frac{d^2\ell d^2\ell'}{\ell^2(\ell')^2}(2 - e^{-i\underline{\ell}\cdot\underline{x}} - e^{i\underline{\ell}\cdot\underline{x}})$$

$$\cdot(2 - e^{-i\underline{\ell}'\cdot\underline{x}'} - e^{i\underline{\ell}'\cdot\underline{x}})f(\ell,\ell,'Y) \tag{53}$$

where the onium-onium scattering amplitude given by (50) and (53) is illustrated in Fig.11. It is useful to write the Y-dependence of f as

$$f(\underline{\ell},\underline{\ell},'Y) = \int \frac{d\omega}{2\pi i}e^{\omega Y} f_\omega(\underline{\ell},\underline{\ell}') \tag{54}$$

where the ω-integration goes parallel to the imaginary axis from $\omega_i = a - i\infty$ to $\omega_f = a + i\infty$ with $a > 0$. Note that

$$f_\omega(\ell,\ell') = \int_0^\infty dY e^{-\omega Y} f(\ell,\ell'Y) \tag{55}$$

so that (55) is a Laplace transform with (54) the inverse Laplace transform. The value of a is chosen so that the Y-integral in (55) converges at $Y = \infty$.

$f_\omega(\underline{\ell},\underline{\ell}')$ obeys the BFKL equation

$$\omega f_\omega(\underline{\ell},\underline{\ell}') = \delta(\ell^2 - \ell'^2)\delta(\phi - \phi') + \frac{\alpha C_A}{\pi^2}\int \frac{d^2k}{(\underline{k}-\underline{\ell})^2}\left[f_\omega(\underline{k},\ell') - \frac{\ell^2 f_\omega(\ell,\ell')}{k^2 + (\underline{k}-\underline{\ell})^2}\right] \tag{56}$$

where ℓ and ϕ are the polar coordinates of $\underline{\ell}$. All derivations of the BFKL equation are very difficult. Perhaps the most straightforward to follow is the one given in Ref.18. There is no scale in (56). (We take α as a fixed coupling in (56) since running coupling effects have additional powers of α which are not compensated by powers of Y, and so are beyond the leading logarithmic approximation considered here.) It is natural to write[8]

$$f_\omega(\underline{\ell},\underline{\ell}') = \sum_n \frac{\psi_n(\underline{\ell})\psi_n^*(\underline{\ell}')}{\lambda_n} \tag{57}$$

where the $\psi_n(\underline{\ell})$ are eigenfunctions of the kernel in (56). From the scaling behavior of f_ω, given by the inhomogeneous term in (56), it is clear that the appropriate eigenfunctions are

$$\frac{1}{(2\pi)}e^{im\phi}(\ell^2)^{\frac{1}{2}-i\nu} \tag{58}$$

with m an integer and ν a real number. Writing

$$f_\omega(\underline{\ell},\underline{\ell}') = \frac{1}{(2\pi)^2}\sum_m e^{im(\phi-\phi')}\int_{-\infty}^{\infty} d\nu\, f_\omega(m,\nu)(\ell^2)^{-\frac{1}{2}-i\nu}(\ell'^2)^{-\frac{1}{2}+i\nu}. \tag{59}$$

Substituting (59) in (56) gives

$$f_\omega(m,\nu) = \frac{1}{\omega - \omega_0(m,\nu)} \tag{60}$$

with

$$\omega_0(m,\nu) = \frac{2\alpha C_A}{\pi}\chi(m,\nu) \tag{61}$$

where

$$\chi(m,\nu) = \psi(1) - \frac{1}{2}\psi(\frac{|m|+1}{2}+i\nu) - \frac{1}{2}\psi(\frac{|m|+1}{2}-i\nu). \tag{62}$$

Problem 8(H).

From (56) and (59) show that (60) follows. Equations (60) to (62) give the solution to the BFKL equation.

When Y becomes large the asymptotic behavior of $f(\underline{\ell},\underline{\ell}',Y)$ is given by the rightmost singularity of $f_\omega(\underline{\ell},\underline{\ell}')$, in ω, as can be seen from (54) where it is clear that one should distort the ω path of integration to the left when Y is large. From (59) to (62) one can see that the rightmost singularity of $f_\omega(\underline{\ell},\underline{\ell}')$, in ω, occurs for the term $m=0$ in (59). We now keep only that term. Then from (54), (59) and (60)

$$f(\underline{\ell},\underline{\ell}',Y) = \frac{1}{(2\pi)^2}\int_{-\infty}^{\infty} d\nu\, e^{\frac{2\alpha C_A}{\pi}\chi(\nu)Y}(\ell^2)^{-\frac{1}{2}-i\nu}(\ell'^2)^{-\frac{1}{2}+i\nu}, \tag{63}$$

with $\chi(\nu) = \chi(0,\nu)$.

Problem 9(M-H). Substituting (63) into (53) show that

$$F(\underline{x}, \underline{x}', Y) = -4\alpha^2 x\, x' \frac{N_c^2 - 1}{N_c^2} \int_{-\infty}^{\infty} d\nu\, (x/x')^{2i\nu} e^{\frac{2\alpha C_A}{\pi} \chi(\nu) Y}.$$

Hint: You may find the integral

$$\int_0^{\infty} d\ell\, \ell^{-2-2i\nu}(1 - J_0(\ell x)) = -x^{1+2i\nu} 4^{-1-i\nu} \frac{\Gamma(-\frac{1}{2} - i\nu)}{\Gamma(\frac{3}{2} + i\nu)}$$

useful.

Problem 10(E).

Use the fact that $\nu = 0$ is a saddle point of $\chi(\nu)$ to evaluate the integral, over ν, for F as given in Problem 9. Use $\chi(\nu) \approx \chi(0) + \frac{1}{2}\nu^2\chi''(0) = 2\ell n\, 2 - 7\zeta(3)\nu^2$ to show that

$$F = -\frac{2\pi\alpha^2 x x' exp\{(\alpha_P - 1)Y - \frac{a}{2}\ell n^2 x/x'\}}{\sqrt{\frac{7}{2}\alpha C_A \zeta(3)Y}}$$

for large Y.$\alpha_P - 1 = \frac{4\alpha C_A}{\pi}\ell n\, 2$ and $a = [7\alpha N_c \zeta(3)Y/\pi]^{-1}$.

Using the result of problem 10 in (50) leads to

$$\sigma = 16\pi R^2 \alpha^2 \frac{N_c^2 - 1}{N_c^2} \frac{e^{(\alpha_P - 1)Y}}{\sqrt{\frac{7}{2}\alpha N_c \zeta(3)Y}} \tag{64}$$

for the total onium-onium cross section as Y becomes large.[6-8] Eq.(64) illustrates the high energy behavior due to the BFKL pomeron, the rightmost ω-singularity of f_ω. For small α and not too large Y the σ given by (64) is much less than the geometric cross-section $4\pi R^2$. However, as Y grows the cross-section becomes much larger than geometric and one would suspect that (64) is no longer reliable. This is indeed the case, however, the present problem does not have a strict Froissart bound so one must be careful in deciding at what energies unitarity corrections must become large. To understand this a little better it is useful to look at high energy onium-onium scattering at a definite relative impact parameter.

The generalization of the F given in problem 10 to scattering at a definite impact separation, $\underline{b}$, of the two onia is[17]

$$F(\underline{x}, \underline{x}', Y, \underline{b}) = -\frac{\pi\alpha^2 x x' \ell n(\frac{b^2}{x\, x'})}{[\frac{7}{2}\alpha N_c \zeta(3)Y]^{3/2} b^2} exp\left\{(\alpha_P - 1)Y - \frac{a}{2}\ell n^2(\frac{b^2}{x\, x'})\right\}, \tag{65}$$

when $b/x, b/x' >> 1$.

Problem 11 (E). Show that

$$\int F(\underline{x}, \underline{x}', Y, \underline{b})d^2 b = F(\underline{x}, \underline{x}', Y).$$

From (65) onium-onium scattering at an impact parameter $b/R >> 1$ is given by substituting (65) into (50) and using (49), however, it is simpler just to view -2F as $\frac{d\sigma}{d^2 b}$ for the scattering of a heavy quark-antiquark pair, a dipole, having separation $\underline{x}$

between the heavy quark and antiquark on a heavy quark-antiquark pair of separation $\underline{x}'$. Thus, $-2F(\underline{x},\underline{x}',Y,\underline{b})$ should not be larger than 1.

Problem 12 (E). Take $x = x' = 2R, \alpha = 1/5, \alpha_P - 1 = \frac{1}{2}$ and $b/R = 2$. Show that $Y \approx 12$ is necessary for $\frac{d^2\sigma}{db^2} = 1$.

The result of problem 12 shows that unitarity corrections are likely not very important over quite a large rapidity region. Unitarity corrections will become important first for $b \approx R$, but we cannot use (65) for such small values of b. In any case the forward scattering amplitude is dominated by rather large values of b, as compared to R, so that for the total onium-onium cross section unitarity corrections will not be important until Y is quite large.

Finally, from (65) and problem 11 it is clear that distances much larger than R are important in onium-onium scattering. The values of b which dominate the integral in problem 11 are clearly given by $\ln^2 b^2/4R^2 \approx 2/a$, where we have set $x = x' = 2R$, which gives

$$b = 2Re^{\sqrt{\frac{7}{2}\frac{C_A}{\pi}\zeta(3)Y}}. \tag{66}$$

The b given by (66) is the diffusion radius, the radius to which gluons have evolved in the high energy scattering.

Problem 13 (M). Show that

$$\frac{\alpha(b) - \alpha(R)}{\alpha(b)} = \frac{\sqrt{14\frac{\alpha C_A}{\pi}\zeta(3)Y}}{\ln(1/R^2\Lambda^2)}$$

when b is given by (66). Show that $\frac{\alpha(b)-\alpha(R)}{\alpha(b)} << 1$ is equivalent to $Y << \frac{\ln^2(1/R^2\Lambda^2)}{14\frac{\alpha C_A}{\pi}\zeta(3)}$.

From the result in problem 13, one sees that for $R\Lambda$ very small one has a very wide range of rapidities over which the fixed coupling approximation is valid. In particular, by choosing $R\Lambda$ sufficiently small $\frac{d^2\sigma}{db^2}$ becomes large for rapidities satisfying the constraint given in problem 13 for running coupling effects to be small. This means that the unitarity problem associated with the rapid growth of cross-sections from the BFKL pomeron can be studied, and solved, in the fixed coupling approximation.

5. FINDING THE BFKL POMERON

The most urgent problem in small-x physics is to measure the BFKL pomeron, that is to measure $\alpha_P - 1$. In principle, there are known processes where this can be done both in deep inelastic lepton-nucleon scattering and in proton-proton collisions, however so far the relevant measurements have not yet been done in anything like a definitive manner. From our earlier discussion, it is clear that the small-x behavior of νW_2 is not necessarily determined by BFKL dynamics. It might be that BFKL dynamics does account for much of the growth of νW_2 at small-x, and if that can be established, for example, by a careful study of the final states associated with small-x events,[19] it would simplify the experimental understanding of small-x physics. However, even if it turns out that the growth in νW_2 is unrelated, or only partly related, to BFKL dynamics that does not mean that the BFKL pomeron cannot be experimentally studied at present high energy accelerators. There are specific measurements that can be made that focus on BFKL evolution, and it is to a description of those processes that we now turn.

5.1 Two-Jet Inclusive Production in Hadron-Hadron Collisions

In principle, it is possible to describe single jet inclusive cross-sections in deep inelastic lepton-nucleon collisions and two-jet inclusive cross-sections in hadron-hadron collisions in terms of sthe BFKL pomeron for certain kinematic regimes of the produced jets.[20-23] The single jet inclusive measurement has been discussed here already[11] so let me describe the two-jet inclusive measurement.[20] The process is $proton(p_1) + proton(p_2) \rightarrow jet(k_1) + jet(k_2)+$ anything. For explicitness, consider the center of mass of the collision with p_{1+} and p_{2-} being the large components of the momenta of the colliding particles. Let $k_{1+} = x_1 p_{1+}$ and $k_{2-} = x_2 p_{2-}$ and define the cross-section

$$\sigma_2(s, Q^2, x_1, x_2) = \int d^2 k_1 d^2 k_2 \Theta(\underline{k}_1^2 - Q^2)\Theta(\underline{k}_2^2 - Q^2)\frac{x_1 x_2 d\sigma}{dx_1 dx_2 d^2 k_1 d^2 k_2} \tag{67}$$

where $s = (p_1 + p_2)^2$ and the differential cross-section on the right-hand of (67) is the two-jet inclusive differential cross-section. Using factorization and defining the gluon-gluon cross-section $\hat{\sigma}$ one has

$$\sigma_2 = x_1 P(x_1, Q^2) x_2 P(x_2, Q^2)\hat{\sigma}(Y, Q^2) \tag{68}$$

where

$$xP(x, Q^2) = xG(x, Q^2) + \frac{4}{9}x \sum_f (q_f(x, Q^2) + \bar{q}_f(x, Q^2)) \tag{69}$$

with $Y = \ell n \hat{s}/Q^2$ and $\hat{s} = (k_1 + k_2)^2 \approx x_1 x_2 s$. We suppose Y is large. $\hat{\sigma}$ can be described in perturbative QCD with the relevant coupling being $\alpha(Q^2)$. At lowest order the process is illustrated in Fig. 12 and one has

$$\hat{\sigma} = \int_{Q^2}^{\infty} dt \frac{d\hat{\sigma}}{dt} = (\frac{\alpha C_A}{\pi})^2 \int_{Q^2}^{\infty} dt \frac{\pi^3}{2t^2} = (\frac{\alpha C_A}{\pi})^2 \frac{\pi^3}{2Q^2} \tag{70}$$

where the $\frac{d\hat{\sigma}}{dt}$ in (70) is the Born term for gluon-gluon wide angle elastic scattering. Eqs.(68) and (70) are the usual parton distributions times hard scattering term that are familiar in jet physics. The approximation that has been made is to treat $\frac{d\hat{\sigma}}{dt}$ for quark-quark, quark-gluon and gluon-gluon scattering as identical except for the Casimir factors appearing in (69). This is a good approximation when Y is large.

Normally, one would correct (70) by taking the next term in $\alpha(Q)$ in $\hat{\sigma}$ and taking parton distributions through next-to-leading order. This is what is called a second order formalism for jet-production. However, when Y is large this may not be a good approximation because higher orders of α in $\hat{\sigma}$ can be compensated by powers of Y so that when $\alpha Y \geq 1$ one should resume all powers of αY. Such a resummation is just the calculation of the leading logarithmic series which is solved by the BFKL equation. In leading logarithmic approximation we may view the process as illustrated in Fig.13. If one writes

$$\hat{\sigma}(Y, Q^2) = (\alpha C_A)^2 \int \frac{d^2 k_1 d^2 k_2}{\underline{k}_1^2 \underline{k}_2^2}\Theta(\underline{k}_1^2 - Q^2)\Theta(\underline{k}_2^2 - Q^2)f(\underline{k}_1, \underline{k}_2, Y) \tag{71}$$

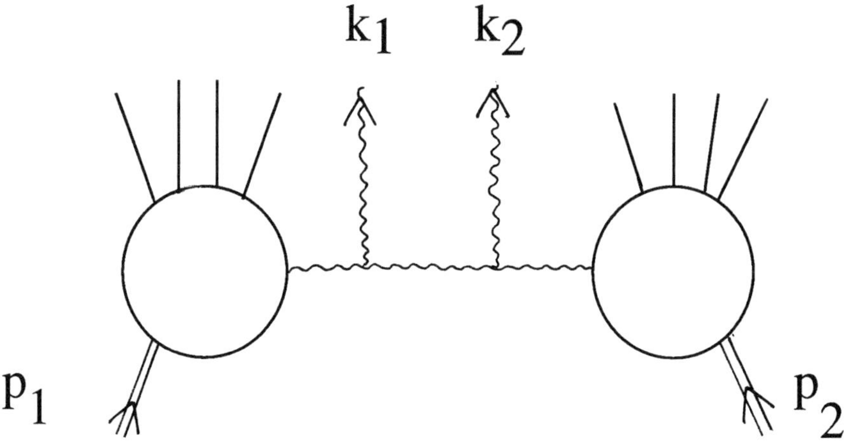

12. Two-jet Production in a Hadron-Hadron Collision at Lowest Order in the Hard Scattering.

then the f in (71) is the same as in (53). Using (54) and (59) to (61) one finds

$$\hat{\sigma}(Y, Q^2) = \frac{(\alpha C_A)^2}{Q^2} \int \frac{d\nu}{1 + 4\nu^2} e^{2\frac{\alpha C_A}{\pi} \chi(\nu) Y} \tag{72}$$

which has an asymptotic behavior

$$\hat{\sigma}(Y, Q^2) \sim \left(\frac{\alpha C_A}{\pi}\right)^2 \frac{\pi^3}{2Q^2} \frac{e^{(\alpha_P - 1)Y}}{\sqrt{\frac{7}{2}\alpha C_A \zeta(3) Y}}. \tag{73}$$

Comparing (73) with (70) one sees that the last factor on the right-hand side of (73) is the enhancement factor due to BFKL dynamics. The gluon-gluon hard scattering cross-section given in (73) and the onium-onium cross section given in (64) have exactly the same Y-dependence. Comparing these two expressions it is clear that two-jet inclusive scattering defines an effective radius proportional to $1/Q$ the minimum allowed momentum of the measured jets.

Eqs.(68) and (73) give the leading logarithmic formula for two-jet inclusive production in hadron-hadron collisions. The ideal way to test BFKL evolution would be to measure $\alpha_P - 1$ in a ramping run at Fermilab. If x_1 and x_2, along with Q^2, were fixed and s increased by increasing the accelerator energy the complete energy dependence of the cross section would reside in the last factor in (73). It appears very hard to measure $\alpha_P - 1$ at a fixed energy[24] setting because x_1 and x_2 also vary as Y varies. Finally, we would not expect the normalization of the cross section given by (68) and (73) to be reliable until higher order corrections are done.

There is a similar measurement which may be possible at HERA.[21-23] In this case, one need only measure one jet associated with a deep inelastic event. The transverse momentum should be on the order of Q, the photon virtuality of the deep inelastic scattering. The quantity which replaces Y is $\ell n\, x_1/x$ with x_1 the jet's longitudinal

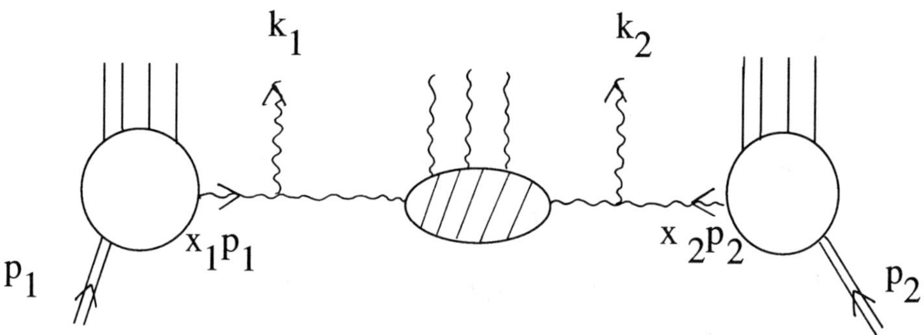

13. Two-jet Inclusive Production in a Hadron-Hadron Collision.

momentum fraction, as a fraction of the proton momentum, and with x the usual Bjorken x. The advantage that HERA has is that for fixed x_1 one may vary x within fixed beam settings. Of course, because of limited energy, $\ell n\, x_1/x$ at HERA cannot be as large as Y at Fermilab. For a further discussion of this process see the talk of De Roeck at this school.[11]

5.2 Rapidity Gaps or "Almost" Gaps.[24–26]

Consider high energy and large momentum transfer quark-quark scattering, $\frac{d\hat{\sigma}}{dt}$, where $Y = \ell n(-s/t)$ with s the center of mass energy squared and t the invariant momentum transfer squared. The process is shown in Fig.14 where the gluon-gluon scattering part of that graph is the generalization of the f, given in (56), to the non-forward direction. The cross section takes the form

$$\frac{d\hat{\sigma}}{dt} = (\frac{\alpha C_F}{\pi})^4 \frac{\pi^3}{(N_c^2 - 1)^2} \left| \int d^2\underline{k}_1 d^2\underline{k}_2 f^q(\underline{k}_1, \underline{k}_2, Y) \right|^2 \tag{74}$$

with f^q the non-forward BFKL amplitude.[8] In lowest order perturbation theory

$$f_0^q(\underline{k}_1, \underline{k}_2, Y) = \frac{\delta(\underline{k}_1 - \underline{k}_2)}{\underline{k}_1^2(\underline{q} - \underline{k}_1)^2} \tag{75}$$

which leads to a divergent integral in (74). However, the infrared region is softened in the asymptotic form of the BFKL solution and one obtains[25]

$$\frac{d\hat{\sigma}}{dt} = (\alpha C_F)^4 \frac{\pi^3}{4t^2} \frac{e^{2(\alpha_P - 1)Y}}{[\frac{7}{2}\alpha C_A \zeta(3) Y]^3} \tag{76}$$

in the leading logarithmic approximation. The corresponding formula for gluon-gluon scattering is obtained by replacing C_F in (76) with C_A.

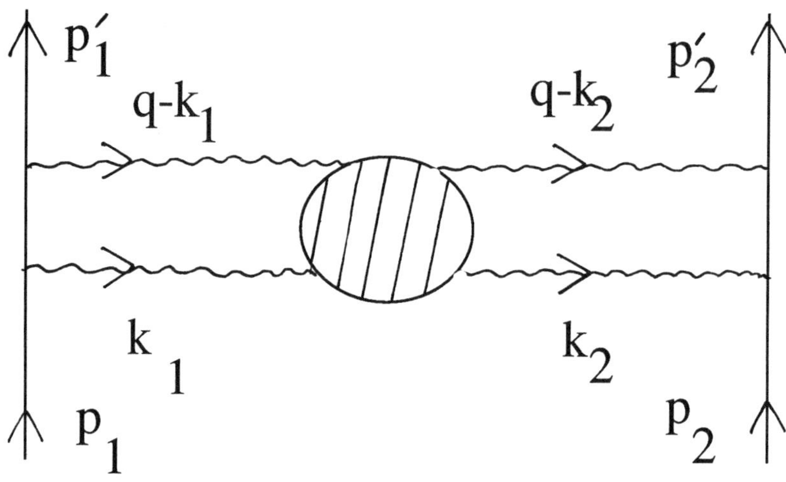

14. Quark-Quark Scattering at High Energy and at Large Momentum Transfer.

The above formula for parton-parton high energy scattering can furnish the basis for the presence of large rapidity gap events bounded by jets in a high energy hadron-hadron collision[24] as illustrated in Fig.15. At first glance, one might think that the cross section for two-jet production in a hadron-hadron collision, with no particles produced in the rapidity interval between the jets, would be given simply by the hard scattering cross section (76) times the parton distributions giving the flux of colliding partons much as in (68). However, the present process is not an inclusive process since one requires that no particles be produced between the two measured jets. Because the process is not inclusive the QCD factorization "theorem" does not work and there is no simple formula describing this process. Stated in more physical terms the above reaction can only take place if the spectator quarks, and gluons, in the two colliding hadrons do not interact, presumably in a soft way, to produce particles which would fill in the rapidity gap. The cross section given by (76) includes the requirement of a rapidity gap in the active quark-active quark scattering, but does not include the suppression coming from the lack of interaction between spectators. One sometimes writes the two-jet, along with a gap between the two jets, cross section as[24]

$$\frac{x_1 x_2 d\sigma}{dx_1 dx_2 dt} = x_1 P(x_1, Q^2) x_2 P(x_2, Q^2) \frac{d\hat{\sigma}}{dt} < S^2 > \tag{77}$$

where $\frac{d\hat{\sigma}}{dt}$ is as given in (76), with $Y = \ell n(\frac{x_1 x_2 s}{-t})$, and where now

$$xP(x, Q^2) = \frac{81}{16} xG(x, Q^2) + x \sum_f (q_f(x, Q^2) + \bar{q}_f(x, Q^2)) \tag{78}$$

with $-t = Q^2$. $< S^2 >$ represents the probability that the spectator partons not interact, the survival probability. $< S^2 >$ is expected to lie between 0.05 and 0.2.[24,27] One can considerably improve the situation, at least theoretically, by requiring not that a complete gap exist between the two measured jets but that no hard particles

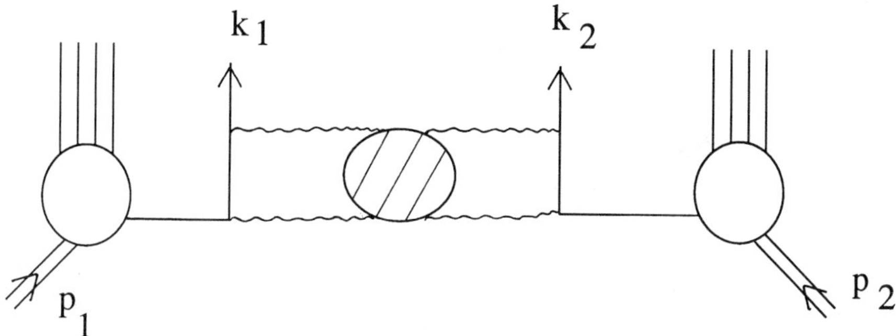

15. Two-jet Production in a Hadron-Hadron Collision. The Color singlet BFKL pomeron connects the two jets allowing a rapidity gap to appear between the two jets.

be produced in the rapidity interval between the two jets.[26] In this case soft spectator interactions are allowed, factorization is recovered, and one can set $< S^2 >$ to 1 in (77).

Again, the best way to measure $\alpha_P - 1$ here would again be in a ramping run at Fermilab where x_1 and x_2, in (77), could be held fixed while Y increases. Recently, a similar process has been discussed for a HERA measurement where diffractive electro-production of vector mesons at large momentum transfer with a rapidity gap between the vector meson and the recoil jet is required.[28] In this case, $< S^2 >= 1$ since a highly virtual photon has no spectator partons. However, the conclusion of Ref.28 is that in the HERA energy regime the non-forward gluon-gluon scattering amplitude will not have reached the asymptotic form where $\alpha_P - 1$ appears.

REFERENCES

1   C.G. Callan and D.J. Gross, Phys. Rev.D8, 4383 (1973).

2   J.D. Bjorken, Phys. Rev.179, 1547 (1969).

3   Yu. L. Dokshitzer, JETP 73, 1216 (1977).

4   V.N. Gribov and L.N. Lipatov, Sov.J. Nucl. Phys. 15, 78 (1972).

5   G. Altarelli and G. Parisi, Nucl. Phys. B126, 298 (1977).

6   E.A. Kuraev, L.N. Lipatov and V.S. Fadin, Sov. Phys. JETP 45, 199 (1978).

7   Ya. Ya. Balitsky and L.N. Lipatov, Sov.J. Nucl. Phys. 28, 22 (1978).

8   L.N. Lipatov in "Perturbative Quantum Chromodynamics," ed., A.H. Mueller, *World Scientific*, Singapore 1989.

9   H1 Collaboration: I. Abt. et al., Nucl. Phys. B407, 515 (1993).

10   ZEUS Collaboration: M. Derrick et al., Phys.Lett. B316, 412 (1993).

11   A. De Roeck, (these proceedings).

12   M.Glück, E. Reya, A. Vogt, Phys.Lett. B306, 391 (1993).

13   A.D. Martin, W.J. Stirling and R.G. Roberts, Phys.Lett. B306, 145 (1993).

14   L. V. Gribov, E.M. Levin and M.G. Ryskin, Phys. Rep.100C, 1 (1983).

15   A.H. Mueller, Nucl. Phys. B415, 373 (1994).

16   A.H. Mueller and B. Patel, Nucl. Phys.B 425, 471 (1994).

17   A.H. Mueller, CU-TP-640 (to be published in Nuclear Physics B).

18   T. Jaroszewicz, Acta. Phys. Polon.B11, 965 (1980).

19   J. Kwiecinski, A.D. Martin, P.J. Sutton and K. Golec-Biernat, Durham preprint DTP/94/08 (1994).

20   A.H. Mueller and H. Navelet, Nucl.Phys. B282, 727 (1987).

21   W.-K. Tang, Phys. Lett. 278B, 363 (1992).

22   J. Bartels, A. De Roeck and M. Loewe, Z. Phys. C54, 635 (1992).

23   J. Kwiecinski, A. Martin and P.J. Sutton, Phys. Lett. 278B, 254 (1992);

Phys. Rev. D46, 921 (1992).

24   J.D. Bjorken, Int.J. Mod. Phys.A7, 4189 (1992).

25   A.H. Mueller and W.-K. Tang,Phys.Lett. B284, 123 (1992).

26   V. Del Duca and W.-K. Tang, Phys. Lett. B312, 225 (1993).

27   E. Gotsman, E.M. Levin and U. Maor, Phys. Lett. B309, 199 (1993).

28   J.R. Forshaw and M.G. Ryskin (in preparation).

HIGHER ORDER QCD CORRECTIONS

Z. Kunszt
Institute of Theoretical Physics, ETH,
Zurich, Switzerland

INTRODUCTION

QCD is a renormalizable non-Abelian gauge theory with color SU(3) gauge group and color triplet quarks as matter fields[1]. It has two fundamental properties: *asymptotic freedom*[2] and *color confinement*. QCD as a renormalizable field theory formally can be studied in perturbation theory around the Fock vacuum state of free quarks and gluons in terms of an effective coupling constant. We know from the data[3] that the effective coupling constant $\alpha_S = g_s^2/4\pi$ is about 0.12 at $Q = 90\,\mathrm{GeV}$ (in the $\overline{\mathrm{MS}}$ scheme). Its value increases with decreasing the scale Q (*asymptotic freedom*) such that at the mass scale of low lying hadrons it reaches the strong coupling regime. Qualitatively one can say that the perturbative description may appear to be a good approximation for phenomena in which the relevant momentum transfer scale is clearly above the proton mass.

But even for reactions with large momentum transfer Q the various correlations may become sensitive to soft and collinear configurations of gluons and/or quarks giving terms of order $(\alpha_s log(Q/m_q))^n$ where m_q denotes a light quark mass. The presence of these terms is not allowed in a perturbative treatment. The applications of perturbative QCD are limited to phenomena where such terms are either cancelled or do not occur. Fortunately, the infrared structure of perturbative QCD is relatively well understood[4,5,6,7]. Two fundamental theorems valid in all orders in perturbation theory reveal some of the basic properties of the mass and soft singularities. First we have a cancelation theorem which states that in simple inclusive reactions, such as for example the total cross section of e^+e^- annihilation into quarks and gluons, the soft and collinear contributions cancel (KNL theorem)[8]. In this case, only one high momentum transfer scale is relevant, the effective coupling becomes small and it is expected that the cross section can reliably be calculated in power series of the effective coupling.

The second theorem is a factorization theorem[9,10,5,4] which is valid in all orders in perturbation theory as well. In the infinite momentum frame the hadrons are considered as beams of free partons (quarks and gluons) which carry some fractions of the momenta of the parent hadrons. Collinear splittings of initial partons (a long distance effect) in

Frontiers in Particle Physics: Cargèse 1994
Edited by M. Lévy *et al.*, Plenum Press, New York, 1995

perturbation theory result in terms of large logarithms of order $\ln Q/m_q$ where m_q denotes (light)quark mass and Q is the large momentum scale of the process. The factorization theorem asserts that these singular terms can be factorized into process independent universal functions therefore can be absorbed as renormalization factors into the initial state number density functions of quarks and gluons. The KNL theorem and the factorization theorem, constitutes the theoretical basis of the description of scattering processes of hadrons in perturbative QCD. The factorization of the collinear contributions depends on a factorization scale therefore parton densities become scale dependent. This scale dependence, however, is calculable in perturbative QCD: it is given by the Altarelli-Parisi (DGLAP) evolution equation. The factorization theorem and the QCD improved parton model is described in great detail by A. Mueller in this volume[12].

A fundamental difficulty of applying perturbative QCD to the description of high energy scattering processes is the problem of hadronization. The measurements are made in terms of hadrons while in perturbative QCD the predictions are given for reactions involving quarks and gluons. Since non-perturbative phenomena occur at large distances, on general ground, one expects that in reactions dominated by short distance processes the non-perturbative contributions are power suppressed: they give rise to small corrections proportional to $(m/Q)^n$ where m is the proton mass and Q is the scale of the hard process. In simple cases this behavior can be derived assuming the validity of operator product expansion[11] (deep inelastic scattering, e^+e^- annihilation into hadrons *etc.*).

Motivated by these general arguments, in applications of perturbative QCD it is assumed that for infrared safe inclusive quantities, the measured values obtained in terms of hadrons, up to small power corrections, are equal with the same quantities calculated in terms of partons. By infrared safe quantities we mean those measurable-s which are insensitive to collinear splittings of partons and/or emission of soft gluon.

The more accurate results of recent high energy scattering experiments require more precise theoretical predictions. Within the framework of perturbative QCD one can systematically improve the accuracy of the predictions by calculating more and more higher order corrections. In the last years impressively large number of higher order corrections have been calculated. Several of these calculations are rather complex and their successful completions required numerous technical improvements.

In this lecture I describe briefly the theoretical concepts which allow us to obtain quantitative predictions for higher energy scattering processes with the help of perturbative QCD, and I review some of the most important new techniques which allowed to obtain complex higher order corrections with acceptable labour. My selection of the vast number of possible examples is subjective and it reflects partially the author's experience. Section 2 is a collection of comments on Feynman rules, ultraviolet counter terms, regularization and renormalization schemes. In Section 3 the derivation of the KNL theorem using the concept of pinch singularities and Landau equations is given. It serves the purpose to illustrate the origin and nature of infrared singularities. Extension of the KNL theorem to jet production is also discussed. Section 4 is devoted to the next-to-next-to leading order description of the e^+e^- annihilation cross section and the measurements of α_S from the LEP data. The limitations and inherent ambiguities of the perturbative treatment is discussed in terms of Borel transforms. Section 5 describes the fundamental formula of the QCD improved parton model giving the physical cross sections of hadron-hadron hard scattering processes to any fixed order of perturbative QCD. Its use is illustrated on the example of jet production. Finally Section 6 is devoted to a brief discussion of some of the basic technical tools (helicity method, use of supersymmetry, dimensional reduction, background gauge etc.) used and invented recently for calculating next-to-leading order corrections to four, five and six leg amplitudes.

RENORMALIZATION

In this section I briefly summarize the building blocks of the ultraviolet renormalization procedure necessary for the calculations of higher order QCD corrections.

Feynman rules

In perturbative QCD we consider correlation functions in the Fock space of free colored quarks and gluons. The derivation of the Feynman rules follows standard Wick-theorem pattern with two specific features. First, due to non-Abelian gauge symmetry the gauge fields have self interactions. Second, gauge fixing requires the addition gauge fixing term and Fadeev-Popov ghost terms to the Lagrangian[13,14,15]. If background gauge is used there are additional vertices coupled to external lines of the background gauge field[16].

Regularization and Renormalization

Straightforward use of Feynman rules in higher orders of the coupling constants leads to divergent integrals in the ultraviolet region. Renormalization is needed to make the theory well defined. For renormalizable theories the appearance of the divergences is simple and universal[17]. They can be absorbed into factors which just renormalize the wave functions of the quark and gluon fields, the coupling constant and the quark mass parameters. The isolation of the divergences requires some regularization of the divergent integrals. The procedure of renormalization becomes algebraically significantly simpler if the regularization scheme respects the symmetries of the bare Lagrangian In QCD, dimensional regularization[18] is particularly convenient since it is a gauge invariant and Lorenz invariant. An additional advantage is that it can also be used to regularize the soft and collinear singularities[19] appearing in the loop and phase space integrals at intermediate steps. In dimensional regularization the integrals which are divergent in four dimensions are carried out in different dimensions where the integrals are finite and the singularities of the integrals are exhibited as poles in ϵ with analytic continuation into $d = 4 - 2\epsilon$ dimensions. As an illustration let us consider a typical integral

$$\int \frac{d^d k}{(2\pi)^d} \frac{1}{[-k^2 + C - i\epsilon]^m} = i(4\pi)^{-2+\epsilon}[C - i\epsilon]^{2-m-\epsilon} \frac{\Gamma(m - 2 + \epsilon)}{\Gamma(m)},$$

where $\epsilon = (4 - d)/2$ and $\Gamma(x)$ is the gamma-function. This integral is divergent if $m = 2$ and it has a pole as $\epsilon \to 0$. The pole terms are subtracted and absorbed into the renormalization factor. The subtraction of a singular piece, however, is ambiguous up to a finite term therefore a unique definition of renormalized Green functions requires some renormalization prescription. In other words the calculational rules of renormalized perturbation theory becomes complete provided the algorithm for the calculation of the counter terms is also uniquely defined. The most widely used renormalization prescription is the $\overline{\text{MS}}$ scheme, a practical version of the minimal subtraction scheme in which case only the pole terms are subtracted.

The counter terms are generated with the help of the renormalization factors Z_i[2,7,15,20]. The renormalized gluon, quark and Fadeev-Popov ghost fields (G, q, c) and the renormalized parameters (g, m_i, λ) are defined as

$$
\begin{aligned}
G^i_\mu &= Z_3^{-1/2} G^{(0)}_\mu, \quad q_a = Z_2^{-1/2} q^{(0)}_a, \\
c_a &= \tilde{Z}_3^{-1/2} c^{(0)}_a, \quad \lambda = Z_3^{-1} \lambda^{(0)}, \\
g(\mu) &= \mu^{-\epsilon} Z_1^{-1} Z_3^{3/2} g^{(0)}(\epsilon), \quad m = Z_m^{-1} m^{(0)},
\end{aligned}
\tag{1}
$$

where λ is the gauge fixing parameter, c^i denotes the ghost field (complex scalar field obeying Fermi statistics). The bare Lagrangian is decomposed into the renormalized Lagrangian and counter terms. The algorithm of renormalized Feynman rules is very simple: we calculate first the renormalization factors from primitively divergent diagrams, then we construct the counter terms iteratively generating additional Feynman rules for higher order diagrams. The sum of all diagrams (including counter terms) for a given physical process defines the finite renormalized transition amplitude.

Renormalization invariant S-matrix

The definition of the renormalized coupling introduces a renormalization scale. In dimensional regularization this scale appears via factoring the physical dimension of the coupling constant into a dimensionful mass parameter μ and a dimensionless coupling constant

$$
g^{(0)} \to g^{(0)} \mu^{-\epsilon} \, .
$$

In next-to-leading order, the ultraviolet divergent terms given by one loop integrals $C g^2 \mu^{-2\epsilon} / \epsilon$ are cancelled by the MS counter term $-C g^2 / \epsilon$ and we get finite remaining μ-dependent logarithmic terms of type $-2C \ln \mu$. In fixed order of perturbation theory the theoretical prediction for physical cross sections appear to depend on this arbitrary unphysical scale parameter μ. The numerical value of the coupling constant extracted from the measured value of the cross section will also depend on μ. The renormalized Lagrangian with counter terms, however, is equal to the bare Lagrangian which does not contain any unphysical mass parameter, therefore, physical quantities like the S-matrix are invariant with respect to the change of the renormalization scale parameter μ

$$
\mu^2 \frac{dS}{d\mu^2} = 0.
$$

The total derivative with respect to μ can be written as

$$
\left(\mu^2 \frac{\partial}{\partial \mu^2} + \beta(\alpha_S) \frac{\partial}{\alpha_S} - \gamma_m \sum_i \frac{\partial}{\partial m_i} \right) S = 0 \, .
\tag{2}
$$

This equation is called the renormalization group equation[17]. It is valid to all orders of perturbation theory. Equation (2) tells us that the μ dependence of the cross section is cancelled by the μ dependence of the renormalized coupling constants determined by the β and γ functions

$$
\mu^2 \frac{d\alpha_s(\mu)}{d\mu^2} = \beta(\alpha_S), \quad \alpha_S = \frac{g^2}{4\pi} \, ,
\tag{3}
$$

and

$$
\mu^2 \frac{dm_j(\mu)}{d\mu^2} = -\gamma_m(\alpha_S) m_j \, .
\tag{4}
$$

In any fixed order of perturbation, however, eq. (2) is valid only in perturbative sense. If the perturbative series is truncated at order α_S^n than the μ dependence will cancel only in this order and the calculated cross section will have remaining μ

dependence of order $\alpha_S{}^{n+1}$. In general, the scale ambiguity reflects the size of the theoretical error of fixed order of perturbation theory given by uncalculated higher order terms.

If we choose the renormalization scale equal to Q the typical scale of the physical process, the potentially dangerous logarithmic terms $\ln^n Q/\mu$ will become harmless. Therefore the effective coupling constant have to be defined around the physical scale Q. In view of the relatively large value of α_S the choice of the scale of the coupling constant is an important issue when we hope to correctly estimate the cross sections in the Born approximation.

This structure remain valid also for quark mass parameters. As a phenomenologically relevant example I recall the calculation of the partial width of the Higgs boson into bottom-anti-bottom pair[21]. The Higgs coupling is proportional to the bottom mass and therefore in the Born approximation one should use running bottom mass value renormalized at the Higgs mass and not the pole mass value. The difference is numerically significant and the correctness of this argument can be explicitly demonstrated by calculating the next-to-leading corrections.[22]

The running coupling constant

The rate of change of the coupling constant with μ is given by the β function (3). The beta function is known up to next-to-next-to-leading order (NNLO) accuracy

$$\beta(\alpha_S) = -\alpha_S{}^2 b_0 - \alpha_S{}^3 b_1 - \alpha_S{}^4 b_2 + \mathcal{O}(\alpha_S{}^5) \tag{5}$$

The first two coefficients of the beta function[2,23] b_0 and b_1 are renormalization scheme independent while the coefficient b_2 is scheme dependent. The value of b_2 was obtained in [24,25] in the $\overline{MS}$ scheme. We note that the calculations are conveniently performed in the background gauge[16]. The great technical advantage of this gauge is that similarly to the case of QED the coupling constant renormalization is given by the vacuum polarization diagrams of the gauge boson propagator which leads to a dramatic reduction of the complexity of the calculation in higher orders. Since b_0 is positive,

$$b_0 = (11 N_C - 2 n_f)/12\pi \tag{6}$$

with increasing the value of the scale, the effective coupling becomes smaller and smaller. This is the celebrated *asymptotic freedom* of QCD which allows for weak coupling description of short distance phenomena.

The first order differential equation (3) defines the running coupling constant up to one initial value. A good reference point is obtained with the introduction of the so called Λ parameter of QCD. Since the β function is known only perturbativly it is most natural to solve (3) iteratively. The general form of the iterative solution is

$$\alpha_S(\mu, n_f, \Lambda) = a_{00}\frac{1}{t} + a_{11}\frac{\ln(t)}{t^2} + a_{10}\frac{1}{t^2} + a_{22}\frac{\ln^2(t)}{t^3} + a_{21}\frac{\ln(t)}{t^3} + a_{20}\frac{1}{t^3} + \mathcal{O}(1/t^4), \tag{7}$$

where $t = \ln\left(\mu^2/\Lambda^2\right)$. This general perturbative structure of the expansion is invariant with respect to the change of the parameter Λ . This ambiguity, is conventionally eliminated by requiring that

$$a_{10} = 0 \ . \tag{8}$$

The measured numerical value of the lambda parameter therefore depends in addition to the regularization scheme and renormalization prescription also on the initial value convention (8). Finally we note that special attention has to be paid to its dependence on the number of quark flavors n_f appearing explicitly in the coefficients b_0, b_1 and b_2. The measured value of the quark mass parameters depends also on a similar series of conventions.

INFRARED SAFE QUANTITIES

Perturbative QCD may be used to predict the values of physical quantities which are not sensitive to effects coming from long distance interactions of quarks and gluons. If one wants to see the range of applicability of perturbative QCD it is fundamentally important to analyses and understand the origin and nature of perturbative infrared singularities[26,5].

The study of the infrared behavior is important also in QED since asymptotic charged particles are characterized by inclusive measures over soft photons integrated over finite angular and energy resolution of the measurement. In these inclusive quantities the soft and collinear singularities cancel between virtual and Bremsstrahlung contributions and multiple soft photon radiation are re-summed with the Bloch-Nordsieck method. In fixed order of perturbation theory the remaining corrections due to soft emission are of double logarithmic type $(\alpha \ln \Delta\Theta/m_e \ln \Delta E/m_e)^n$. Where $\Delta\Theta$ and ΔE denote the angular and energy resolution of the measurement for charged particles. Nearly collinear emission of a hard photon gives rise to large single logarithmic corrections of the order $\alpha^n \ln^n E/m_e$.

The physical meaning of the finiteness of inclusive quantities in QCD is less clear. Quarks and gluons are confined and therefore the resolution parameters of the experiments are defined in terms of hadrons while in perturbative QCD calculations they are defined in terms of partons. If the resolution parameters, however, are unable to resolve distance scales below the confinement radius one expects that up to small power corrections the perturbatively calculated physical quantities can directly be compared with the experiment. The validity of this assumption have been successfully tested.

KNL theorem, Landau equations and pinch surfaces

Let us consider e^+e^- annihilation into quarks and gluons. The total cross section is determined by the imaginary part of the vacuum polarization tensor of the electromagnetic current

$$\Pi_{\mu\nu}(q) = -i \int d^4x e^{iqx} < 0|T J_\mu(x) J_\nu(0)|0 > = \left(g_{\mu\nu} q^2 - q_\mu q_\nu \right) \Pi(q^2). \tag{9}$$

Due to unitarity the total cross section is related to the imaginary part of $\Pi(q^2)$

$$\sigma(e^+e^- \to \text{hadrons}) = \text{Im}\Pi(q^2)/q^2. \tag{10}$$

According to the KNL theorem[8] calculating $\Pi(q^2)$ in perturbative QCD the soft and collinear singular contributions coming from individual diagrams cancel in their sum. The proof of the theorem is relatively simple. In perturbation theory the amplitudes are defined in terms of Feynman diagrams built from propagators, vertices and loop integrals. The integrands are rational functions with denominators given by the product of propagator denominator factors. In performing the loop integrals singular contributions can emerge due to bad ultraviolet behavior. These divergences are controlled by the ultraviolet renormalization procedure and are cancelled by counter terms. Divergences may appear, however, also due to vanishing denominator factors. Let us consider the

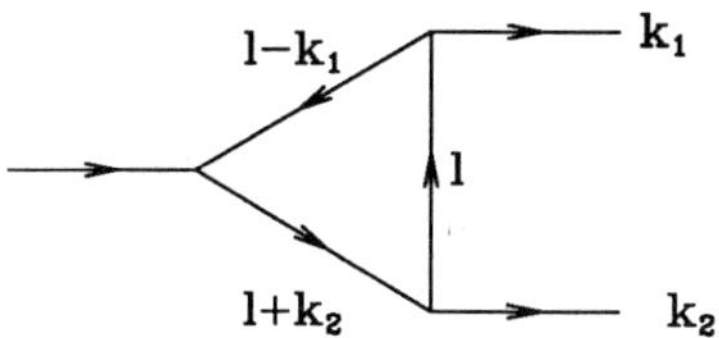

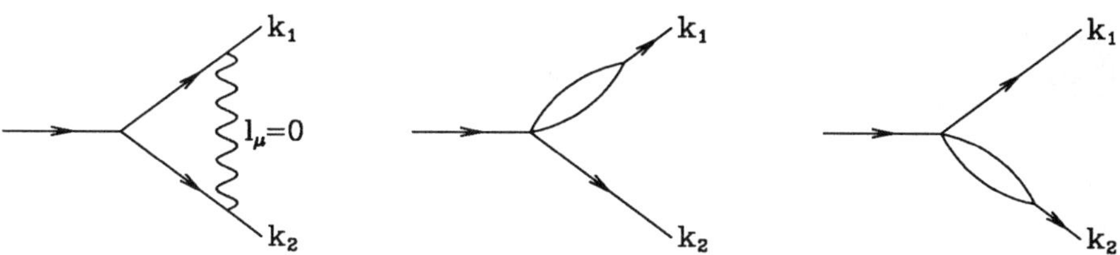

Fig. 1: Triangle diagram and its reduced diagrams defining the pinch surfaces.

triangle diagram depicted in Fig. 1 with two on-shell lines (k_1, k_2)

$$I_\Delta(k_1, k_2) = \int d^4l \int_0^1 \frac{da_1 da_2 da_3 \delta(1 - \sum_{i=1}^3 a_i)}{D(a_i, l, k_1, k_2)^3} \tag{11}$$

where

$$D(a_i, l, k_i) = a_1(l - k_1)^2 + a_2 l^2 + a_3(k_2 + l)^2 + i\epsilon. \tag{12}$$

and we used standard Feynman parameter representation for the denominator factors given by the propagators[15].

The necessary condition for divergences is $D = 0$. This condition, in general, does not guarantee yet that the integral is divergent since integration contour can be deformed away from the $D = 0$ region. The contour can not be deformed, however, if it is pinched between two degenerate poles or if the singularities are at the endpoint of the contours. Since the denominator D is quadratic function in l the conditions for pinch singularities are

$$D(a_i, l) = 0 \quad \text{and} \quad \frac{\partial}{\partial l_\mu} D(a_i, l_\mu) = 0. \tag{13}$$

These equations (known as Landau equations[15]) represent the necessary and enough conditions for the existence of the integration regions in variables l and a_i leading to singularities. In the case of our simple example we can easily see that that equations (13) are fulfilled if

$$a_i = 0 \quad \text{or} \quad q_i^2 = 0 \quad a_i \geq 0 \quad i = 1, 2, 3 \tag{14}$$

and

$$\sum_{i=1}^3 a_i q_{i\mu} = 0 \quad q_1 = l - k_1, \, q_2 = l, \, q_3 = l + k_2. \tag{15}$$

We can classify the three solutions in more physical terms as follows
(i) the loop momentum is soft: $l_\mu = 0$, $a_1 = a_3 = 0$,
(ii) the loop momentum is collinear with k_1: $a_3 = 0$, $a_1 = \lambda$, $l_\mu = -\lambda k_{1\mu}$,
(iii) the loop momentum is collinear with k_2: $a_1 = 0$, $a_3 = \lambda$, $l_\mu = \lambda k_{1\mu}$.

Infrared singularities come from the regions where the loop momentum is either soft or is collinear with one of the massless external legs. The surfaces in the space of loop momenta and Feynman parameters defined by the solutions of eq.(13) are called pinch surfaces.

For more complicated diagrams the Landau equations and their solutions have the same form: equations (14) remain true for every lines while equations (15) will hold for each closed loop. We obtain I+4L equation for (I-1+4L) unknowns where I denotes the number of internal lines and L the number of the loops. Coleman and Norton[27] pointed out that these solutions can be represented by the physical picture of real space time particles moving around a network with on-shell momentum conservation in the vertices and with conserved values $a_i q_i^\mu$ for closed loops $\sum_i a_i q_i^\mu = 0$. A diagram has infrared singularities if the external momentum configurations allow for such on-shell jet lines. Momentum conservation for three particle kinematics in the vertices along the network allows only collinear splitting and soft emission. All the lines which are hard have to be shrunk to a point (they have $a_i = 0$). The remaining lines can be grouped into jet-lines of collinear particles which join into the hard bubble and are connected by soft lines.

The triangle diagram with two massless external lines has one soft range and two collinear range. The matrix element of the electromagnetic current between two quark states in leading order of α_S is given by a triangle diagram therefore it has infrared singular terms. The vacuum polarization, however, has no massless external line therefore it can not have jet lines. Furthermore the electromagnetic current is colorless hence the external line can not emit soft lines. As a result vacuum polarization diagrams have no pinch surface and therefore they are free from infrared and soft singularities. But then their imaginary parts are also singularity free which is the KNL theorem: in the value of the total cross-section $\sigma(e^+ e^- \to$ hadrons) the contributions from soft and collinear regions cancel each others. In this argument unitarity plays the important role: in the sum over all possible final state in inclusive quantities the contribution of pinch surfaces cancel each other. Unitarity is consistent with perturbation theory therefore the cancelation theorem is valid in any fixed order. For this reason infrared singularities of loop corrections are cancelled by infrared singularities coming from gluon Bremsstrahlung and collinear final state configurations.

Infrared safe jet cross sections

The interesting feature of the derivation of the KNL theorem using pinch surfaces and unitarity is that it can be generalized to jet production.*

Qualitatively the argument goes as follows [28,15]. Let us consider a limited region of phase space where the particles are either soft (their energies are less than some energy resolution parameter ϵ) or have momenta within two back-to-back cones (with half opening angles δ)[29]. Consider the pinch surfaces of all Feynman diagrams with external lines lying within this two jet like regions. The pinch surfaces are defined by jet lines joining the final state particles connected by soft bubbles exactly like in the

---

*Jets are qualitatively defined as nearly collinear beam of particles. The hadronic final states in $e^+ e^-$ annihilation into hadrons can be classified into two jet, three jet, four jet *etc.* final states. The probabilities of these topologically different final states decrease with the number of the jets as given by powers of α_S.

case of the total cross section calculation. The only new feature is that now all jet lines lie within the cone. As a result if we restrict the loop integrals to the regions where the space components of the momenta lie within the same cones the singularity structure remain the same. Hence if we define our field theory in momentum space such that the three momenta are restricted to be within the two cones defined above with the help of ϵ and δ the cancelation theorem remains valid. Lorenz invariance gets lost but unitarity and the Landau equations remain the same.

The structure of the jet lines allowed by the external momenta will not change if an extra soft gluon is emitted or a collinear parton is splitted into to other collinear partons. As a result the cancelation theorem expected to remain valid for inclusive quantities defined with use of some resolution parameters consistent with the structure of the jet lines.

$\Gamma(Z \to \textbf{HADRONS})$

This section summarizes the higher order QCD corrections calculated to the simplest infrared safe physical quantity the total hadronic width of the Z boson. We review all the relevant ingredients which enter into the theoretical prediction for $\Gamma(Z \to$ hadrons). The limitations inherent in the use of perturbative methods are illuminated with the discussion of non-Borel summability of the perturbative series due to infrared renormalons.

NNLO corrections and measurement of α_S

The precise measurement of the ratio

$$R_Z = \frac{\Gamma(Z^0 \to \text{hadrons})}{\Gamma(Z^0 \to e^+ e^-)} \tag{16}$$

at LEP gives one of the most crucial test of QCD[30,31,32]. If QCD is correct, the value of α_S extracted from this measurement should be the same as the value obtained in other experiments such as three jet production in $e^+ e^-$ annihilation and deep inelastic scattering. This test of QCD is regarded as the best and cleanest because the theoretical ambiguities are best controlled for this quantity[33]. The hadronization corrections are suppressed by four power of the mass of the Z-boson M_Z and the perturbative calculation is most complete: it is known up to next-to-next-to leading order accuracy together with possible heavy quark mass corrections. Simple analicity arguments relate R_Z to the behavior of the current correlation functions at short distances therefore we do not need to use complicated arguments about good jet definitions or about factorization of long-distance effects. We only need the KNL theorem which assures that in the asymptotic limit $s = q^2 \to \infty$ the imaginary part of the vacuum polarization of the weak current is free from mass and soft singularities.

The hadronic width is decomposed into the sum of the contribution from the weak vector and weak axial vector neutral current

$$\Gamma_h = \Gamma_h^V + \Gamma_h^A = M_Z \, \text{Im}\Pi(M_Z^2 + i0^+) \tag{17}$$

In leading order of the weak coupling the QCD correction have the form[34,35]

$$\Gamma_h^V = \frac{G_F M_Z^3}{8\sqrt{2}\pi} \left(\sum_{i=1}^{5} v_i^2\right) \left[1 + \frac{\alpha_s}{\pi} + 1.409(\frac{\alpha_s}{\pi})^2 - 12.8(\frac{\alpha_s}{\pi})^3\right]$$

$$+ \frac{G_F M_Z^3}{8\sqrt{2}\pi}\left(\sum_{i=1}^{5} v_i\right)^2\left[-0.41(\frac{\alpha_s}{\pi})^3\right]$$

and

$$\Gamma_h^A = \Gamma_h^{NS} + \Gamma_h^S$$
$$\Gamma_h^{NS} = \frac{G_F M_Z^3}{8\sqrt{2}\pi}\left(\sum_{i=1}^{5} a_i^2\right)\left[1 + \frac{\alpha_s}{\pi} + 1.409(\frac{\alpha_s}{\pi})^2 - 12.8(\frac{\alpha_s}{\pi})^3\right]$$
$$\Gamma_h^S = \frac{G_F M_Z^3}{8\sqrt{2}\pi}\left[(\frac{\alpha_s}{\pi})^2(-\frac{37}{12} + \ln\frac{M_Z^2}{m_t^2}) + (\frac{\alpha_s}{\pi})^3\left[-18.7 + \frac{31}{18}\ln\frac{M_Z^2}{m_t^2} + \frac{23}{12}\ln^2\frac{M_Z^2}{m_t^2}\right]\right],$$

where the axial vector contributions are decomposed into singlet and non-singlet flavor contributions; a_i and v_i denote the vector and axial vector neutral current couplings

$$a_q^2 = 1 \qquad v_q = 2I_q^{(3)} - 4Q_q\sin^2\Theta_w \tag{18}$$

G_F is the Fermi coupling, and m_t denotes the mass of the top quark.

At this accuracy of the QCD corrections the higher order electroweak corrections should also be taken into account. There are non-negligible corrections due to

(i) $\mathcal{O}(G_F^2 m_Z^4)$ corrections to the ρ-parameter and to the $Z\bar{b}b$ vertex[36]

(ii) $\mathcal{O}(G_F\alpha_s m_t^2)$ corrections to the Z-boson self-energies and to the $Z\bar{b}b$ vertex[37,38]

(iii) $\mathcal{O}(\alpha_s^2 m_t/M_Z)$ dependent corrections plus $\mathcal{O}(\alpha_s^2 m_b^2/M_Z^2)$ corrections to the axial Z^0 coupling in the triangle diagrams[39].

The theoretical prediction has uncertainties due to the large ambiguities in the value of the mass of the Higgs boson and in the value of the mass of the top quark.

These additional corrections in general can not be written in the form

$$R_Z = R_Z^{(0)}(1 + \delta_{QCD}) \tag{19}$$

where $R_Z^{(0)}$ denotes the pure electroweak result. There is no exact factorization formula since the vector and axial part receive different QCD and different electroweak corrections[40]. The final analysis of course coded all this corrections correctly into a computer code. It is rather impressive to see the amount of theoretical input needed for the theoretical description of this quantity provided we would like to control the accuracy of the calculation for R_Z at the level 0.05% accuracy. From the latest measured value at LEP

$$R_Z^{\exp} = 20.795 \pm 0.040 \tag{20}$$

taking into account all the corrections listed above one obtains[†]

$$\alpha_S = 0.124 \pm 0.006(\exp.) \quad \pm 0.004(\text{theor.}) \tag{21}$$

The remaining theoretical ambiguity coming from m_t, M_H dependence and from the remaining scale dependence of the QCD corrections. It was a dream for us for a long time to get a value of α_S from the "gold plated" quantity R_Z with such an accuracy. It is a very beautiful result from LEP.

An independent determination of α_S in NNLO is provided by the measurement of the hadronic width of the τ lepton. This quantity is as simple theoretically as the

---

[†]More details can be found in the lecture of E. Martinez in this volume.

hadronic width of the Z boson since it is also related to the the current correlation function. But here we have one important difference: the scale of the process given by the tau mass 1.78 GeV is very low, the non-perturbative corrections are more important and they have to be controlled more precisely. The hadronic branching ratio is decomposed as

$$R_\tau = \frac{B(\tau \to \nu_\tau + \text{hadrons})}{B(\tau \to \nu_\tau l \overline{\nu}_l)} = R_\tau^{(0)} \left(1 + \delta_{\text{pert.}} + \delta_{\text{non-pert.}}\right) \tag{22}$$

The non-perturbative corrections are estimated to be $\delta_{\text{non-pert.}} = -0.02 \pm 0.01$. This theoretical estimate has to be still further scrutinized. Although this value is quite small it is comparable to the size of $\alpha_S{}^3$ corrections since at the scale m_τ the value of α_S is rather large (≈ 0.3). The experimental value of the hadronic width of the tau lepton[3] is known about 1% accuracy giving the value for α_S

$$\alpha_S(M_Z) = 0.122 \pm 0.002(\text{exp.}) \pm 0.004(\text{theor.}) \tag{23}$$

in agreement with the value obtained from the measurement of the hadronic width of the Z boson.

Nature of QCD perturbation theory

In the previous sections using perturbative QCD we could predict physical cross sections in power series of the coupling constant. Calculating higher order corrections we apparently increased the accuracy of the theoretical predictions. This is very important for the quantitative success of the prediction of perturbative QCD. One can not improve the prediction of the theory, however, beyond a certain limit since the QCD series for R_Z is not convergent, even worse it is not Borel summable[41,42]. This is not unexpected since non-perturbative topological configurations can lead to corrections not analytic in α_S at $\alpha_S = 0$. Furthermore it is not meaningful to try to improve the accuracy of perturbation theory beyond an accuracy given by the power corrections coming from non-perturbative effects. At the Z-mass assuming that $\alpha_S{}^n \approx (1\,\text{GeV}/M_Z)^4$ this gives very weak constraint $n < 10$.

The argument for non-Borel summability of perturbative QCD can be best understood on the QED example provided by the perturbative calculation of the electron anomalous magnetic moment[43] A_e. The expansion of the magnetic moment in α is

$$A_e = \sum_{n=0}^{\infty} a_n \left(\frac{\alpha}{\pi}\right)^n \tag{24}$$

Let us consider the one loop vertex corrections but with self energy corrections inserted n-times in the virtual photon line. Using dispersion relation satisfied by the vacuum polarization amplitude the contribution of vacuum polarization insertion to the anomalous magnetic moment can be given as

$$A_e^{\text{vac.pol.}} = \frac{\alpha}{\pi} \int_0^1 dx(1-x)\left(-\Pi\left[\frac{-x^2 m^2}{1-x}\right]\right) \tag{25}$$

where for n bubble insertion

$$\Pi(t) = -\left(-\Pi_2(t)\right)^n \tag{26}$$

where $\Pi_2(t)$ is the standard second order vacuum polarization

$$\Pi_2(t) = \frac{\alpha}{\pi}\left(\frac{8}{9} - \frac{1}{3}\kappa^2 + \frac{1}{6}(3 - \kappa^2)\ln\frac{\kappa-1}{\kappa+1}\right) \tag{27}$$

with $\kappa = \sqrt{1 - 4m_e^2/t}$. For t negative $\Pi_2(t)$ is negative definite. Assuming the same mass for the particle in the internal loop we can write

$$-\Pi_2\left[\frac{-x^2m^2}{1-x}\right] = \frac{\alpha}{\pi}f(x) \tag{28}$$

where $f(x)$ can be derived from Π_2. The contributions of vacuum polarization bubbles to a_n become

$$a_n = \int_0^1 dx(1-x)[f(x)]^n . \tag{29}$$

These integrals can explicitly be evaluated[43]. For large n the integral is dominated with the region $x = 1$ and one obtains the asymptotic behavior

$$a_n = \frac{1}{2}n!\frac{1}{6^n}e^{-10/3} + \mathcal{O}(1/n) \tag{30}$$

We see that for large n the expansion coefficients are all positive and grow as $n!$. Such a series is not convergent but even not Borel summable for $\alpha > 0$. We recall that the Borel transform of a series

$$A(u) = u + \sum_1^{\infty} a_n u^{n+1} \tag{31}$$

is defined by forming with the coefficients a_n a new series

$$A_{\mathcal{B}}(z) = \sum_0^{\infty} \frac{a_n}{n!}z^n \tag{32}$$

with $a_0 = 1$. Using the integral representation of the factorial

$$n! = \int_0^{\infty} dt e^{-t}t^n \tag{33}$$

we obtain for the original function $A(u)$ the integral representation

$$A(u) = \int_0^{\infty} dz e^{-z/u} A_{\mathcal{B}}(z) \tag{34}$$

The use of the Borel transform is motivated by the idea that even if the coefficients a_n grow like $n!$, as it is expected in field theory, the Borel transform may still be defined and as a result the integral representation may exist. For the perturbative series of the the anomalous magnetic moment $a_n = C(1/6)^n n!$ then $A_{\mathcal{B}}=C'\frac{1}{1-z/6}$ and the integral is divergent since the integrand has a pole at $z = 6$ along the real axis. Such a perturbative series is called Borel non-summable. If the coefficient had an alternating sign then the position of the pole would be at $z = -6$ and the corresponding Borel sum would exist.

The Borel non-summability of the perturbative series of the electron anomalous moment is related to the Landau pole of the QED running coupling constant. One can easily see that with interchanging the sum over the bubble contribution and the remaining integral defining a_n. For large n the integral is dominated with contribution from the region around $x = 1$ and it can be evaluated with saddle point method using the variable change $x = 1 - e^{-u}$ and one obtains that the the asymptotic behavior is determined by the integral

$$A_e \approx \frac{\alpha}{\pi} \int_0^{\infty} \frac{m^4 dt}{t^3} [-\Pi_2(t)]^n \tag{35}$$

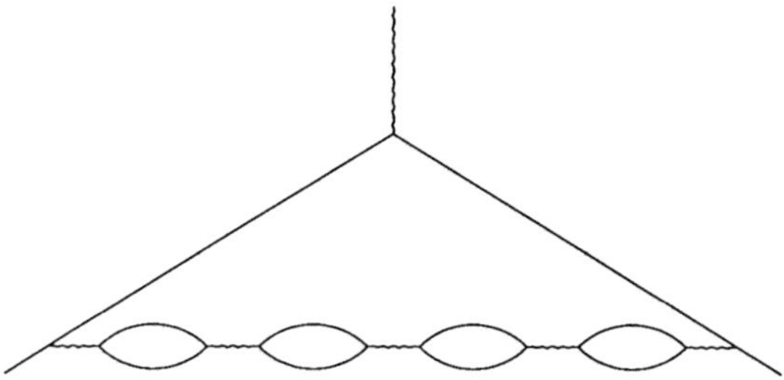

Fig. 2: Fermion loop (bubble) contributions to the photon self-energy in vertex diagram contributing to the electron anomalous magnetic moment.

with

$$-\Pi_2(t) \approx -\bar{b}_0 \alpha \ln t/m^2 \tag{36}$$

with $\bar{b}_0 = -1/3\pi$ the first coefficient of the beta function of α using the same normalization as we used for QCD. Introducing $u = t/m^2$ as new variable and replacing the the coefficients with the asymptotic form also at low orders we get

$$A_e \approx \frac{\alpha}{\pi} \sum_0^\infty \int_0^\infty \frac{du}{u^3} (-b_0 \alpha \ln u)^n = \frac{\alpha}{\pi} \int_0^\infty \frac{du}{u^3} \frac{1}{1 + \bar{b}_0 \alpha \ln u} . \tag{37}$$

The integrand is proportional to the running coupling constant

$$\alpha(k^2) = \frac{\alpha}{1 + b_0 \alpha \ln k^2/m^2} .$$

The integration has to be carried out over k^2. At the scale where the denominator is vanishing the running coupling has a pole (Landau pole) and the integral can not be carried out without regularizing this pole contribution. This means QED can not be defined uniquely in perturbation theory. With further change of variable $2\alpha \ln u = e^z$ this integral representation of A_e gets the form of a Borel representation

$$A_e(\alpha) = \frac{1}{2\pi} \sum_{n=0}^1 \int_0^\infty dz e^{-z/\alpha} (-\bar{b}_0 z/2)^n = \frac{1}{2\pi} \int_0^1 dz e^{-z/\alpha} \frac{1}{1 + \bar{b}_0/2z} . \tag{38}$$

We see that the Borel integral is divergent at $z = -2/\bar{b}_0 = 6\pi$. The divergence is due to the Landau pole of the running coupling constant and the singularity is called ultraviolet renormalon. In QCD one can carry out a similar analysis but since the sign of b_0 is opposite at large k^2 the integrand is well behaved it has no Landau singularity. In QCD, however, the behavior of the running coupling constant give rise to singularities in the infrared region. In general we can say that if a simple dimensionless physical quantity receives contribution from vacuum polarization insertions as given by the integral

$$A(\alpha(Q^2)) \approx \int dk^2 F(k^2)\alpha(k^2) \tag{39}$$

with running coupling constant

$$\alpha(k^2) \approx \frac{\alpha}{a + b_0\alpha(Q^2)\ln k^2/Q^2}, \qquad (40)$$

than this integrand has a Landau pole either in the infrared or in the ultraviolet region depending on the sign of b_0. The actual position of the singularities depends on the asymptotic behavior of $F(k^2)$ as $k^2 \to \infty$ or as $k^2 \to 0$. Expanding the integrand in $\alpha(Q^2)$ and forming its Borel transform we obtain Borel representation for A.

In summary, QCD perturbation theory is not Borel summable since contribution from bubble diagrams effectively equivalent to carry out loop integrals with running coupling constant depending on the loop integration variable. This effect appears as a singularity of the Borel transform along with the positive real axis. This singularity is called infrared renormalon. The resummation of perturbation theory can only be achieved provided we give some instruction how this singular contribution should be regularized. The Borel transform is ambiguous for $b > 2/b_0$. Regularizing the singularity and assuming convergent behavior for large b will in general lead to power corrections. This can be seen from equation (38) replacing $1/\alpha_S$ with its asymptotic value $b_0 \ln Q^2/\Lambda^2_{\mathrm{QCD}}$.

When we calculate infrared safe quantities in higher orders of perturbation theory via loop integrals these quantities receive contributions from long distance small virtuality regions. The contributions from these regions in the case of infrared safe quantities, however, are suppressed with certain powers of the inverse of the typical large scale of the problem. In high orders of perturbation theory the sensitivity to this region will grow leading to divergent non-Borel summable series. If we regularize the infrared renormalons the coefficients of the perturbative series will change. Clearly the nature of the failure of perturbative QCD in higher orders is sensitive to non-perturbative effects[41,42].

While all these arguments are based on the contributions of vacuum polarization it is likely that these contribution can not be cancelled by contributions of more complicated Feynman diagrams therefore it provides to us a qualitative understanding of the nature of the perturbative approximation to QCD.

JET PRODUCTION IN NEXT-TO-LEADING ORDER

NLO corrections to three jet production in e^+e^- annihilation

The value of R_Z in leading order is independent from α_S hence the extraction of α_S requires very high precision experimental input (we needed less than 5% accuracy on an effect of less than 10%). This problem can be avoided by studying hard gluon Bremsstrahlung which is proportional to α_S in leading order and it typically leads to three jet configurations for the final state hadrons. As a result detailed QCD tests can be performed at e^+e^- colliders by studying infrared safe three jet like quantities. Jet studies do not suffer from low statistics data but the theoretical uncertainties become larger. The hadronization (non-perturbative) corrections are sizeable: they are only suppressed by the first power of the annihilation energy. The data can not be compared directly with the predictions of perturbative QCD: they require corrections for hadronization effects. These effects are estimated using shower Monte Carlo programs[44,45,46,47]. Although the experimental method for these corrections is scrutinized by various consistency checks it is not possible to obtain a rigorous bound on the size of the power corrections. Therefore the measurement of α_S from three jet like quantities contains more theoretical uncertainty than the measurement of α_S from R_Z.

The perturbative calculation of any three jet measure is based on the next-to-leading order matrix elements of three parton production in $d = 4 - 2\epsilon$ dimension and on the leading order four-parton matrix element squared in d-dimension calculated by Ellis Ross and Terrano [48]. With this analytic input any infrared safe three jet like quantity can be calculated in next-to-leading order accuracy. The next-to-leading order corrections are given by the interference between the leading order term and the virtual corrections to $q\bar{q}g$ production ($d\sigma^{\mathrm{virt}}$) and by the order α_S^2 four parton production ($d\sigma^{\mathrm{4par}}$). After ultraviolet renormalization the cross-section for any inclusive quantity calculated in $d = (4 - 2\epsilon)$ dimension has the form

$$d\sigma(\epsilon, s, ..) = d\sigma^{\mathrm{virt}}(\epsilon, s, ..) + d\sigma^{\mathrm{4par}}(\epsilon, s, ...) \tag{41}$$

where both the virtual and Bremsstrahlung contributions have soft and collinear singular pieces as $1/\epsilon^2$, $1/\epsilon$ pole terms. For an infrared safe inclusive quantity these pole terms cancel each other. The singular terms of the four parton final states arise from the phase space integral over soft and collinear contributions carried out in d-dimension. Fortunately the matrix element squared of four parton production becomes very simple in the singular regions. For these simplified expressions the integration can be carried out analytically giving the same pole terms ($d\sigma^{\mathrm{pole}}$) as the full expression. As a result the cancelation of the singularities can be arranged analytically in the sum

$$d\sigma^{\mathrm{virt}} + d\sigma^{\mathrm{pole}}.$$

In this way local subtraction term can be defined for the four parton contribution. With local counter terms

$$d\sigma^{\mathrm{4par}} - d\sigma^{\mathrm{pole}}$$

is finite the regularization parameter ϵ can be set to zero and it can be subject of numerical evaluation[49].

Let us briefly summarize the method of analytic cancelation of the singular terms. $\left|A^{(4)}(p_i, k)_{\mathrm{real}}\right|^2$ denotes the spin and color averaged value of the squared amplitudes of tree diagrams for the process

$$e^+ + e^- \rightarrow q(p_1) + \bar{q}(p_2) + g(p_3) + g(k) \tag{42}$$

calculated in $d = 4 - 2\epsilon$ dimension. The singularities arise from regions where either k is soft or k is collinear with one of the momenta p_i ($i = 1, 2, 3$).

In the collinear limit $k \rightarrow z p_1$, for example, we obtain the factorized form

$$\left|A^{(4)}(p_1, p_2, p_3, k)_{\mathrm{real}}\right|^2_{\mathrm{collinear}} = \left|A^{(3)}(p_1, p_2, p_3)\right|^2 \frac{1}{p_1 \cdot k} \tilde{P}_{g/q}(z, \epsilon) \tag{43}$$

where $\left|A^{(3)}\right|^2$ denotes the amplitude square of the leading order process

$$e^+ + e^- \rightarrow q(p_1) + \bar{q}(p_2) + g(p_3) \tag{44}$$

in Born approximation and in d-dimension. $\tilde{P}$ is the Altarelli-Parisi splitting function[4] in d-dimension defined only for $z < 1$. The k-dependence of the amplitude is contained in the collinear factor $1/p_1 \cdot k$ and in the z-dependence of $\tilde{P}_{g/q}$.

In the soft region $k^\mu \rightarrow 0$ the singular behavior can also be obtained very simply. Due to a general theorem, in Feynman gauge (and also in axial gauge), the singular contributions are given by diagrams where the soft line connects external legs[50]. These contributions can easily be calculated[51,52]. For example when the soft line connects the quark and the hard gluon lines we obtain

$$|A(p_i,k)_{\text{real}}|^2_{\text{soft,qg}} = A^{(3)\dagger}(p_i)^c_{\alpha\gamma} \frac{t^a_{\alpha\beta} p_1 \cdot p_3}{p_1 \cdot k p_3 \cdot k} T^a_{bc} A^{(3)}(p_i)^b_{\gamma\beta} \tag{45}$$

where t^a and T^a denote the color matrices in the fundamental and adjoint representations, respectively, and the color labels of the matrix element of $q\bar{q}g$ production are also indicated. This contribution is different from the standard Born term since the color components, as a result of soft gluon exchange, got rotated. But the explicit evaluation of this contribution as trivial as the evaluation of the Born term of the leading order process. The k-dependence of the amplitude square in the soft limit is completely given by the eikonal factors

$$\frac{p_i \cdot p_j}{p_i \cdot k p_j \cdot k} \quad (i \neq j).$$

If the kinematical constraints defining our physical quantity is insensitive to collinear splitting or to the emission of a soft line the singular region contributions of the Bremsstrahlung process is also independent from those kinematical constraint. As a result the phase space integrals leading to the singular terms are universal and have the same form for all infrared safe inclusive quantities. Their cancelation can be achieved analytically without making any reference to the definition of the three jet like infrared safe physical quantity to be calculated. After arranging the cancelation of the singular contributions the remaining integrals can be easily carried out numerically[49,53].

The definition of a three jet like quantity, as well as the definition what is a jet, is genuinely ambiguous. This explains the large number of shape variables proposed in the literature. For sake of illustration we recall the definition of thrust. It is defined as sum of the lengths of the longitudinal momenta of the final state particles relative to the axis $\vec{n}_{\text{thrust}}$ chosen to maximize this sum[54]

$$\mathcal{T} = \max \frac{\sum_i |\vec{p}_i \cdot \vec{n}|}{\sum_i |\vec{p}_i|} \tag{46}$$

where i runs over all the final state particles. For two parton final states we have $\mathcal{T} = 1$, for three particle final states $\mathcal{T}$ is in the region

$$1 \geq \mathcal{T} \geq \frac{2}{3}, \tag{47}$$

for four particle final states we get

$$1 \geq \mathcal{T} \geq \frac{1}{\sqrt{3}}, \tag{48}$$

while for an arbitrary number of particles

$$1 \geq \mathcal{T} \geq \frac{1}{2}. \tag{49}$$

The thrust distribution is thus discontinuous with respect to the multiplicity in the final state. Since the multiplicity is an infrared sensitive quantity, some amount of smearing is necessary around this kinematical boundaries. The leading order cross-section is

$$\frac{1}{\sigma_0}\frac{d\sigma}{d\mathcal{T}} = \frac{\alpha_S(\mu)}{2\pi(1-\mathcal{T})}\frac{4}{3}\left[\frac{2(3\mathcal{T}^2 - 3\mathcal{T} + 2)}{\mathcal{T}}\log\left(\frac{2\mathcal{T}-1}{1-\mathcal{T}}\right) - 3(3\mathcal{T}-2)(2-\mathcal{T})\right]. \tag{50}$$

Three jet like quantities may be defined directly in terms of jets. In this case we must use and explicit jet definition. The most popular jet definitions are provided by successive jet cluster algorithms. They give the most commonly used methods for defining and reconstructing jets at e^+e^- colliders. It was originally introduced by the JADE group[55]. Such algorithms are iterative, beginning with a list of jets that are just the observed particles. (In a perturbative calculation, one begins with a list of partons instead.) At each stage of the iteration, one considers two jets i and j as candidates for combination into a single jet according to the value of a dimensionless "jettiness" variable y_{ij}, which may be, for example,

$$y_{ij} = \frac{2E_i E_j (1 - \cos\theta_{ij})}{s} . \tag{51}$$

The pair i, j with the smallest value of y_{ij} is combined first. When two jets are combined the four-momentum of the new jet is determined by a combination formula, which may be, for example,

$$p^\mu = p_i^\mu + p_j^\mu . \tag{52}$$

After this joining, there is a new list of jets. The process continues until every remaining y_{ij} is larger than a preset cutoff, y_{cut}. In this way, each event is classified as containing two, three, four ... jets, where the number of jets depends on the cutoff y_{cut} chosen. The success of this and similar algorithms is mainly due to the fact that the hadronization of the parton final states can be shown to have, on average, little influence on the jet rates[55,56,49]. Other versions of the this type of jet definition are obtained by modifying either the variable used to define jettiness or/and the recombination algorithm rates[56].

In next-to-leading order the distributions of three jet measures are given as a second order polynomial in α_S. Let us denote such a quantity with X. The cross-section can be conveniently given as

$$\frac{1}{\sigma_0}\frac{d\sigma}{dX} = \frac{\alpha_S(\mu)}{2\pi} A_X(X) + \left(\frac{\alpha_S(\mu)}{\pi}\right)^2 \left[A_X(X) 2\pi b_0 \log(\mu^2/S) + B_X(X)\right] \tag{53}$$

where b_0, b_1 are the coefficients of the beta function of α_S and $A(x)$ and $B(x)$ are scale independent functions. Their values are tabulated for many quantities in ref. [49]. Cross-section (53) manifestly satisfies the renormalization group equation to order $\mathcal{O}(\alpha_S^3)$

$$\frac{d}{d\mu^2}\left(\frac{d\sigma}{dX}\right) = 0 + \mathcal{O}(\alpha_S^3) . \tag{54}$$

The size and sign of the corrections can be rather different for the various jet measures and the corrections are usually rather large (30%-40% at LEP). The study the remaining scale dependence of the result gives a rough estimate on the size of the remaining theoretical error due to uncalculated higher order corrections. Some principles have been advocated for the best choice of the scale [57,58,59].

According to the minimal sensitivity (MS) principle[57] the optimal choice is the scale where eq. (54) is satisfied exactly. If a physical cross-section is parameterized as

$$d\sigma = C\alpha_S^n (1 + r\alpha_S)) \tag{55}$$

then the MS principle gives the relation

$$\mu_{\text{opt}} = Q e^{-\frac{r}{2nb_0} - \frac{1}{2(n+1)b_1}} \tag{56}$$

If the correction r is large and positive the optimal scale is much smaller than Q, if the correction is large and negative the optimalization scale is much larger than Q. Clearly the optimalized scale is in one-to-one correspondence with the size of the next-to-leading order corrections.

Let us summarize the general features of NLO jet studies in e^+e^- annihilation.

i) Even at the scale M_Z the NLO corrections are rather large. Typically they give $\approx$ 30%-40% corrections. As a result the remaining scale dependence is rather large[49,44,45].

ii) Fixed order perturbation theory fails at the boundary values. For example at $\mathcal{T} \approx 1$ there are large contributions of order $\mathcal{O}(\alpha_S{}^n \ln^{2n}(1 - \mathcal{T}))$. These logarithmically enhanced terms can be resummed together with the first sub-leading terms. Such a resummation extends the range of appiicability of the perturbative result. Detailed descriptions of the resummation of such terms exists by now for many shape variables[60,61]. The study of this so called Sudakov resummation helped also to improve the jet finding algorithm. As was pointed out in ref. [62], the jet fractions defined using the formula for y_{ij} given in eq.(55) do not exhibit the usual Sudakov exponentiation from multiple soft gluon emission, despite having an effective expansion in $\alpha_S \log^2(y_{\text{cut}})$. A modified algorithm was proposed which satisfied the criterium to allow for this resummation without destroying other attractive feature of the original algorithm[63].

iii) Hadronization corrections are non-negligible and are of order $1\,\text{GeV}/Q$. Even this estimate is still optimistic since the coefficient may become large depending on the jet resolution parameter. Recent studies of possible effects of infrared renormalons and model studies of hadronization confirm this expectations[64,65].

iv) In view of the large perturbative corrections and of the large hadronization corrections it would be interesting to know the NNLO corrections for jet production. This is a very difficult calculation which can not be carried out with straightforward applications of existing technical tools[66]. Part of the corrections come from the NLO corrections to four jet production. This calculation is feasible and work is in progress.

vi) The value of α_S obtained from measuring three jet like quantities a LEP is competitive and consistent[45] with the value obtained from the measurement of α_S from R_Z

$$\alpha_S = 0.122 \pm 0.002(\text{exp.}) \pm 0.005(\text{th.}) \tag{57}$$

NLO studies of jet production at hadron colliders

We describe briefly the fundamental formula of the QCD improved parton model for processes involving initial hadrons[5], and we discuss the structure of the soft and collinear singularities appearing in the NLO corrections[52,7,20].

Factorization theorem for initial state collinear singularities. In simple hard scattering processes involving hadrons in the initial state, the cancelation theorem for jet production is incomplete: there are remaining infrared singularities due to collinear splitting of the incoming partons. These contributions are controlled fortunately by a factorization theorem. According to this theorem the initial state collinear singularities are universal and can be factored into parton distribution functions for the incoming hadrons[9,5]. This theorem gives the required consistency condition for the validity of the parton picture in which the cross-sections of the large momentum transfer reactions can be calculated in terms of parton densities and parton scattering cross-sections in a factorized form. This theorem allows to calculate higher order corrections to physical cross-section $d\sigma$ of any hard scattering process as given by the parton model. The cross-section in the collision of hadrons A and B with incoming momenta p_A and p_B are obtained by folding the initial parton number densities $f_{a/A}(x_A,\mu)$ of momentum fraction X_A and of scale μ with the so called finite *hard scattering* cross-section $d\hat{\sigma}$ defined in terms of partons

$$d\sigma(p_A,p_B,...) = \sum_{a,b} \int_0^1 dx_A \int_0^1 dx_B f_{a/A}(x_A,\mu) f_{b/B}(x_B,\mu) d\hat{\sigma}_{a,b}(x_A p_A, x_B p_B, \mu, \alpha_S(\mu)) \ .$$

(58)

This formula is valid for three type of final states final states of i) inclusively produced lepton-pairs, W, Z bosons; ii) inclusively produced jets of hadrons iii) the mixed states of final states of the previous two cases[‡]

In leading order of the strong coupling constant α_S the short distance cross-section $d\hat{\sigma}$ is equal to the Born cross-section of the corresponding parton level process. The jets are identified with the final state partons. In higher order the definition of the hard scattering cross-section contains some kinematical constraints defining the jets. Since in higher orders the parton scattering cross-sections are singular we have to define precisely the finite hard scattering cross-section $d\hat{\sigma}$. In d-dimension the cross section of parton-parton scattering is defined by a set of Feynman diagrams. After adding the the ultraviolet renormalization counter terms the ultraviolet divergences cancel. This ultraviolet finite parton cross section, however, contains soft and collinear singularities. These singularities are also regularized in d-dimension. According to the factorization theorem in infrared safe quantities the final state collinear singularities and the soft singularities cancel and the remaining initial collinear singularities of the bare parton cross-section $d\sigma^{(bare)}$ can be factored into singular *process independent* perturbative "parton in a parton" splitting functions $\Gamma_{i/a}$. Since the initial collinear singularities are universal we can define universal counter terms to cancel them. The remaining finite cross section is called *finite short distance cross section $d\hat{\sigma}$*. The counter term is not unique: two schemes are used in the literature the DIS and the $\overline{\mathrm{MS}}$ schemes. The short distance cross section will have a remaining dependence on the scale parameters μ introduced by dimensional regularization. This scale can be interpreted as the scale of separating the long distance contributions represented by the parton number densities from the short distance contributions given by the remaining finite hard scattering cross section. This scale is clearly unphysical hence its effect must cancel in the physical cross section $d\sigma(p_A,p_B,..)$

$$\frac{d\sigma}{d\mu} = 0 \ .$$

(59)

In fixed order of perturbation theory this equation holds only up to the accuracy or the truncated perturbative expansion. The μ dependence of the partonic short distance cross section $d\hat{\sigma}$ is cancelled by the μ dependence of the parton densities as described by the Altarelli-Parisi evolution equations[5]. The uniquely defined counter terms for the initial state collinear singularities also summarize the conventions entering in the definition of the value of the next-to-leading order Altarelli-Parisi splitting functions. Clearly we must use the same conventions for the definition of the hard scattering

---

[‡]For inclusively produced single hadron or photon final states factorized fragmentation functions also appear in the cross-section formula.

cross sections and for the definition of the next-to-leading order AP kernels in order to obtain scheme independent answer for the physical cross sections. The $\overline{\text{MS}}$ scheme and the so called DIS scheme are the most generally accepted schemes. If a calculation is performed in a different scheme one should work out the transition functions which changes the result from one scheme to the other. This step is unavoidable if the parton number densities are extracted from a general fit to several hard scattering processes analyzed in NLO accuracy.

Analytic cancelation of the soft and collinear singularities. We shall describe the structure of the soft and collinear singularities appearing in next-to-leading order calculation on the example of inclusive jet production[67,68,69]. The analytic cancelation is carried out using the subtraction method[52,70]. We note that that an alternative method the so called slicing method[71,72] can also be used. Let us denote the cross section inclusive one-jet production by $\mathcal{I}$. At next-to-leading order, $\mathcal{I}$ is a sum of two terms,

$$\mathcal{I} = \mathcal{I}[2 \to 2] + \mathcal{I}[2 \to 3], \tag{60}$$

where $\mathcal{I}[2 \to n]$ is the part of the cross section defined by $2 \to n$ parton scattering processes. The jet is equal to either one of the final state parton or the combination of two final state partons. According to the factorization theorem, the physical cross section in the QCD improved parton model for hadron-hadron scattering is a folding between the parton densities and the hard-scattering cross section:

$$\mathcal{I}[2 \to n] = \sum_{a,b} \int_0^1 dx_A \int_0^1 dx_B f_{a/A}(x_A, \mu) f_{b/B}(x_B, \mu) d\hat{\sigma}_{a,b}(x_A p_A, x_B p_B, \mu, \alpha_s(\mu)). \tag{61}$$

In eq. (61), $d\hat{\sigma}_{a,b}$ is the hard-scattering cross section for the process $a + b \to j_1 + \ldots + j_n$. It is defined as a product of the flux factor and the integral of the squared matrix element over the phase space of the final state particles

$$d\hat{\sigma}_{a,b} = \frac{1}{n!} \sum_{j_1,\ldots,j_n} \frac{1}{2x_a x_b s} \tag{62}$$

$$\int d\mathcal{P}^{(n)}(\mathbf{p}_{j_i}) \, \mathcal{S}_n(p_{j_i}^\mu) \langle |\mathcal{M}(a + b \to j_1 + \ldots j_n)|^2 \rangle (2\pi)^d \delta^d \left(p_a^\mu + p_b^\mu - \sum_{i=1}^n p_{j_i}^\mu \right),$$

where $d\mathcal{P}^{(n)}$ denotes the n-body phase space integral and $\mathcal{S}_n(p_{j_i}^\mu)$ is the so called measurement function that defines the infrared-safe physical quantity. The counting factors $1/n!$ are present when all partons are treated indistinguishable and we sum over the possible parton types. For the cancelation of the soft and final state collinear singularities, it is important that the physical measurement, represented by the functions $\mathcal{S}$, are "infrared safe." This means that one obtains the same measured result whether or not a parton splits into two collinear partons and whether or not one parton emits another parton that carries infinitesimal transverse momentum. A physical quantity that is designed to look at short distance physics should have this property, otherwise it will be sensitive to the details of parton shower development and hadronization. The mathematical requirements for $\mathcal{S}_2$ and $\mathcal{S}_3$ are as follows. First, $\mathcal{S}_3$ should reduce to $\mathcal{S}_2$ when two of the outgoing partons become collinear:

$$\begin{aligned}
\mathcal{S}_3(p_1^\mu, (1 - \lambda)p_2^\mu, \lambda p_2^\mu) &= \mathcal{S}_2(p_1^\mu, p_2^\mu) \\
\mathcal{S}_3((1 - \lambda)p_1^\mu, p_2^\mu, \lambda p_1^\mu) &= \mathcal{S}_2(p_1^\mu, p_2^\mu) \\
\mathcal{S}_3(\lambda p_1^\mu, (1 - \lambda)p_1^\mu, p_2^\mu) &= \mathcal{S}_2(p_1^\mu, p_2^\mu)
\end{aligned} \tag{63}$$

for $0 \leq \lambda \leq 1$. Second, $\mathcal{S}_3$ should reduce to $\mathcal{S}_2$ when one of the partons becomes parallel to one of the beam momenta, which we will denote by p_A^μ and p_B^μ:

$$
\begin{aligned}
\mathcal{S}_3(p_1^\mu, p_2^\mu, \lambda p_A^\mu) = \mathcal{S}_3(p_1^\mu, p_2^\mu, \lambda p_B^\mu) &= \mathcal{S}_2(p_1^\mu, p_2^\mu) \\
\mathcal{S}_3(p_1^\mu, \lambda p_A^\mu, p_2^\mu) = \mathcal{S}_3(p_1^\mu, \lambda p_B^\mu, p_2^\mu) &= \mathcal{S}_2(p_1^\mu, p_2^\mu) \\
\mathcal{S}_3(\lambda p_A^\mu, p_1^\mu, p_2^\mu) = \mathcal{S}_3(\lambda p_B^\mu, p_1^\mu, p_2^\mu) &= \mathcal{S}_2(p_1^\mu, p_2^\mu) \, .
\end{aligned}
\tag{64}
$$

For the $[2 \to 2]$ process, the square of the matrix element — summed over final spins and colors and averaged over initial spins and colors — has the following perturbative expansion

$$
\langle |\mathcal{M}(a + b \to j_1 + j_2|^2 \rangle = \frac{g^4}{\omega(a)\omega(b)} \left\{ \psi^{(4)}(\vec{a}, \vec{p}) + 2g^2 \left(\frac{\mu^2}{Q^2}\right)^\varepsilon c_\Gamma \psi^{(6)}(\vec{a}, \vec{p}) + \mathcal{O}(g^4) \right\},
\tag{65}
$$

where we denote $\vec{a} = (a, b, j_1, j_2)$, $\vec{p} = (p_a^\mu, p_b^\mu, p_{j_1}^\mu, p_{j_2}^\mu)$ and $\omega(a)$ represents the number of spin and color states of a parton type a where $\psi^{(4)}(\vec{a}, \vec{p})$ is the d-dimensional Born term and $\psi^{(6)}(\vec{a}, \vec{p})$ is the contribution of the next-to-leading order contributions

$$
2g^6 \left(\frac{\mu^2}{Q^2}\right)^\varepsilon c_\Gamma \psi^{(6)}(\vec{a}, \vec{p}) \equiv \sum_{\text{hel}} \sum_{\text{col}} [\mathcal{M}^* \mathcal{M}]_{\text{NLO}}.
\tag{66}
$$

where c_Γ

$$
c_\Gamma = \frac{1}{(4\pi)^{2-\varepsilon}} \frac{\Gamma^2(1-\varepsilon)\Gamma(1+\varepsilon)}{\Gamma(1-2\varepsilon)}
\tag{67}
$$

is a ubiquitous prefactor.

In ref. [52], using the results of Ellis and Sexton[73], the following structure has been found for the next-to-leading order term:

$$
\begin{aligned}
\psi^{(6)}(\vec{a}, \vec{p}) = \; &\psi^{(4)}(\vec{a}, \vec{p}) \left\{ -\frac{1}{\varepsilon^2} \sum_n C(a_n) - \frac{1}{\varepsilon} \sum_n \gamma(a_n) \right\} \\
&+ \frac{1}{\varepsilon} \sum_{m<n} \log\left(\frac{2p_n \cdot p_m}{Q^2}\right) \psi_{mn}^{(4,c)}(\vec{a}, \vec{p}) \\
&+ \psi_{\text{NS}}^{(6)}(\vec{a}, \vec{p}),
\end{aligned}
\tag{68}
$$

where $\psi_{mn}^{(4,c)}(\vec{a}, \vec{p})$ are the color-linked Born squared matrix elements in d dimensions as defined in the case of $e^+ e^-$ annihilation in the previous section and $\psi_{\text{NS}}^{(6)}(\vec{a}, \vec{p})$ represents the remaining finite terms. The sum over m and n runs from one to four. In eq. (68), $C(a)$ is the color charge of parton a and the constant $\gamma(a)$ represents the contribution from virtual diagrams to the Altarelli-Parisi kernel. Specifically,

$$
C(g) = N_c, \; \gamma(g) = \frac{1}{2}\beta_0,
\tag{69}
$$

$$
C(q) = \frac{V}{2N_c}, \; \gamma(q) = \frac{3V}{4N_c}.
\tag{70}
$$

The contribution of the $2 \to 3$ parton scattering process should give exactly the same pole terms with opposite sign. This can be demonstrated analytically. The idea of the analytic cancelation of the singularities is the same as in the case of jet production in $e^+ e^-$ annihilation: we can construct simple local counter terms for the squared matrix element of the $2 \to 3$ process such that subtracting these terms the subtracted matrix elements are free from soft and collinear singularities. The local subtraction terms can easily be obtained as soft and collinear limits of the $2 \to 3$ matrix elements and have the same structure as the ones found in e^+e^- annihilation. In these limits the jet definition function of the $2 \to 3$ process S_3 will become equal to the jet definition function of the $2 \to 2$ process S_2 . The additional integration appearing in the calculation of the contribution of the $2 \to 3$ process with respect to the calculation of the contribution of the $2 \to 2$ process can be easily carried out over the simple pole terms and eikonal factors of the limiting functions. The sum of the counter terms for the initial state singularities, the contributions of the virtual corrections and the pole contributions defined by the soft and collinear limits all the singularities is finite and the integrations over the subtracted matrix element squared of the $2 \to 3$ processes which can be carried out numerically are also finite.

One technical observation: the local counter terms can be calculated directly without calculating the full matrix elements. Since the singularities are completely controlled by these terms only these terms have to be known in d-dimensions. Due to the cancelation of the soft and collinear singularities, however, we do not need to know the full d-dependence of the local subtraction terms. This leads to considerable simplification.

With this method efficient Monte Carlo program could be built which can be used to calculate any one or two jet like inclusive quantity at NLO accuracy[74,76,75]. I note that recently the slicing method was also successfully coded into a general purpose Monte Carlo program two calculate one jet, two jet quantities at NLO order[77,78]

The very good data on three jet production at hadron colliders and four jet production in e^+e^- annihilation call for the calculation of the next-to-leading order corrections to these processes. It would be interesting to perform a a quantitative NLO test of QCD in case of processes with more complex final states. Knowing the NLO corrections for inclusive three jet production it becomes possible to extract α_S from jet production at hadron collier with measuring the ratio of three jet to two jet production in a narrow transverse energy interval

$$R_{3/2} = \frac{\Delta \sigma_{3jet}}{\Delta \sigma_{2jet}} \approx C_0 \alpha_S (1 + c_1 \alpha_S) \tag{71}$$

It is expected that the coefficients C_0 and C_1 in this ratio will have very weak dependence on the value of the parton densities therefore α_S can reliably be extracted.

IMPROVED METHODS FOR LOOP CORRECTIONS

At moderate jet energies the production rate of jets at hadron collider is very high, hence, although the final state is dominated by two jet production, the UA1, UA2[79,80], CDF[81,82] and D0[83] collaborations could also observe the production of 3, 4, 5 and 6 jets[80] with a rate suppressed by increasing powers of α_S

$$d\sigma^{2jet} : d\sigma^{3jet} : ... : d\sigma^{6jet} \approx \alpha_S^2 : \alpha_S^3 : ... : \alpha_S^6 \tag{72}$$

and could compare their data with the QCD prediction[84,85]. We note that five and six jet production as well as four jet production in association with a W boson give important background to top production at Fermilab[86].

The study of multijet production offers at the moment qualitative test of the predictions of the QCD improved parton model up to exotic order $\mathcal{O}(\alpha_S^6)$. The theoretical calculation at this order, even in the Born approximation, is very complicated since the number of the Feynman diagrams $\mathcal{N}$ grows with the increase of the number of the final state jets n_J as a factorial

$$\mathcal{N} \propto (n_J + 2)^{n_J} , \quad n_J \to \infty . \tag{73}$$

In Table 1 I listed the number of Feynman graphs for the production of up to seven gluon in gluonic QCD (flavor number $n_f = 0$)[88].

n_J	2	3	4	5	6	7
N	4	25	220	2485	34300	559405

Table 1: The number of Feynman graphs for $gg \to n_J g$ in pure gluonic QCD.

Fortunately new methods have been found to calculate tree diagrams which allowed to evaluate *exactly* even the $8g$ process given by 34300 Feynman diagrams[89]. The most important new technique is the use of the so called helicity method developed by the CALCUL collaboration for QED and reviewed in the recent book by Gastmans and Wu[90]. The power of the helicity method could be fully exploited in QCD with a crossing symmetric formulation[91] using single reference momentum in the gluon polarization [92,93] and decomposing the color structure consistent with the duality property of the amplitudes in the tree approximation[94]. The decomposition of the helicity amplitudes into color subamplitudes allowed to construct recursion relations among amplitudes of increasing number of external legs. The tree level color decomposition and the duality properties emerge quite naturally from string theories in the zero slope limit of an open string amplitude[95].

$N = 1$ and $N = 2$ supersymmetry could be used to calculate amplitude of processes involving quarks from the amplitudes of processes involving gluons,gluinos and scalar particles [96,97]. An exact formula has been found for the so called maximally helicity violating amplitude (conjectured by Parke and Taylor[98] and proved in ref. [94]) valid for any numbers of gluons. Helicity conservation and the general pole structure in two and three body channels gives very strong restrictions on the helicity amplitudes at tree level. In most of the cases one obtains surprisingly short formulae. The new technique also applies to processes in which gauge bosons are involved. A summary of these developments with many references can be found in the excellent review article by Mangano and Parke[99].

Recently it was found that the new technical improvements found in the calculation of tree amplitudes lead to significant simplifications also in the case of the calculation of loop corrections. The feasibility of such a calculation was demonstrated first by Bern and Kosower[100] with deriving the one loop radiative corrections to the four gluon helicity amplitudes from string theory. Subsequently it was pointed out that for the feasibility of the calculation the helicity method, the use of dimensional reduction [101,102,103,104] is very important. One interesting result obtained in the string theory derivation is that subleading color subamplitudes can be constructed from the leading color subamplitudes[105]. Similarly to the case of Born amplitudes supersymmetry significantly reduces the matrix problem of calculating subprocesses of quarks and gluons. I also note that interesting universal structure was found for one loop multi-parton amplitudes in the limit when two of the external legs become collinear[106].

I restrict myself here only to describe very briefly the main ideas of the helicity method and the Ward identities given by supersymmetry.

Helicity Method

The calculation of the jet cross section is based on the spin and color averaged matrix elements squared of the transition amplitude of the contributing subprocesses (see eq. (68)). According to the helicity method one first calculates the matrix elements for a definite helicity configuration. Since helicity states are orthogonal to each other the spin summed amplitude squared is obtained simply as incoherent sum

$$\sum_{\lambda} |M_n(\lambda)|^2$$

where the sum runs over 2^n helicity configuration of an n-parton amplitude. There are several advantages over the standard Dirac trace method.

i) Using parity and charge conjugation symmetry the number of the independent helicity amplitudes is greatly reduced.

ii) The wave function of external quarks and gluons can be described completely in terms of massless Dirac spinors of definite helicity

$$\hat{p}u(p,\pm) = 0, \quad u(p,\pm) = \frac{1}{2}(1 \pm \gamma_5), \quad u(p,\pm)^c = u(p,\mp) \tag{74}$$

where the upper index c denotes charge conjugation. The normalization of the spinors is chosen such that

$$< p\pm |\gamma_\mu| p\pm >= 2p_\mu, \qquad |p\pm >\equiv u(p,\pm) \, . \tag{75}$$

The massless spinors have a number of property which greatly simplify the calculations. We note only a few of them

$$< p\pm |k\pm >= 0, \quad |< p\pm |k\mp >|^2 = 2pk, \quad < p\lambda|p\lambda' >= 0 \, . \tag{76}$$

iii) The polarization vectors of the gluons can also be given in terms of massless spinors with a single reference vector k^μ

$$\epsilon^\pm(p,k)_\mu = \pm\frac{< p\pm |\gamma_\mu| k\pm >}{\sqrt{2} < k\mp |p\pm >}, \quad k^2 = 0, \tag{77}$$

$$\epsilon^\pm(p,k)_\mu \gamma^\mu = \frac{\sqrt{2}}{< k\mp |p\pm >}(|p\mp >< k\mp| + |k\pm >< p\pm|) \, . \tag{78}$$

Therefore all the terms in the amplitudes are proportional to some inner products $< p\lambda|k\lambda' >$ which vanish if $\lambda = \lambda'$ giving a substantial reduction in the number of the contributing terms. Because of gauge invariance the result is independent from the choice of the reference momentum k_μ. A clever choice sets large blocks of terms equal to zero giving significant reduction of the contributing terms.

iv) The subamplitudes appearing in the color decomposition have many important properties. They are gauge invariant, they are invariant under the cyclic and anti-cyclic permutation of the gluon variables. They have simple soft and collinear limits. They can be constructed by recursive relations discovered by Berends and Giele[94]. In this way one can prove the validity of Park-Taylor formula. With the help of these recursion relation one could calculate the exact eight gluon amplitudes[89] as well as the W plus six parton amplitudes[87].

v) In terms of spinor inner products the amplitudes have much less terms. This is well illustrated with the following example. For massless four momenta p_i, $(i = 1, ..2n)$ the trace

$$\frac{1}{2}\text{Tr}(\hat{p_1}\hat{p_2}..\hat{p_{2}n}(1 + \gamma_5)) = < p_1 - |p_2+ >< p_2 + |p_3- > ... < p_{2n} + |p_1- >$$

is given by a single term. The traditional trace method, however, generates an exponentially increasing large expression for increasing values of n.

Supersymmetry Ward identities

Let us consider $N = 1$ supersymmetric $SU(3)$ Yang-Mills theory with helicity states $g^\pm(p)$, $\lambda^\pm(p)$ for gluons and gluinos, respectively, and with supersymmetry generator Q_α. The gluons and gluinos are in the adjoint octet representations. This theory is different from QCD with one quark flavor since the quarks belong to the fundamental representation of color $SU(3)$. At tree level the Feynman diagrams are the same only the color factors are different. Using the commutation relations with $Q(\eta) = Q_\alpha\eta^\alpha$ where η is an anticommuting auxiliary spinor

$$\begin{aligned}
[Q(\eta), g^\pm(p)] &= \mp\Gamma^\pm(p,\eta)\lambda^\pm(p) \\
[\overline{Q}(\eta), \lambda^\pm(p)] &= \mp\Gamma^\mp(p,\eta)g^\pm(p)
\end{aligned} \tag{79}$$

where

$$\Gamma(p,\eta) = (\Gamma(,p,\eta))^* = \bar{\eta}u_-(p). \tag{80}$$

The supersymmetry Ward identities are obtained[107] simply applying these commutation relations to the identities

$$0 = < 0|[Q, \Pi_{i=1}^n a_i]|0 > \tag{81}$$

where a_i denotes gluino or gluon creation and annihilation operators. If we consider the special case $\Pi_{i=1}^n a_i = g_1^- g_2^- g_3^+ \lambda_4^+$, for example, we get that

$$< p_1 p_2 >< 0|l_1^- g_2^- g_3^- |0 > = < p_4 p_2 >< 0|g_1^- g_2^- g_3^+ g_4^+ |0 > \tag{82}$$

which is an exact relation between a helicity amplitude of four gluon scattering and a helicity amplitude of the scattering of two gluinos and two gluons. After decomposition in color we can obtain from "quark" amplitudes the "gluon" amplitudes in ref. (97) in this way the six gluon amplitudes have been derived without any new calculation from the analytic results obtained for the two quark four gluon amplitudes. In next-to-leading order one should correct for internal loops and one should use supersymmetric regularization such as dimensional reduction.

New results for five parton one loop amplitudes

The use of helicity method, dimensional reduction and string theory method allowed recently to calculate the helicity amplitudes of all $2 \to 2$ and $2 \to 3$ parton processes in next-to-leading order. First the four gluon one loop amplitudes have been obtained using string theory[100]. Then the NLO amplitudes of all the other $2 \to 2$ processes[108] have been calculated with application of dimensional reduction, helicity method and supersymmetry. The one loop corrections to five gluon amplitudes have been derived by string theory method[105] while the two gluon three quark[109,110] and one gluon four quark amplitudes[111] have been obtained with combinations of several new methods.

Acknowledgements

I would like to thank Professors R. Gastmans and J.-M. Gerard for a very pleasantly organized Summer Institute.

REFERENCES

1. M. Gell-Mann, *Acta Physica Austriaca, Suppl.* **IX**(1072)733;
   H. Fritzsch and M. Gell-Mann, XVI International Conference on High Energy Physics, Batavia, Vol. II p.135 (1972);
   H. Fritzsch, M. Gell-Mann and H. Leutwyler, *Phys. Lett.* **47B** (1973) 365.
2. D. J. Gross and F. Wilczek, *Phys. Rev. Lett.* **30** (1973) 1343; *Phys. Rev.* **D8** (1973) 3633;
   H. D. Politzer, *Phys. Rev. Lett.* **30** (1973) 1346.
3. M.Martinez, Lecture in this volume
4. G. Altarelli, *Phys. Rep.* **81** (1982) 1; *Ann. Rev. Nucl. Part. Sci.* **39** (1989) 357.
5. J. C. Collins and D. E. Soper, *Ann. Rev. Nucl. Part. Sci.* **37** (1987) 383;
   J. C. Collins, D. E. Soper and G. Sterman, in Perturbative QCD, ed. A.H. Mueller (World Scientific 1989)
6. A. H. Mueller, *Phys. Rep.* **73** (1981) 237;
7. R. K. Ellis, Proceedings, Santa Fe Tasi-87, (1987). FERMILAB-CONF-88/60-T, May 1988. 60pp.
8. T. Kinoshita, *J. Math. Phys.* **3** (1965) 56;
   T. D. Lee and M. Nauenberg *Phys. Rev.* **133** (1964) 1549
9. D. Amati and G. Veneziano, *Nucl. Phys.* **B140** (1978) 54; R .K. Ellis, H. Georgi, M. Machacek, H. D. Politzer and G. G. Gross *Nucl. Phys.* **B152** (1979) 285; A. V. Efremov and A. V. Radyushkin, *Theor. Math. Phys.* **44** (1980) 17; S. Libby and G. Sterman, *Phys. Rev.* **D18** (1978) 3252; A. Mueller, *Phys. Rev.* **D18** (1978) 3705.
10. G. Bodwin, *Phys. Rev.* **D31** (19;) 2616 *ibid.* **D34** (1986) 3932;
    J. C. Collins , D. E. Soper and G. Sterman, *Nucl. Phys.* **B261** (1985) 104; *ibid.* **B308** (1988) 833.
11. K. Wilson, *Phys. Rev.* **179** (1969) 1699;
    W. Zimmermann, *Commun. Math. Phys.* **15** (1969) 208; *Ann. Phys. (NY)* **77** (1970) 536,570.
12. A. H. Mueller, this volume.
13. E. Abers and B. W. Lee, *Phys. Rep.* **9** (1973) 1.
14. E. L. Fadeev and A. A. Slavnov , Gauge Fields, The Benjamin/Cummings Publishing Company, 1980
15. G. Sterman, An introduction to quantum field theory, Cambridge Univ. Press, 1993,
16. L. F. Abbott, *Nucl. Phys.* **B185** (1981) 189 .
17. J. C. Collins, Renormalizaton, Cambridge Univ. Press (1984).
18. G. 't Hooft and M. Veltman, *Nucl. Phys.* **B44** (1972) 189 .
19. R. Gastmans and R. Meuldermans, *Nucl. Phys.* **B63** (1973) 277 .

20. Z. Kunszt, Proceedings of the 1990 Theoretical Advanced Study Institute in Elementary in Particle Physics Boulder, Colorado, Eds.: M. Cvetic and P. Langacker, World Scientific, Singapur.

21. Z. Kunszt and W.J. Stirling, in Proceedings of the Large Hadron Collider Workshop, Aachen, 1990 (G. Jarlskog and D. Rein eds.), Vol. II, p. 428.

22. S.G. Gorishnii, A.L. Kataev, S.A. Larin, L.R. Surguladze, *Phys. Rev.* **D43** (1991) 1633.

23. W. E. Caswell, *Phys. Rev. Lett.* **33** (1974) 244; D.R.T. Jones *Nucl. Phys.* **B75** (1974) 531.

24. O. V. Tarasov, A. A. Vladimirov, and A. Yu, Zharkov, *Phys. Lett.* **B93** (1980) 429.

25. S.A. Larin and J.A.M. Vermaseren, *Phys. Lett.* **B303** (1993) 334

26. M. Ciafaloni, in Perturbative QCD, ed. Mueller, World Scientific, 1985

27. S. Coleman and R.E. Norton, *Nuovo Cimneto* **38** (1965) 438

28. G. Sterman, *Phys. Rev.* **D17** (1978) 2773; *Phys. Rev.* **D17** (1978) 2789.

29. G. Sterman and S. Weinberg, *Phys. Rev. Lett.* **39** (1977) 1436.

30. T. Hebbeker, *Phys. Rep.* **217** (191992) 69

31. S. Bethke, in Proc. of the Aachen Conf. QCD - *20 Years Later*, eds. P.M. Zerwas and H.A. Kastrup, World Scientific, Singapur,1993.

32. B.R. Webber, Plenary talk at the 27th Int. Conf. on High Energy Physics, Glasgow, July 1994, Cavendish-HEP-94/15 (1994).

33. . Zee, *Phys. Rev.* **D8** (1974) 4038.

34. S.G. Gorishnii, A.L. Kataev and S.A. Larin, *Phys. Lett.* **B259** (1991) 144; L.R. Surguladze and M.A. Samuel *Phys. Rev. Lett.* **66** (1991) 560, 2416(E).

35. S.A. Larin, T. van Ritbergen and J.A.M. Vermaseren, *Phys. Lett.* **B320** (1994) 159.

36. R. Barbieri *et al. Phys. Lett.* **B288** (1992) 95.

37. A. Djoudi, C. Verzegnassi, *Phys. Lett.* **B195** (1987) 265;
F. Halzen and B. A. Kniehl, *Nucl. Phys.* **B353** (1991) 567

38. J. Fleischer, O.V. Tarasov, F. Jegerlehner and P. Raczka, *Phys. Lett.* **B293** (1992) 437.

39. K.G. Chetyrkin, J.H. Kuehn and A. Kiatkowski, *Phys. Lett.* **B282** (191992) 221

40. T. Hebbekker, M. Martinez, G. Passarino and G.Quast, *Phys. Lett.* **B331** (1994) 165.

41. G. 'tHooft, in "The Whys of Subnuclear Physics", Erice 1977, ed. Zichichi, Plenum, New York.

42. A. H. Mueller, In "QCD – Twenty Years Later", Aachen, Vol.1.pp. 162, ed. H. A. Kastrup and P. M. Zerwas, World Scientific, 1992.

43. B. Lautrup, *Phys. Lett.* **69B** (1977) 109.

44. S. Bethke, in Proc. of the 26th Int.Conf. on High Energy Physics, Dallas, 1992, ed. J. Sanford, AIP New York, 1993, p. 81.

45. S. Catani, in Proc. of the Int. Europhysics' Conf. on High Energy Physics, HEP-93, Marseilles, Eds. J.Carr and M. Perrottet (Editions Frontiers, Gif-sur-Yvette,1994).

46. G. Marchesini and B.R. Webber, *Nucl. Phys.* **B310** (1988) 461.

47. T. Sjöstrand and M. Bengsston, *Comput. Phys. Commun.* **43** (1987) 367.

48. R. K. Ellis, D. A. Ross and A. E. Terrano, *Nucl. Phys.* **B178** (1981) 421 .

49. Z. Kunszt, P. Nason, G. Marchesini and B. Webber, in "*Z* Physics at LEP1", CERN Yellow Report 89-08 (1989), Vol.1.

50. G. Jr. Grammer and D. R. Yennie *Phys. Rev.* **D8** (1973) 4332; R. Tucci, *Phys. Rev.* **D32** (1985) 945.

51. A. Bassetto, M. Ciafaloni and G. Marchesini, *Phys. Rep.* **100** (1983) 202 .

52. Z. Kunszt and D. E. Soper, *Phys. Rev.* **D46** (1992) 192 .

53. W. T. Giele and E. W. N. Glover, *Phys. Rev.* **D46** (1992) 1980 .

54. E. Fahri, *Phys. Rev. Lett.* **39** (1977) 1587.

55. JADE collaboration: S. Bethke *et al.*, *Phys. Lett.* **213B** (1988) 235.

56. S. Bethke, Z. Kunszt, D. E. Soper and W. J. Stirling, *Nucl. Phys.* **B370** (1992) 310.

57. P.M. Stevenson, *Nucl. Phys.* **B150** (1979) 357 357.

58. G. Grunberg, *Phys. Lett.* **95B** (1980) 70.

59. S. J. Brodsky, G.P. Lepage and Mackenzie, *Phys. Rev.* **D28** (1983) 228.

60. J.C. Collins and D. E. Soper, *Nucl. Phys.* **B197** (1982) 446; J. Kodaira and L. Trentadue, *Phys. Lett.* **294B** (1992) 431.

61. S. Catani *et al.*, *Phys. Lett.* **263B** (1991) 491; *Phys. Lett.* **B295** (1992) 269; *Nucl. Phys.* **B377** (1992) 445.

62. N. Brown and W.J. Stirling, *Phys. Lett.* **252B** (1990) 657.

63. S. Catani, Yu. L. Dokshitzer and B. R. Webber, *Phys. Lett.* **285B** (1992) 291.

64. B. R. Webber, preprint, Cavendish-HEP-94/7, hep-ph/9408222

65. G. P. Korchemsky and G. Sterman, preprint, ITP-SB-94-50 (1994), hep-ph/9411211.

66. "New Techniques for Calculating Higher Order QCD Corrections, Proc. ETH Workshop, Zürich, 1992, Ed. Z. Kunszt preprint ETH-TH/93-01.

67. S. D. Ellis, Z. Kunszt and D. E. Soper, *Phys. Rev.* **D40** (1989) 2188.

68. S. D. Ellis, Z. Kunszt and D. E. Soper, *Phys. Rev. Lett.* **64** (1990) 2121.

69. F. Aversa, M. Greco, P. Chiappetta and J. Ph. Guillet, *Phys. Rev. Lett.* **65** (1990) 401; *Zeit. Phys.* **C49** (1991) 459.

70. M. L. Mangano, P. Nason, G. Ridolfi, *Nucl. Phys.* **B373** (1992) 295.

71. H. Baer, J. Ohnemus and J.F. Owens, *Phys. Rev.* **D40** (1989) 2844.

72. W. T. Giele, E. W. N. Glover and D. Kosower, *Nucl. Phys.* **B403** (1993) 633 .

73. R. K. Ellis and J. Sexton, *Nucl. Phys.* **B269** (1986) 445.

74. S. D. Ellis, Z. Kunszt and D. E. Soper, in Proc. 1991 International Symposium on Lepton and Photon Interactions at High Energies, Geneva, July, 1991

75. S. D. Ellis, Z. Kunszt and D. E. Soper *Phys. Rev. Lett.* **69** (1992) 3615; hep-ph 9208249

76. S. D. Ellis, Z. Kunszt and D. E. Soper *Phys. Rev. Lett.* **69** (1992) 1496.

77. W. Giele, E. W .N. Glover and D. A. Kosower, *Phys. Lett.* **B339** (1994) 181

78. W. Giele, E. W .N. Glover and D. A. Kosower, *Phys. Rev. Lett.* **73** (1994) 2019

79. R. K. Ellis and W. G. Scott, Contribution to the volume *Proton-Antiproton Collider Physics* eds. G. Altarelli and L. Di Lella, World Scientific (1988).

80. P. Lubrano, Proc. Les Rencontre Phys. de la Vallée Aosta, LaThuile, 1990, (Editions Frontieres, Gif sur Yvette, Ed. M. Greco).

81. F. Abe *et al.*, *Phys. Rev. Lett.* **62** (1989) 613; *ibid* **62** (1989) 3020.

82. CDF Collaboration (F. Abe, *et al.*) *Phys. Rev.* **D47** (1993) 4857.

83. Studies of jet production with the D0 detector. By D0 Collaboration (Harry Weerts, for the collaboration), FERMILAB-CONF-94-035-E, Jan 1994. 18pp. Presented at 9th Topical Workshop on Proton - Anti-proton Collider Physics, Tsukuba, Japan, 18-22 Oct 1993.

84. Z. Kunszt and W. J. Stirling, *Phys. Lett.* **176B** (1986) 263.

85. Z. Kunszt and W. J. Stirling, *Phys. Rev.* **D37** (1988) 2439.

86. F. Berends, H. Kuijf, B. Tausk and W. Giele, *Nucl. Phys.* **B357** (1991) 32;

87. W. Giele, E. Glover, and D. A. Kosower, *Nucl. Phys.* **b403** (1993) 633.

88. R. Kleiss and H. Kuijf, *Nucl. Phys.* **312B** (1989) 616.

89. F. A. Berends, W. T. Giele and H. Kuijf, *Phys. Lett.* **232B** (1989) 266.

90. R. Gastmans and T.T. Wu, International Series of Monographs on Physics, Vol. 80 (Clarendon Press, Oxford, 1990) xvi + 648 pages

91. J. F. Gunion and Z. Kunszt, *Phys. Lett.* **161B** (1985) 333.

92. R. Kleiss and W. J. Stirling, *Nucl. Phys.* **B262** (1985) 235.

93. Z. Xu, Da-Hua Zhang and L.Chang, *Nucl. Phys.* **B292** (1987) 392.

94. F. A. Berends and W. Giele, *Nucl. Phys.* **B306** (1988) 759.

95. M. Mangano, S. Parke, and Z. Xu, *Nucl. Phys.* **B298** (1988) 653;
D. A. Kosower, B.-H. Lee, and V. P. Nair, *Phys. Lett.* **201B** (1988) 85.

96. S. Parke and T. Taylor, *Phys. Lett.* **157B** (1985) 81.

97. Z. Kunszt, *Nucl. Phys.* **B271** (1986) 333.

98. S. Parke and T. Taylor, *Phys. Rev. Lett.* **56** (1986) 2459.

99. M. L. Mangano and S. J. Parke, *Phys. Rep.* **200** (1991) 301.

100. Z. Bern and D. A. Kosower, *Nucl. Phys.* **B379** (1992) 451.

101. W. Siegel, *Phys. Lett.* **84B** (1979) 193.

102. D. M. Capper, D. R T. Jones, P. van Nieuwenhuizen, *Nucl. Phys.* **B167** (1980) 479.

103. W. Siegel, *Phys. Lett.* **94B** (1980) 37.

104. G. Altarelli, G. Curci, G. Martinelli and S. Petrarca, *Nucl. Phys.* **B187** (1981) 461.

105. Z. Bern L. Dixon, and D. A. Kosower, *Phys. Rev. Lett.* **70** (1993) 2677.

106. Z. Bern L. Dixon,D. C. Dunbar and D. A. Kosower, *Nucl. Phys.* **B425** (1994) 217.

107. M. T. Grisaru and H. N. Pendleton, *Nucl. Phys.* **B124** (1977) 81.

108. Z. Kunszt, A. Signer and Z. Trócsányi *Nucl. Phys.* **B411** (1994) 397.

109. Z. Bern L. Dixon, and D. A. Kosower, SLAC-PUB-6663, hep-ph/9409393 (1994).

110. Z.Kunszt, A. Signer and Z. Trócsányi, in preparation.

111. Z.Kunszt, A. Signer and Z. Trócsányi, *Phys. Lett.* **B336** (1994) 529.

NON-PERTURBATIVE QCD ON THE CONTINUUM : SOLVING THE DYSON-SCHWINGER EQUATIONS

Andrew J. Gentles

Theoretical High-Energy Physics Group
University of Southampton
Southampton S017 1BJ
U.K.

1 INTRODUCTION

Quantum Chromodynamics (QCD) is almost universally believed to provide us with a correct description of the strong interactions between quarks and gluons. However, most of its successes have come in the realm of perturbative calculations. Only relatively recently have non-perturbative approaches such as Lattice Gauge Theory, QCD sum-rules and Dyson-Schwinger equations (DSEs) begun to provide reliable and accurate predictions for the infrared behaviour of the theory. This is an important enterprise: QCD should be able to predict correctly such quantities as decay rates and to explain the phenomena of chiral symmetry breaking and confinement.

The DSEs are a natural way to approach non-perturbative field theory. They are the equations of motion of the continuum theory, being exact relations between full n-point Green functions. Unfortunately they comprise an infinite set of coupled nonlinear integral equations which must be truncated to be solved. In contrast to Lattice QCD, the approximations made are not systematic and thus their effect is *a priori* hard to determine. However they are less numerically intensive and relate directly to the continuum. The hope is that the truncations can be controlled and improved upon to the extent that the DSE formalism can provide a complementary and computationally competitive companion to lattice methods.

We begin by outlining the derivation of DSEs and present the equations relevant to gauge theories. After briefly reviewing the huge progress made in QED, we then apply the formalism to QCD - in particular to a study of dynamical chiral symmetry breaking. Finally we examine future prospects and challenges for the approach.

Frontiers in Particle Physics: Cargèse 1994
Edited by M. Lévy *et al.*, Plenum Press, New York, 1995

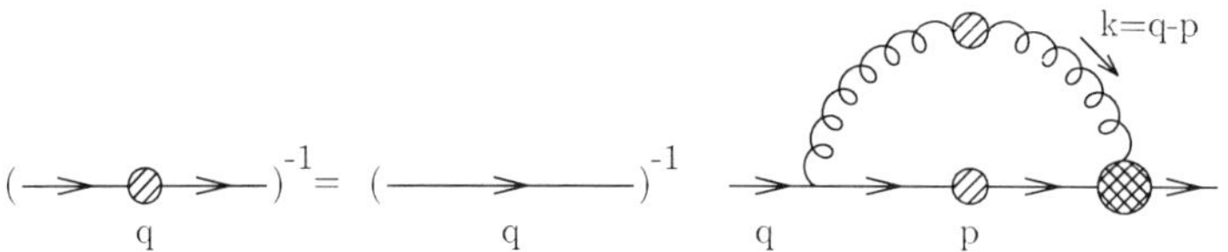

Figure 1: The Dyson-Schwinger equation for the quark propagator.

2 DERIVING THE DYSON-SCHWINGER EQUATIONS

The DSEs for a field theory can be derived in a number of ways, the simplest of which is in the context of the Path Integral formalism [1]. This involves the generalisation of Ehrenfest's theorem from real analysis to the functional integral - namely that, given suitable boundary conditions, the integral of a derivative is zero [2]. The functional derivative can be commuted with the functional integral and acts upon the exponentiated action. This can be rewritten as a functional differential equation for the generating functional of the theory. Following a Legendre transformation, taking increasing numbers of derivatives generates an infinite hierarchy of relations between n- and $(n+1)$-point proper Green functions. These are effectively the Euler-Lagrange equations of the theory - although unlike their more familiar counterparts they take the form of coupled functional differential equations.

For example, considering the integral of the derivative of the gauge field A^μ leads to an equation for the gauge-boson propagator, whilst the same technique applied to the fermion field $\bar\psi$ gives an equivalent expression for the fermion propagator. In gauge theories, especially non-Abelian ones such as QCD, the DSEs are complicated. We will concentrate on the fermion equation which relates the complete fermion propagator to the complete gauge-boson propagator and fermion-boson vertex. The diagrammatic form of this is shown in Figure 1 where hatched circles indicate full propagators and vertices.

In what follows, we will assume that the form of the gluon propagator is known. In general the gluon equation involves not only the 2- and 3-point gluon vertices, but also the 4-point gluon, fermion-ghost and fermion-gluon vertices. In QED (and in axial gauges for QCD) the situation is simpler due to the absence of ghost fields - a feature which we shall exploit later.

In order to make any progress, the fermion DSE must be closed – i.e. we must make some assumption for the fermion-gauge vertex. The effects of such approximations are the subject of the next section. A comprehensive and up-to-date review can be found in Ref. [3] and references therein.

3 THE FERMION-GAUGE BOSON VERTEX
3.1 Rainbow Approximation

The simplest possible ansatz is merely to replace the full vertex, Γ^ν by the bare one γ^ν. This is often termed the rainbow approximation. In general such truncations have unsatisfactory consequences:

- the answers obtained are not gauge invariant

- we do not satisy the requirements of multiplicative renormalisability.

It is unacceptable for physical quantities such as masses to depend on the gauge in which we do our calculations. The fact that this dependence is not small leaves us with little confidence in the whole procedure. Nevertheless, such methods have been used frequently in, for example, technicolour calculations [4].

When solving the gluon equation for QCD in covariant gauges, one also neglects the effect of ghost fields, directly violating the unitarity and gauge invariance of the theory. In defence of this there is the assertion that at the one-loop level, the contribution of ghost loops (in Landau gauge) is of order 10%. However, we cannot foretell in advance what the magnitude of their non-perturbative contributions will be.

3.2 Ball-Chiu Ansatz

In axial gauges and in QED, the Ward-Slavnov-Taylor identity between the fermion propagator and fermion-gauge boson vertex reduces to the simpler Ward-Takahashi identity (WTI). This is an exact relation between the propagator, $S_F(q)$, and vertex, $\Gamma^\nu(q,p)$:

$$k_\nu \Gamma^\nu(q,p) = S_F^{-1}(q) - S_F^{-1}(p) \cdot \tag{1}$$

It is convenient to split the vertex into longitudinal and transverse parts i.e. $\Gamma^\nu = \Gamma_L^\nu + \Gamma_T^\nu$. Ball and Chiu have shown how this can be solved for the longitudinal part of the vertex, the transverse part remaining undetermined [5]. The first step in the solution is to rewrite Γ^ν in the form

$$\Gamma_L^\nu(q,p) = \left[g^{\mu\nu} - \frac{q^\mu q^\nu}{q^2} \right] \Gamma_\mu(q,p) + \frac{q^\nu}{q^2} \left[S_F^{-1}(q) - S_F^{-1}(p) \right], \tag{2}$$

where we have added and subtracted a term proportional to $q^\mu q^\nu$ and then used the WTI to replace $q^\nu \Gamma_\nu$ by its expression in terms of the inverse propagators. This determines Γ^ν up to an unknown transverse part. To this can be added any piece Γ_T^ν which satisfies $k_\nu \Gamma_T^\nu = 0$. However there is an additional constraint as the WTI has the differential limit

$$\Gamma^\nu(p,p) = \frac{\partial S_F^{-1}(p)}{\partial p_\nu}, \tag{3}$$

which is the original Ward Identity. Not all forms of Γ_T will satisfy this – it must remove the $1/q^2$ singularities generated by its substitution into equation (2), so that

$$\Gamma_T^\nu(q,p) = \left(g^{\mu\nu} - \frac{q^\nu q^\mu}{q^2} \right) \frac{\partial S_F^{-1}(p)}{\partial p^\nu} \cdot \tag{4}$$

As an example, in massless QED with the fermion propagator $S(p) = \mathcal{F}(p^2)/\not{p}$ (where $\mathcal{F}$ is the fermionic wavefunction) a solution to (1) is

$$\Gamma_L^\nu(q,p) = \frac{1}{2} \left[\frac{1}{\mathcal{F}(q^2)} + \frac{1}{\mathcal{F}(p^2)} \right] \gamma^\nu + \frac{1}{2} \left[\frac{1}{\mathcal{F}(q^2)} - \frac{1}{\mathcal{F}(p^2)} \right] \frac{(q+p)^\nu (\not{q} + \not{p})}{p^2 - q^2} \tag{5}$$

Armed with this improved approximation to the vertex, we can proceed to re-solve the fermion equation. There have been a number of careful studies done in QED (see for example Ref.[6]). The conclusions which can be drawn from these are essentially

- still highly gauge dependent

- do not satisy multiplicative renormalisability

- very different from the rainbow approximation in the infrared

To be able to make meaningful physical statements we need to do better.

3.3 Multiplicative Renormalisability (MR)

A crucial advance in recent years has been the observation that MR powerfully constrains the transverse part of the fermion-gauge boson vertex [7]. We can calculate the one-loop corrections to the vertex in QED and, by solving the renormalization group equations in the $q^2/p^2 \to \infty$ limit, obtain the asymptotic behvaviour of the transverse vertex. In covariant gauge, with gauge parameter ξ, the RG-improved one-loop vertex in the leading logarithm approximation is

$$\Gamma^\nu_{LL}(q,p) = \gamma^\nu \left[1 - \frac{\alpha\xi}{4\pi} ln\left(\frac{q^2}{\Lambda^2}\right)\right] - \frac{\alpha}{4\pi}[\not{p}\gamma^\nu\not{q} + (\xi - 1)q^\nu\not{p}]\frac{1}{q^2} ln\left(\frac{q^2}{p^2}\right), \qquad (6)$$

where Λ is the ultraviolet cutoff and α the gauge coupling constant. If we subtract from this the solution to the WTI of equation (5) then we have the $q^2/p^2 \to \infty$ limit of Γ_T. The tensor structure can be extended and symmetrised in q and p to find an ansatz for Γ_T which automatically satisfies MR to all orders in leading *and* next-to-leading logarithms. The form suggested in Ref. [7] is

$$\Gamma^\nu_T(q,p) = \frac{1}{2}\left(\frac{1}{\mathcal{F}(q^2)} + \frac{1}{\mathcal{F}(p^2)}\right)\frac{(q^2 - p^2)\gamma^\nu - (q + p)^\nu(\not{q} - \not{p})}{d(q^2,p^2)}. \qquad (7)$$

$d(q^2,p^2)$ is not uniquely determined but should be analytic and free of kinematic singularities.

Studies in QED have indicated that the solutions to the Dyson-Schwinger equations obtained with a vertex specified by (5) and (7) are still gauge dependent, but less so - at least in covariant gauges over a wide range of gauge parameters [8].

4 THE QUARK PROPAGATOR IN QCD

Having looked briefly at investigations of QED, we now turn our attention to the question of QCD and in particular the phenomenon of dynamical chiral symmetry breaking . This is an interesting area, both in its own right and because chiral symmetry breaking is believed to be connected intimately with confinement. Although axial gauges have been used before in DSE studies [9, 10], they have been neglected recently, due to the explicit breaking of Lorentz covariance and because their use introduces kinematical singularities. However, they have the distinct advantage that the ghost fields are decoupled and the full Ward-Slavnov-Taylor identities can be replaced by the WTI. The quark-gluon vertex can then be approximated by the form used in QED.

4.1 QCD in Axial Gauge

The axial gauge is fixed by demanding that the gauge field A^μ satisfies the condition $n_\mu A^\mu = 0$ where n is some fixed four-vector. This has the unfortunate side effect of complicating the Feynman rule for the gluon propagator. The latter is also the source of the much maligned kinematical singularities, as it contains terms with $k.n$ in the denominator. In general the quark propagator is the sum of four scalar functions of momentum, F, G, H and I in the form $S(p) = (\not{p}F + G) + \not{n}(\not{p}H + I)$. However, the functions H and I can be absorbed into F and G by means of the choice $p.n = 0$. Axial gauge has a number of other subtleties which are not discussed here. The most important of these is that in performing some of the angular integrations inherent in the DSE, we are forced to take a Principal Value prescription.

4.2 DSE for the Quark Propagator

The DSE is most conveniently expressed as

$$\not{p} - \Sigma(p^2) = \not{p}\mathcal{F}(p^2) - \frac{iC_f\alpha_s}{4\pi^2} \int d^4k \, \gamma^\mu S_F(q)\Gamma^\nu(q,p)\mathcal{F}(p^2)D_{\mu\nu}(k^2), \tag{8}$$

where the full fermion propagator is now written as $S(p) = \mathcal{F}(p^2)/(\not{p} - \Sigma(p^2))$. By alternately taking the trace of equation (8) and the trace after multiplying throughout by $\not{p}$, we obtain a pair of coupled nonlinear integral equations for the functions $\mathcal{F}$ and Σ. It is reasonable to assume that the complete gluon propagator should have the same spin structure as the bare one. We can attempt therefore to parameterise it by a single scalar function $Z(k^2)$ which is determined by the DSE for the gluon propagator (modulo whatever approximations we make in order to solve it). With this in mind we write

$$D_{\mu\nu}(k^2) = -\frac{Z(k^2)}{k^2} \left(g_{\mu\nu} - \frac{k_\mu n_\nu + k_\nu n_\mu}{k.n} + n^2 \frac{k_\mu k_\nu}{(k.n)^2} \right). \tag{9}$$

Hereafter we shall specialise to $\Sigma = 0$, the case of massless fermions with no explicit chiral symmetry breaking term. Over what range of values of α_s can we obtain a solution ? Before answering this there remains the task of renormalization.

4.3 Renormalization

The renormalization of DSEs is generally very awkward. In axial gauge however, the renormalization constants Z_1 and Z_2 occuring in the QCD Lagrangian are equal and the only nontrivial renormalization is for $\mathcal{F}(p^2)$. This is achieved if we can write $\mathcal{F}(p^2) = \mathcal{F}_m\mathcal{F}_R(p^2)$, where $\mathcal{F}_R(p^2)$ is now a finite quantity. $\mathcal{F}_m$ is an infinite constant which is determined by requiring that there be some point in momentum space, μ^2, at which $\mu^2\mathcal{F}_R(\mu^2) = 1$.

After tracing the quark equation (8), we are left with an equation for $\mathcal{F}$ which reads (suppressing factors of α_s and C_F which occur in front of all of the integrals)

$$1 = \mathcal{F}(p^2) - \mathcal{F}(p^2) \int \xi_1(p^2, k^2)Z(k^2)\, dk^2 - \int \mathcal{F}(k^2)\xi_2(p^2, k^2)Z(k^2 - p^2)\, dk^2 . \tag{10}$$

The functions ξ_1 and ξ_2 are complicated polynomials in p^2 and k^2 which are generated by the angular integrations. The first of these is well behaved apart from an integrable singularity at $k^2 = p^2$. In deriving the integral equations we made the choice $p.n = 0$ and the singularity in ξ_1 therefore occurs when $k.n = 0$. The integral over ξ_2 produces a logarithmic divergence and this is what necessitates our renormalization. The procedure is straightforward - by writing equation (10) at the point $p^2 = \mu^2$ we can extract the renormalization constant $\mathcal{F}_m$ with the result that the integral equation to be solved for $\mathcal{F}_R(p^2)$ becomes

$$1 - \int \xi_1(\mu^2, k^2) Z(k^2)\, dk^2 \;=\; \mathcal{F}_R(p^2) - \mathcal{F}_R(p^2) \int \xi_1(p^2, k^2) Z(k^2)\, dk^2$$
$$- \int \left(\xi_2(p^2, k^2) - \xi_2(\mu^2, k^2) \right) \mathcal{F}_R(k^2) Z(k^2)\, dk^2 \,. \quad (11)$$

4.4 Dynamical Chiral Symmetry Breaking

There are numerous methods of solution for integral equations of the type in question [11]. The procedure we adopt is to apply a Gauss-Legendre quadrature to reduce the problem to that of solving the set of linear equations $\mathcal{K}\vec{\mathcal{F}} = \vec{s}$. Here $\mathcal{K}$ is the quadrature matrix obtained from discretising the right-hand side of (11) and $\vec{s}$ is a vector representative of the left-hand side. The quantity $\vec{\mathcal{F}}$ is a vector of values $\mathcal{F}_R(p_i^2)$ at the Gaussian abscissae p_i^2. This can be taken as the input to a Liouville-Neumann iteration sequence to produce a more accurate, smoother solution. Both of these techniques give a unique well-defined solution as long as the left-hand side of (11), which we will refer to as $\phi(p^2)$, is non-zero for all values of p^2. If $\phi^2 = 0$ at any point, the problem becomes ill-defined and no unique solution exists - in fact the function $\mathcal{F}_R(p^2)$ develops an imaginary part. At $\alpha_s = 0$ it is clear that $\phi(p^2) = 1$. As α_s increases, $\phi(p^2)$ decreases until it reaches zero, at which point the solution is lost. This leads us to conclude that there exists a critical coupling α_c above which there is no real solution to equation (11) - in other words chiral symmetry is dynamically broken. Incorporating the explicit breaking term Σ stabilises the situation as it makes an opposite contribution to $\phi(p^2)$. As α_s increases, the magnitude of $\Sigma(p^2)$ must increase in order to maintain a solution.

In their original study of the gluon sector in axial gauge, Baker, Ball and Zachariasen found a highly singular gluon propagator which behaved like $1/p^4$ in the infrared [9]. Phenomenologically one would like to have softer behaviour than this. It has been found that the nonlinearity of the gluon equation does indeed admit other solutions which are softer than that found by BBZ [12]. The result of explicitly solving our quark equation for such a propagator, with the form ($K = k^2/\mu^2$)

$$Z(K) = \frac{K}{0.88 K^{0.22} - 0.95 K^{0.86} + 0.59\, ln(2.1K + 4.1)} \quad (12)$$

is shown in Figure 2. At $\alpha_s = 0$ the wavefunction is unity, corresponding to free propagation of quarks and $\mathcal{F}_R(p^2)$ matches the RG improved perturbative result at high momentum as it must do. As expected there is a critical value $\alpha_c \approx 1.4$ above which the solution disappears and the quarks acquire a dynamically generated mass. Approaching the critical coupling from below, the solution begins to oscillate with increasing magnitude. In fact the situation is analogous to a forced oscillator with damping. The size of the damping is $\phi(p^2)$ which vanishes at α_c.

5 CONCLUSION AND OUTLOOK

We have seen that given sensible truncations of the DSE we can find solutions for the quark propagator in QCD which exhibit dynamical chiral symmetry breaking. It should be stressed of course that this conclusion is valid only in the context of the approximations made. We have not and *cannot* at this time solve the DSEs for full QCD. However, it is to be hoped that further developments will allow an increasingly close approach to this ideal situation.

An obvious next step is to approach the quark and gluon equations simultaneously to find a fully consistent solution set. Recent developments in QED have shown that

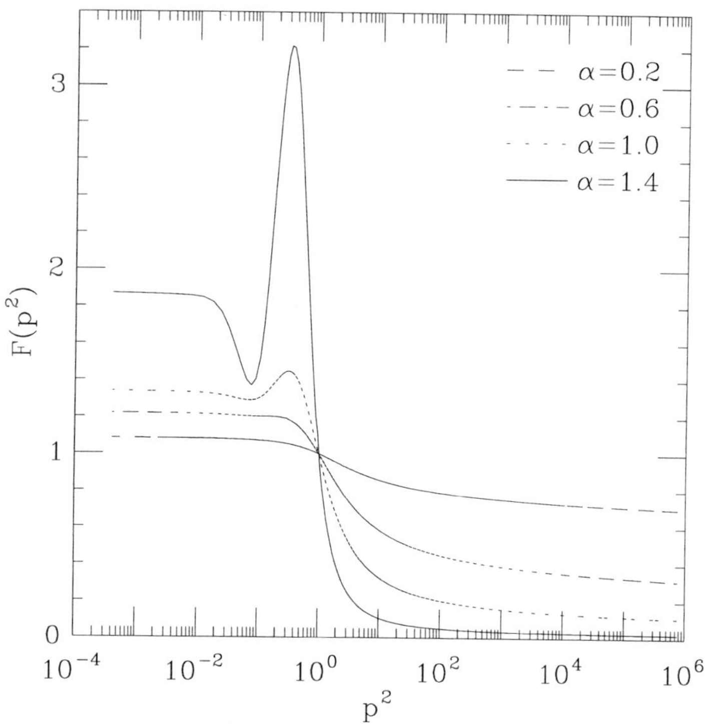

Figure 2: Behaviour of the fermion wavefunction $\mathcal{F}(p^2)$ at various values of α_s

the vertex (7) is a special case of restoring gauge invariance to the theory, superseding multiplicative renormalisability [8]. In the gluon sector, only the longitudinal part of the triple-gluon vertex is usually employed. In principle we can apply the same techniques which we used to constrain the transverse part of the quark-gluon vertex to get a more realistic equation for the gluon propagator.

Finally, a word about confinement. We have resisted the temptation to make any comment about this important subject because it is not known what behaviour of the propagators leads to the inability of quarks to propagate freely to large distances. Indeed, it seems likely that confinement is a collective property of quark-gluon dynamics, which cannot be addressed in terms of individual particles. Much work still needs to be done before we can claim to understand the mechanism responsible.

Acknowledgements

It is a pleasure to express my gratitude to Prof. D. A. Ross (University of Southampton) and Dr. J-R. Cudell (McGill University) with whom the work in section 4 was carried out [13]. I would also like to thank the University of Southampton and the U.K. Particle Physics and Astronomy Research Council for providing the funding which enable me to attend Cargèse 1994.

References

[1] C. Itzykson and J-B. Zuber. *Quantum Field Theory*, chapter 9. John Wiley and Sons, 1980.

[2] J. C. Collins. *Renormalization*, pages 13–18. Cambridge University Press, 1984.

[3] C. D. Roberts and A. G. Williams. *hep-ph 9403224, to appear in Prog. Part. Nucl. Phys.*, 1994.

[4] A. A. Kamli and D. A. Ross. *Phys. Lett.*, 255B:285–289, 1991.

[5] J. S. Ball and T-W. Chiu. *Phys. Rev. D.*, 22:2542–2549, 1980.

[6] D. C. Curtis and M. R. Pennington. *Phys. Rev. D.*, 48:4933–4939, 1993.

[7] D. C. Curtis and M. R. Pennington. *Phys. Rev. D.*, 42:4165–4169, 1990.

[8] A. Bashir and M. R. Pennington. *hep-ph 9407350 (to appear in Phys. Rev.) and this volume.*

[9] J. S. Ball M. Baker and F. Zachariasen. *Nucl Phys*, B186:531, 560, 1981.

[10] J. S. Ball and F. Zachariasen. *Phys. Lett.*, 106B:133, 1981.

[11] K. Kondo. *Integral Equations.* Oxford University Press.

[12] J-R. Cudell and D. A. Ross. *Nucl Phys*, B359:247–261, 1991.

[13] A. J. Gentles J-R. Cudell and D. A. Ross. *hep-ph 9407220, submitted to Nucl. Phys. B.*, 1994.

ON THE NEW METHOD OF COMPUTING
TWO-LOOP MASSIVE DIAGRAMS

Andrzej Czarnecki

Institut für Theoretische Teilchenphysik
Universität Karlsruhe
D-76128 Karlsruhe, Germany
e-mail: ac@ttpux2.physik.uni-karlsruhe.de

INTRODUCTION

The improving precision of experiments in the high energy physics motivates theoretical studies of quantum corrections to various processes. In the two-loop approximation this is connected with great computational difficulties, especially if there are
several mass scales involved in the process, which is a typical situation in the case
of electroweak or mixed chromodynamic and electroweak corrections. Recently a new
method has been proposed for the evaluation of scalar two-loop vertex and propagator
functions [1, 2]. It has also been shown that a similar approach works even for the four-
point functions [3]. In this talk I present a few examples which illustrate the principle
of this method.

The aim will be to obtain a double integral representation which is suitable for
numerical evaluation. In the following section I will derive it for a special case of the
vertex function with zero momentum transfer. The next section refers to the general
case of a planar vertex function with space-like values of external momenta, and the last
one shows an example of dealing with ultraviolet divergent diagrams. The examples of
two-loop functions to be considered in this paper are depicted in Fig. 1(a,b,c).

VERTEX FUNCTION AT ZERO MOMENTUM TRANSFER

While ref. [2] describes the general method of computing the two-loop vertex function, here the method is illustrated with the special case of zero-momentum transfer
which is in fact a two-point function. The principle remains the same, but the computation becomes much simpler and it is easy to write down explicit formulas. The
diagram and numbering of lines is depicted in Fig. 1(b). The four momenta in the rest

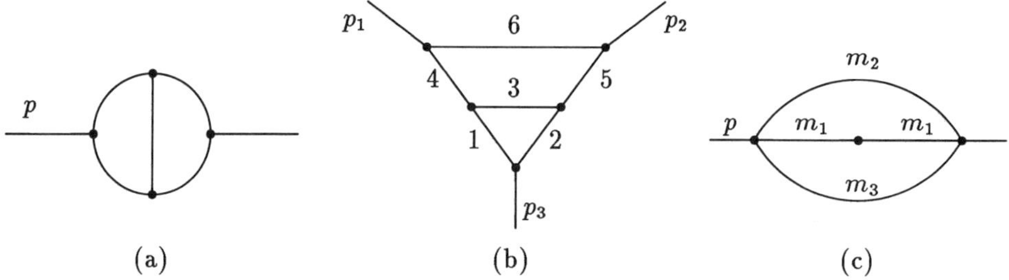

(a)   (b)   (c)

Figure 1: Examples of two-loop diagrams: a) Master two-point diagram, b) Planar vertex, c) Sunrise diagram with a zero-momentum insertion

frame of the external particle are (p_i are outgoing)

$$
\begin{aligned}
p_2^\mu &= -p_1^\mu = (q,0,0,0), \\
l^\mu &= (l_+ + l_-, l_+ - l_-, \vec{l}_\perp), \\
k^\mu &= (k_+ + k_-, k_+ - k_-, \vec{k}_\perp),
\end{aligned}
\tag{1}
$$

and the momentum k runs along the lines (4,3,5,6) and l along (1,3,2). We shall see later that one obtains different, but equivalent, formulas if one uses a different routing of the momenta. The two-loop function of Fig. 1(b) is

$$
V_0(q^2) = \iint d^4k\, d^4l \frac{1}{P_1 P_2 P_3 P_4 P_5 P_6}.
\tag{2}
$$

With the present choice of momenta, and with $s \equiv l_\perp^2$ and $t \equiv k_\perp^2$, the explicit form of the propagators is

$$
\begin{aligned}
P_{1,2} &= (2l_+ - q)(2l_- - q) - s - m_{1,2}^2 + i\eta, \\
P_3 &= 4(l_+ + k_+)(l_- + k_-) - s - t - 2\sqrt{st}\,z - m_3^2 + i\eta, \\
P_{4,5} &= (2k_+ + q)(2k_- + q) - t - m_{4,5}^2 + i\eta, \\
P_6 &= 4k_+ k_- - t - m_6^2 + i\eta.
\end{aligned}
\tag{3}
$$

There is only one propagator, P_3, through which both internal momenta flow, and z denotes the cosine of the angle between the two perpendicular momentum vectors $\vec{k}_\perp$ and $\vec{l}_\perp$. Integrations over the two angular variables describing the absolute and relative configuration of $\vec{k}_\perp$ and $\vec{l}_\perp$ can now be performed and we obtain from (2)

$$
V_0(q^2) = -4\pi^2 \int dk_+ dl_+ ds\, dt\, dk_-\, dl_- \frac{1}{\sqrt{A^2 - B^2}} \frac{1}{P_1 P_2 P_4 P_5 P_6},
\tag{4}
$$

with

$$
\begin{aligned}
A &= 4(l_+ + k_+)(l_- + k_-) - s - t - m_3^2 + i\eta, \\
B^2 &= 4st.
\end{aligned}
\tag{5}
$$

The integrations over the k_- and l_- are done with help of contour integrals. It turns out that the singularities in P_4 and P_5 do not contribute, while P_1, P_2 and P_6 contribute only for k_+ and l_+ lying in a triangular region T in the $k_+ l_+$ plane delimited by the lines

$$
\begin{aligned}
k_+ + l_+ &= 0, \\
k_+ &= 0, \\
l_+ &= \frac{q}{2}.
\end{aligned}
\tag{6}
$$

The function V_0 becomes now

$$V_0(q^2) = 2\pi^4 \iint_T \frac{dk_+ dl_+}{k_+(2l_+ - q)} \int_0^\infty dt \int_0^\infty ds$$
$$\times \left. \sum_{\{i,j\}=\{1,2\},\{2,1\}} \frac{1}{P_i P_4 P_5} \frac{1}{\sqrt{A^2 - B^2}} \right|_{k_- \to k_6, l_- \to l_j}, \tag{7}$$

where

$$k_6 = \frac{t + m_6^2}{4k_+}, \qquad l_{1,2} = \frac{s + m_{1,2}^2}{2(2l_+ - q)} + \frac{q}{2}. \tag{8}$$

Substituting the explicit formulas for the propagators one obtains

$$V_0(q^2) = \frac{8\pi^4}{q^2(m_2^2 - m_1^2)} \iint_T dk_+ dl_+ \frac{k_+}{2l_+ - q} \int_0^\infty \frac{dt}{(t + t_4)(t + t_5)}$$
$$\times \int_0^\infty ds \left(\frac{1}{\sqrt{(at + b_2 + cs)^2 - 4st}} - \frac{1}{\sqrt{(at + b_1 + cs)^2 - 4st}} \right), \tag{9}$$

with

$$a = \frac{l_+}{k_+},$$
$$b_{1,2} = 2(k_+ + l_+)\left(\frac{m_6^2}{2k_+} + \frac{m_{1,2}^2}{2l_+ - q} + q \right) - m_3^2,$$
$$c = \frac{2k_+ + q}{2l_+ - q},$$
$$t_{4,5} = \frac{2k_+}{q}\left\{ (2k_+ + q)\left(\frac{m_6^2}{2k_+} + q \right) - m_{4,5}^2 \right\}. \tag{10}$$

For the sake of simplicity let us assume that q^2 lies below all thresholds so that the function V_0 is real. In such case the integration over s is elementary

$$\int_0^\infty ds \left(\frac{1}{\sqrt{(at + b_2 + cs)^2 - 4st}} - \frac{1}{\sqrt{(at + b_1 + cs)^2 - 4st}} \right)$$
$$= \frac{1}{c} \ln \frac{t(1 - ca) - cb_2}{t(1 - ca) - cb_1}, \tag{11}$$

and in the integration over t one encounters dilogarithms,

$$\mathrm{Li}_2(x) = -\int_0^x dy \, \frac{\ln|1 - y|}{y}. \tag{12}$$

The final result is

$$V_0(q^2) = \frac{4\pi^4}{q(m_2^2 - m_1^2)(m_5^2 - m_4^2)} \int_{-q/2}^0 \frac{dk_+}{2k_+ + q} \int_{-k_+}^{q/2} dl_+$$
$$\times \left\{ \ln \frac{t_4}{t_5} \ln \frac{t_2}{t_1} + \mathrm{Li}_2\left(1 - \frac{t_5}{t_2}\right) - \mathrm{Li}_2\left(1 - \frac{t_4}{t_2}\right) - \mathrm{Li}_2\left(1 - \frac{t_5}{t_1}\right) + \mathrm{Li}_2\left(1 - \frac{t_4}{t_1}\right) \right\}, \tag{13}$$

with $t_{1,2} = -cb_{1,2}/(1 - ca)$.

The complication which arises in the case of the non-zero momentum transfer consists in the fact that the propagators $P_{1,2}$ after substitution of the appropriate value for l_- do not have the simple form $\pm(m_2^2 - m_1^2)$ but retain dependence on the variables l_+ and s. This leads to a more complicated form for the integrations over s and t, and the final formula contains dilogarithms as well as Clausen functions. There are also two additional residues which contribute and each contribution in general comes from a different triangle in the $k_+ l_+$ plane.

Since the final two integrations over k_+ and l_+ are to be done numerically it is useful to have an alternative formula which provides a cross check and a test of accuracy. Such formula can be derived by choosing the internal momenta in such way that k runs through the lines $(1, 2, 5, 6, 4)$ and l through $(1, 2, 3)$.

With this choice the propagators are

$$
\begin{aligned}
P_{1,2} &= (2k_+ + 2l_+ + q)(2k_- + 2l_- + q) - s - t - 2\sqrt{st}z - m_{1,2}^2 + i\eta, \\
P_3 &= 4l_+ l_- - s - m_3^2 + i\eta, \\
P_{4,5} &= (2k_+ + q)(2k_- + q) - t - m_{4,5}^2 + i\eta, \\
P_6 &= 4k_+ k_- - t - m_6^2 + i\eta.
\end{aligned}
\tag{14}
$$

There are now two propagators which depend on z: P_1 and P_2, but only one combination of propagators $(P_3$ and $P_6)$ whose singularities contribute to the contour integrations over k_- and l_-. The triangular region of the integration over k_+ and l_+ is now limited by the lines

$$
\begin{aligned}
k_+ + l_+ &= -\frac{q}{2}, \\
l_+ &= 0, \\
k_+ &= 0.
\end{aligned}
\tag{15}
$$

After the integration over the angular variables and over k_- and l_- we get

$$
V_i(q^2) = \frac{\pi^4}{m_2^2 - m_1^2} \int \frac{dk_+ \, dl_+}{k_+ l_+} \int_0^\infty \frac{dt}{P_4 P_5}
$$
$$
\times \int_0^\infty ds \left(\frac{1}{\sqrt{A_1'^2 - B^2}} - \frac{1}{\sqrt{A_2'^2 - B^2}} \right) \Bigg|_{k_- \to k_6, l_- \to l_3}
\tag{16}
$$

with

$$
k_6 = \frac{t + m_6^2}{4k_+}, \qquad l_3 = \frac{s + m_3^2}{4l_+},
$$
$$
A_{1,2}' = (2k_+ + 2l_+ + q)\left(\frac{t + m_6^2}{2k_+} + \frac{s + m_3^2}{2l_+} + q \right) - s - t - m_{1,2}^2.
\tag{17}
$$

The integrations over s and t proceed in exactly the same way as in the previous calculation. Finally, we can make the shift $l_+ \to l_+ - q/2$. It turns out that this change of variables not only makes the region of integration in the $k_+ l_+$ plane equal to the triangle T defined in (6), but the whole formula for V_i becomes almost the same as formula (13), the only difference being the coefficients b_i, which in the present case are

$$
b_{1,2}' = 2(k_+ + l_+) \left(\frac{m_6^2}{2k_+} + \frac{m_3^2}{2l_+ - q} + q \right) - m_{1,2}^2.
\tag{18}
$$

156

The equivalence of the two formulas can be checked after integrating over k_+ and l_+. It provides an excellent cross check for the numerical calculation.

In practical calculations one can encounter a mass configuration in which $m_1 = m_2$. In this case the formula simplifies:

$$V_0(q^2, m_1 = m_2) = -\frac{4\pi^4}{q^2(m_5^2 - m_4^2)}$$
$$\times \int_{-q/2}^0 dk_+ \int_{-k_+}^{q/2} dl_+ \frac{k_+}{k_+ + l_+} \left(\frac{1}{t_5 - t_0} \ln \frac{t_5}{t_0} - \frac{1}{t_4 - t_0} \ln \frac{t_4}{t_0} \right), \qquad (19)$$

where $t_0 = -cb_1'/(1 - ca)$.

SPACE-LIKE EXTERNAL MOMENTA

If the external momenta have space-like values the computation of the propagator and vertex diagrams is greatly simplified since the internal particles do not become on-shell. In particular we can easily check the analytical results obtained for the two-point and the planar three-point functions when all internal particles are massless. While the result for the two-point function (see Fig. 1(a)) has been know for long time [4, 5], the much more complex formulas for the vertex functions (of both planar and crossed topologies) have been obtained only very recently [6]. We present here numerical evaluation of the vertex function with all internal masses equal m and space-like external momenta. In the limit $m \to 0$ we reproduce the result of [6].

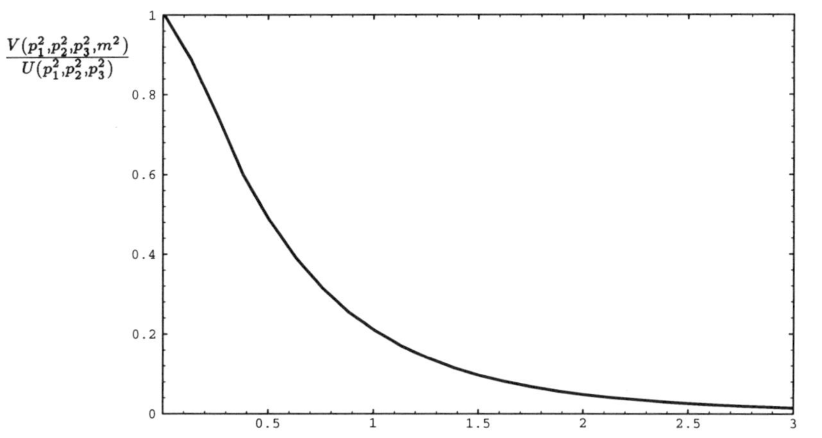

Figure 2: Vertex function for space-like external momenta $p_1^2 = -1$, $p_2^2 = -4$, $p_3^2 = -25$. All internal masses are equal m and the function is divided by the result of Ussyukina and Davydychev which corresponds to the vertex with $m = 0$.

For the numerical calculation it is convenient to choose such reference frame that the external outgoing momenta become (according to the notation of Fig. 1(b))

$$\begin{aligned}
p_1^\mu &= (e, q_1, 0, 0), \\
p_2^\mu &= (-e, q_2, 0, 0), \\
p_3^\mu &= (0, -q_1 - q_2, 0, 0).
\end{aligned} \qquad (20)$$

Repeating the calculations described in the previous section we arrive at a double integral representation which is easy to evalulate numerically. Fig. 2 shows the ratio of the vertex function

$$V(p_1^2, p_2^2, p_3^2, m^2) = \int\int \frac{d^4k d^4l}{P_1 P_2 P_3 P_4 P_5 P_6} \tag{21}$$

to the value of the vertex at zero internal masses

$$U(p_1^2, p_2^2, p_3^2) = \left(\frac{i\pi^2}{p_3^2}\right)^2 \Phi^{(2)}\left(\frac{p_1^2}{p_3^2}, \frac{p_2^2}{p_3^2}\right), \tag{22}$$

where the function $\Phi^{(2)}$ has been derived in [6]. In formula (21) P_i denote propagators defined analogously to the formula (3)). For the purpose of numerical calculation we choose one arbitrary configuration of external momenta $p_1^2 = -1$, $p_2^2 = -4$, $p_3^2 = -25$. We see that for very small masses the ratio of the two formulas becomes unity which confirms the analytical result of Ussyukina and Davydychev.

DIVERGENT INTEGRALS

The method of calculation of two-loop diagrams described here is limited to the four-dimensional space. In dealing with divergent integrals we first have to find another diagram with the same divergent part but simple enough to be computed analytically. The difference of the two diagrams can then be calculated in four dimensions and the final result is obtained by adding the analytical formula for the simpler diagram. Such procedure, based on the representation of the two-point functions proposed in [7], has been described in [8]. In the present section I illustrate an analogous procedure in the framework of the representation which works for both two- and three-point functions, with the example of the sunrise diagram with a zero momentum insertion in one of the propagators (Fig. 1(c)).

Considerable effort has been recently devoted to the investigation of this diagram. Asymptotic expansions have been derived in the papers [9, 10], and explicit expressions in terms of generalized hypergeometric, or Lauricella, functions were obtained in ref. [11]. The same diagram has also been analyzed in [12]. It has been noted in [13] that a general two-loop diagram can formally be expressed as a sunrise diagram with masses and the external momentum being functions of Feynman parameters over which one can integrate numerically. The latter reference gave a very convenient one-dimensional integral representation for this diagram.

The value of the sunrise diagram is

$$S(p^2, m_1, m_2, m_3) = (\pi e^{\gamma_E})^{2\omega} \int\int d^D k d^D l \frac{1}{P_1^2 P_2 P_3} \tag{23}$$

with γ_E being Euler's constant and

$$\begin{aligned}
P_1 &= (l + k + p)^2 - m_1^2 + i\eta, \\
P_2 &= l^2 - m_2^2 + i\eta, \\
P_3 &= k^2 - m_3^2 + i\eta, \tag{24}
\end{aligned}$$

and since the sunrise diagram is ultraviolet divergent we have to compute it in $D \equiv 4 - 2\omega$ dimensions.

It has been shown in the previous sections that the triangular regions over which one has to perform the final two integrations numerically are determined only by the values of external momenta, and are independent of masses of particles inside the diagram. Therefore it is convenient to choose for the subtraction a diagram which differs from the diagram we are interested in only by the values of internal masses. In the present case we choose a diagram with vanishing m_2 and m_3 which can be computed analytically

$$S(p^2, m_1, 0, 0) = -\pi^4 \left[\frac{1}{2\omega^2} + \frac{1}{2\omega}(1 - 2\ln m_1^2) - \frac{1}{2} + \frac{\pi^2}{4} \right.$$
$$\left. + \ln m_1^2 \left(\ln m_1^2 - 1 \right) + \mathrm{Li}_2 \left(\frac{p^2}{m_1^2} \right) + \frac{p^2 - m_1^2}{p^2} \ln \left(\frac{m_1^2 - p^2}{m_1^2} \right) \right] + O(\omega) \tag{25}$$

and the value of a diagram with arbitrary masses can be expressed by

$$S(p^2, m_1, m_2, m_3) = S(p^2, m_1, 0, 0) + \Delta(p^2, m_1, m_2, m_3) \tag{26}$$

where $\Delta(p^2, m_1, m_2, m_3) \equiv \Delta$ is free from divergences and can be computed using our method. For simplicity we only consider the case of $p^2 < m_1^2$ where both diagrams are real.

After the integration over angular variables as in (4) and over k_- and l_- with help of contour integrals we obtain

$$\Delta = \pi^4 \iint_T \frac{dk_+ dl_+}{k_+ l_+} \iint_0^\infty ds\,dt \left(\frac{A}{(A^2 - B^2)^{3/2}} - \frac{A_0}{(A_0^2 - B^2)^{3/2}} \right), \tag{27}$$

with

$$A = (2l_+ + 2k_+ + p)\left(\frac{m_2^2 + s}{2l_+} + \frac{m_3^2 + t}{2k_+} + p \right) - m_1^2 - s - t + i\eta$$
$$\equiv at + b + cs$$
$$B^2 = 4st \tag{28}$$

and the subscript 0 means that we take $m_2 = m_3 = 0$. The region of k_- and l_- integration is a triangle T delimited by the lines $k_+ = 0$, $l_+ = 0$ and $l_+ + k_+ = -p/2$. The integrations over s and t are easy

$$\int_0^\infty ds \frac{at + b + cs}{[(at + b + cs)^2 - 4st]^{3/2}} = \frac{1}{(1 - ac)t - bc} \qquad \text{for } a, b, c < 0 \tag{29}$$

and finally we arrive at

$$\Delta = -4\pi^4 \int_{-p/2}^0 dk_+ \int_{-k_+}^0 \frac{dl_+}{p(2l_+ + 2k_+ + p)} \ln \frac{p(2l_+ + 2k_+ + p) - m_1^2}{\left(p + \frac{m_2^2}{2l_+} + \frac{m_3^2}{2k_+} \right)(2l_+ + 2k_+ + p) - m_1^2}. \tag{30}$$

Thus we have found a double integral representation of the sunrise diagram. One of the k_+, l_+ integrations can still be carried out, and since the argument of the logarithm is a polynomial of the second degree the result will in general involve dilogarithms of complex arguments even below the threshold. For the purpose of numerical evaluation it may be convenient to work with a double-integral, but explicitly real representation.

ACKNOWLEDGMENTS

I thank D. Broadhurst, K.G. Chetyrkin, and A.I. Davydychev for discussion and advice, and B. Krause and M. Steinhauser for checking some of the formulas. I am very grateful to the organizers of the Cargèse Summer Institute for the opportunity to take part in this great event. I thank Graduiertenkolleg Elementarteilchenphysik at the University of Karlsruhe for support.

References

[1] D. Kreimer, Phys. Lett. **B292** (1992) 341.

[2] A. Czarnecki, U. Kilian, and D. Kreimer, *New representation of two-loop propagator and vertex functions*, hep-ph/9405423, in press in Nucl. Phys. B.

[3] D. Kreimer, *A short note on two-loop box functions*, hep-ph/9407234.

[4] J.L. Rosner, Ann. Phys. **44** (1967) 11.

[5] K.G. Chetyrkin and F.V. Tkachov, Nucl. Phys. **B192** (1981) 159.

[6] N.I. Ussyukina and A.I. Davydychev, Phys. Lett. **B298** (1993) 363; Yad. Fiz. **56** (1993) 172; Phys. Lett. **B332** (1994) 159.

[7] D. Kreimer, Phys. Lett. **B273** (1992) 277.

[8] F.A. Berends and J.B. Tausk, Nulc. Phys. **B421** (1994) 456.

[9] A.I. Davydychev and J.B. Tausk, Nucl. Phys. **B397** (1993) 123.

[10] A.I. Davydychev, V.A. Smirnov, and J.B. Tausk, Nucl. Phys. **B410** (1993) 325.

[11] F.A. Berends, M. Buza, M. Böhm, and R. Scharf, Zeit. Phys. **C63** (1994) 227. S. Bauberger, F.A. Berends, M. Böhm, and M. Buza, hep-ph/9409388.

[12] F.A. Lunev, *On evaluation of two-loop self-energy diagram with three propagators*, hep-th/9408161.

[13] A. Ghinculov and J.J. van der Bij, *Massive two-loop diagrams: the Higgs propagator*, hep-ph/9405418.

PRECISION TESTS
OF THE STANDARD MODEL

Manel Martinez

Institut de Fisica d'Altes Energies (IFAE)
Edifici Cn
Universitat Autonoma de Barcelona
E-08193 Bellaterra (Barcelona)
Spain

1 Introduction

In the last few years, High Energy Physics has advanced in an unprecedented manner towards the detailed probing of the Standard Model of electroweak interactions. A whole bunch of high precision measurements have been performed in several laboratories improving largely what was available just few years ago. The experimental accuracy reached is such that tests at the quantum level of the Electroweak theory have become possible. In this case, unlike in precision tests of QED such as the measurement of $g - 2$, the radiative corrections are the door for new physics since, radiative corrections in the electroweak theory are also sensitive to particles with masses far beyond the range of direct production. This fact makes particularly important these measurements. For instance, the analysis of the data has enabled already the inference of the elusive top quark mass (confirmed by the recent direct observation at Fermilab) and might start giving some insight into the symmetry breaking sector.

Probably the most important reason for this big improvement is the success of e^+e^- machines at the Z resonance. In the case of LEP, an impressively good performance has provided the experiments with high Luminosity and precise knowledge of the beam energy.

Concerning the LEP detectors, their adequate design and their good performance together with the clean background conditions have allowed the understanding of the data to the level of the statistical precision or better already almost from the very beginning. Typical systematic uncertainties in event selections are at the few per mille level. Nevertheless, in the last years, most of the experiments have upgraded their detectors to achieve even higher performances for some applications. In the case of SLC, substantial improvements have been achieved in increasing the the electron longitudinal polarization and calibrating it. This has allowed a complementary determination of the effective electroweak mixing angle with an accuracy which matches the most precise measurements at LEP.

A part from the introduction and the conclusions, this report is organized in three sections which correspond to three different goals:

- Recall (and try to clarify) the Standard Model language used to describe, compile and analyze precision electroweak measurements. This is done in section two.

- Give a brief description of the most relevant measurements performed so far emphasizing the main ideas behind them and their limitations and discussing the latest results. This is the subject of section three.

- Used these measurements to analyze some assumptions and extract some relevant free parameters of the theory, to test their consistency within the framework of the MSM and, assuming it, to infer the value of some of its basic Lagrangian parameters. Section four deals with this subject.

2   Standard Model language for Z physics.

In this section, we are going to give a short review of the most relevant theoretical elements to understand the physics contents and relevance of the electroweak precision measurements. Given the experimental dominance of the Z resonance observables, special attention will be given to the discussion of the theoretical predictions for Z physics. In addition, the actual theoretical language used in practice to handle these measurements will be introduced, justified and discussed in some detail. For this, a choice in the theoretical framework is mandatory. Ours, follows the way these concepts were introduced in reference [1], [1] because, for historical reasons, this is the approach which has been closer to the development of the experimental analysis, though most of the ideas discussed here have a clear parallelism in other approaches and the conclusions are equivalent [2].

2.1   The input constants: G_F.

Within the MSM, any measurement can be predicted in terms of a small set of input parameters. In the on-mass-shell renormalization scheme, this set consists of the electromagnetic coupling constant, the masses of all the particles in the theory and the Cabibbo-Kobayashi-Maskawa fermionic mixing angles. Nevertheless, to make predictions for precision measurements, it seems natural to use instead, as input parameters, the most precise quantities known so far. The two most precise measurements related to the electroweak coupling constants are

$$\alpha(0) = \frac{1}{137.0359895(61)}$$
$$G_F = 1.166389(22) \times 10^{-5} \; GeV^{-2} \tag{1}$$

the first one is the electromagnetic coupling constant measured at very low q^2 (Thompson limit) and the second one is the Fermi coupling constant, obtained from the analysis of the muon decay using the Fermi interaction language. The detailed analysis of this second input constant is important not just because it plays a crucial

---

[1] In that reference, the on-mass-shell renormalization scheme is used and the calculations are performed in the t'Hooft-Feynman gauge

role in the prediction of precision measurements but also because it illustrates the need of higher orders to analyze the present data and because it helps in introducing the meaning and properties of the main pieces of the weak radiative corrections, which are a fundamental issue of this report.

In the Fermi language, the muon lifetime τ_μ can be predicted in terms of the coupling G_F thorough the expression

$$\frac{1}{\tau_\mu} = \frac{G_F^2 m_\mu^5}{192\pi^3} \left(1 - \frac{8m_e^2}{m_\mu^2}\right) \left[1 + \frac{\alpha}{2\pi}(1 + \frac{2\alpha}{3\pi} \log \frac{m_\mu}{m_e})(\frac{25}{4} - \pi^2)\right] \tag{2}$$

where the first term comes from the quartic interaction strength, the second is a phase space correction and the third one comes from the QED corrections. Given the high accuracy of the τ_μ measurement, this equation can be considered in practice, the actual definition of G_F.

If instead of the Fermi language, one uses the Standard Model language at Born level to predict the muon lifetime measurement, then G_F is equivalent to the product of the W boson couplings times the W boson propagator at $q^2 \to 0$ (see fig. 1), namely

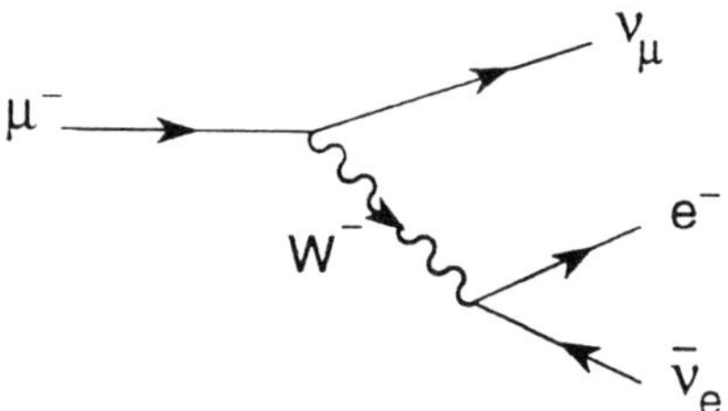

Figure 1: Feynman diagram for the muon decay at Born level.

$$G_F = \frac{\pi\alpha}{\sqrt{2}} \frac{1}{M_W^2 s_W^2} \tag{3}$$

where

$$s_W^2 \equiv 1 - \frac{M_W^2}{M_Z^2} \tag{4}$$

so that, in practice, we can still use the on-mass-shell scheme if we substitute M_W (the least known gauge boson mass) by G_F. This is conceptually what is done in most of the calculationnal approaches used in practice. Defining

$$A_0 \equiv \frac{\pi\alpha}{\sqrt{2}G_F} = (37.2802(3) \ GeV)^2 \tag{5}$$

and knowing $M_Z = 91.190$ GeV one can compute M_W

$$\begin{aligned} M_W &= \frac{M_Z}{\sqrt{2}} \left[1 + (1 - \frac{4A_0}{M_Z^2})^{1/2}\right]^{1/2} \longrightarrow 80.942 \ GeV \\ s_W^2 &= \frac{1}{2} \left[1 - (1 - \frac{4A_0}{M_Z^2})^{1/2}\right] \longrightarrow 0.2121 \end{aligned} \tag{6}$$

Nevertheless, as we shall see later, the direct data on M_W gives $M_W = 80.23 \pm 0.18$ GeV, that is, about 4 sigma off from the above prediction. If the neutrino-nucleon data is used in addition, then the best experimental determination of s_W^2 is $s_W^2 = 0.2247 \pm 0.0025$ at about 5 sigma from the above prediction. The conclusion of these comparisons is that the Born Standard Model language is not accurate enough to describe the data and therefore, since the Standard Model is a renormalizable theory, we must include higher order contributions to correct the above expressions.

At one loop, expression 3 becomes

$$G_F = \frac{\pi\alpha}{\sqrt{2}}\frac{1}{M_W^2 s_W^2}(1 + \Delta r) \tag{7}$$

being

$$\Delta r = \frac{\Re\hat{\Sigma}_W(0)}{M_W^2} + \frac{\alpha}{4\pi s_W^2}\left(6 + \frac{7 - 4s_w^2}{2s_W^2}\log c_W^2\right) \tag{8}$$

where the first term accounts for the renormalized W self energy correction while the second one corresponds to the rest of corrections (vertex and boxes). Since in the self energy corrections all kind of heavy particles may show up virtually, in practice the size of Δr depends on all the constants of the Lagrangian and, in particular, on the still unknown (or badly known) top mass and Higgs mass. From the above best experimental determination of s_W^2 one gets

$$s_W^2 = 0.2247 \pm 0.0025 \implies \Delta r = 0.042 \pm 0.008$$

and, by looking to figure 2 we can see that, for values of the top quark mass in a "reasonable" range, the prediction is in perfect agreement with the measured values. Therefore, the one loop Standard Model language describes well these measurements.

In the on-mass-shell renormalization scheme used here, the actual behaviour of Δr cannot be directly analyzed in terms of the contribution of the different unrenormalized diagrams which contribute. This simple connection gets obscured by the counterterm subtraction. Instead, Δr is customarily splitted into pieces which have different conceptual origin:

$$\Delta r = \Delta\alpha + \Delta r_W = \Delta\alpha - \frac{c_W^2}{s_W^2}\Delta\rho + \Delta r_{REM} \tag{9}$$

where the meaning and properties of the different pieces is as follows:

Photon vacuum polarization: $\Delta\alpha$

$\Delta\alpha$ describes the change in the electric charge coupling from $q^2 = 0$ to $q^2 = M_Z^2$:

$$\alpha(M_Z^2) = \frac{\alpha(0)}{1 - \Delta\alpha} \tag{10}$$

being

$$\Delta\alpha = \Pi_0^\gamma(0) - \Re\Pi_0^\gamma(M_Z^2) \tag{11}$$

which within the Standard Model is

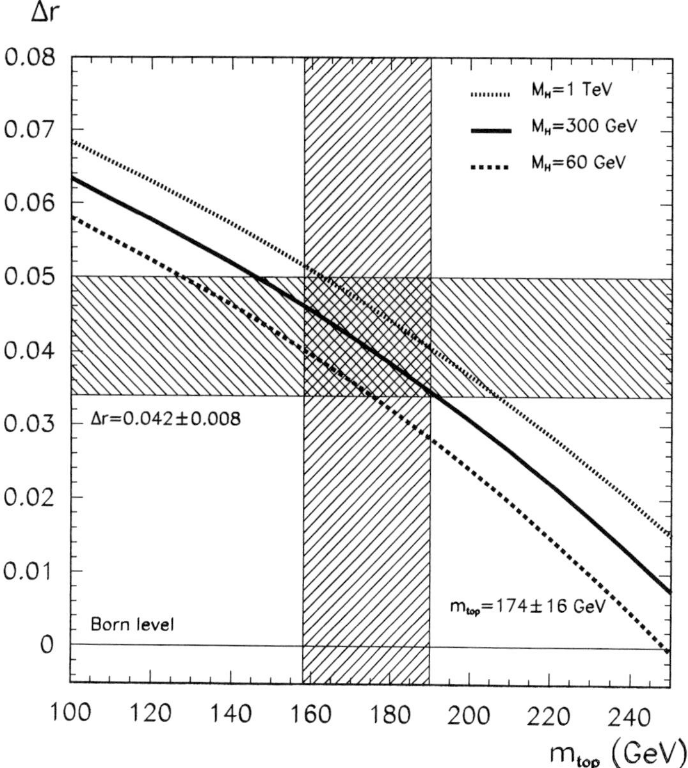

Figure 2: MSM prediction for Δr as a function of m_t for different values of M_H together with the experimental determination band $\Delta r = 0.042 \pm 0.008$ and CDF's claim $m_t = 176 \pm 16$ GeV.

$$\Delta\alpha = \frac{\alpha}{3\pi} \sum_l Q_l^2 \left(\log \frac{M_Z^2}{m_l^2} - \frac{5}{3} \right) + \Delta\alpha_{HADS} + \dots \tag{12}$$

where the first term is the contribution for charged leptons, $\Delta\alpha_{HADS}$ the contribution for quarks and the dots indicate remaining (small) bosonic contributions.

Given the expression for leptons, it is obvious that $\Delta\alpha$ is dominated by the contribution of light particles and *it remains unchanged by new physics*. In the case of quarks, since the actual masses to be used in this expression are not well determined, $\Delta\alpha_{HADS}$ is in practice computed via dispersion relations using the experimental data on hadronic e^+e^- cross sections, $R^\gamma = \sigma_{HAD}^\gamma / \sigma_\mu^\gamma$:

$$\Delta\alpha_{HADS} = -\frac{\alpha M_Z^2}{3\pi} \Re \int_{4m_\pi^2}^{\infty} ds \, \frac{R^\gamma}{s(s - M_Z^2 - i\epsilon)} \tag{13}$$

The published value obtained with this procedure [3] was so far $\Delta\alpha_{HADS} = 0.0288 \pm 0.0009$ which leads to $\alpha^{-1}(M_Z^2) = 128.79 \pm 0.12$[2] and this is the number used thorough all this review. A later update, published by F.Jegerlehner in a general review article

---

[2]This value, contains just the light fermion contributions (all fermions but the top quark). This is the standard way to refer to it.

[4], which included also Cristal Ball data was $\Delta\alpha_{HADS} = 0.0282\pm0.0009 \Rightarrow \alpha^{-1}(M_Z^2) = 128.87 \pm 0.12$.

Very recently, few new studies which claim to improve technically that procedure and include in addition all the most relevant presently available low energy data (mainly Cristal Ball), coincide in quoting a value of $\Delta\alpha_{HADS}$ significatively smaller. For instance Swartz [5] claims $\Delta\alpha_{HADS} = 0.02666 \pm 0.00075 \Rightarrow \alpha^{-1}(M_Z^2) = 129.08 \pm 0.10$ while Martin and Zeppenfeld [6] claim $\alpha^{-1}(M_Z^2) = 128.99\pm0.06$ and argue that the difference among these two numbers might be due to their different reliance on $R(QCD)$.

So far the difference between the published results and the new ones is not yet completely understood but, since this change has some relevance in several of the aspects discussed in the last part of this review, its consequences will be discussed there.

Quantum corrections to the ρ parameter: $\Delta\rho$

The ρ parameter is defined as the relation between the neutral and charged current strength at $q^2 = 0$. For a minimal Higgs sector[3], at tree level

$$\rho_0 = \frac{M_W^2}{\cos^2\theta_W M_Z^2} = 1 \tag{14}$$

but after computing one loop corrections $\rho = \rho_0 + \Delta\rho$ being

$$\Delta\rho = \frac{\Sigma^Z(0)}{M_Z^2} - \frac{\Sigma^W(0)}{M_W^2} + 2\frac{s_W}{c_W}\frac{\Sigma^{\gamma Z}(0)}{M_Z^2} \tag{15}$$

which within the Standard Model is

$$\Delta\rho = \frac{\sqrt{2}G_F}{16\pi^2} \sum_f N_{c,f}\Delta m_f^2 + \dots \tag{16}$$

being f all fermion doublets, $N_{c,f}$ their possible number of colours and $\Delta m_f^2 = |m_{f1}^2 - m_{f2}^2|$ the doublet mass splitting. $\Delta\rho$ is negligeable for light fermions but large for heavy fermions with a light iso-doublet partner. Therefore, the largest contribution is, by far:

$$\Delta\rho \sim \Delta\rho_{tb} = \frac{\sqrt{2}G_F}{16\pi^2}3m_t^2 \tag{17}$$

which amounts to about 1% for $m_t = 175$ GeV. $\Delta\rho$ is sensitive to all kind of $SU(2)_L$ multiplets which couple to gauge bosons and exhibit large mass splitting and hence, *it is very sensitive to new physics.*

Remainder corrections: Δr_{REM}

In addition to the terms included in the two previous corrections, there are other non-leading (but not negligeable) contributions for light fermions:

$$\Delta r_{REM}^f \simeq \frac{\alpha}{4\pi s_W^2}\left(1 - \frac{c_W^2}{s_W^2}\right)\frac{N_{c,f}}{6}\log c_W^2 \tag{18}$$

which yield ~ 0.0015 for leptons and ~ 0.0040 for quarks, and logarithmic terms for heavy fermions:

---

[3]This also applies to a Higgs sector with any number of Higgs doublets

$$\Delta r_{REM}^t \simeq \frac{\sqrt{2}G_F M_W^2}{16\pi^2}\left\{2\left(\frac{c_W^2}{s_W^2}-\frac{1}{3}\right)\log\frac{m_t^2}{M_W^2}+\dots\right\} \tag{19}$$

In Δr at one loop the leading Higgs contribution is logarithmic due to the accidental $SU(2)_R$ symmetry of the Higgs sector in the MSM which implies $\rho_0 = 1$ (Veltman screening).The leading contribution (provided that $M_H >> M_W$) is:

$$\Delta r^{Higgs} \simeq \frac{\sqrt{2}G_F M_W^2}{16\pi^2}\left\{\frac{11}{3}\left(\log\frac{M_H^2}{M_W^2}-\frac{5}{6}\right)\right\} \tag{20}$$

The structure discussed in this section for Δr is similar to the one that one can observe in the corrections to the different precision measurements that we will discuss later. All these corrections can be decomposed into a piece which can be identified as $\Delta\alpha$ with some coefficient in front, another which is $\Delta\rho$ with a different coefficient and the remainder which accounts for the rest. We will discuss this decomposition for the Z observables in a forecoming section. In the case of Δr, table 1 summarizes the basic features discussed in this section.

Table 1: Summary of the leading dependences on m_t and M_H of the different contributions to Δr.

	$\Delta\alpha$	$\Delta\rho$	Δr_{REM}	Δr
m_t	-	$\sim m_t^2$	$\sim \log m_t$	$\sim m_t^2$
M_H	-	$\sim \log M_H$	$\sim \log M_H$	$\sim \log M_H$

2.2 $\quad e^+e^- \to f\bar{f}$ beyond tree level.

In the MSM, at tree level the process $e^+e^- \to f\bar{f}$, $f \neq e$ is described by the sum of two s-channel amplitudes [4] (see fig. 3) For photon exchange it can be written as:

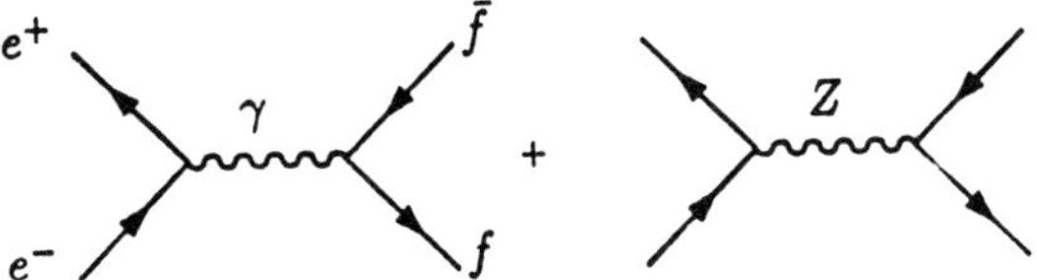

Figure 3: Tree level diagrams for the process $e^+e^- \to f\bar{f}$, $f \neq e$.

$$\mathcal{A}_\gamma = e^2\frac{Q_e Q_f}{s}\gamma_\mu \otimes \gamma^\mu \tag{21}$$

---

[4]The Higgs exchange contribution, though also possible at tree level, is customarily forgotten because of its tiny size due to the low mass of the possible final state fermions.

and for Z exchange as:

$$
\begin{aligned}
\mathcal{A}_Z &= \frac{e^2}{4s_W^2 c_W^2}\frac{\left[\gamma_\mu(I_3^e - 2Q_e s_W^2) - \gamma_\mu\gamma_5 I_3^e\right] \otimes \left[\gamma^\mu(I_3^f - 2Q_f s_W^2) - \gamma^\mu\gamma_5 I_3^f\right]}{s - M_Z^2} \\
&= e^2\frac{\left[\gamma_\mu(v_e - a_e\gamma_5)\right] \otimes \left[\gamma^\mu(v_f - a_f\gamma_5)\right]}{s - M_Z^2}
\end{aligned}
\tag{22}
$$

$$
\begin{aligned}
v_f &\equiv \frac{1}{2s_W c_W}(I_3^f - 2Q_f s_W^2) \\
a_f &\equiv \frac{1}{2s_W c_W}I_3^f
\end{aligned}
\tag{23}
$$

where a short notation for bilinear spinor combination is used:

$$
A_\mu \otimes B^\mu = [\bar{v}_e A_\mu u_e] \cdot [\bar{u}_f B^\mu v_f].
\tag{24}
$$

and the charge and weak isospin assignment are given in table 2.

Table 2: Charge and isospin assignment for the possible final state fermion types.

	ν	e	u	d
Q_f	0	-1	2/3	-1/3
I_3^f	1/2	-1/2	1/2	-1/2

Alternatively, if the tree level G_F relation

$$
\frac{e^2}{4s_W^2 c_W^2} = \sqrt{2}G_F M_Z^2
\tag{25}
$$

is used, then the Z exchange amplitude can be written as

$$
\mathcal{A}_Z = \sqrt{2}G_F M_Z^2\frac{\left[\gamma_\mu(g_{v_e} - g_{a_e}\gamma_5)\right] \otimes \left[\gamma^\mu(g_{v_f} - g_{a_f}\gamma_5)\right]}{s - M_Z^2}
\tag{26}
$$

$$
\begin{aligned}
g_{v_f} &\equiv (I_3^f - 2Q_f s_W^2) \\
g_{a_f} &\equiv I_3^f
\end{aligned}
\tag{27}
$$

Like in the previous section, the accurate description of the precision measurements requires dressing these amplitudes with higher order contributions. At one loop, these contributions can be classified into two groups: photonic and non-photonic corrections. The first one includes all contributions in which a photon line is added to the born diagrams and the second group includes the rest. This separation is specially important for neutral current processes in which the non-photonic corrections at one-loop level separate naturally from the photonic ones forming a gauge-invariant subset. This fact enables the separate study of these corrections which, as we shall see have very different properties and relevance in the present discussion.

2.2.1 Photonic corrections.

The photonic corrections near the Z pole are very large for many of the precision observables. As we shall see, they distort noticeably the shape of their energy dependence and hence their size depends strongly on the actual energy. In addition, given the presence of real photon emission, they depend also strongly on the experimental cuts applied to analyze the data and therefore, their detailed evaluation is linked to the specific experimental analysis used.

Nevertheless, the inclusion of photon lines does not add more physics than just QED and therefore, the physics interest of photonic corrections is rather limited. In general, the strategy applied to deal with these correction consists in unfolding them as accurately as possible from the observed measurements to recover the non-photonic measurements. In this short review we will limit ourselves to recall the most important conceptual features of photonic corrections for the analysis of precision measurements.

For s-channel lineshapes, at one loop, photonic correction can be classified into three infrared-finite gauge-invariant set of diagrams:

- Final state radiation (FSR). contribution from diagrams in which a photonic line is attached to the final state fermion line (see fig.4a). In this set of contributions, the infrared divergence which shows up in the real photon emission when the photon energy vanishes, cancels the infrared divergence present in the interference between the Born amplitude and the one in which a virtual photon is attached to the final state vertex (vertex correction). In the total cross section and the forward-backward asymmetry, if just lose detection cuts are applied, the correction is

$$\delta_{FSR} = \frac{3\alpha(s)}{4\pi} Q_f^2$$

  (positive for the cross section and negative for the forward-backward asymmetry) which for leptons amounts to just $\sim 0.17\%$.

- Initial-Final state interference. Contribution from the interference between the diagrams in which a real photonic line is attached either to the initial or the final fermionic lines (see fig.4b). In this set of contributions, as before, the infrared divergence which shows up in the real photon emission when the photon energy vanishes, cancels the infrared divergence present in the interference between the Born amplitude and the one in which a virtual photon links the initial and the final state fermion lines (box correction). This contribution, unlike the previous one, depends on $\cos\theta$ and its analytic form is rather involved. Anyway, to give a feeling on its size, for instance, in the cross section for hadrons, if lose detection cuts are applied, the correction amounts to just $\sim 0.02\%$

- Initial state radiation (ISR). Contribution from diagrams in which a photonic line is attached to the initial state fermion line (see fig.4c). In this set of contributions, there is a cancellation among infrared divergences like in the case of the final state corrections. These corrections near the Z pole are very large and of paramount importance for the precision measurements and therefore we will concentrate in their discussion in the following.

Two understand why ISR corrections are so important and to handle them, the physical picture of structure functions results very useful. In that picture, the colliding electrons are though as composite objects inside which, parton electrons are dressed

final state corrections:

real photons

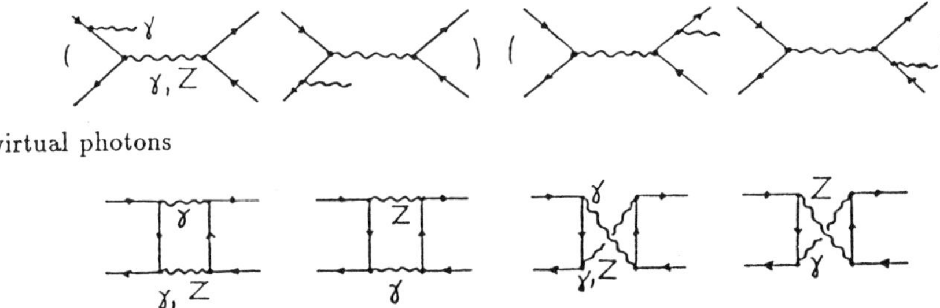

virtual photons

initial-final state interference corrections:

real photons

virtual photons

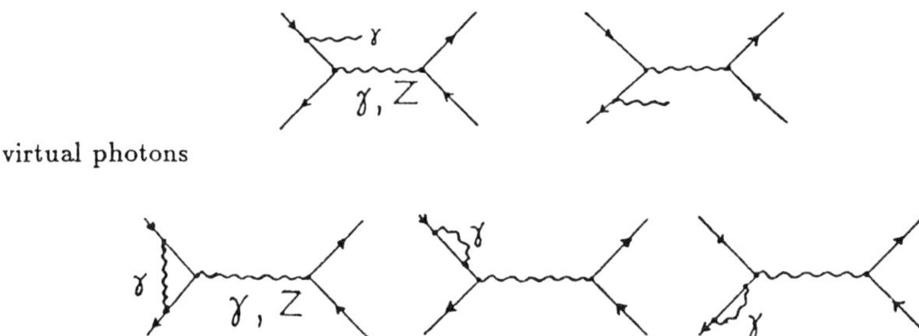

initial state corrections:

real photons

virtual photons

Figure 4: Photonic corrections to the $e^+e^- \to f\bar{f}$, $f \neq e$ process.

by photons with which they share the beam's momentum. These partons collide in a "hard scattering" which is described by the amplitudes without photonic corrections. By using this simple picture, two important effects of ISR can be easily understood, namely:

1. The Z resonance acts as a natural photon energy cut-off which decreases drastically the cross section. Out of all the beam's collisions, only those in which the actual parton energies is close enough to the Z mass will produce a "hard

scattering" event. Therefore, the Z resonance acts as a monocromator of the parton energy since only those with the right energy annihilate into a Z boson. Therefore, the existence of ISR decreases drastically the actual probability of Z production (cross section). In fact, at $\mathcal{O}(\alpha)$

$$\sigma(s) \simeq \sigma_{NP}(1 + \delta_1 + \beta \log x_M)$$

where β is the coefficient of the infrared term and acts as the actual coupling strength of photonic radiation if there is any energy cut-off

$$\beta = \frac{2\alpha}{\pi}(\log \frac{s}{m_e^2} - 1) \sim 0.11$$

δ_1 the non-infrared part

$$\delta_1 = \frac{3}{4}\beta + \frac{\alpha}{\pi}(\frac{\pi^2}{3} - \frac{1}{2}) \sim 0.09$$

and x_M the maximum photonic energy scaled to the beam energy. Near the Z pole the maximum photonic energy is limited to about a Z width and so

$$x_M \sim \frac{\Gamma_Z}{M_Z} \implies \beta \log x_M \simeq -0.40$$

so that the infrared term dominates by far and the final total correction is of about 30%.

2. The hard photon emission shifts the actual "hard scattering" energy and hence distorts noticeably the energy dependence of the observables. The effective "hard scattering" energy is $s' \leq s$ due to the energy carried by the hard photons. For instance, the shift in the resonance peak position is

$$\Delta M_Z \sim \frac{\beta\pi}{8}\Gamma_Z \simeq 110 MeV$$

Above M_Z the energy shift is large ("radiative return" to the Z peak) whereas below M_Z the energy shift is small and soft photon emission dominates. This results in an asymmetric distortion of the energy dependence.

Given their large size, the pure one loop calculation of ISR corrections is clearly insufficient to match the experimental precision. Therefore, the calculation of the two loop terms as well as the study of the procedure to resum to all orders the infrared contributions (exponentiation) were attacked before LEP started operation. The outcome of this work was that the photonic corrections to the e^+e^- annihilation near the Z pole are very accurately known [7]. Several approaches to handle higher orders, based on different physical pictures and different technical implementations (inductive exponentiation, structure functions, YFS,...) have been developed and their results compare well [7]. Figure 5 illustrates the effects of ISR corrections in the different Z observables that we will discuss in this report.

In practice, following the picture of the structure function approach, the ISR corrections to the cross sections are accurately taken into account by convoluting a photon energy structure function with the "hard scattering" cross sections: [5]

---

[5]The index A denotes the conditions in which the cross section is computed. For the observables discussed in this report, $A = $ forward, backward, left, right or total.

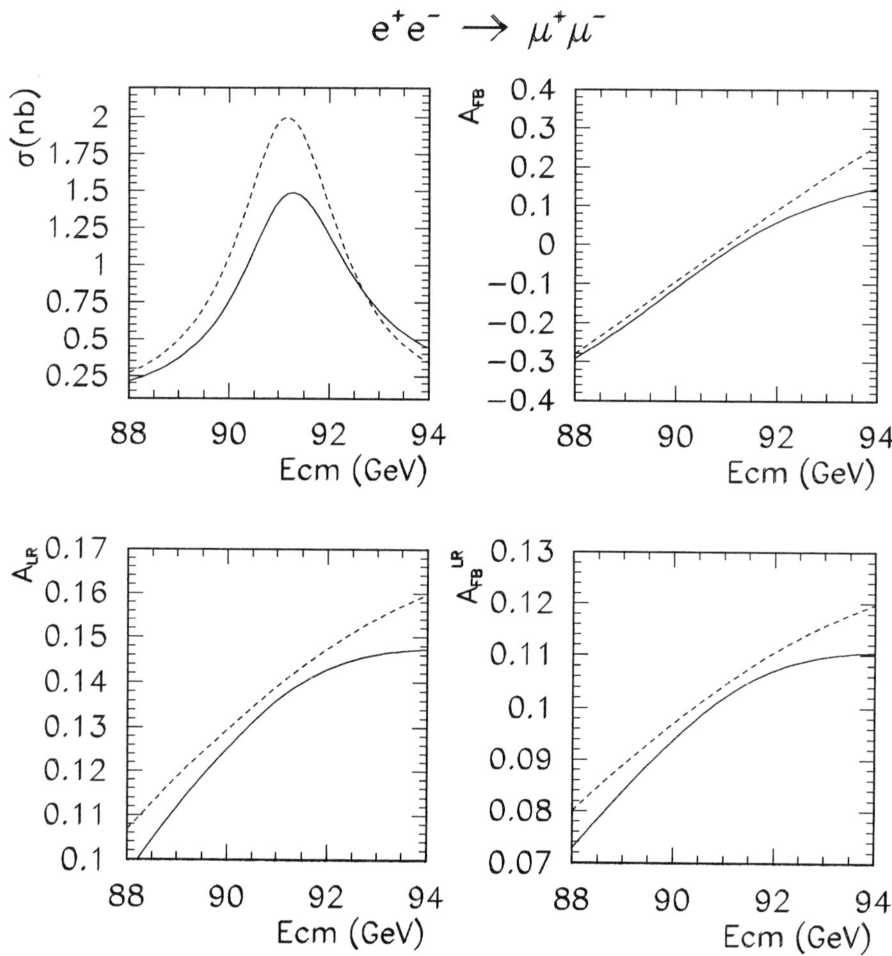

Figure 5: Distortion due to ISR corrections on the energy dependence of some of the main Z observables discussed in this report. The dashed line is the ISR unfolded prediction whereas the solid line is the complete one.

$$\sigma_A(s) = \int_{4m_f^2}^{s} ds' H_A(s,s') \hat{\sigma}_A(s') \tag{28}$$

where $H_A(s,s')$ is the so-called radiator function which typically is computed up to $\mathcal{O}(\alpha^2)$ and includes soft photon exponentiation and $\hat{\sigma}$, the so-called reduced cross section is the Born cross section dressed by non-photonic corrections.

2.2.2 Non-photonic corrections.

Non-photonic corrections are also large but, in contrast with photonic ones, they do not depend on experimental cuts. On top of that, they contain relevant information on the non-energetically available elements of the theory, so they are the ones that allow the detailed test of the quantum structure of the Standard Model and the search for new physics by performing precision measurements

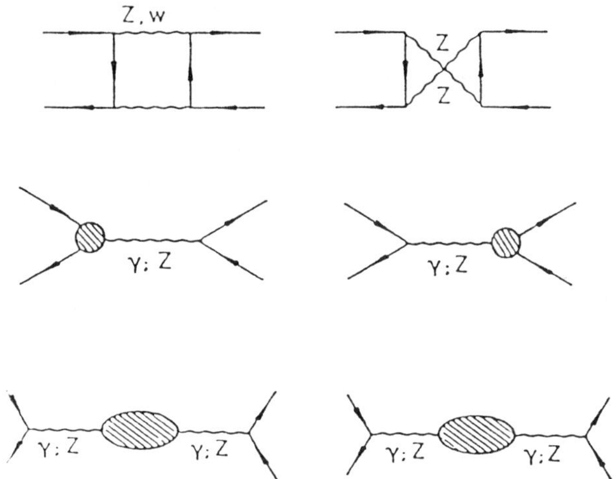

Figure 6: Non-photonic corrections to the $e^+e^- \to f\bar{f}$, $f \neq e$ process.

At one loop, these contributions can be classified into three types, namely box corrections, vertex corrections and vacuum polarization corrections, (see fig. 6) in such a way that one can just write

$$\hat{\sigma}(s) = \sigma_0(1 + \Delta_{BOX}(s) + \Delta_{VERTEX}(s) + \Delta_{VACUUM}(s)) \tag{29}$$

The box corrections are very small [6] near the Z peak due to their non-resonant structure. Their size depends on θ and therefore they depend on the the actual observable studied but their typical order of magnitude is $\Delta_{BOX}(s) \leq 0.1\%$. The vertex corrections are small but not negligeable. For all fermions (but the b quark) their typical size is of $\Delta_{VERTEX} \leq 1\%$. For the b quark, the situation is different: the correction can be rather larger depending on the actual value of the top mass, its isospin partner, which directly enters the vertex contributions (see fig.7) giving a correction of the form

$$\Delta_{VERTEX}^b = -\frac{20}{13}\frac{\alpha}{\pi}\left(\frac{m_t^2}{M_Z^2} + \frac{13}{6}\log\frac{m_t^2}{M_Z^2}\right) + \ldots \tag{30}$$

so that, for instance, for $m_t = 175$ GeV then $\Delta_{VERTEX}^b \simeq 2.4\%$.

Finally, the vacuum polarization corrections (also called oblique corrections, propagator corrections and self-energies) are the largest ones (typically $\Delta_{VACUUM} \leq 10\%$) and will be discussed in some more detail in the following.

It has been shown by several groups that in four fermion processes, the matrix element squared including non-photonic corrections can be rewritten keeping a Born-like structure by defining running effective complex parameters [8, 9, 10].

---

[6]Since these three types of contributions do not constitute gauge-invariant subsets, their size can be different for different gauge choices. In our discussion, the t'Hooft-Feynman gauge as been taken.

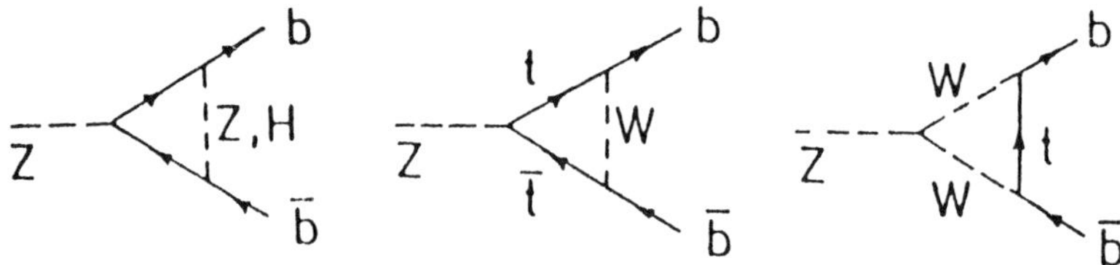

Figure 7: Vertex corrections for Z decay into b quarks.

Initial-final state factorizable corrections such as self-energies and vertex corrections can be easily absorbed by redefining the Born couplings as we shall see. Concerning non-factorizable corrections such as boxes, two different approaches exist:

- Absorbing them also in the definition of the effective parameters [10]. The price to pay is that some effective parameters become not just a function of s but also of $\cos\theta$ and, in addition, the Born-like structure is somewhat spoiled by the presence of effective parameters which do not show up in the pure Born approach.

- Keeping them out from the definition of the effective parameters [9]. They must be included as explicit corrections afterwards. This approach has the advantage of being very simple and producing a set of effective parameters which depend only on s and have a clear Born interpretation. Nevertheless, the price to pay is that, in this case, the effective parameters are defined in a gauge non-invariant way so that attention should be paid to the gauge choice. [7].

For the sake of simplicity we will follow the second approach to define the meaning of the effective couplings.

Universal effective parameters.

In the 't Hooft-Feynman gauge the dominant corrections are by far, the vector boson vacuum polarizations. Since these corrections do not depend of the species of the external fermions they are, in fact, universal (process independent). By applying Dyson equations, the one-loop fermionic self-energies can be resumed to all orders dressing the propagators. The three 1-loop one-particle-irreducible (1PI) dressed propagators showing up in Z processes, can be absorbed in three universal parameters [9]:

- the photon renormalized self energy $\Pi^\gamma(s)$ is absorbed in an effective coupling constant $\alpha(s)$, defined as:

$$\bar{\alpha}(s) = \frac{\alpha_0}{1 + \Pi^\gamma(s)} \tag{31}$$

---

[7]For a while, this observation prevented the theoretical community from accepting the usefulness of the effective parameter approach which, at the end, has been the one chosen by the experimental community to perform the measurements. In fact, if the predictions are computed in a gauge in which the non-absorbed corrections are numerically irrelevant (as is the case for the 't Hooft-Feynman gauge) the calculations using these effective parameters produce, in fact, numerically gauge-invariant results

It is important to stress that, since $\Pi^\gamma(s)$ is a complex function, so is $\bar{\alpha}(s)$. Nevertheless, since the imaginary part of $\bar{\alpha}(s)$ is small compared with the real one, its main effect happens in observables which are sensitive to phase differences between photon and Z exchange diagrams such as the forward-backward charge asymmetry.

- the way the γZ mixing is treated is slightly more complicated. Since the γZ mixing does not show up at tree level, to keep the Born structure, it must be absorbed in the neutral current coupling parameters. Then the neutral current is redefined as

$$C_\mu^Z = \gamma_\mu(v_f - a_f\gamma_5) + \gamma_\mu Q_f \frac{\Pi_{\gamma Z}(s)}{1 + \Pi^\gamma(s)} = \frac{1}{2s_W c_W}\left[\gamma_\mu(I_3^f - 2Q_f \bar{s}_W^2(s)) - I_3^f \gamma_\mu \gamma_5\right]$$

being

$$\bar{s}_W^2(s) = s_W^2(1 + \Delta\kappa(s)) \tag{32}$$

the complex universal effective weak mixing angle, where

$$\Delta\kappa(s) = -\frac{c_W}{s_W}\frac{\Pi^{\gamma Z}(s)}{1 + \Pi^\gamma(s)} \tag{33}$$

- finally, the Z self energy has to be absorbed into a third parameter. The problem is that we have already used the only two tree level parameter (α and s_W) to absorb vacuum polarization corrections. The way out is the following: lets first consider the corrected Z propagator with the overall coupling constant that will come from the initial and final state particle couplings

$$\frac{e^2}{4s_W^2 c_W^2}\frac{1}{s - M_Z^2 + \Re(\Sigma_Z(s)) + i\Im(\Sigma_Z(s))} = \frac{e^2}{4s_W^2 c_W^2}\frac{1}{1 + \Pi^Z(s)}\frac{1}{s - M_Z^2 + i\frac{\Im(\Sigma_Z(s))}{1+\Pi^Z(s)}}$$

where

$$\Pi^Z(s) = \frac{\Re(\Sigma_Z(s))}{s - M_Z^2}$$

Therefore, the factor $\frac{e^2}{4s_W^2 c_W^2}\frac{1}{1+\Pi^Z(s)}$ which multiplies the Breit-Wigner like propagator, can be considered as the effective strength of the purely weak interactions and an appropriate way of introducing a running parameter to account for it is recalling the tree level G_F relation:

$$\frac{e^2}{4s_W^2 c_W^2} = \sqrt{2}G_F M_Z^2(1 - \Delta r)\rho_0$$

where ρ_0 is the tree level ρ parameter which in the Minimal Standard Model is exactly 1. Therefore we can write

$$\frac{e^2}{4s_W^2 c_W^2}\frac{1}{1+\Pi^Z(s)} = \sqrt{2}G_F M_Z^2 \rho(s)$$

being

$$\bar{\rho}(s) = \rho_0(1 + \Delta\rho(s)) \tag{34}$$

the universal effective ρ parameter, where

$$\Delta\rho(s) = \frac{1-\Delta r}{1+\Pi^Z(s)} - 1 \tag{35}$$

$\Delta\rho(s)$ and hence $\bar{\rho}(s)$ are real quantities by definition, since the imaginary part of the Z vacuum polarization will be treated separately. It is important to stress that this $\Delta\rho$ is numerically and conceptually different from the one introduced when discussing Δr. The one introduced there accounted for the ratio of W to Z vacuum polarization corrections at $q^2 \cong 0$, whereas the one introduced now is more complex since accounts for the ratio of the whole Δr correction (which in spite that is dominated by the W vacuum polarization at $q^2 \cong 0$, includes also sizable QED corrections) to the real part of the derivative of the Z vacuum polarization at $q^2 = s$. Nevertheless, the coefficient of the dominant m_t terms is the same.

Finally, it can be shown [9] that the imaginary part of the Z self energy can be interpreted, through the use of the optical theorem as

$$\frac{\Im(\Sigma_Z(s))}{1+\Pi^Z(s)} = \bar{\Gamma}_Z(s) \cong \frac{s}{M_Z}\bar{\Gamma}_Z(M_Z^2)$$

where $\bar{\Gamma}_Z(s)$ stands for the Born total Z decay width in terms of effective couplings.

Summarizing, for Z physics, the vacuum polarization corrections can be properly included by writing the Born amplitudes in terms of the effective running couplings $\bar{\alpha}(s)$, $\bar{\rho}(s)$ and $\bar{s}_W^2(s)$ as:

$$\begin{aligned}
\bar{A} =\ & Q_e Q_f \frac{4\pi\bar{\alpha}(s)}{s}[\gamma_\mu \otimes \gamma^\mu] + \sqrt{2}G_F M_Z^2 \bar{\rho}(s)\frac{1}{s - M_Z^2 + is\frac{\bar{\Gamma}_Z}{M_Z}} \\
& [\gamma_\mu(I_3^e - 2Q_e\bar{s}_W^2(s)) - \gamma_\mu\gamma_5 I_3^e] \otimes [\gamma^\mu(I_3^f - 2Q_f\bar{s}_W^2(s)) - \gamma^\mu\gamma_5 I_3^f]
\end{aligned}$$

This representation of the amplitudes is accurate for the calculation of any observable [8] at the percent level. It is worth mentioning also that at this level, the definition of the effective amplitudes in the most popular electroweak libraries [11, 20] conceptually agree.

---

[8] Exception made of the observables for b quarks for which, as pointed out before, vertex corrections play an important role

Flavour-dependent effective parameters.

The fact that the accuracy for some Z observables reaches the permile level requires the consideration of the next level of corrections, namely the weak vertex ones. As we have seen, in the case of b-quark final state, these corrections have a quadratic dependence on the top mass and exceed the percent level for realistic m_t, so they become as relevant as the vacuum polarization ones.

These corrections, unlike the vacuum polarization ones, depend explicitly on the species of the external fermions and therefore are flavour dependent. Hence, they can be absorbed into effective parameters at the price of making them flavour dependent, that is, having a set of effective parameters for every fermion species.

As of the photon exchange amplitude, the current including vertex corrections at one loop can be written as:

$$C_\mu^\gamma = \left[\gamma_\mu \left(Q^f + F_{V\gamma f}(s) - F_{A\gamma f}(s)\gamma_5 \right) \right]$$

where $F_{V\gamma f}$ and $F_{A\gamma f}$ are the complex vector and axial photon formfactors (see for instance [9]).

In the case of the weak vertex corrections for the Z amplitudes, at one loop we can introduce them through the use of complex form factors modifying the Born axial and vector couplings in the currents [9]:

$$C_\mu^Z = \left[\gamma_\mu \left(v_f + F_{VZf}(s) \right) - (a_f + F_{AZf}(s))\gamma_5) + \gamma_\mu Q_f \frac{\Pi_{\gamma Z}(s)}{1 + \Pi^\gamma(s)} \right]$$

where $F_{V\gamma Z}$ and $F_{A\gamma Z}$ are the complex vector and axial Z formfactors and the γZ mixing correction has also been included explicitly. This can be rewritten as

$$\left(1 + \frac{F_{AZf}(s)}{a_f} \right) \frac{1}{2 s_W c_W} [\gamma_\mu (I_3^f - 2Q_f \sin^2 \theta_{eff}^f(s)) - I_3^f \gamma_\mu \gamma_5]$$

being

$$\sin^2 \theta_{eff}^f(s) = s_W^2 (1 + \Delta\kappa(s) + \Delta\kappa_f(s)) \tag{36}$$

the flavour-dependent complex effective weak mixing angle, where

$$\Delta\kappa_f(s) = -\frac{1}{Q_f} \frac{c_W}{s_W} \left(F_{VZf}(s) - \frac{v_f}{a_f} F_{AZf}(s) \right) \tag{37}$$

is the flavour-dependent vertex correction.

After this algebra, the effective strength of the purely weak interactions becomes

$$\frac{e^2}{4 s_W^2 c_W^2} \frac{1}{1 + \Pi^Z(s)} \left(1 + \frac{F_{AZe}(s)}{a_e} \right) \left(1 + \frac{F_{AZf}(s)}{a_f} \right) = \sqrt{2} G_F M_Z^2 (\rho_e(s)\rho_f(s))^{\frac{1}{2}}$$

being

$$\rho_f(s) = \rho_0 (1 + \Delta\rho(s) + \Delta\rho_f(s)) \tag{38}$$

the flavour-dependent complex effective ρ parameter, where

$$\Delta\rho_f(s) = \left(1 + \frac{F_{AZf}(s)}{a_f} \right)^2 - 1 \tag{39}$$

is the flavour-dependent complex vertex correction.

With the introduction of these complex flavour-dependent effective parameters, the Z exchange amplitude can simply be written as:

$$\mathcal{A}_Z = \sqrt{2}G_F M_Z^2 \frac{1}{s - M_Z^2 + is\frac{\Gamma_Z}{M_Z}}[\gamma_\mu(g_{V_e}(s) - g_{A_e}(s)\gamma_5)] \otimes [\gamma^\mu(g_{V_f}(s) - g_{A_f}(s)\gamma_5)] \quad (40)$$

where the complex effective vector and axial couplings are defined as [9]

$$\begin{aligned}
g_{V_f}(s) &= \sqrt{\rho_f}(I_3^f - 2Q_f \sin^2\theta_{eff}^f) \\
g_{A_f}(s) &= \sqrt{\rho_f}I_3^f
\end{aligned} \quad (41)$$

2.3   Leading behaviour of E.W. corrections.

We have seen that beyond tree level, the dominant non-photonic corrections where the vacuum polarizations, which where connected to the improved Born parameters in the following way:

$$\begin{aligned}
\Pi_\gamma &\to \Delta\alpha \\
\Pi_W &\to \Delta r \simeq \Delta\alpha - \frac{c_W^2}{s_W^2}\Delta\rho + \ldots \\
\Pi_{\gamma Z} &\to \bar{s}_W^2(s) \simeq s_W^2 + c_W^2\Delta\rho + \ldots \\
\Pi_Z &\to \bar{\rho}(s) \simeq 1 + \Delta\rho + \ldots
\end{aligned} \quad (42)$$

$$(43)$$

where also their leading behavior in the basic pieces discussed in the section on G_F,is shown. Therefore, all the E.W. vacuum polarizations in the MSM (but $\Delta\alpha$) have a leading sensitivity to heavy particles given by $\Delta\rho$. Therefore, all what we can extract from our precision E.W. measurements is always basically the combination of m_t and M_H which shows up in $\Delta\rho$ and hence, there is no possible disentangling of m_t and M_H in leading corrections in vacuum polarizations.

Concerning the vertex corrections, as we have seen they are subdominant and flavour-dependent. In this corrections the dependence on M_H is negligeable (due to the fact that the external fermions are light) and the dependence on m_t is only relevant in the vertex correction for $Z \to b\bar{b}$, ΔV_b. Therefore, measuring Γ_b could help in disentangling m_t from M_H in E.W. corrections.

So, as we see in any case, there are four relevant (non-trivial) loop contributions entering the precision E.W. measurements.

In models other than the MSM the $\Delta\rho$ entering Π_W, $\Pi_{\gamma Z}$ and Π_Z could get different contributions from hidden (heavy) physics. Therefore, it is desirable having a language allowing to extract these informations from the data as if they could, indeed, behave in a completely independently way. This allows:

- checking for physics beyond the S.M. in precision measurements and

---

[9]Note the use of uppercase V and A subindices to distinguish the effective couplings from the Born ones.

- compiling easily several measurements to check for consistency within the S.M.

Several alternatives have been suggested [13], among which the most popular is the ε language [14]. The basic reasons for this choice are the following:

- the εs define deviations with respect to the Born (plus QED plus QCD) predictions, and hence measure directly loop effects,

- they are defined thorough direct quantities measured experimentally and

- they are a complete set of four quantities.

To define the εs, first a "Born sinus" s_0 is introduced as

$$s_0^2 c_0^2 = \frac{\pi \alpha(M_Z)}{\sqrt{2} G_F M_Z} \tag{44}$$

and then the defining measurements taken are to be:

$$\frac{M_W}{M_Z} \;\to\; \Delta r_W$$

$$\Gamma_l \;\to\; g_{A_l} = -\frac{1}{2}\left(1 + \frac{\Delta \rho_l}{2}\right)$$

$$A_{FB}^{0,l} \;\to\; \sin^2 \theta_{eff}^{lept} = s_0^2(1 - \Delta \kappa')$$

$$\Gamma_b \;\to\; g_{A_b} = -\frac{1}{2}\left(1 + \frac{\Delta \rho_l}{2}\right)(1 + \varepsilon_b) \tag{45}$$

so that the quantities $\Delta r_W, \Delta \rho_l, \Delta \kappa'$ and ε_b reflect the deviations from the Born expectations for these measurements. Since new physics is easier to disentangle if not masked by large, conventional m_t effects, the *varepsilons* are defined keeping $\Delta \rho_l$ and ε_b while trading Δr_W and $\Delta \kappa'$ for two quantities with no contributions of order $G_F m_t^2$:

$$\varepsilon_1 \;=\; \Delta \rho_l$$

$$\varepsilon_2 \;=\; c_0^2 \Delta \rho_l + \frac{s_0^2}{c_0^2 - s_0^2} \Delta r_W - 2 s_0^2 \Delta \kappa'$$

$$\varepsilon_3 \;=\; c_0^2 \Delta \rho_l + (c_0^2 - s_0^2)\Delta \kappa' \tag{46}$$

By doing so, within the MSM one has:

$$\varepsilon_1 \;=\; \frac{3 G_F}{8\pi^2 \sqrt{2}} m_t^2 - \frac{3 G_F M_W^2}{4\pi^2 \sqrt{2}} \frac{s_W^2}{c_W^2} \log \frac{M_H}{M_Z} + \ldots$$

$$\varepsilon_2 \;=\; -\frac{G_F M_W^2}{2\pi^2 \sqrt{2}} \log \frac{m_t}{M_Z} + \ldots$$

$$\varepsilon_3 \;=\; \frac{G_F M_W^2}{12\pi^2 \sqrt{2}} \log \frac{M_H}{M_Z} + \ldots$$

$$\varepsilon_b \;=\; -\frac{G_F}{4\pi^2 \sqrt{2}} m_t^2 + \ldots \tag{47}$$

so that basically ε_1 has the dominant m_t and M_H dependences ($\Delta \rho$), ε_2 contains the logarithmic m_t dependence, ε_3 is mainly sensitive to $\log(M_H/M_Z)$ (but about 3 times less sensitive than $\varepsilon 1$) and ε_b has the quadratic sensitivity to m_t from the b-vertex correction.

2.4 Predictions for Z observables.

By using the effective coupling language, accurate predictions for Z observables can be formulated with just some small modifications with respect to the pure Born description. These predictions include basically the energy dependence of the total cross section (lineshape) and the different asymmetries. In the following we are going to review their formulation to introduce the concepts needed to understand their experimental analysis. The actual treatment of photonic corrections which, as mentioned before, is strongly linked to the experimental analysis, will be discussed during the presentation of the experimental results.

As we have seen, near the Z pole, the process $e^+e^- \to f\bar{f}$ being $f \neq e$ can be precisely described by the sum of a s-channel photon exchange amplitude and a s-channel Z boson exchange amplitude with the effective couplings discussed in the previous sections. The photon exchange amplitude is non-resonant and, at that energy, is strongly suppressed by the photon propagator whereas the Z exchange amplitude is resonant and constitutes the largest contribution.

2.4.1 Total cross sections.

As we have seen, for the dominant contribution, the Z exchange, the amplitude in the effective parameter language can be written as

$$A_Z = \sqrt{2}G_F M_Z^2 \mathcal{P}_Z(s)(J_{Ze}J_{Zf}^*)_s \tag{48}$$

where $\mathcal{P}_Z(s)$ is the Z propagator and the last term are the current contraction

$$
\begin{aligned}
\mathcal{P}_Z(s) &= \frac{1}{s - M_Z^2 + is\Gamma_Z/M_Z} \\
(J_{Zf}^\mu)_s &= [\bar{u}_f \gamma^\mu (g_{V_f}(s) - g_{A_f}(s)\gamma_5)v_f]
\end{aligned} \tag{49}
$$

whereas in the same language, the Z decay rate into a fermion pair, leaving aside final state corrections, can be expressed as

$$\Gamma(z \to f\bar{f}) = N_C^f \sqrt{2}G_F M_Z^2 \frac{1}{3}\frac{1}{16\pi M_Z}\frac{3}{4}M_Z^2(g_{V_f}^2 + g_{A_f}^2) \tag{50}$$

where N_C^f stands for the number of colours, $1/3$ comes from the polarization average, $1/(16\pi M_Z)$ comes from phase space integration and the last term comes from $(J_{Zf}J_{Zf}^*)_{s=M_Z^2}$.

If $m_f^2 << s$ then, in good approximation, $(J_{Ze}J_{Zf}^*)_s \propto s$ and therefore, we can write

$$(J_{Zf}J_{Zf}^*)_s = \frac{s}{M_Z^2}\Gamma(Z \to f\bar{f})\frac{3 \cdot 16\pi M_Z}{N_C^f \sqrt{2}G_F M_Z^2} \tag{51}$$

So that the matrix element squared can be written as

$$
\begin{aligned}
|A_Z|^2 &= N_C^f(\sqrt{2}G_F M_Z^2)^2 |\mathcal{P}_Z(s)|^2 (J_{Zf}J_{Zf}^*)(J_{Ze}J_{Ze}^*) \\
&= (3 \cdot 16\pi M_Z)^2 |\mathcal{P}_Z(s)|^2 \frac{s}{M_Z^2}\Gamma(Z \to e^+e^-)\Gamma(Z \to f\bar{f})
\end{aligned} \tag{52}
$$

and therefore, since the integrated cross section is

$$\sigma_Z(s) = \frac{1}{4}\frac{1}{16\pi s}|A_Z|^2 \tag{53}$$

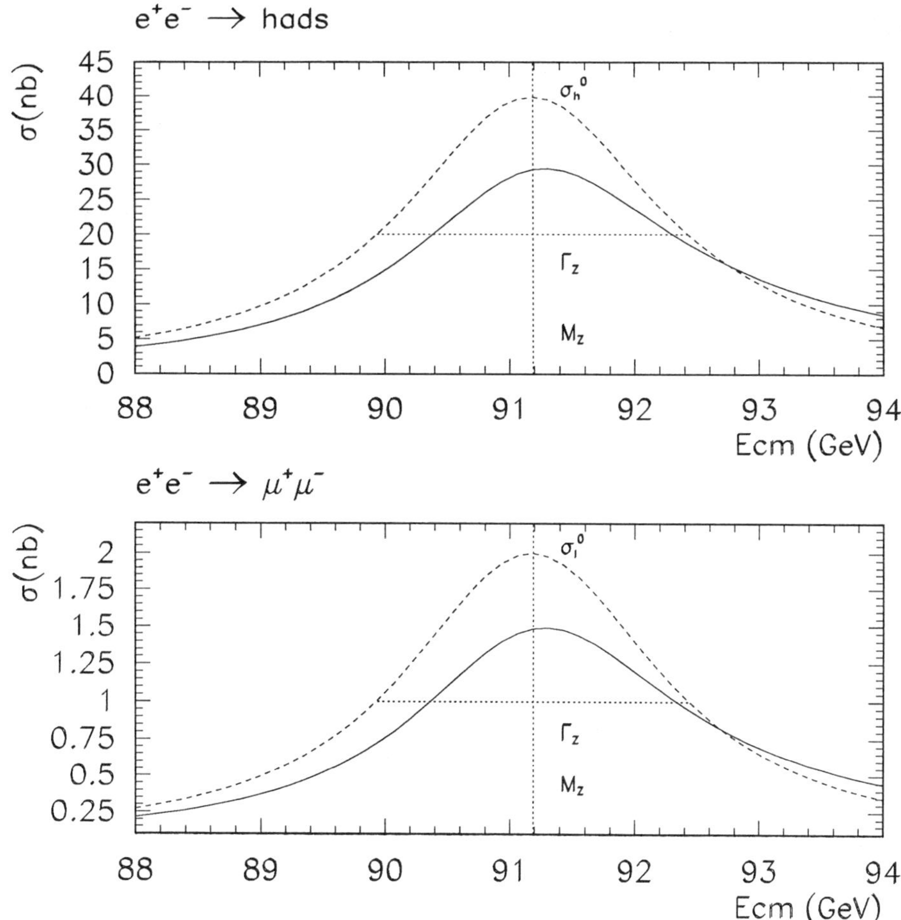

Figure 8: Theoretical predictions for the total cross section. The dashed line is the ISR unfolded prediction whereas the solid line is the complete one.

one gets

$$\sigma_{Z,f}(s) = \frac{12\pi}{M_Z^2} \frac{s}{(s - M_Z^2)^2 + \left(s\frac{\Gamma_Z}{M_Z}\right)^2} \Gamma_e \Gamma_f \tag{54}$$

where Γ_f stands for $\Gamma(Z \to f\bar{f})$. This equation relates directly the total cross section to the Z partial decay widths and hence, constitutes the basic expression used to describe the lineshape in an almost-model independent way, since very few assumptions have been applied in this deduction.

For $s = M_Z^2$ this expression simplifies to

$$\sigma_{Z,f}(M_Z^2) = \frac{12\pi}{M_Z^2} \frac{\Gamma_e \Gamma_f}{\Gamma_Z^2} = \frac{12\pi}{M_Z^2} BR(Z \to e^+ e^-) BR(Z \to f\bar{f}) \tag{55}$$

181

and is the so-called *peak cross section* $\sigma_f^0 = \sigma_{Z,f}(M_Z^2)$. By introducing this concept, eq. 54 can be rewritten as

$$\sigma_{Z,f}(s) = \sigma_f^0 \frac{s\Gamma_Z^2}{(s - M_Z^2)^2 + \left(s\frac{\Gamma_Z}{M_Z}\right)^2} \tag{56}$$

from this expression, it is clear that by studying $\sigma_{Z,f}(s)$ one can measure (see fig.8):

- M_Z, the "perturbative" Z mass, which gives basically the resonance peak position,

- G_Z, the Z width, which gives basically the resonance width,

- σ_f^0, the peak cross section, which gives basically the resonance height.

Given the clear connection between these parameters and the geometrical properties of the resonance, this parametrization allows a simple understanding of the data.

For hadron final state, since $BR(Z \to hads) \sim 70\%$, the analysis has a high statistical power and therefore, the hadron lineshape is the one that in practice determines M_Z, Γ_Z and σ_h^0.

For charged leptons final state, since $BR(Z \to lepts) \sim 9\%$, the statistical power is much lower and, in practice, only σ_e^0 is extracted, using for that M_Z and Γ_Z from the hadron lineshape analysis. Given the fact that

$$\sigma_e^0 = \frac{12\pi}{M_Z^2} \frac{\Gamma_e^2}{\Gamma_Z^2} \tag{57}$$

then from σ_e^0 one obtains Γ_e and since

$$\Gamma_e = \frac{G_F M_Z^3}{6\pi\sqrt{2}} (g_{V_e}^2 + g_{A_e}^2) \left(1 + \frac{3\alpha}{4\pi}\right) \tag{58}$$

that means that, for $g_{V_e}^2 << g_{A_e}^2$, the lepton lineshape measures basically $g_{A_e}^2$ and hence $\Delta\rho_e$

2.4.2 Asymmetries.

By using the effective coupling language, the differential cross section for colliding e^- with longitudinal polarization p can simply be written as ($\theta = \angle(e^-, f)$, $\mu_f = m_f^2/s$, $N_{c,f}$ = number of colours for fermion f):

$$
\begin{aligned}
\frac{d\sigma}{d\cos\theta}(s, \cos\theta; p) = {} & \frac{\pi\alpha^2(s)}{2s} N_{c,f} \sqrt{1 - 4\mu_f} \times \\
& \left\{ (1 + \cos^2\theta)\, G_1(s) + 4\mu_f\, G_2(s) \sin^2\theta + 2\cos\theta\, G_3(s) \right. \\
& \left. + p\left[(1 + \cos^2\theta)\, H_1(s) + 4\mu_f\, H_2(s) \sin^2\theta + 2\cos\theta\, H_3(s)\right] \right\}
\end{aligned} \tag{59}
$$

with

$$
\begin{aligned}
\chi_{\gamma Z}(s) &= F_G(s) \frac{s(s - M_Z^2) + s\Gamma_Z/M_Z \Im(\Delta\alpha)}{(s - M_Z^2)^2 + s^2\Gamma_Z^2/M_Z^2} \\
\chi_{ZZ}(s) &= F_G^2(s) \frac{s}{(s - M_Z^2)^2 + s^2\Gamma_Z^2/M_Z^2}
\end{aligned} \tag{60}
$$

$$
\begin{aligned}
G_1(s) &= Q_e^2 Q_f^2 &&+ 2Q_e Q_f g_{V_e} g_{V_f} \chi_{\gamma Z}(s) &&+ (g_{V_e}^2 + g_{A_e}^2)(g_{V_f}^2 + g_{A_f}^2 - 4\mu_f g_{A_f}^2)\chi_{ZZ}(s) \\
G_2(s) &= Q_e^2 Q_f^2 &&+ 2Q_e Q_f g_{V_e} g_{V_f} \chi_{\gamma Z}(s) &&+ (g_{V_e}^2 + g_{A_e}^2)g_{V_f}^2 \chi_{ZZ}(s) \\
G_3(s) &= && 2Q_e Q_f g_{A_e} g_{A_f} \chi_{\gamma Z}(s) &&+ 4g_{V_e} g_{A_e} g_{V_f} g_{A_f} \chi_{ZZ}(s) \\[8pt]
H_1(s) &= && 2Q_e Q_f g_{A_e} g_{V_f} \chi_{\gamma Z}(s) &&+ 2g_{V_e} g_{A_e}(g_{V_f}^2 + g_{A_f}^2 - 4\mu_f g_{A_f}^2)\chi_{ZZ}(s) \\
H_2(s) &= && 2Q_e Q_f g_{A_e} g_{V_f} \chi_{\gamma Z}(s) &&+ 2g_{V_e} g_{A_e} g_{V_f}^2 \chi_{ZZ}(s) \\
H_3(s) &= && 2Q_e Q_f g_{V_e} g_{A_f} \chi_{\gamma Z}(s) &&+ 2(g_{V_e}^2 + g_{A_e}^2)g_{V_f} g_{A_f} \chi_{ZZ}(s)
\end{aligned}
$$

being

$$
F_G(s) = \frac{G_F M_Z^2}{2\sqrt{2}\pi\alpha(s)} \tag{61}
$$

So that the unpolarized total cross section is

$$
\sigma(s) = \frac{4\pi\alpha^2(s)}{3s} N_{c,f} \sqrt{1 - 4\mu_f}\,(G_1(s) + 2\mu_f G_2(s)) \tag{62}
$$

Forward-backward charge asymmetry A_{FB}.

It is defined as

$$
A_{FB}(s) = \frac{\sigma(\cos\theta > 0) - \sigma(\cos\theta < 0)}{\sigma(s)} \tag{63}
$$

where θ is the azimuthal angle of the the outgoing fermion and therefore, by using equations 59 and 62:

$$
A_{FB} = \frac{3}{4}\sqrt{1 - 4\mu_f}\,\frac{G_3(s)}{G_1(s) + 2\mu_f G_2(s)} \simeq \frac{3}{4}\frac{G_3(s)}{G_1(s)} \tag{64}
$$

Near the Z pole, as we have seen, the total cross section $\sigma(s)$ is, by far, dominated by the pure Z-exchange term and therefore, defining

$$
A_f \equiv \frac{2g_{V_f} g_{A_f}}{(g_{V_f}^2 + g_{A_f}^2)} \tag{65}
$$

one can write

$$
A_{FB}(s) \cong \frac{3}{4} A_e A_f + (s - M_Z^2)\frac{3}{4F_G(s)}\frac{2Q_e Q_f g_{A_e} g_{A_f}}{(g_{V_e}^2 + g_{A_e}^2)(g_{V_f}^2 + g_{A_f}^2)} \tag{66}
$$

so that for $\sqrt{s} = M_Z$

$$
A_{FB}(M_Z^2) = \frac{3}{4} A_e A_f \equiv A_{FB}^{0,f} \implies \text{ Peak F-B asymmetry} \tag{67}
$$

Therefore, by studying $A_{FB}(M_Z^2)$ one measures $A_e A_f$.

- For $f = l$ then, assuming universality, one measures A_e^2, but since for leptons $g_{V_e} << g_{A_e}$ then

$$A_f \equiv \frac{2g_{V_f}/g_{A_f}}{(g_{V_f}^2/g_{A_f}^2) + 1} \cong 2\frac{g_{V_f}}{g_{A_f}} = 2(1 - 4\sin^2 \theta_{eff}^{lept})$$

so that the peak asymmetry measures directly the effective sinus. Since $A_e \sim 0.15$ then the peak asymmetry $A_{FB}^{0,l}$ is a very small quantity ~ 0.015. On top of that, since the energy slope is $\mathcal{O}(g_{A_e}^2)$ its energy dependence is large and therefore, the precise knowledge of $s - M_Z^2$ is important and ISR corrections distort very noticeably the observed asymmetry.

- For $f = q$ the one measures $A_e A_q$ but, since for quarks $g_{V_q} \sim g_{A_q}$ then the quark coupling is large ($A_c \sim 0.7$ and $A_b \sim 0.9$) and insensitive to the effective sinus (see fig. 9) so that one basically measures A_e and, since the dependence is linear, the sensitivity to the effective sinus is larger than in the lepton asymmetry case. On top of that, the peak asymmetry $A_{FB}^{0,q}$ is large and therefore, in this case the energy dependence is much less relevant.

Left-right polarization asymmetry A_{LR}.

It is defined as

$$A_{LR}(s) = \frac{\sigma(p = +1) - \sigma(p = -1)}{\sigma(s)} = -A_{POL}(s) \tag{68}$$

where p is the initial state polarization (first equality) or the final state one (second equality). Therefore, by using equation 59, one can write

$$A_{LR}(s) = \frac{H_1(s)}{G_1(s)}$$
$$A_{POL}(s) = -\frac{H_3(s)}{G_1(s)} \tag{69}$$

so that in the region $\sqrt{s} \sim M_Z$

$$A_{LR}(s) \cong A_e + (s - M_Z^2)\frac{1}{F_G(s)} \frac{2Q_e Q_f g_{A_e} g_{V_f}}{(g_{V_e}^2 + g_{A_e}^2)(g_{V_f}^2 + g_{A_f}^2)} \tag{70}$$

For $A_{POL}(s)$ the equation is the same by interchanging $e \leftrightarrow f$ and multiplying times -1. Therefore for $\sqrt{s} = M_Z$

$$A_{LR}(M_Z^2) = A_e \equiv A_{LR}^0 \implies \text{Peak L-R asymmetry}$$
$$A_{POL}(M_Z^2) = -A_\tau \equiv P_\tau \implies \text{Tau polarization} \tag{71}$$

where in the second equation, for reasons that we will discuss later, we have taken $f = \tau$. Given the fact that $A_e \sim 0.15$ then A_{LR}^0 and P_τ are large and, since in addition the slope of the energy dependence is $\mathcal{O}(g_A g_V)$, the energy dependence is small and therefore, the effect of ISR is also very small.

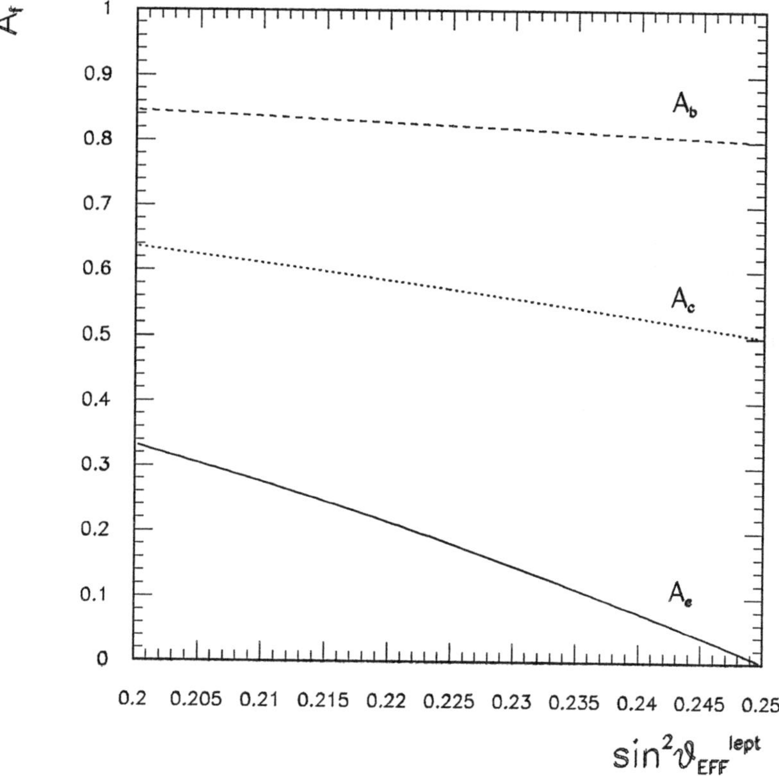

Figure 9: Sensitivity of A_f to $\sin^2 \theta_{eff}^{lept}$ for quarks and leptons.

Forward-backward left-right polarization asymmetry A_{LR}^{FB}.

It is defined as

$$A_{LR}^{FB}(s) = \frac{1}{\sigma(s)} \times \{\sigma(p = +1, \cos\theta > 0) - \sigma(p = -1, \cos\theta > 0)$$
$$- \quad \sigma(p = +1, \cos\theta < 0) + \sigma(p = -1, \cos\theta < 0)\}$$
$$= -A_{POL}^{FB}(s) \tag{72}$$

where θ is the azimuthal angle of the the outgoing fermion and, as before, p is the initial state polarization (first equality) or the final state one (second equality). Therefore, by using equation 59, one can write

$$A_{LR}^{FB}(s) = \frac{3}{4} \frac{H_3(s)}{G_1(s)}$$
$$A_{POL}^{FB}(s) = -\frac{3}{4} \frac{H_1(s)}{G_1(s)} \tag{73}$$

It is worth noticing that, since $H_1(s)$ is equal to $H_3(s)$ under the exchange $e \leftrightarrow f$, then the following equations are verified:

$$
\begin{aligned}
A^{FB}_{LR}(s) &= -\frac{3}{4} A_{POL}(s) \\
A^{FB}_{POL}(s) &= -\frac{3}{4} A_{LR}(s)
\end{aligned}
\tag{74}
$$

so that, for instance, the tau (final state) forward-backward polarization asymmetry measures precisely the same coefficient ratio as the initial state left-right polarization asymmetry. In the region $\sqrt{s} \sim M_Z$

$$
A^{FB}_{LR}(s) \cong \frac{3}{4} A_f + (s - M_Z^2) \frac{3}{4 F_G(s)} \frac{2 Q_e Q_f g_{V_e} g_{A_f}}{(g_{V_e}^2 + g_{A_e}^2)(g_{V_f}^2 + g_{A_f}^2)}
\tag{75}
$$

For $A^{FB}_{POL}(s)$ the equation is the same by interchanging $e \leftrightarrow f$ and multiplying time -1. Therefore for $\sqrt{s} = M_Z$

$$
\begin{aligned}
A^{FB}_{LR}(M_Z^2) &= \frac{3}{4} A_f \equiv A^{FB;0,f}_{LR} \Longrightarrow \text{Peak F-B L-R asymmetry} \\
A^{FB}_{POL}(M_Z^2) &= -\frac{3}{4} A_e \equiv P^{FB}_\tau \Longrightarrow \text{F-B Tau polarization}
\end{aligned}
\tag{76}
$$

where, like in the previous case, in the second equation, for reasons that we will discuss later, we have taken $f = \tau$. The same observations than in the previous case hold now, namely: given the fact that $A_e \sim 0.15$ then $A^{FB;0,f}_{LR}$ and P^{FB}_τ are large and, since in addition the slope of the energy dependence is $\mathcal{O}(g_A g_V)$, the energy dependence is small and therefore, the effect of ISR is also very small.

Figure 10 shows the exact predictions for the different asymmetries presented and the features discussed can be clearly observed.

2.5   Theoretical uncertainties.

During the last few years, a remarkable theoretical effort in trying to match the big experimental progress has happened. Calculations of the leading terms in m_t and M_H from genuine two loop E.W. corrections for Δr and Z physics, and of the QCD corrections to the leading one loop E.W. terms have been performed and checked by several groups. Also studies on how the resumation of known leading one-loop terms should be done, and of the interplay between QCD and E.W. corrections have been incorporated into the E.W. libraries used by the experimentalists and further improvement is still ongoing (for a clear review see [2]).

Moreover, an attempt to quantify the present theoretical limitations in the interpretation of the measurements within the MSM has recently taken place [2]: given the high precision of the experimental measurements, a question in order is what is the precision of the theoretical predictions with which they have to be confronted. Two different sources can be the origin of uncertainties in the theoretical predictions:

- On the one hand, the uncertainty coming from the precision of the input parameters used (masses and couplings constants), some of which are intended to be determined from the precision data and some of which are taken from other, low energy, experiments.

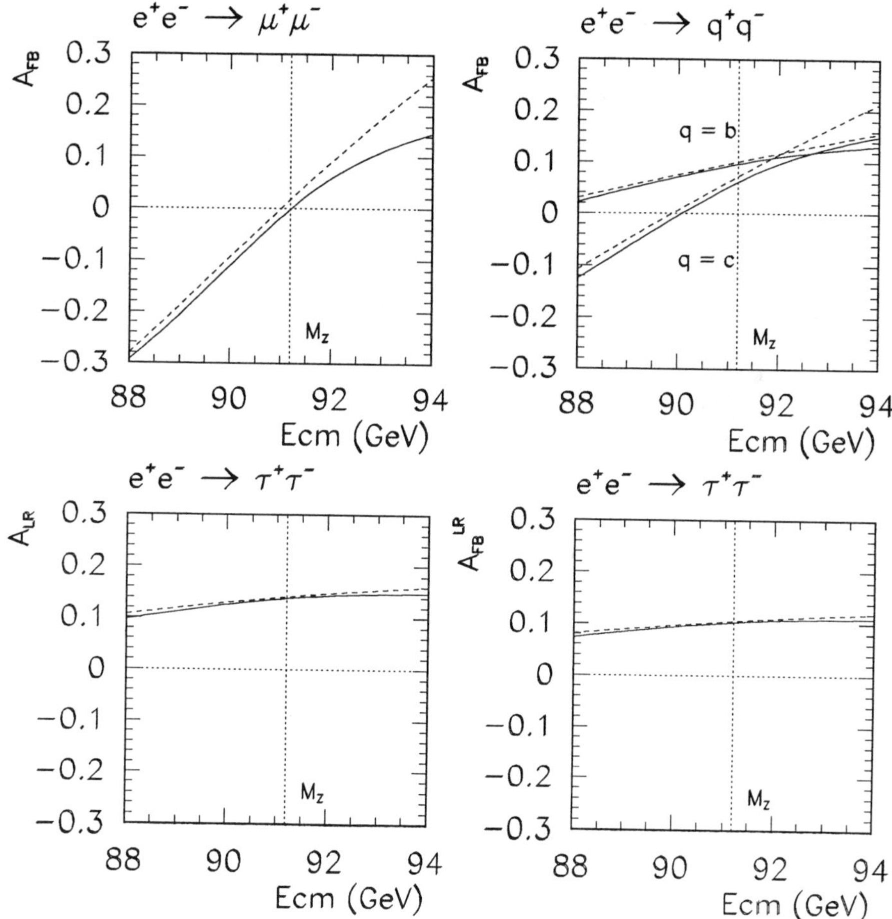

Figure 10: Theoretical predictions for the different asymmetries: lepton forward-backward charge asymmetry, quark forward-backward charge asymmetry, left-right polarization asymmetry and forward-backward left-right polarization asymmetry. The dashed line is the ISR unfolded prediction whereas the solid line is the complete one.

- On the other hand, the uncertainty coming from the limitation of the calculation itself. Since we use predictions computed up to some order in the perturbative expansion, our predictions have some limited precision. Customarily the precision is quoted from the estimation of the next missing order in the expansion.

Concerning the first source, the analysis of these uncertainties shows [2] that the present precision in the knowledge of $\Delta(\alpha^{-1}(M_Z^2)) = 0.12$, (linked to the experimental precision in the low energy $e^+e^- \rightarrow hads$ data used to estimate the hadronic contribution through dispersion relations) is still, one of the most important limitations[10],

---

[10]To illustrate this fact, the effect of the uncertainty in $\alpha(M_Z)$ is shown explicitly in the MSM

and specially for the interpretation of the effective weak mixing angle, $\sin^2 \theta_{eff}^{lept}$. As discussed already, recent evaluations claim that, by using all the presently available low energy data and an improved analysis methode, the error in $\Delta(\alpha^{-1}(M_Z^2))$ might be reduced to about 0.06 [6].

Concerning the second source, several calculations, implementing different renormalization schemes, different gauge choices and different estimations on missing terms, have been compared in great detail to be able to disentangle the technical precision (linked to the existence of programming "bugs" or approximations in the computer codes) from the true difference due to the diversity of choices. This second difference is an estimator of missing orders and thus enables the quotation of some sort of theoretical uncertainty.

In practice, in each code options have been setup allowing to produce the predictions for all the precision observables for a variety of choices. One given choice, agreed among the different groups, is used to compare the codes while, by changing internally the options, each code is able to estimate the uncertainty in its own predictions.

Comparisons have been made at different prediction levels, matching the way the actual experimental measurements are analyzed, which, for most of the precision measurements, consists in two steps:

- first, deconvolution from the direct measurements ("real-observables") of well established effects like photon emission, QED contributions,... to lead to the extraction of so-called electroweak parameters ("pseudo-observables") with almost-model-independent expressions,

- second, interpretation of these electroweak parameters in terms of basic Lagrangian parameters.

Therefore, the determination of the theoretical uncertainties has also been splitted in two parts:

- The precision in the prediction, within the MSM, of the electroweak parameters. The study of the different possible contributions, shows as main sources for this uncertainty [2]: the missing purely electroweak higher orders (estimated by comparing results obtained with resumations using different prescriptions on how to handle higher orders), the higher orders coming from the interplay between electroweak and QCD corrections (factorization or not of these corrections) and the actual technical precision of the calculations (due to simplifications to make programs faster, different practical implementations and programming bugs).

- The precision in the transformation from direct measurements to electroweak parameters. The main limitation in this step is the knowledge of photon emission corrections.

The estimates of theoretical uncertainties are highly subjective and their values partly reflect the internal philosophy of the actual implementation of radiative corrections in a given code. The main conclusions of these studies have been the following [2]:

- The differences between results of different codes are small compared to existing experimental uncertainties.

---

predictions given in the figures of this report for the different measurements.

- At present the most promising are measurements of g_V/g_A in various P- and C-violating asymmetries and polarizations.

- The real bottleneck for improved theoretical accuracy in g_V/g_A is presented by the uncertainty of the input parameter $\alpha(M_Z^2)$.

- In many cases the one-loop approximation in the electroweak gauge coupling is adequate enough at the present level of experimental accuracy. Anyway, a complete evaluation of the sub-leading corrections, $\mathcal{O}(G_F^2 M_Z^2 m_t^2)$ would greatly reduce the uncertainty that we observe, one way or the other, for all observables.

- In case the next generation of experiments at LEP 1 and SLC would improve considerably the accuracy (which is a problem not only of statistics but mainly of systematics) the full program of two-loop electroweak calculations should be carried out.

3 Electroweak precision measurements.

In this section, we are going to give a short review of the most relevant precision electroweak measurements performed so far. Most of them come from the study of the Z resonance at the LEP machine and therefore, we are going to pay an special attention to the description of the most relevant characteristics of the instruments used: the LEP collider and its detectors.

The results presented here come, for most observables, from the data presented during the 1994 summer conferences, in the case of LEP, the dominant measurements, this corresponds to all data taken between 1989 and 1993 by the four LEP experiments: almost two million hadronic Z decays and about two-hundred thousand charged-lepton Z decays per experiment. Most of the results discussed are still preliminary although they have been already presented in public [15].

3.1 Measurements at the Z resonance.

In the last few years, the e^+e^- colliders LEP and SLC have proven to be crucial tools to explore the Standard Model and in particular the electroweak theory to a level which is sensitive to its quantum structure. Many very high precision measurements performed by the four LEP collaborations and the SLD experiment allow today the determination of important parameters of the theory with an unprecedented level of accuracy.

3.1.1 The instruments.

The LEP machine.

The LEP machine is an e^+e^- storage ring of 27 Km. of circumference. It is the largest collider ever built and the reason for its size is the synchrotron radiation, which is proportional to the square of the inverse mass of the particle and to the fourth power of its energy. It is negligible for protons, but fierce for electrons. It is also inversely proportional to the square of the radius of the machine, and so becomes manageable at high energy only if the radius is sufficiently large. In the latest LEP operation mode, $N_b = 8$ bunches of electrons are driven into collision against 8 bunches of positrons, their average current is about 1 mA and their crossing rate is of about 11 microseconds. The beams collide in four interaction regions in which four detectors are located: ALEPH, DELPHI, L3 and OPAL. The typical beam lifetime reached was of about 12 hours.

The machine luminosity can be expressed in terms of the bunch characteristics as

$$\mathcal{L} = \frac{N_{e^+} N_{e^-} f_{rev}}{4\pi \sigma_x \sigma_y} N_b$$

where N_e are the number of particles per bunch (typically of about 10^{12} at LEP and limited by the so-called beam-beam effect), f_{rev} is the revolution frequency (about 10^4 at LEP) and σ_x and σ_y are the beam transverse sizes (about 200 and 10 microns respectively at LEP). This luminosity so far has reached the value of $\mathcal{L} = 1 - 2 \times 10^{31} \, cm^{-2} \, s^{-1}$. Several times intrinsic limitations of the machine luminosity have been overcome by a thorough study of its behavior and by the creativity of the machine physicists, who have been constantly trying to improve the machine by applying new ideas, like for example the Pretzel scheme or bunch trains. This enabled the experiments to collect $1.5 \cdot 10^7$ visible Z decays by the end of 1994 and leaves scope for a significant increase of the integrated luminosity before the start of LEP-II.

The machine energy, so far (LEP-I phase), has been set near the Z mass (about 91 GeV) and is expected to increase above the W pair threshold (about 180 GeV) in the forecoming years (LEP-II phase). The determination of the LEP beam energy constitutes the highest precision measurement performed at LEP so far and has direct consequences in the precision of important electroweak quantities as we shall see.

The method presently used to measure the beam energy, takes advantage of the fact that, under favourable conditions, transverse beam polarization can be naturally built up due to the interaction of the electrons with the magnetic guide field (Sokolov-Ternov effect). The number of spin precessions in one turn around the ring ("spin tune") is

$$\nu = \frac{g_e - 2}{2} \frac{E_{beam}}{m_e} = \frac{E_{beam}(GeV)}{0.4406486(1)}$$

where g_e is the gyromagnetic constant and m_e is the electron mass. The spin precession frequency is then equal to

$$f_{prec} = \nu f_{rev}$$

with f_{rev} the revolution frequency being in typical conditions of $f_{rev} = 11245.5041(1)$. Resonant depolarization is produced by using a sweeping kicker magnet which produces an exciting field perpendicular to the beam axis and in the horizontal plane, when $f_{spin-kick} = f_{prec}$, that is, when the exciting field is in phase with the precession. This calibration methode has an intrinsic precision of few hundred KeV and requires a time of about 4 hours per calibration.

Nevertheless, since just about 2 calibrations per week are, in practice, feasible, this means that these precise measurements, have to be extrapolated to the whole running time by correlating them with the energy measurements performed by using some reference magnets. The scatter in this correlation, depends on the stability of the machine energy and is affected by several variables, such as the status of the radiofrequency cavities, the temperature and humidity in the LEP tunnel and the distortions of the ring length, for instance due to the tidal forces of the sun and the moon, which change the circumference of the machine by just ~ 1 mm, but affects the beam energy at the few MeV level [16] (see fig. 11). The final precision of the measurement improves as these effects are understood and, even thought at present is of about 4 MeV, it is expected to be brought down to about 1.5 MeV [17].

The LEP detectors.

As mentioned, there are four detectors operating at LEP. All four detectors, though rather different in practice, follow similar conceptions in their design. Their inner

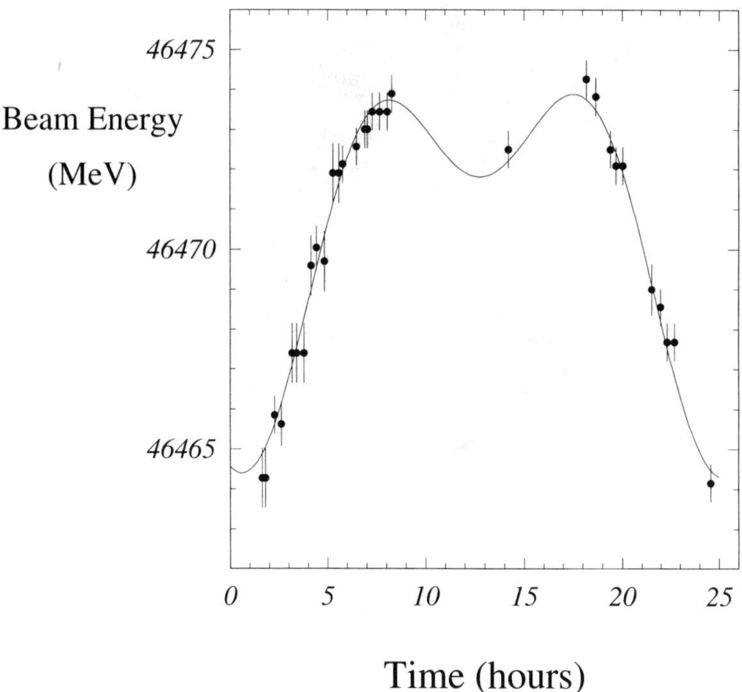

Figure 11: Effect of the tidal forces of the moon and the sun on the LEP energy. The dots represent the energy measured with resonant depolarization, while the solid line is the prediction.

volumes are devoted to perform accurate tracking which use either proportional wire or semiconductor techniques or both. The tracking volume is surrounded by calorimeters, which are in two layers, an inner layer which measures electromagnetic, and an outer layer which measures hadronic energy. The whole is surrounded by wire chambers to detect the penetrating muons. In figure 12 one of such detectors is shown. Given the fact that the techniques used in each detectors are rather different, we refer the reader to their published descriptions to get a detailed discussion [18].

In discussing the detector aspects of the experiments it should be noted that their adequate design and their good performance together with the clean background conditions have allowed the understanding of the data to the level of the statistical precision or better already almost from the very beginning. Typical systematic uncertainties in event selections are at the few per mille level. Nevertheless, in the last years, most of the experiments have upgraded their detectors to achieve even higher performances for some applications. Especially two detector improvements of this kind deserve mentioning:

- The luminometers have been upgraded by installing Silicon-Tungsten calorimeters or by improving the tracking capabilities. This has enabled the experiments to master the detector systematics in the luminosity determination below the per mille level, a limit unconceivable just a few years ago.

- The installation and steady improvement of microvertex detectors has decreased

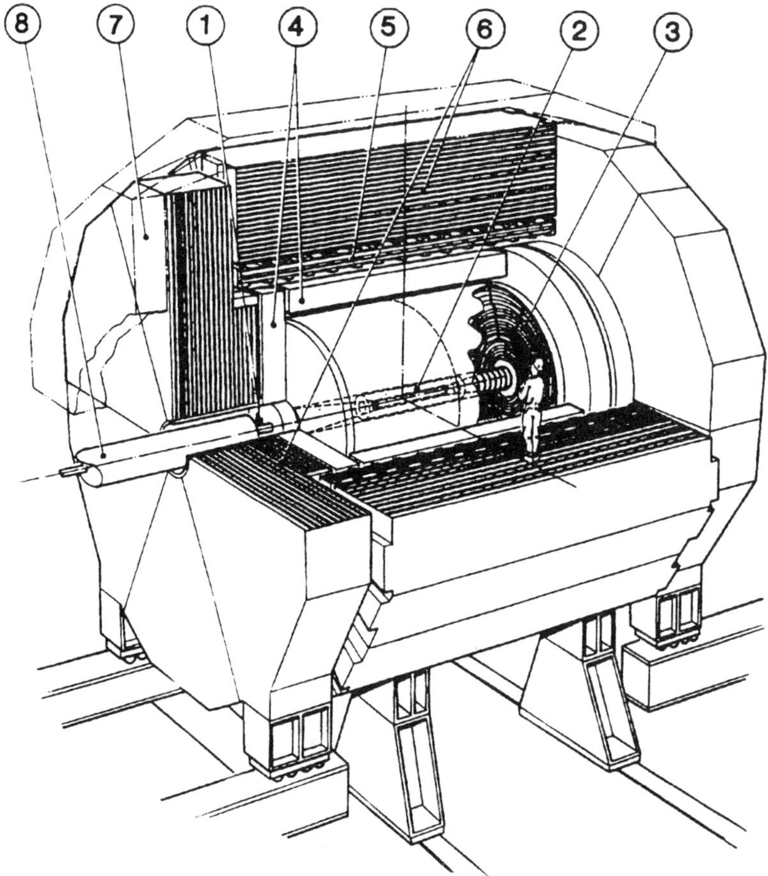

Figure 12: Schematic view of the ALEPH detector at LEP. (1) Luminosity Monitor. (2) Silicon Microvertex Detector and Inner Tracking Chamber. (3) Time Proportional Chamber. (4) Electromagnetic Calorimeter. (5) Superconducting Coil. (6) Hadronic calorimeter. (7) Muon Chambers. (8) Focusing Quadrupoles.

drastically the uncertainties in electroweak measurements with heavy flavors. These measurements play now a very important role when analyzing the implications of electroweak precision measurements.

The actual measurement of the luminosity, given its impact on the LEP precision measurements, deserves some more explanation. The experimental cross section for the production of any kind of final state σ is determined in practice by counting the number of observed events of such kind recorded during some given period N_s divided by the integrated luminosity of the machine during the same period $L = \int \mathcal{L}dt$, say

$$\sigma = \frac{N_s - N_b}{\epsilon L}$$

being N_b the events from background processes and ϵ the detection efficiency. Instead of using the expression which gives the instantaneous luminosity as a function of the beam characteristics, the LEP experiments determine the luminosity by counting the number of events coming from a process with very well known cross section, such the small angle Bhabha scattering ($e^+e^- \to e^+e^-$). The non-electromagnetic contributions to this process are small and its cross section for small angles is very high, namely

$$\sigma_{Bhabha} \sim \frac{16\pi\alpha^2}{s} \left(\frac{1}{\theta^2_{min}} - \frac{1}{\theta^2_{max}} \right)$$

and all the LEP detectors have specialized small angle calorimeters (typically $\theta_{min} \sim 25$ mrad), the luminosity monitors, to study it. From the above expression it follows directly that the precise knowledge of the detector inner edge radius is one of the fundamental milestones from the experimental point of view. Given the fact that, with the upgrade of the luminosity monitors, the experimental systematic uncertainty is now below one per mille, presently the limitation in the knowledge of the luminosity comes from the calculation of the prediction of the Bhabha cross section which, although being basically a QED problem, technically is a rather difficult task and is just known a the 2 per mille level.

From the experimental point of view, one of the main reasons for the high accuracy of the measurements performed at LEP is the cleanliness of the events which allows an easy and precise recognition of the events.

The most common decay of the Z, accounting for around 65 % of the decays, is into a quark and its antiquark, which fragment producing hadron jets with typically of the order of ten charged particles per quark. A typical hadronic (quark) decay of the Z is shown in fig. 13a. The typical background levels for the selection of this process is of about 0.1 % and the detection efficiency of close to 100 %. It is in general not possible to determine the quark flavor of a particular event: up, down, strange, charm, or bottom. However, in a certain number of cases it is possible, especially for the b quark, using certain particularities of the jets. Recently b quark identification has improved dramatically with the help of a new detection method based on semiconductor strip detectors with spacial resolution measured in microns, which have permitted reconstruction of the b decay vertex at distances from the interaction point typically of one millimeter.

About ten percent of the decays are to the charged leptons of the three families, a third of this to each one. Typical backgrounds in these process are about 0.1 - 1 % and efficiencies larger than 90 % within the detector acceptance. An example of electron decay is shown in 13b. The two tracks emitted in opposite directions look as one, because the particles are emitted back to back to conserve momentum. Even though the tracks look straight on the scale of the picture, their curvatures are measured with a precision of 2.5They identify themselves as electrons by the shapes and the magnitudes of their showers in the electro-magnetic calorimeter. Fig. 13c shows the decay into mu leptons. The tracks are similar to those of electrons, however the calorimetric signatures are very different: the muons penetrate both calorimeters and leave only a characteristic small amount of energy in these. Fig. 13d shows the decay into a positive and a negative tau lepton. The electron is stable, and the muon, although

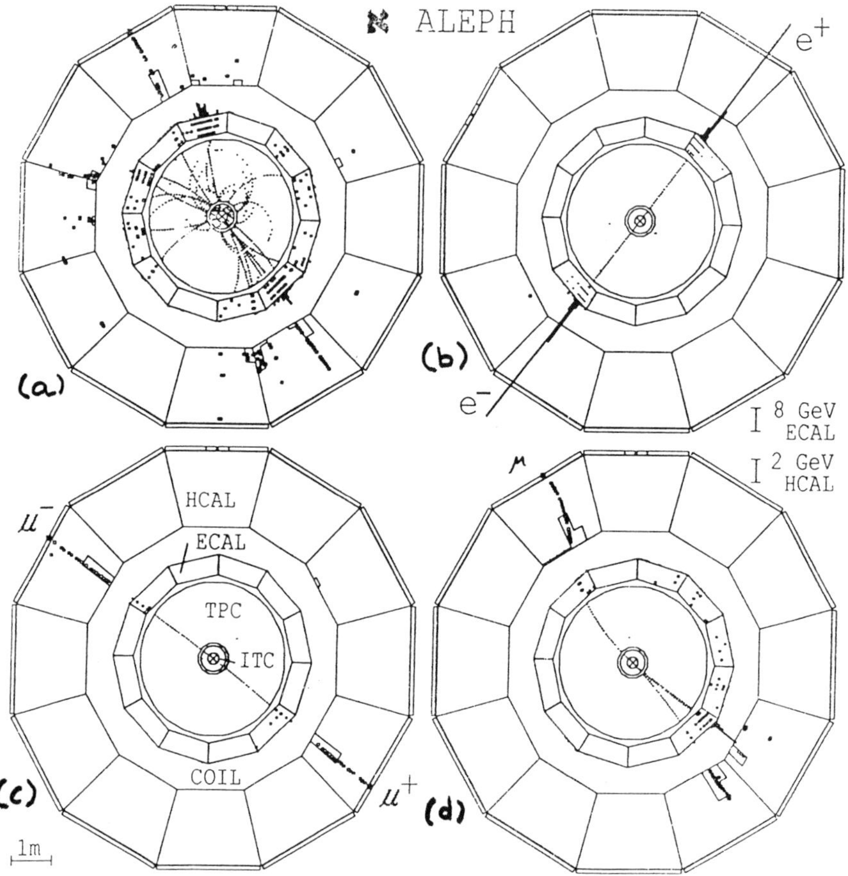

Figure 13: Examples of typical decays of the Z, as seen in one of the four LEP detectors. (a) The decay into quark-antiquark pair, with two back to back hadron jets. (b) Decay into electron and positron. The two back to back tracks starting from the center look as one. Both particles deposit all their energy in the electromagnetic calorimeter. (c) Decay to a muon pair. The back to back tracks traverse the calorimeters and register in the outer muon detection layers. (d) Decay to tau leptons. The tau decays in the beam pipe. The positive tau produces one charge secondary, in this case a muon, the negative tau decays into three pions.

unstable, has a decay length, at the LEP energy, of the order of 300 km, and so decays only extremely rarely within the apparatus. The tau however decays with a typical path of only a millimeter or two. It therefore decays within the beam pipe and only its decay products are observed. These contain usually one charged track, more rarely three, very seldom as many as five. The tracks may identify themselves as hadrons or leptons. The decay of a Z into tau's is shown in fig. 13d. The main purpose of showing these four typical events is to give some feeling for the clarity and simplicity of the primary data. Finally, about 20 % of the Z's decay to neutrinos. These events are not observed at all, since neutrinos pass the apparatus without a trace.

The results presented in the following section corresponds to the total collected data at LEP which amounts to some 7.1 million hadronic Z decays and about 780 thousand leptonic Z decays [15].

Cross sections.

The measurement of the cross sections at the Z pole allows the determination of the Z partial widths into visible channels and the analysis of its energy dependence determines directly the Z mass and total width. From these measurements, the invisible partial width can be derived and hence, the number of light neutrino species can be determined.

Cross sections are measured exclusively for charged leptons (e,μ and τ), heavy quarks (c and b), and, through the use of radiative hadronic events, for u-type and d-type quarks. They are also measured inclusively for hadrons. The clean selections for leptons and hadrons, enable the detailed study of the energy dependence of the cross section (lineshape analysis) whereas for heavy quarks and radiative hadronic events, since the tag is more complicated, only the partial widths have been studied so far.

Lineshape analysis.

As we have seen, the cross section for the production of a fermion pair $f\bar{f}$ can be written as

$$\sigma_f(s) = \int_{4m_f^2}^{s} ds' H(s, s')\hat{\sigma}_f(s') \tag{77}$$

where $H(s, s')$ is the total cross section radiator function which takes care of initial state radiation corrections and the reduced cross section $\hat{\sigma}$ is written as

$$\hat{\sigma}_f(s) = \sigma_f^0 \cdot \frac{s\Gamma_Z^2}{\left(s - M_Z^2\right)^2 + \left(s\Gamma_Z/M_Z\right)^2} + (\gamma - Z) + |\gamma|^2 \tag{78}$$

with σ_f^0 being the peak cross section, $\Gamma_Z = \Gamma_h + \Gamma_e + \Gamma_\mu + \Gamma_\tau + \Gamma_{inv}$ the total Z width and M_Z is the perturbative mass (not to be confused with the S-matrix mass). This parameterization assumes the validity of QED for the photon exchange part and also takes from the Minimal Standard Model the interference between the photon- and Z-mediated amplitudes [11]. This interference is very small around the Z pole. In the case of Bhabha scattering, $f = e$, one can either subtract for the data the t-channel contributions, also taken from the theory or add them to the previous expression. The cross section at the peak can be written in turn in terms of the Z mass and width and the Z partial widths to the initial state Γ_e and the final state Γ_f:

$$\sigma_f^0 = \frac{12\pi}{M_Z^2} \cdot \frac{\Gamma_e \Gamma_f}{\Gamma_Z^2} \tag{79}$$

where the partial widths can be written in terms of effective vector and axial couplings of the fermions to the Z:

$$\Gamma_l = \frac{G_F M_Z^3}{6\pi\sqrt{2}} \left({g_{V_l}}^2 + {g_{A_l}}^2\right) \left(1 + \frac{3\alpha}{4\pi}\right) \tag{80}$$

Assuming lepton universality, only four parameters are needed to describe the s-dependence of the hadronic and leptonic cross sections: M_Z, Γ_Z, σ_h^0 and the ratio of hadronic to leptonic partial widths ($R_l = \Gamma_h/\Gamma_l = \sigma_h^0/\sigma_l^0$), where the lepton is taken

---

[11]For some experiments this is only the case for hadronic final states.

to be massless. This choice of parameters, given the fact that they are directly related to geometrical characteristics of the lineshape, has two advantages: on the one hand, their correlation are small, and on the other hand they simplify the task of disentangling common uncertainties:

- M_Z and G_Z are the only measurements in the energy scale and, as we shall see, they are strongly affected by common LEP energy uncertainties,

- σ_h^0 is the only lineshape observable in which the overall normalization, and hence the luminosity measurement uncertainties, enter.

If lepton universality is not assumed, R_l is substituted by three analogous quantities, R_e, R_μ, R_τ.

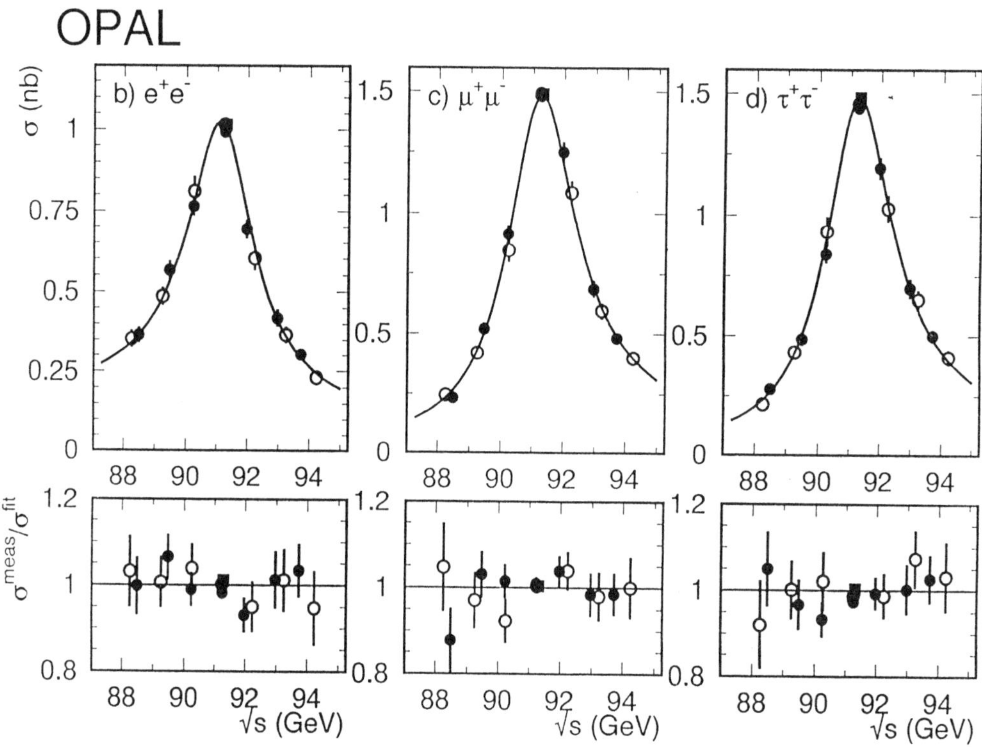

Figure 14: The charged lepton lineshape OPAL data compared to the best fits. For electrons, the lineshape looks different due to the t-channel contributions which here have not been subtracted.

Two computer programs which implement the scheme sketched above have become the standard ones at LEP: MIZA [19] is used by the ALEPH collaboration; ZFITTER [20] by DELPHI, L3 and OPAL. At the current level of experimental precision, the results obtained with both of them are equivalent. Figure 14 shows the measurements of the three leptonic lineshapes and the best fits as obtained by the OPAL

196

collaboration. In this case, the electron lineshape data has not been corrected to subtract the t-channel contributions but instead, these have been added to the s-channel lineshape expressions used in the fit.

Z mass.

The Z mass is the most precise single measurement performed at LEP. The results by the four experiments are shown in fig. 15 where it can be seen that the measurement uncertainty is systematics-dominated and comes from the preliminary estimation of ± 0.004 GeV uncertainty in the absolute energy scale of the machine.

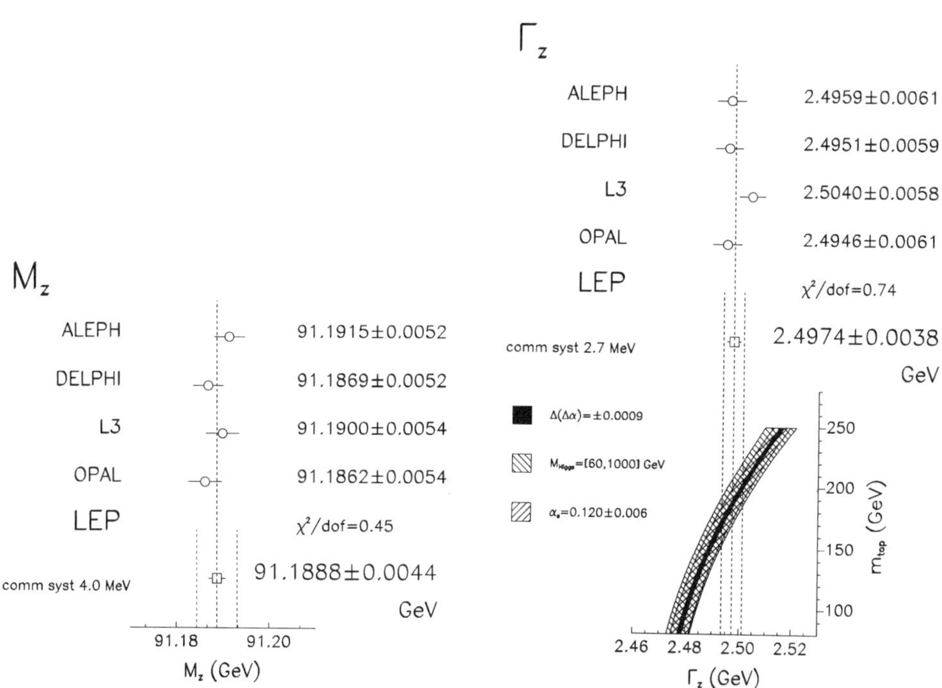

Figure 15: The Z mass and width measured by the four experiments together with the mean. The bands show the MSM prediction as a function of the top quark mass. The numerical values given include all uncertainties while the error bars include only non-common uncertainties to allow a visual inspection of the actual agreement between them. Unless explicitly stated, the same convention applies to all the figures of this kind in this report.

Z width.

The Z width is the only lineshape direct observable sensitive to the top quark mass. Fig. 15 shows the LEP results together with the MSM prediction as a function of m_t. In this measurement, statistical and systematical uncertainties are presently of the same order. The main systematic comes from the uncertainty in the difference in LEP energy between scan points. During 1993 the resonant depolarization technique was used in the three scan points and this has brought this error down to a preliminary estimation of 2.7 MeV which might improve down to about 1.7 MeV [17]. The second most important source of systematics comes from the uncertainty in the background from non-resonating processes like two-photon collisions. The current error is close to

2 MeV per experiment but incoherent among them.

Hadronic peak cross section.

The measurements by the four collaborations of the hadronic peak cross section, σ_h^0, are shown in fig. 16. As can be seen, this measurement is already dominated by the common systematical error due to the theoretical uncertainty in the low-angle Bhabha cross section. The uncertainty used has been 0.25% but recent claims reduce it to around 0.15% [21]. The rest of systematical uncertainties (the knowledge of the efficiency and background of the hadron selection, which contributes about 0.2% per experiment, and the experimental uncertainty in the measurement of the absolute luminosity which, after the upgrade in the luminosity set-ups of some detectors is better than 0.1%) are uncorrelated and can probably be improved.

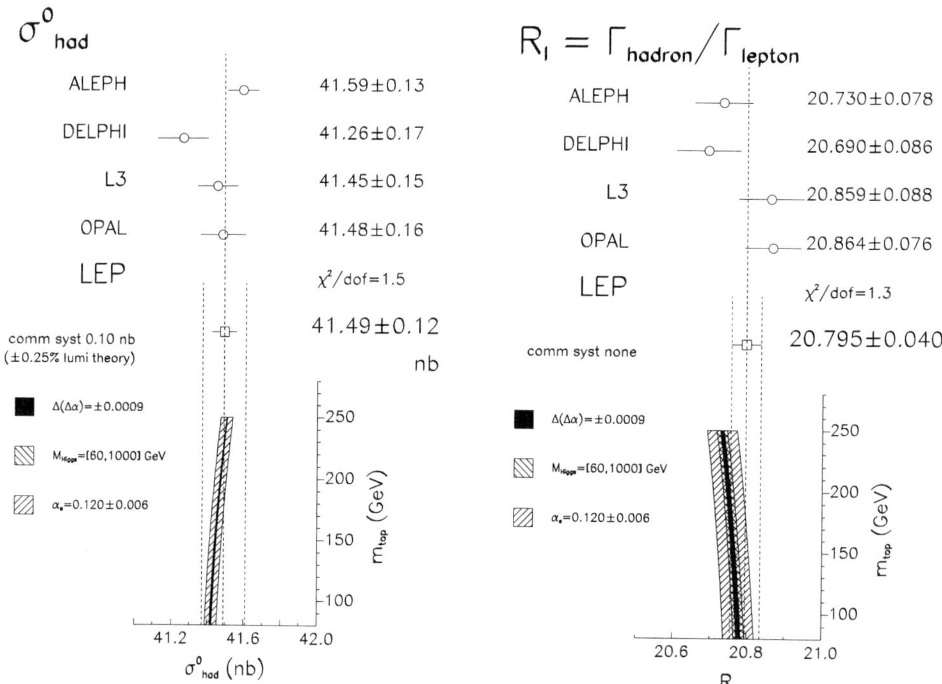

Figure 16: The hadronic peak cross section and its ratio to the leptonic one measured by the four experiments together with the mean and the MSM prediction as a function of the top quark mass.

Hadronic over leptonic width ratio R_l.

The results of the measurements of R_l, the ratio of the hadronic to the leptonic Z partial widths are given in fig. 16. The experimental systematic error is dominated by the knowledge of the efficiencies and backgrounds in the leptonic channels: about 0.5% per experiment but uncorrelated among them. On top of this, there is a common error coming from the t-channel correction in the electron channel, which contributes 0.1% to the error in R_l. This uncertainty is directly due to the lack of a full $\mathcal{O}(\alpha^2)$ Monte Carlo event generator for Bhabha scattering.

Heavy quark partial widths

The measurement of the Z decay width into b hadrons is especially important because, within the Minimal Standard Model, it receives a vertex correction involving the top quark which is absent from any other final state. By taking the ratio $R_b = \Gamma_b/\Gamma_h$, most of the vacuum polarization corrections depending on the top quark and the Higgs mass cancel out, and one is left with the following approximate expression already discussed:

$$R_b \simeq R_d \cdot \left[1 - \frac{20}{13} \frac{\alpha}{\pi} \left(\frac{m_t^2}{M_Z^2} + \frac{13}{6} \log \frac{m_t^2}{M_Z^2} \right) \right] \tag{81}$$

Therefore, R_b has a singular rôle since its accurate measurement should provide a determination of m_t independent on M_H, which is something that none of the rest of precision measurements can do. Anyway, the effect of the top quark vertex corrections is only of order 2% for a top mass of 150 GeV. Therefore only a precise measurement, to better than 1%, is useful to get information on the top mass. With the new preliminary measurements just made available by the LEP collaborations the overall error has reached a very interesting 0.9%

The relatively large b mass (~ 4.7 GeV) and lifetime ($\sim 1.5 \times 10^{-12}$ s), makes possible the use of its decay kinematics to have the largest identification efficiency and purities among all the quarks. Three methods have been used to tag b events at LEP:

- Lepton tag: It uses high p, high p_t leptons from b decays. High purity can be achieved but one has to pay for the small semileptonic b branching ratio.

- Event shape tag: High mass, high momentum b mesons or baryons give raise to particular event shapes which have been used to tag b events with high efficiency although rather modest purity. Recognition has been optimized using Neural Network techniques.

- Lifetime tag: The long lifetime of the b quark can be used, using silicon microvertex detectors, to tag b events by looking for tracks not coming from the Z production vertex. This is currently the best performing method with both high purity and efficiency.

The main systematic errors come from the evaluation of the efficiency and the background of the selection. The best option is to try to use data to estimate both. In the case of the efficiency, the techniques mentioned above can be used to tag only one hemisphere and look at the other one to measure the tag efficiency. Similar techniques could also be used for the backgrounds. The results from the four collaborations are shown in fig. 17.

The c quarks can be tagged in two different ways:

- using b tag techniques (lepton tag, event shape) extending them to the lower p and p_t regions and fitting then simultaneously the b and c information,

- through the reconstruction on charmed meson decays. The cleanest one is $D^{*+} \to D^0 \pi^+$ and then $D^0 \to K^- \pi^+$ and its charge conjugate, because the low $D^* - D^0$ mass difference produces the signature of a soft π acompaning a D^0 with opposite charge to the K.

In either case, the efficiencies and purities are much lower than for b quarks and the dependence on external input for the product and decay branching ratios very

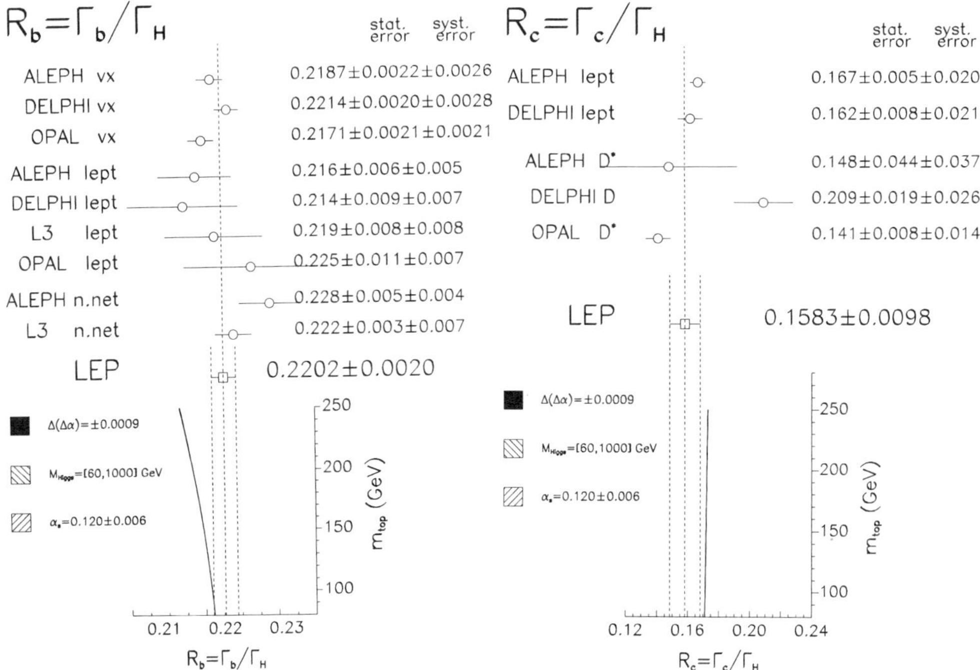

Figure 17: The ratios of the Z partial widths to b and to c hadrons to the one to all hadrons measured by the four experiments together with the mean and the MSM prediction as a function of the top quark mass. The error bars reflect the full uncertainty.

important. The results from the four collaborations are shown in fig. 17, where one can see that the agreement between the different measurements of both quantities is excellent. Usually the selection cuts are chosen to optimize the ratio between statistical and systematical uncertainties and, in this sense, these measurements are not yet systematically limited. At any rate, the weight of correlated systematical uncertainties is very important in the final errors and therefore, the proper averaging of the measurements from the different experiments and using different techniques is non-trivial. A detailed discussion, which also applies to the heavy flavour asymmetries to be presented later, can be found in [22]. One relevant conclusion of such study is that, given the contamination of b events into the c sample and viceversa the correlation between the final averages of R_b and R_c is as big as ~ -0.4.

For both measurements, the agreement with the MSM prediction is modest. In the case of R_b, the current central value of the mean corresponds to a negative value for m_t^2 and it is more than 1.5σ higher than the MSM for $m_t > 150$ GeV, as favoured by the other electroweak measurements. In the case of R_c the measurement is about 1.5σ lower than the MSM prediction, which is extremely precise.

Given the anticorrelation of both measurements, due to the intercontamination of both samples, it makes sense trying to check which is the discrepancy of their sum with respect to the MSM prediction. The result is

$$R_b + R_c = 0.3785 \pm 0.0092$$

to be compared to the MSM prediction for $m_t = 175$ GeV which is $R_b + R_c = 0.3878$, so that the difference is just about one sigma. Therefore, with the present data, the individual discrepancies could be explained if, for any reason, the border defining which heavy flavor events should be classified as b and which as c would be placed inconsistently in data and in theory.

Γ_u, Γ_d partial widths.

For light quarks, the use of radiative events with photons coming from final state radiation, has been advocated as a technique to disentangle u-type from d-type partial widths. The photon is supposed to probe the quark charge and therefore, should see $+2/3$ for u-type quarks and $-1/3$ for d-type quarks. The naif expectation is therefore:

$$\Gamma_{q\bar{q}\gamma} \sim (8\Gamma_u + 3\Gamma_d) \quad \text{and} \quad \Gamma_{q\bar{q}} \sim (2\Gamma_u + 3\Gamma_d) \tag{82}$$

and hence, the combination of both measurements should allow the determination of Γ_u and Γ_d. The present LEP average (coming from DELPHI, L3 and OPAL) is:

$$\Gamma_{u-like} = 244 \pm 39 \text{ MeV} \quad \text{and} \quad \Gamma_{d-like} = 419 \pm 27 \text{ MeV}$$

in good agreement with the MSM predictions

$$\frac{1}{2}(\Gamma_u + \Gamma_c) = 297 \pm 3 \text{ MeV} \quad \text{and} \quad \frac{1}{3}(\Gamma_d + \Gamma_s + \Gamma_b) = 381 \pm 3 \text{ MeV}$$

Nevertheless, there is not yet universal agreement about the precise validity of the argument: does really the photon probe directly the electric charge of the primordial quark or the one of its fragmentation products ?. In fact the QCD corrections estimated with Monte Carlo are of about a 1.7 factor and therefore some experiments suggest using these events just as a QCD model test.

3.1.2   Asymmetries

The measurement of the different asymmetries near the Z pole provides direct determinations of the effective weak mixing angle, $\sin^2 \theta_{eff}^{lept}$, [12] defined via the ratio of the effective vector and axial lepton couplings to the Z:

$$\sin^2 \theta_{eff}^{lept} = \frac{1}{4}\left(1 - \frac{g_{V_l}}{g_{A_l}}\right) \tag{83}$$

Forward-backward asymmetries are measured for all tagged flavors (e,μ, τ,c and b) and inclusively for hadrons (jet charge asymmetry). For leptons the expected forward-backward asymmetry at the Z pole is very small ($\sim 1.5\%$) due to the smallness of the lepton vector coupling to the Z whereas it changes very rapidly with energy. Therefore, the precise determination of the peak asymmetry requires a proper handling of the energy dependence (mainly of the difference between the measurement energy $\sqrt{s_i}$ and M_Z). Because of that, the asymmetries measured at different energies are fitted together with the lineshape data to extract the peak asymmetry. For heavy flavors, the fact that the expected asymmetries are large ($\sim 10\%$ for b and $\sim 7\%$ for c), makes

---

[12] It should be remarked that the angle presented here and used in the following is defined via the ratio of the charged lepton couplings: the angle determined from quark final states is (slightly) corrected to this definition.

their energy dependence much less relevant than for leptons and then it is properly accounted for by applying a correction just at the end, as we shall see.

At LEP, since the beams are unpolarized, only final state polarization asymmetries can be measured. Among the leptons, only the tau decay inside the detectors and their polarization can be inferred from the momentum distribution of their decay products. The expected tau polarization asymmetry is large ($\sim 15\%$) and its energy dependence is very small so that in practice its proper handling only requires a small correction. The expected forward-backward polarization asymmetry is also large ($\sim 11\%$) and its energy dependence is also very small.

Lepton Forward-Backward Asymmetry

The forward-backward asymmetry is defined as

$$A_{FB} = \frac{\sigma_F - \sigma_B}{\sigma_F + \sigma_B}, \tag{84}$$

where F and B indicates the forward or backward hemisphere. Normally it is obtained by fitting the measured angular distribution to the formula

$$\frac{d\sigma^l}{d\cos\theta}(s) = \frac{3}{8}\sigma^l(s) \cdot \left(1 + \cos^2\theta + \frac{8}{3}A^l_{FB}(s)\cos\theta\right). \tag{85}$$

In the case of e^+e^- final state, the t-channel contribution is either subtracted from the observed asymmetry or added to the previous expression.

Once the different $A^l_{FB}(s_i)$ are obtained, they are fitted together with the lineshape data to get the the lineshape parameters mentioned above and the peak asymmetry, $A^{0,l}_{FB}$:

$$\begin{aligned} A^{0,l}_{FB} &= \frac{3}{4}A_e A_l \\ A_f &= \frac{2g_{V_f}/g_{A_f}}{1 + \left(g_{V_f}/g_{A_f}\right)^2}. \end{aligned} \tag{86}$$

from which the effective weak mixing angle is measured. The results of the four collaborations are shown in fig. 18 where it can be seen that the agreement between the experiments (and, in particular, between ALEPH and OPAL) is not excellent. The main error is still statistical. Experimental systematics can only come from simultaneous charge and forward-backward asymmetries in the detector, which are bound to be very small. The knowledge of the beam energy contributes a non-negligible 0.0008 to $\Delta A^{0,l}_{FB}$, although this can be improved.

Jet charge Forward-Backward Asymmetry.

Given the lack of techniques to tag efficiently the light quark flavors, most of the LEP experiments compute inclusively the forward-backward asymmetry in all hadronic events by estimating the average quark charge via a momentum-weighted mean of charges of the hadrons belonging to each quark's jet. This way, the inclusive forward-backward asymmetry for the actual Z decay mixture of quark flavors can be extracted. The precise definition of the observable is the following:

$$\langle Q_{FB}\rangle = \left\langle \frac{\sum_F q_i p^\kappa_{iL}}{\sum_F p^\kappa_{iL}} - \frac{\sum_B q_i p^\kappa_{iL}}{\sum_B p^\kappa_{iL}} \right\rangle = cA_e \sum_q \delta_q A_q \frac{\Gamma_q}{\Gamma_h} \tag{87}$$

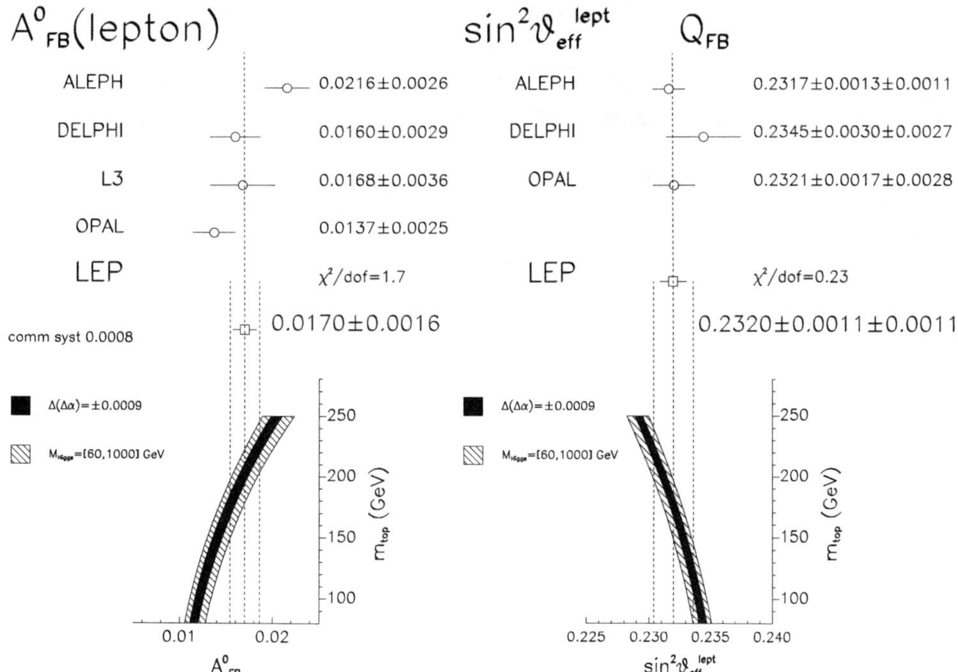

Figure 18: The leptonic forward-backward asymmetry and the effective weak mixing angle obtained from the quark forward-backward asymmetry measured by the LEP experiments together with the mean and the MSM prediction as a function of the top quark mass.

where c is a normalization constant, p_L is the longitudinal momentum along the thrust axis, the quantity δ_q is called the charge separation and measures the difference between the two hemisphere charges for a given flavour and κ is an arbitrary weight power, chosen to optimize δ_q. For b quarks, δ_q can be obtained from data, using semileptonic events, for instance. For the rest, it is obtained from Monte Carlo, although work is in progress to get the c quark contribution also from data. The uncertainty in the determination of δ_q is the dominant systematic error, which in turn, dominates by far the total error.

The measurements from three of the four collaborations are shown in fig. 18. The resulting uncertainty in $\sin^2\theta_{eff}^{lept}$ is competitive with the other measurements, although its still strong dependence on the Monte Carlo modelling for light quarks, together with the discrepancies among the experiments concerning the criteria to estimate the Monte Carlo uncertainty, might limit further improvements.

Heavy Flavour Forward-Backward Asymmetries

The heavy quark forward-backward asymmetries provide presently the most precise determination of the effective weak mixing angle at LEP. What is measured is

$$A_{FB}^{0,q} = \frac{3}{4}A_e A_q.$$

(88)

where A_q is $\sim 0.66, 0.93$ for c- and b-type quarks, respectively and depends only mildly on $\sin^2 \theta_{eff}^{lept}$. Therefore, the asymmetry is quite large and mainly sensitive to the $\sin^2 \theta_{eff}^{lept}$ dependence of A_e. The main difficulties measuring the forward-backward asymmetry for quark final states are the flavour identification and the charge assignment.

The techniques presently used to identify b and c quarks have already been discussed in section 3. For b quarks, the charge assignment is done in two different ways: either the charge of a high p, high p_t lepton from semileptonic decays which identify the b events is also used to extract the charge of the parent quark, or the lifetime information in one hemisphere is used to tag the event while the weighted mean charge (jet charge) in the other hemisphere measures the quark charge. The two methods lead to samples almost completely statistically independent. The systematical errors differ as well: in the first method, the knowledge of the lepton purities and of the semi-leptonic branching ratios is crucial; in the second, the charm background in the b sample is the main worry.

For c quarks, the charge is extracted either from the lepton charge, like for b's, in global fits using the low p_t leptons from semileptonic decays or, from the charge of the $D^{*\pm}$ meson. Results using these methods are shown in fig. 19 for b and c quarks. The

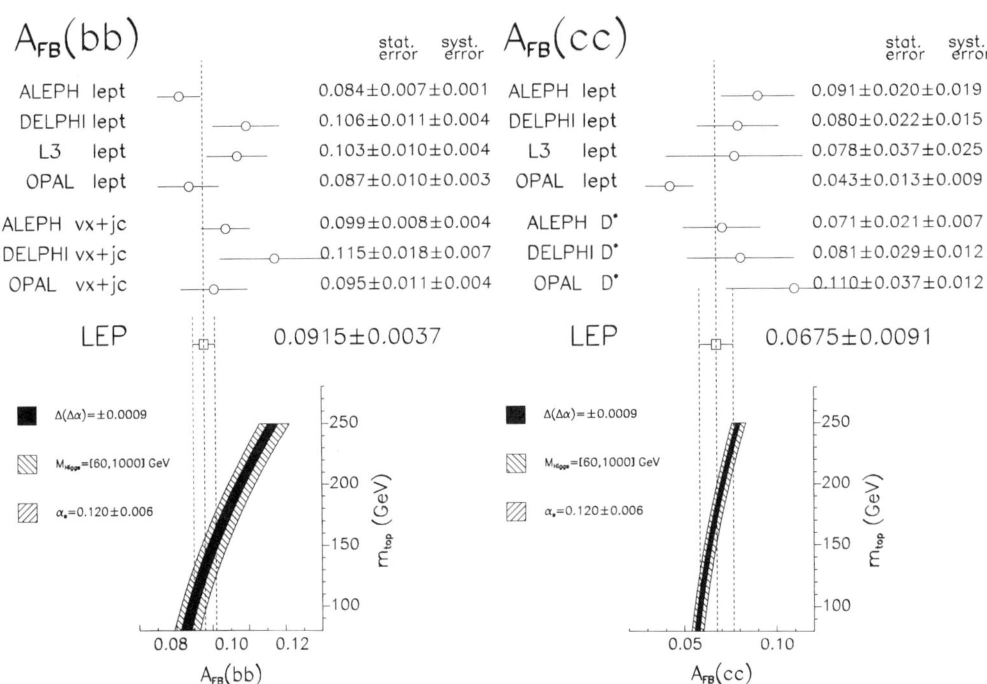

Figure 19: The b and c quark forward-backward asymmetry measured by the four experiments. The error bars reflect the full uncertainty

results given in fig. 19 correspond to the "row" asymmetries measured at the average LEP energy and therefore, they have to be corrected for QED, QCD and energy effects to obtain $A_{FB}^{0,q}$ as appearing in eq. 88. After the corrections, the results are

$$\begin{aligned}
A_{FB}^{0,b} &= 0.0967 \pm 0.0038 \\
A_{FB}^{0,c} &= 0.0760 \pm 0.0091
\end{aligned} \tag{89}$$

Tau Polarization

The measurement of the final state longitudinal polarization asymmetry for tau leptons:

$$< P_\tau >= \frac{\sigma_R - \sigma_L}{\sigma_R + \sigma_L} = -A_\tau, \tag{90}$$

where $\sigma_{R(L)}$ is the integrated cross section for right (left) handed tau's, provides a means to measure the tau couplings to the Z directly. Since the tau decay inside the detectors, their helicity information can be obtained by using their parity-violating weak decays as a spin analyzer. The main decays are used: to electron, muon, pion, rho, and a_1. Maximal sensitivity is obtained for the semileptonic modes (pion and rho) since then, just one neutrino escapes detection. In this case, the decay angle of the hadron in the tau rest frame (or equivalently its scaled energy) is measured and the integrated tau polarization is extracted from

$$\frac{1}{N}\frac{dN}{dx} = 1 + \alpha < P_\tau > \cos \theta^* \tag{91}$$

where α is a sensitivity coefficient linked to the spin of the hadron. For pions (s=0) $\alpha = 1$ and for rho and a_1 (s=1) $\alpha < 1$ but can be improved by studying the hadron helicity through the analysis of its decay products.

The results are given in fig. 20. The systematics are in this case comparable to the statistical errors and their reduction would require a lot of effort in the understanding of the calorimeters. In the π channel they come from the knowledge of the $\pi^-\pi^0$ background and of the energy dependence of the pion detection efficiency. In the ρ channel, the dominating uncertainty comes from the separation of the neutral and charged pions, because the their energy difference is used to measure the ρ decay angle.

Tau Polarization Forward-Backward Asymmetry

By measuring the tau polarization as a function of the the tau production angle, θ one can write

$$P_\tau(\cos \theta) = -\frac{A_\tau + A_e \cdot \frac{2\cos\theta}{1+\cos^2\theta}}{1 + A_e A_\tau \cdot \frac{2\cos\theta}{1+\cos^2\theta}} . \tag{92}$$

From this expression its is apparent that, while the integrated polarization measures A_τ (as seen in the previous section), its forward-backward asymmetry measures the electron coupling, A_e.

The results are shown in fig. 20. The analysis of the forward-backward polarization asymmetry is more complicated than the extraction of the integrated polarization because not only the L-R dependence of efficiencies, intercontaminations and backgrounds has to be known but also its angular dependence, which requires computing all these numbers for different bins in $\cos \theta$. Nevertheless, the error is still mainly statistical, since most systematic effects cancel out when computing the forward-backward asymmetry. The significantly smaller error claimed by ALEPH is due to its advantage in reaching $|\cos \theta| = 0.9$ compared to 0.7 for the other experiments. Figure 21 shows the dependence on $\cos \theta$ of the tau polarization as measured by ALEPH together with the best fit when universality is or is not assumed.

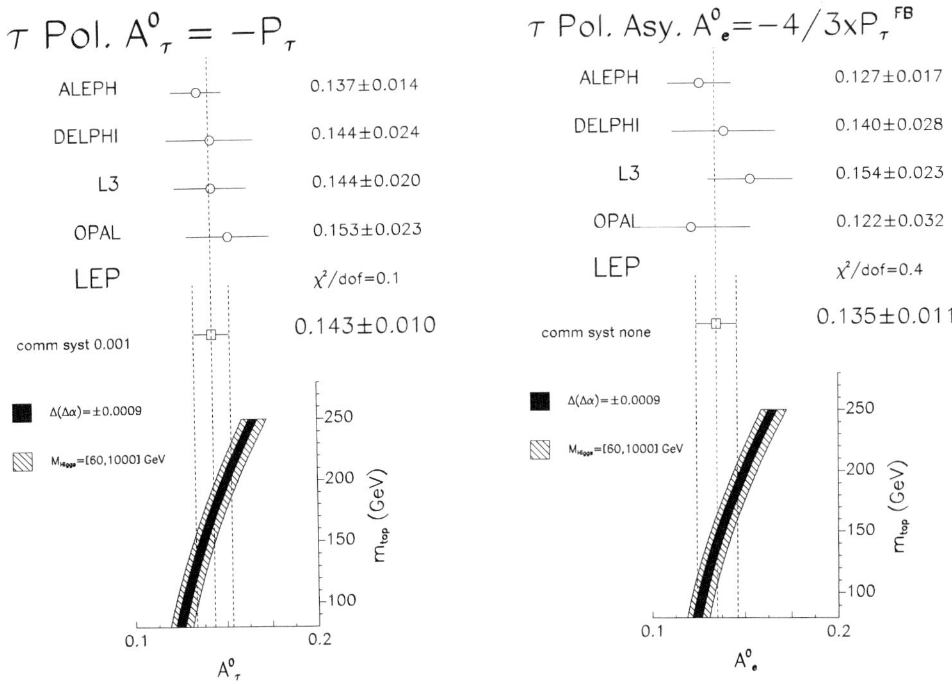

Figure 20: The tau polarization and its forward-backward asymmetry measured by the four experiments together with the mean and the MSM prediction as a function of the top quark mass.

Left-right Polarization Asymmetry

The SLD experiment measures the initial state longitudinal polarization asymmetry by using an strongly polarized electron beam colliding against an unpolarized positron beam at the SLC machine (see figure 22). This asymmetry is defined as:

$$A_{LR} = \frac{\sigma_L - \sigma_R}{\sigma_L + \sigma_R} \simeq A_e \tag{93}$$

where $\sigma_{R(L)}$ stands now for the integrated cross section for right (left) handed electrons, and the last equality is exact modulus small corrections due to ISR and photon exchange. In practice, final state hadronic and tau Z decays calorimetricaly selected, are counted for each of the two longitudinal polarizations of the electron beam and a *measured asymmetry* is defined as:

$$A_m = \frac{N_L - N_R}{N_L + N_R} \tag{94}$$

The extraction of A_{LR} from this direct measurement requires the knowledge of the electron beam polarization P_e. For that, a Compton polarimeter placed at 33 meters downstream of the interaction point is used. This instrument reaches an statistical precision in P_e of about 1 % in 3 minutes of operation while its systematic limitation has been estimated to be 1.3 % and comes mainly from the precision in the polarization

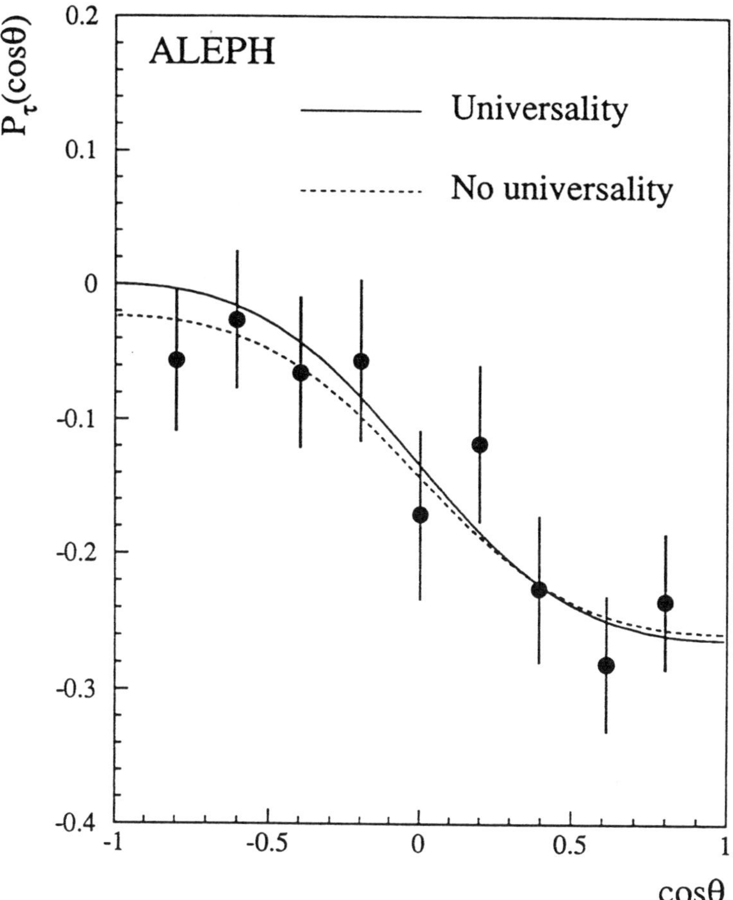

Figure 21: The ALEPH tau polarization measurements as a function of $\cos\theta$ together with the best fits.

of the laser source used to produce the Compton scattering. A Moeller polarimeter placed at the end of the LINAC, before the SLC arcs, is used for cross-checking. The polarization measured in the Compton setup is about 4-5 % smaller, which is consistent with the expected loss due to the polarization transport in the arcs. The average polarization is computed using the Compton polarimeter measurement for every recorded Z event P_i

$$< P_e >= (1 + \xi)\frac{1}{N_Z} \sum_{i=1}^{N_Z} P_i = 0.630 \pm 0.011 \tag{95}$$

where N_Z is the total number of Z events and $\xi = 0.017 \pm 0.011$ is a chromaticity correction due to aberrations in the final focus optics.

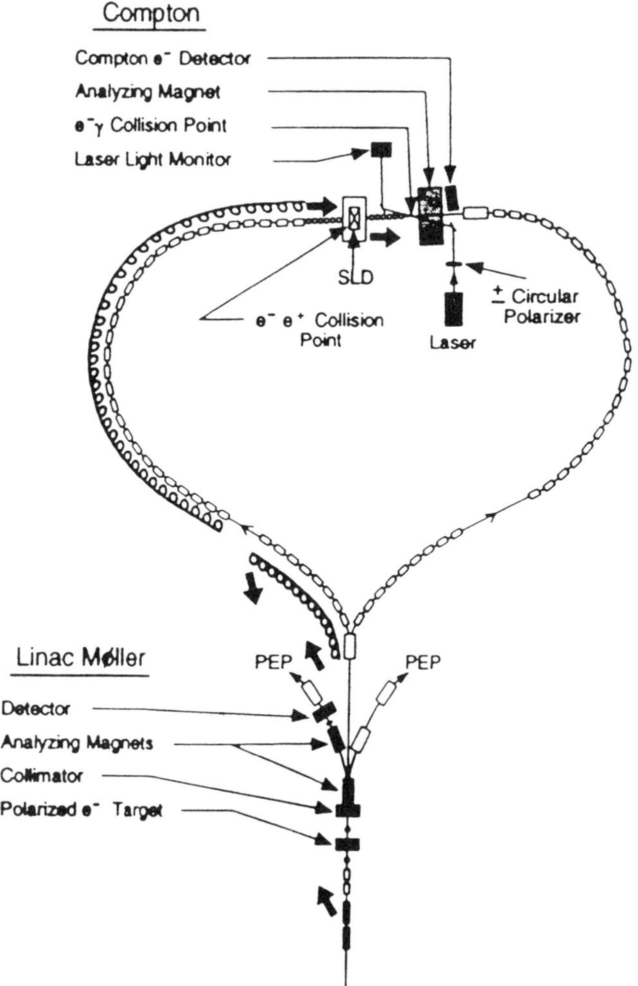

Figure 22: The SLC polarization setup.

Then A_{LR} is computed as

$$A_{LR} = \frac{A_m}{<P_e>} + \Delta A_{LR} \tag{96}$$

where $\Delta A_{LR} = 0.015 \pm 0.012$ is a small correction which accounts for the left-right asymmetry of the background, the efficiencies, the luminosity and the energy. The result for the 93' data obtained by using the previous numbers is

$$A_{LR} = 0.1628 \pm 0.0071(stat.) \pm 0.0028(syst.)$$

so that this measurement is still statistically limited and therefore, a large improvement is expected in the near future. Combining this measurement with the less precise 92' data leads to $A_{LR}^0 = A_e = 0.1637 \pm 0.0075$, corresponding to $\sin^2 \theta_{eff}^{lept} = 0.2294 \pm 0.0010$.

This measurement can be directly compared with the tau forward-backward polarization asymmetry from LEP since, in a rather model independent way, both determine the same quantity: A_e. The SLD measurement $A_e = 0.1637 \pm 0.0075$ turns out to be at about 2.3σ from the LEP forward-backward tau polarization average $A_e = 0.135 \pm 0.011$. Given the statistical origin of the uncertainties claimed, this difference should be assigned to an statistical fluctuation.

3.2 Measurement of the W mass.

Real W bosons cannot be produced yet by the existing e^+e^- machines in spite that this is the fundamental goal of the forecoming LEP-II program. They were discovered in $p\bar{p}$ colliders where the energy allows their production but the experimental conditions make their detection and study more difficult. The first measurements came from the UA1 and UA2 experiments at the CERN SPS collider but, so far the most accurate determinations of the W mass come from the 93' runs of the CDF and D0 experiments at the TEVATRON collider at an energy of $\sqrt{s} = 1.8TeV$ with an integrated luminosity of about $20pb^{-1}$.

Most of these experiments tag W events by their leptonic decay. Candidates are requested to have either an electron or a muon with high transverse momentum (typically larger than 25 GeV), high transverse missing energy (typically larger than 25 GeV) and a high reconstructed transverse mass of the W (typically between 60 and 100 GeV) on top of fiducial isolation cuts. Using these events the W mass is extracted from the fit to the observed transverse mass spectrum. For that, an important issue is the actual energy calibration of the detector, which is done in different ways by the different experiments. For instance, CDF uses $J/\psi \to \mu\mu$ events to calibrate the muon momentum and uses a fit to the E/p spectrum of their $W \to e\nu$ candidates to the simulation to calibrate their electron energy scale. D0 selects leptonic Z decay candidates and rescales the invariant mass of the lepton pair to the LEP measurement of M_Z. In addition, one needs the evaluation of the background and the modelling of the transverse momentum of the hadronic recoil jet in the presence of the underlying event.

The dominant common source of uncertainty among all the experiments is of about 100 MeV and is due to the structure functions used. The combination of the results from UA2, CDF and D0 leads to [23]

$$M_W = 80.23 \pm 0.18 GeV$$

The accuracy of this measurement in the collider experiments is expected to increase with the increase of statistics up to about 60 MeV and with the operation of LEP-II up to few ten MeV.

3.3 Neutrino-nucleon scattering.

In fix target experiments with neutrino beams, the ratio R_ν defined as

$$R_\nu = \frac{\sigma_{\nu,NC}}{\sigma_{\nu,CC}} \tag{97}$$

where $\sigma_{\nu,NC}$ is the cross section for muonic neutrino scattering in nuclei thorough neutral currents ($\nu_\mu q \to \nu_\mu q$) and $\sigma_{\nu,NC}$ it the one thorough charged currents ($\nu_\mu q \to \mu q'$), provides an additional precision electroweak measurement. Three experiments have provided precise measurements of this ratio: CDHS and CHARM at the CERN neutrino facilities and CCFR at the FERMILAB neutrinos beams.

Within the MSM, this ratio can be written as

$$R_\nu = \frac{M_W^4}{M_Z^4}\frac{1}{2}(1+\epsilon) \tag{98}$$

where ϵ is an small correction

$$\epsilon = \frac{1 - 2s_W^2 + 10/9s_W^4(1+r)}{1 - 2s_W^2 + s_W^4} - 1 \simeq 0.05 \pm 0.003 \tag{99}$$

being

$$s_W^2 = 1 - \frac{M_W^2}{M_Z^2}$$
$$r = \frac{\sigma_{\bar{\nu},CC}}{\sigma_{\nu,CC}} \simeq 0.38 \tag{100}$$

so that R_ν measures basically M_W/M_Z.

Experimentally speaking, neutral currents and charged currents are distinguished by the characteristic penetration of the muons produced in charged currents. In the case of CCFR for instance, for neutral currents, nearly all the events have an "event length", defined as the the penetration depth detected by fired counters (one counter has about 10 centimeters of iron) shorter than about 30 while a large fraction of events originated by charged currents have much larger event length. The systematic uncertainties are dominated by the ν_e contamination in the ν_μ beam (because they originate $\nu_e e$ charged current interactions which are identified as coming from the neutral current ν_μ process), by the target modelling and by the charm production [24].

The present results, expressed as measurements of s_W^2 are:

$$s_W^2 = 0.2295 \pm 0.0035 \text{ (stat.+syst.)} \pm 0.005 \text{ (modell.) CDHS+CHARM}$$
$$s_W^2 = 0.2222 \pm 0.0026 \text{ (stat.)} \pm 0.0035 \text{ (syst.)} \pm 0.005 \text{ (modell.) CCFR}$$

which leads to a final average of $s_W^2 = 0.2256 \pm 0.0047$.

4 Interpretation of the measurements

The above measurements can be directly used to analyze some assumptions and extract some relevant free parameters of the theory. They can also be used to test their consistency among the different measurements within the framework of the MSM and, assuming it, to infer the value of some of its basic Lagrangian parameters.

4.1 Direct results

Tests of lepton universality.

If lepton universality is not assumed, then the measurements of the lepton line-shapes and asymmetries can be used to compare the couplings of the Z to the three charged lepton species. The comparison of the partial width of the Z into e, μ and τ shows perfect consistency with lepton universality (fig. 23) and the average provides a direct constraint to the top mass, because it does not depend on α_s, while Γ_Z does.

A deeper test on lepton universality can be carried out by analyzing simultaneously the information coming from the lepton partial widths and the leptonic forward-backward and tau polarization asymmetries. Then, following eqs.80 and 86 the vector

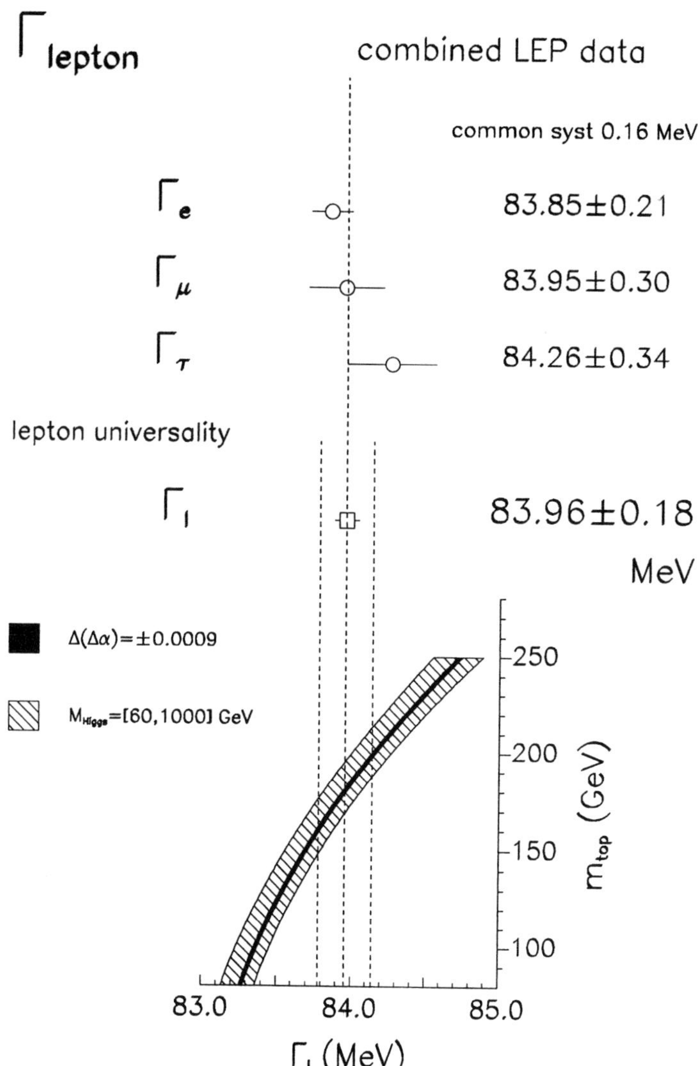

Figure 23: The partial widths measured at LEP for the three charged leptons, with the combined result corrected to a massless lepton.

and axial couplings of the Z to each lepton specie can be disentangled. The LEP results (see table 3 and fig. 24) show perfect agreement with the hypothesis of lepton universality for both, the vector and axial couplings.

$$g_{A_\mu}/g_{A_e}=1.0014 \pm 0.0021 \ , \ g_{A_\tau}/g_{A_e}=1.0034 \pm 0.0023 \ ,$$

$$g_{V_\mu}/g_{V_e} =0.83 \pm 0.16 \qquad , \ g_{V_\tau}/g_{V_e} =1.044 \pm 0.091 \ .$$

By performing a combined analysis of the hadronic and leptonic partial widths and asymmetries, one can directly determine the effective couplings for leptons and quarks [25]. This allows a direct test of universality in the quark sector. Figure 25 shows the

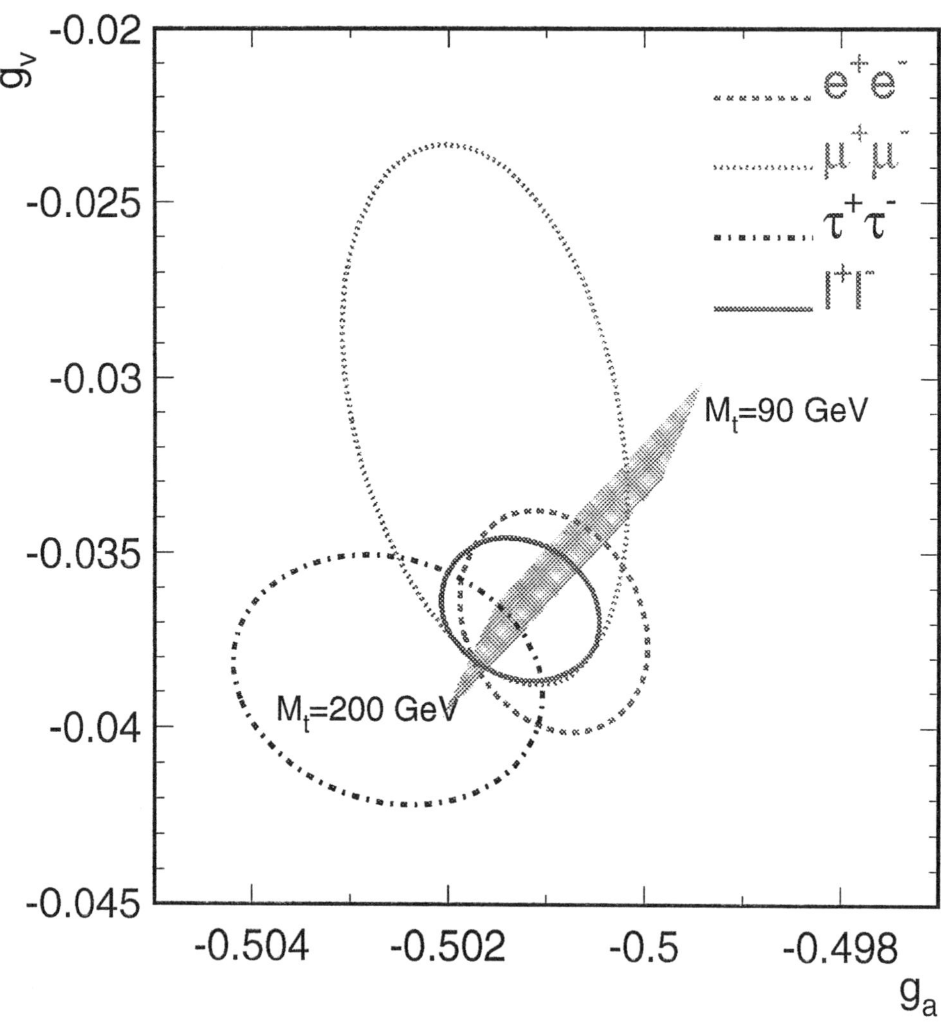

Figure 24: The 68% probability contours in the $g_{V_l} - g_{A_l}$ plane coming from the LEP leptonic measurements compared to the MSM predictions.

effective couplings for all the fermion species. For b and c quarks the one standard deviation contours (39% probability) are shown. The circles for u and d quarks come from the use of the partial widths obtained from the final state radiation in hadronic events and the same happens with the neutrino circle. The small rectangle for l^+l^- corresponds to the size of the enlarged view of lepton couplings given in fig 24 and shows clearly the fact that the precision on lepton couplings is, by far, much higher than for quark couplings.

Quantities derived from ratios.

The ratios of partial widths, due to the cancellation of universal $\Delta\rho$ corrections between numerator and denominator, allow the direct determination of important parameters of the theory without relying too much on the validity of the MSM. Extensions of

Table 3: Results for the leptonic effective vector and axial couplings without and with the assumption of lepton universality.

g_{V_e}	-0.0370 ± 0.0021
g_{V_μ}	-0.0308 ± 0.0051
g_{V_τ}	-0.0386 ± 0.0023
g_{A_e}	-0.50093 ± 0.00064
g_{A_μ}	-0.50164 ± 0.00096
g_{A_τ}	-0.5026 ± 0.0010
g_{V_ℓ}	-0.0366 ± 0.0013
g_{A_ℓ}	-0.50128 ± 0.00054

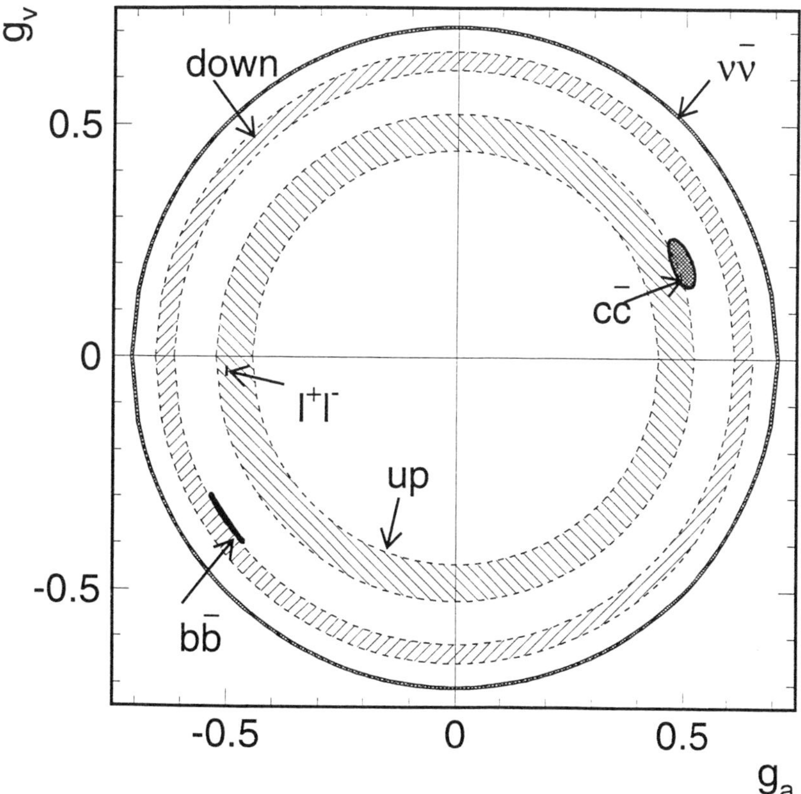

Figure 25: One standard deviation contours for the effective couplings obtained from the analysis of LEP data.

the MSM which would manifest themselves mainly via effects in vacuum polarization would produce very similar predictions for ratios.

As can be seen from fig. 16, R_l depends little on m_t and M_H whereas it has a large dependence on the strong coupling constant $R_l \sim R_l^0(1 + \alpha_s(M_Z)/\pi)$. Therefore it allows a direct determination of $\alpha_s(M_Z)$ with minimum theoretical uncertainties. From the LEP average of R_l and using the formulae suggested in [26], which relate R_l with the QCD prediction, known to $\mathcal{O}(\alpha_s^3)$, one gets

$$\begin{aligned}
\alpha_s(M_Z^2) &= 0.126 \pm 0.006 \pm 0.002_{ew} \pm 0.002_{QCD} \pm 0.003_{m_t,M_H} \\
&= 0.126 \pm 0.007
\end{aligned} \tag{101}$$

where the second and third error reflect uncertainties on the electroweak and QCD parts of the theoretical prediction respectively, and the last one comes from the lack of knowledge on the top quark and Higgs masses

The ratio of the invisible width, Γ_{inv}, to the leptonic width can be derived from the direct lineshape measurements (see fig.26) through the equation:

$$\frac{\Gamma_{inv}}{\Gamma_l} = \sqrt{\frac{12\pi R_l}{M_Z^2 \sigma_H^0}} - R_l - 3(1 + \delta_m) \tag{102}$$

where $\delta_m = -0.0023$ is an small correction which accounts for the tau mass effect. If one assumes that all the invisible width is due to neutrino final states, then one can derive the number of light neutrino species N_ν by writing

$$\frac{\Gamma_{inv}}{\Gamma_l} = N_\nu \cdot \frac{\Gamma_\nu}{\Gamma_l} \tag{103}$$

and taking the ratio Γ_ν over Γ_l from the MSM: $\Gamma_\nu/\Gamma_l = 1.992 \pm 0.003$. It should be noted the small error in the Minimal Standard Model prediction for this ratio, which does not depend on α_s and in which the top and Higgs mass dependences largely cancel. Using the LEP average for $\Gamma_{inv}/\Gamma_l = 5.953 \pm 0.046$ one obtains

$$N_\nu = 2.988 \pm 0.023$$

Since the result favours three species without any doubt, this measurement is actually a test of the MSM, a test of the assumptions made: that all invisible decays are to neutrinos, and of the value Γ_ν/Γ_l. If $N_\nu = 3$ is assumed, the measurement of N_ν can be turned into a measurement of Γ_ν/Γ_l:

$$\frac{\Gamma_\nu}{\Gamma_l} = 1.984 \pm 0.015$$

in good agreement with the MSM prediction and which can also be used to put limits on the mixing of extra neutral bosons to the Z, if one wants to avoid using external information on the strong coupling constant.

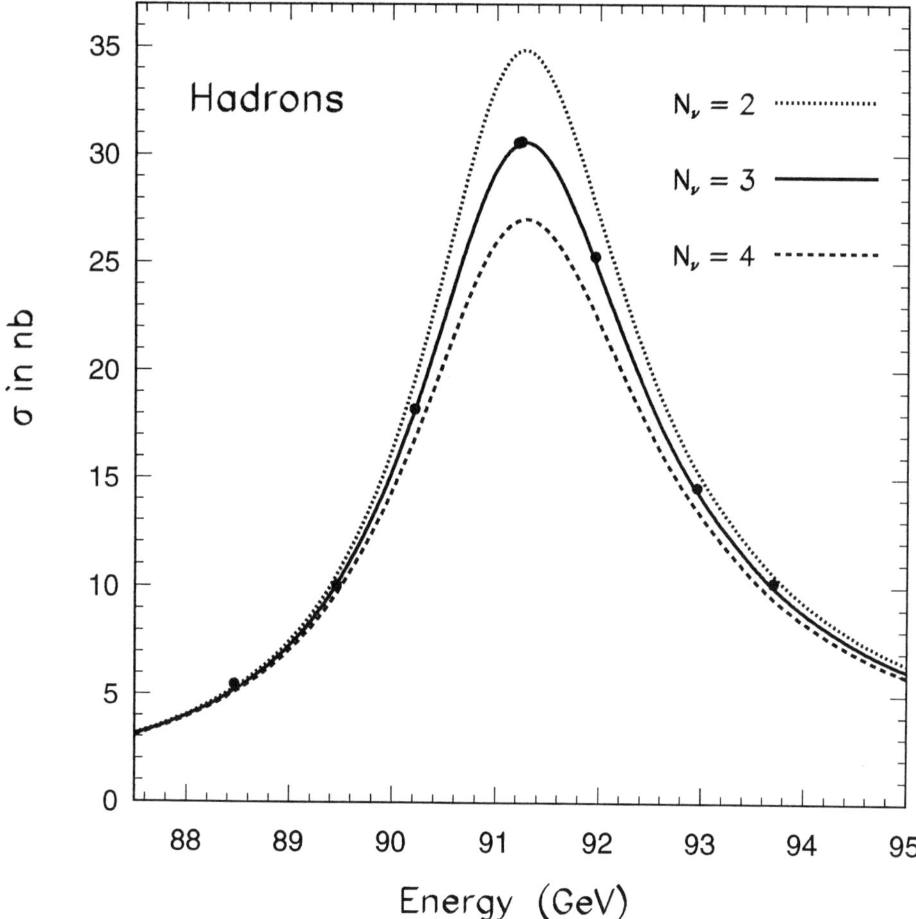

Figure 26: The hadron lineshape measured at ALEPH versus the expectation for $N_\nu = 2, 3$ and 4.

The effective sinus.

Figure 27 shows the compilation of all the values of $\sin^2 \theta_{eff}^{lept}$ obtained from the asymmetry measurements at LEP presented in the previous sections. The SLD determination coming from the measurement of the left-right polarization asymmetry [27] is also shown. Its discrepancy with the average of the LEP measurements alone $((\sin^2 \theta_{eff}^{lept})_{LEP} = 0.2321 \pm 0.0004)$ is of about 2.5 standard deviations. At any rate, the overall agreement of all the measurements is still acceptable and the mean value provides a very precise determination of the effective weak mixing angle, which is very sensitive to the top quark mass.

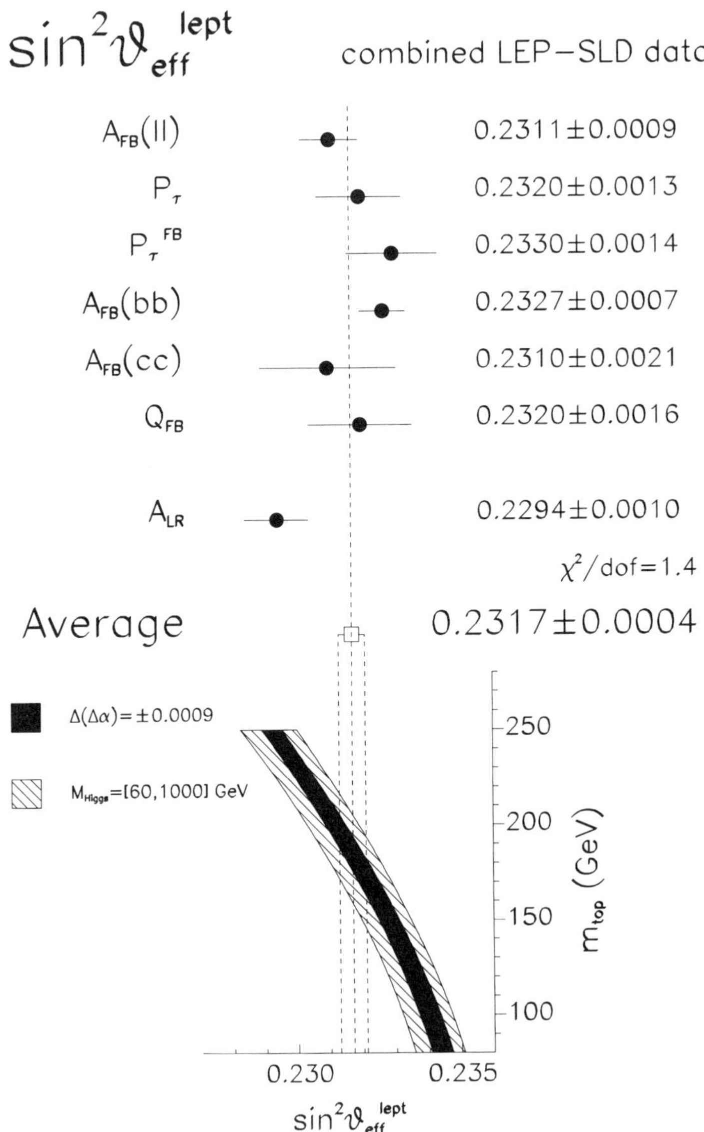

Figure 27: The LEP averages for $\sin^2\theta_{eff}^{lept}$ obtained from the different asymmetry observables discussed in the text, together with SLD measurement and the total mean. This mean is shown as a band in the Γ_b/Γ_{had} versus $\sin^2\theta_{eff}^{lept}$ plane together with the direct determination of Γ_b/Γ_{had} and the band corresponding to the R_l measurement (assuming $\alpha_s(M_Z) = 0.123 \pm 0.006$), compared to the MSM predictions.

4.2 Tests of radiative corrections consistency.

By using specific data sensitive to different electroweak radiative corrections or specific analysis variables such as the ε, one can check for consistency among the measurements within the theory at the quantum level. Since, as we have seen, the number of different leading non-trivial electroweak components of the radiative corrections for the presented observables is 4, in principle one should do a 4-dimensional analysis of the data. Nevertheless, restricting ourselves to the highest precision observables (Γ_l,

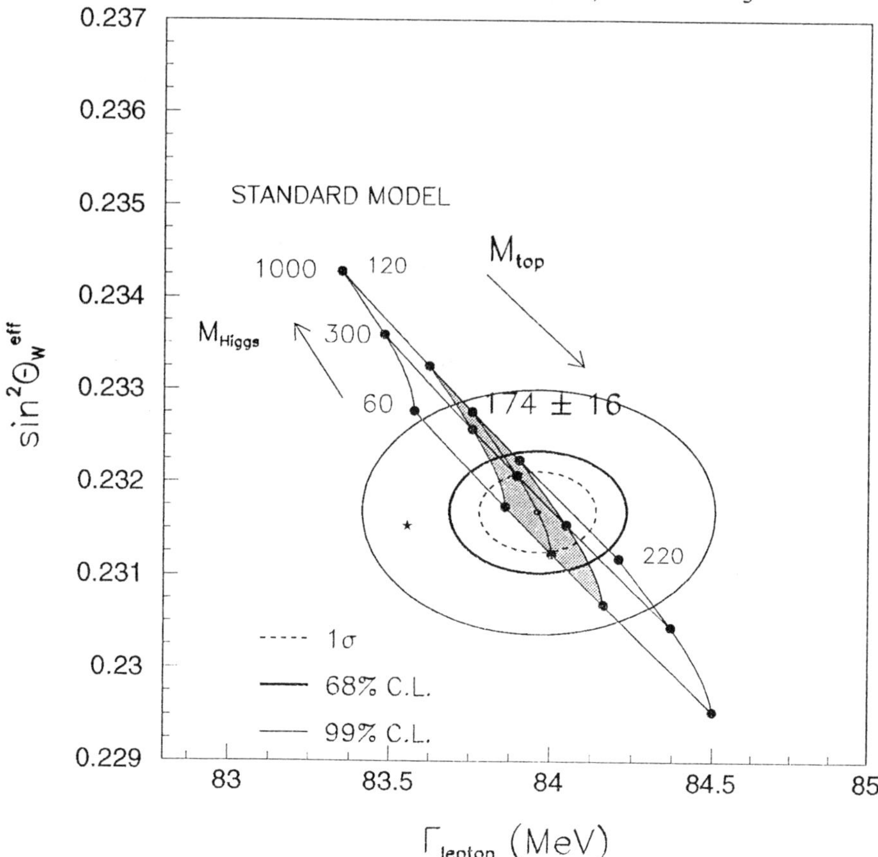

Figure 28: The contours from the direct measurement of Γ_l and $\sin^2 \theta_{eff}^{lept}$ compared to the MSM predictions.

$\sin^2 \theta_{eff}^{lept}$ and Γ_b) we can do a simplified 3-dimensional check from which we will show two projections.

The contours delimited in the Γ_l versus $\sin^2 \theta_{eff}^{lept}$ plain by the direct measurements are shown in figure 28 together with the MSM predictions as a function of the top and Higgs masses. The measurements are consistent with the MSM predictions for top masses in agreement with the CDF claim.

The information in the $\sin^2 \theta_{eff}^{lept}$ versus Γ_b/Γ_{had} plane is shown in figure 29. In this case, a part from the bands showing the direct measurements, a band shows Γ_b as indirectly determined through its contribution to the total hadronic width in R_l and σ_h^0 for instance. Out of these two measurements, R_l is the most powerful at present [28]. The value of Γ_b obtained by means of such an analysis turns out to be in perfect

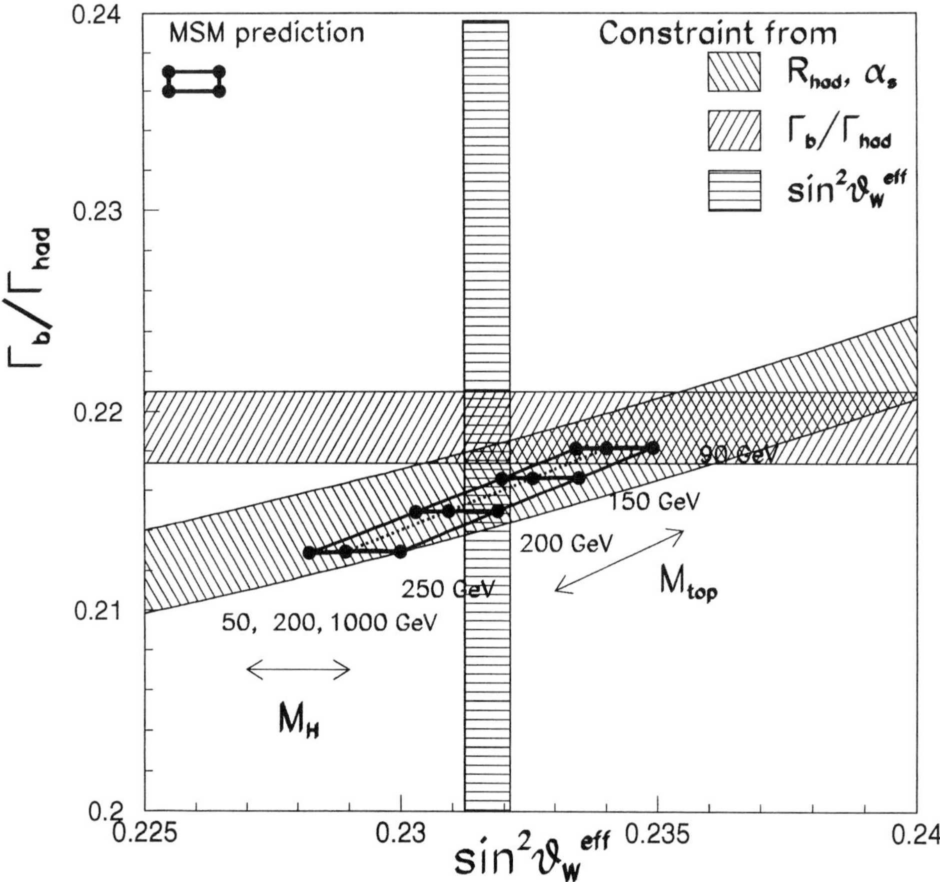

Figure 29: The average for $\sin^2\theta_{eff}^{lept}$ obtained from all the different asymmetry observables discussed in the text shown as a band in the Γ_b/Γ_{had} versus $\sin^2\theta_{eff}^{lept}$ plane together with the direct determination of Γ_b/Γ_{had} and the band corresponding to the R_l measurement (assuming $\alpha_s(M_Z) = 0.123 \pm 0.006$), compared to the MSM predictions.

agreement with the MSM prediction for the presently most favoured m_t value (see fig.29). This fact reinforces the conclusions discussed in the R_b, R_c section.

The results of the analysis of all the LEP and SLC data discussed here, using the ε language, are show in fig.30. There the 1σ (39% c.l.) contours obtained from the measurements for each ε are shown together with the MSM predictions for different values of m_t and M_H. In all these plots it is clear that the Born prediction (corresponding to $\varepsilon_i = 0$) is disfavoured by the data. In the ε_1 v. s. ε_3 plot, is clear that ε_1 is mainly sensitive to m_t and choses a value in the range claimed by CDF while ε_3 is sensitive to M_H and prefers a light Higgs. In the other two figures, ε_b is consistent with its Born expectaction zero $\varepsilon_b = 0$ deviating from the MSM behaviour as we have seen.

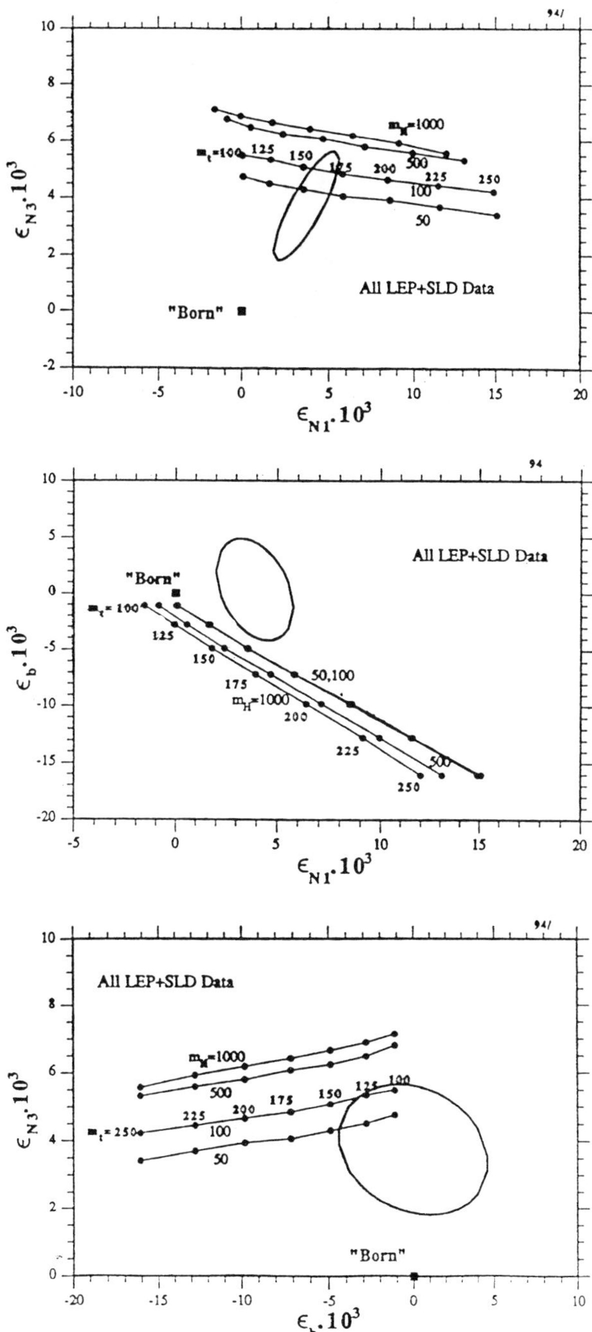

Figure 30: The 39 % c.l. contours obtained from fitting all the data to the ε parameters together with the MSM predictions as a function of m_t and M_H.

4.3 Standard Model Fits

The results presented in the previous sections can be interpreted in the context of the Minimal Standard Model allowing a check of its validity in describing all the observations, and enabling the inference of some of the MSM unknown (or badly known) parameters.

The comparison is made through a fit of the measurements shown in table 4 to their MSM predictions in terms of the top quark mass m_t and the value of $\alpha_s(M_Z)$. The Higgs boson mass is fixed to 300 GeV and, afterwards, varied in the range between 60 and 1000 GeV.

The most up-to-date MSM calculations have been used [11] and their estimated theoretical uncertainties have been also propagated in the fitting procedure, being the uncertainty $\Delta(\Delta\alpha)) = 0.0009$, due to the contribution of light quarks in the photon vacuum-polarization, the one that dominates by far [2].

The results of the fit can be gleaned from table 5. As can be seen in that table, given the current precision of LEP data, non-LEP data does not add very much information. The inclusion of the SLD measurement, however, has a clear effect on the total χ^2 and produces a significant shift on m_t. This is due to the present $\sim 2.5\sigma$ discrepancy with the rest of measurements already discussed. Nevertheless, the quality of all fits is good, and therefore we can claim that the MSM is able to describe well all the measurements discussed at their present (high) precision level. Moreover the m_t value resulting from these fits is in excellent agreement with the direct search evidence reported by CDF of $m_t = 174 \pm 10^{+13}_{-12}$ GeV. Hence, there is evidence, for the first time, that the bulk of non-trivial quantum effects in precision electroweak observables is indeed due to the top quark as predicted by the MSM.

The χ^2 of the fit including all the data increases by around 3.6 when M_H moves from 60 GeV to 1 TeV but this sensitivity is not supported by the expectation (see fig.31a). In fact, this fast χ^2 rise can be traced back to the effect of R_b preferring a very low top quark mass which, given the positive strong correlation among m_t and M_H induced by the rest of measurements, translates into an artificially fast χ^2 rise in the $\log(M_H)$ scale [29]. Therefore, the observed sensitivity is not robust with respect to fluctuations in the input data and has to be taken with care.

If the CDF determination of m_t is used as an additional constraint, then the agreement between the observed and the expected sensitivities becomes good (see fig.31b) almost independently of the choice of the measurements used in the fit. Therefore, the inclusion of m_t in the fit causes the inference of $\log(M_H)$ from the data being robust and hence reliable errors on $\log(M_H)$ can be estimated:

$$\Delta(\log_{10}(M_H/GeV)) \;=\; 0.5 \text{ at } 68\% \text{ C.L.}$$
$$=\; 1. \text{ at } 95\% \text{ C.L.}$$

At any rate, since the measurements are sensitive to $\log(M_H)$ and not directly to M_H, fluctuations in the position of the minimum (even if small in the $\log(M_H)$ scale in comparison to the width of the χ^2 parabola) correspond to very important changes in M_H. Therefore, in spite of the reliability of the present determination of $\log(M_H)$, its actual translation in terms of a measurement of M_H leads to conclusions which strongly depend on the chosen input data and on their fluctuations.

Therefore, being cautious, the only conclusion that should be stressed from fig. 31b is that the data seem to prefer a light Higgs and that the M_H value preferred by the data within the MSM is consistent with the validity of perturbation theory.

Table 4: Summary of measurements included in the combined analysis of Standard Model parameters. Section a) summarizes LEP averages, section b) electroweak precision tests from hadron colliders [23] and νN-scattering [24], section c) gives the result for $\sin^2\theta_{eff}^{lept}$ from the measurement of the left-right polarization asymmetry at SLC [27]. The Standard Model fit result in column 3 and the pulls in column 4 are derived from the fit including all data (Table 5, column 4) for a fixed value of $M_H = 300$ GeV.

	measurement	Correlation matrix	Standard Model fit	pull
a) <u>LEP</u>				
line-shape and				
lepton asymmetries:				
M_Z [GeV]	91.1888 ± 0.0044	1.00	91.1887	0.0
Γ_Z [GeV]	2.4974 ± 0.0038	0.04 1.00	2.4973	0.0
σ_h^0 [nb]	41.49 ± 0.12	0.01 -0.11 1.00	41.437	0.4
R_ℓ	20.795 ± 0.040	-0.01 0.01 0.13 1.00	20.786	0.2
$A_{FB}^{0,\ell}$	0.0170 ± 0.0016	0.04 0.00 0.00 0.01 1.00	0.0153	1.0
τ polarization:				
$\mathcal{A}_\tau$	0.143 ± 0.010		0.143	0.0
$\mathcal{A}_e$	0.135 ± 0.011		0.143	-0.7
b and c quark results:				
$R_b = \Gamma_{b\bar{b}}/\Gamma_{had}$	0.2202 ± 0.0020	1.00	0.2158	2.2
$R_c = \Gamma_{c\bar{c}}/\Gamma_{had}$	0.1583 ± 0.0098	-0.38 1.00	0.172	-1.4
$A_{FB}^{0,b}$	0.0967 ± 0.0038	-0.03 0.10 1.00	0.1002	-0.9
$A_{FB}^{0,c}$	0.0760 ± 0.0091	0.08 -0.07 0.12 1.00	0.0714	0.5
$q\bar{q}$ charge asymmetry:				
$\sin^2\theta_{eff}^{lept}$ from $\langle Q_{FB}\rangle$	0.2320 ± 0.0016		0.2320	0.0
b) $p\bar{p}$ and νN				
M_W [GeV] (CDF, D0 and UA2)	80.23 ± 0.18		80.31	-0.5
$1 - M_W^2/M_Z^2(\nu N)$	0.2256 ± 0.0047		0.2243	0.3
c) <u>SLD</u>				
$\sin^2\theta_{eff}^{lept}$ from $\mathcal{A}_e$	0.2294 ± 0.0010		0.2320	-2.6

Table 6 shows the differences in the best fit to the top mass when, for instance, $\alpha^{-1}(M_Z^2) = 129.01 \pm 0.06$ is used instead of $\alpha^{-1}(M_Z^2) = 128.79 \pm 0.12$. In spite that the central values move sizably, the conclusions about the consistency of the data with the MSM and the good agreement with the CDF m_t determination still hold. Concerning the information on M_H, figure 32 shows that the conclusions on $\Delta(\log_{10}(M_H/GeV))$ remain unchanged whereas the actual minimum is shifted towards higher M_H values. Therefore, in this case, the data does not prefer anymore a light Higgs and at the 95% c.l. M_H is not constrained (it can have heavier masses than the theoretically acceptable).

Table 5: Results of fits to LEP and other data for m_t and $\alpha_s(M_Z^2)$. No external constraint on $\alpha_s(M_Z^2)$ has been imposed. The central values and the first errors quoted refer to $M_H = 300$ GeV. The second errors correspond to the variation of the central value when varying M_H from 60 GeV to 1 TeV.

	LEP	LEP + Collider and ν data	LEP + Collider and ν data + A_{LR} from SLC
m_t (GeV)	$173^{+12}_{-13}\,{}^{+18}_{-20}$	$171^{+11}_{-12}\,{}^{+18}_{-19}$	$178^{+11}_{-11}\,{}^{+18}_{-19}$
$\alpha_s(M_Z^2)$	$0.126 \pm 0.005 \pm 0.002$	$0.126 \pm 0.005 \pm 0.002$	$0.125 \pm 0.005 \pm 0.002$
$\chi^2/(d.o.f.)$	$7.6/9$	$7.7/11$	$15/12$
$\sin^2\theta_{eff}^{\mathrm{lept}}$	$0.2322 \pm 0.0004\,{}^{+0.0001}_{-0.0002}$	$0.2323 \pm 0.0003\,{}^{+0.0001}_{-0.0002}$	$0.2320 \pm 0.0003\,{}^{+0.}_{-0.0002}$
$1 - M_{\mathrm{W}}^2/M_Z^2$	$0.2249 \pm 0.0013\,{}^{+0.0003}_{-0.0002}$	$0.2250 \pm 0.0013\,{}^{+0.0003}_{-0.0002}$	$0.2242 \pm 0.0012\,{}^{+0.0003}_{-0.0002}$
M_{W} (GeV)	$80.28 \pm 0.07\,{}^{+0.01}_{-0.02}$	$80.27 \pm 0.06\,{}^{+0.01}_{-0.01}$	$80.32 \pm 0.06\,{}^{+0.01}_{-0.01}$

Table 6: Results of fits to the whole set of precision data. No external constraint on $\alpha_s(M_Z^2)$ has been imposed. The central values and the first errors quoted refer to $M_H = 300$ GeV. The second errors correspond to the variation of the central value when varying M_H from 60 GeV to 1 TeV.

$\alpha^{-1}(M_Z^2)$	128.79 ± 0.12	129.01 ± 0.06
m_t (GeV)	$178^{+11}_{-11}\,{}^{+18}_{-19}$	$168^{+10}_{-10}\,{}^{+18}_{-21}$
$\alpha_s(M_Z^2)$	$0.125 \pm 0.005 \pm 0.002$	$0.125 \pm 0.005 \pm 0.002$
$\chi^2/(d.o.f.)$	$15/12$	$14/12$

5 Summary

The theoretical language needed to understand the physics contents of precision electroweak measurements has been briefly reviewed. The theoretical meaning of each parameter and its actual connection with the measurements has been discussed.

Using the most relevant Electroweak experimental data accumulated so far the precise determination of several electroweak parameters has been presented. Emphasis has been put in trying to show which measurements may still improve and which are already hitting systematic limits coming from machine energy uncertainty, experimental sources or theoretical limitations.

From the analysis of the precision measurements, conclusions have been extracted about basic ingredients of the theory, such as the numbers of neutrinos, or the value of the strong coupling constant. Moreover, tests on basic assumptions, like universality in the leptonic and the quark sectors, or consistency among the radiative corrections in the different observables, have been discussed.

The confrontation of the measurements with the Minimal Standard Model predic-

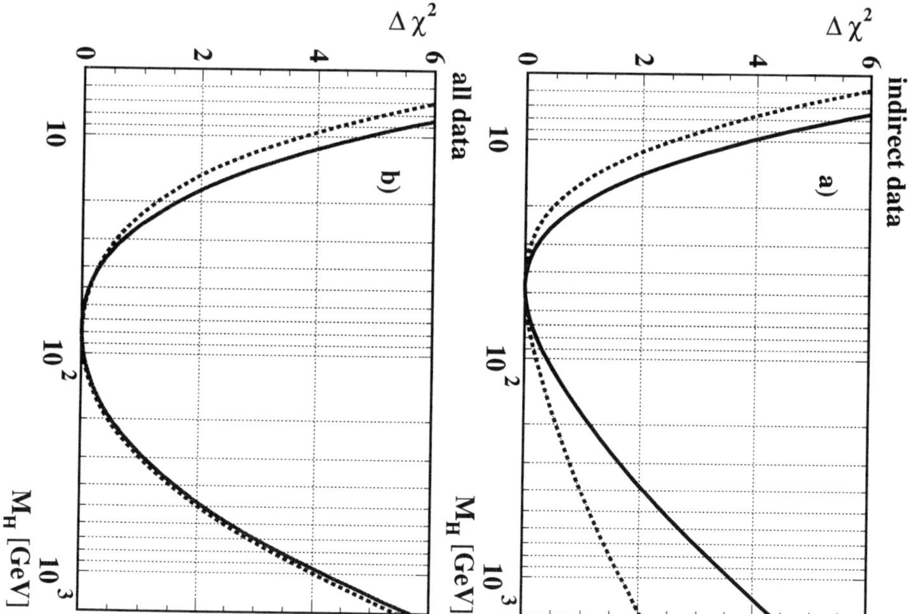

Figure 31: χ^2 vs $\log_{10}(M_\mathrm{H})$ as observed in the data (continuous line) and as predicted from theory (dashed line) using all data except CDF (a) and when the CDF determination of m_t is also included as an additional constraint in the electroweak fits (b).

tions shows perfect agreement at an unprecedent level of accuracy. The interpretation of the measurements in the MSM framework, allows a determination of the top quark mass with a ~ 20 GeV accuracy which agrees with the direct observation evidence by CDF. This agreement constitutes the first direct confirmation of the fact that the the top quark is the responsible for the bulk of Electroweak radiative corrections as predicted in the MSM.

Analyzing the precision Electroweak measurements together with the CDF m_t evidence within the MSM and assuming the applicability of perturbation theory in the Higgs sector, stable errors can be estimated for $\log(M_H)$ for the first time.

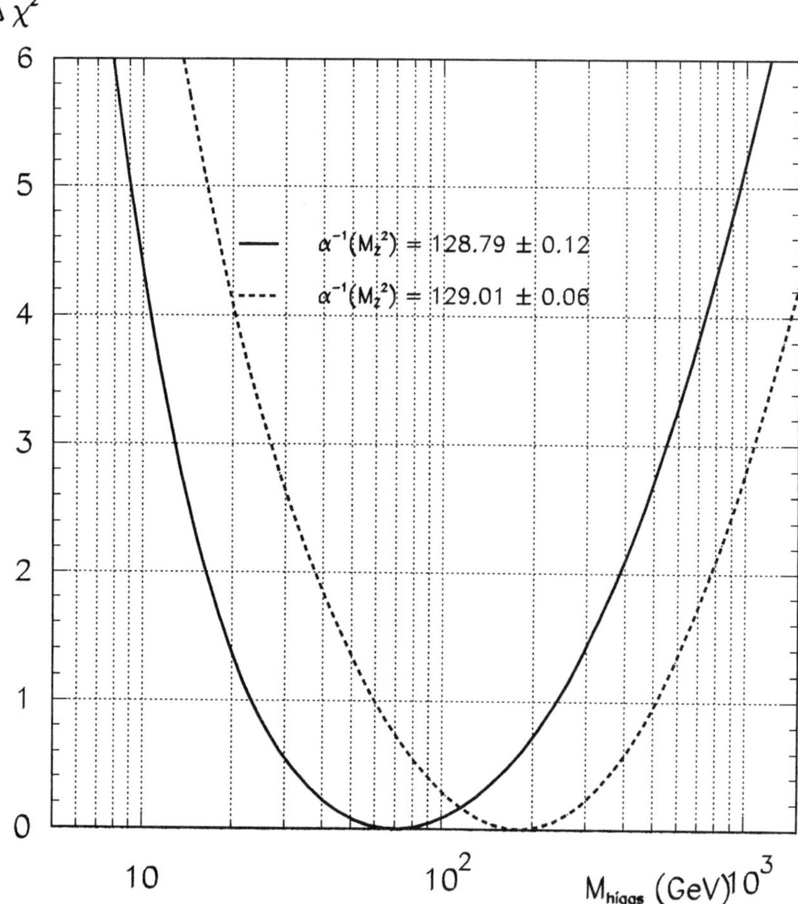

Figure 32: χ^2 vs $\log_{10}(M_{\mathrm{H}})$ using all the precision data plus the CDF determination of m_t as an additional constraint for two different values of $\alpha(M_Z^2)$.

Acknowledgements

I would like to thank the four LEP collaborations for letting me use their unpublished data and specially the members of the "LEP Electroweak Working Group" (LEP-EWWG) for their help in collecting all the relevant information. I am also very grateful to D. Bardin, W. Hollik and G. Passarino for several discussions about higher order corrections and theoretical uncertainties within the framework of the "Precision Calculation Working Group"(PCWG). I want to express my sincere gratitude to R. Gastmans and his team for the excellent organization of the school as well as for the very warm hospitality dispensed to all of us.

References

[1] M. Consoli, W. Hollik and F. Jegerlehner: Proceedings of the Workshop on Z physics at LEP I, CERN Report 89-08 Vol.I, 7

G. Burgers, F. Jegerlehner, B. Kniehl and J. Kühn: the same proceedings, CERN Report 89-08 Vol.I, 55.

[2] D. Bardin et al.,"Precision Calculation Working Group"(PCWG), CERN Yellow report (in press).

[3] H.Burkhardt, F.Jegerlehner,G.Penso and C.Verzegnassi, Z. Phys.**C43** (1989) 497.

[4] F.Jegerlehner, Prog. in Particle and Nucl.Phys. **27** (1991) 1.

[5] M.L.Swartz, SLAC-PUB-6710

[6] A.D.Martin and D.Zeppenfeld, MAD/PH/855

[7] F. A. Berends: Proceedings of the Workshop on Z physics at LEP I, CERN Report 89-08 Vol.I, 89
M. Böhm, W. Hollik: the same proceedings, CERN Report 89-08 Vol.I, 203 and references therein.

[8] B.W.Lynn, High-precision tests of electroweak physics on the Z^0 resonance, Proceedings of the Workshop on Polarization at LEP, CERN 88-06, Sept. 1988, ed. G.Alexander et al., Vol. 1, p.24. at LEP1,

[9] W. Hollik , Radiative Corrections in the Standard Model and their role for precision tests of the Electroweak Theory, Fortschr. Phys. **38** (1990) 165.

[10] D. Bardin, W. Hollik and T. Riemann, Bhabha scattering with higher order weak loop corrections, Z. Phys.**C49** (1991) 485

[11] BHM: Computer code by G. Burgers, W. Hollik and M. Martinez; initially based upon ref. [1]

[12] ZFITTER: Computer code by D. Bardin et al., Z. Phys.**C44** (1989) 493, Nucl. Phys. **B351** (1991) 1; Phys. Lett. **B255** (1991) and CERN-TH 6443/92 (May 1992).

[13] see for instance: M. E. Peskin and T. Takeuchi, Phys. Rev. Lett.**65** (1990) 964,
D. C. Kennedy and P. Langacker, Phys. Rev. Lett.**65** (1990) 2967,
V. Novikov, L. Okun and M. Vysotsky Nucl. Phys.**B397** (1993) 35

[14] G. Altarelli, R. Barbieri and S. Jadach, Nucl. Phys.**B369** (1992) 3,
G. Altarelli, R. Barbieri and F. Caravaglios, Nucl. Phys.**B405** (1993) 3

[15] The LEP Collaborations ALEPH, DELPHI, L3, OPAL and The LEP Electroweak Working Group, CERN-PPE/93-157 and CERN-PPE/94-187.

[16] L. Arnaudon et al., The Working Group on LEP Energy and The LEP Collaborations ALEPH, DELPHI, L3, OPAL, Phys. Lett. **B307** (1993) 187.

[17] The Working Group on LEP Energy, private communication.

[18] ALEPH Coll., D. Decamp et al., Nucl. Instr. and Meth. **A294** (1990) 121
DELPHI Coll., P. Aarnio et al., Nucl. Instr. and Meth. **A303** (1991) 233
L3 Coll., B. Adeva et al., Nucl. Instr. and Meth. **A289** (1990) 35
OPAL Coll., K. Ahmet et al., Nucl. Instr. and Meth. **A305** (1991) 275

[19] M. Martinez, L. Garrido, R. Miquel, J. L. Harton, R. Tanaka, Z. Phys. **C49** (1991) 645.

[20] D. Bardin et al., Z. Phys. **C44** (1989) 493; Nucl. Phys. **B351** (1991) 1; Phys. Lett. **B255** (1991) 290; CERN-TH 6443/92.

[21] B. Ward et al., Contribution to the 27th International Conference on High Energy Physics, Glasgow, Scotland, July 1994; S. Jadach, private communication.

[22] The LEP Electroweak Heavy Flavors Working Group, LEPHF/94-03, July 1994; see also ref. [15]

[23] M. Demarteau et al., *Combining W mass measurements*, CDF/PHYS/2552 and D0NOTE 2115.

[24] C. G. Arroyo et al., CCFR Coll., Columbia University preprint NEVIS R#1498, November 1993.
H. Abramowicz et al., CDHS Coll., Phys. Rev. Lett. **57** (1986) 298; A. Blondel et al., Z. Phys. **C45** (1990) 361.
J. V. Allaby et al., CHARM Coll., Phys. Lett. **B177** (1986) 446; Z. Phys. **C36** (1987) 611.

[25] D. Schaile, Tests of the electroweak theory at LEP, Electroweak Theory, Fortschr. Phys. **42** (1994) 429.

[26] T. Hebbeker, M. Martinez, G. Passarino, G. Quast, Phys. Lett. **B331** (1994) 165.

[27] K. Abe et al., SLD Coll., SLAC-PUB-6456, March 1994, to appear in Physical Review Letters.

[28] A. Blondel, private communication.

[29] F. del Aguila, M. Martinez and M. Quiros, Nucl. Phys. **B381** (1992) 451.

THE TOP ...IS IT THERE?
A SURVEY OF THE CDF AND D0 EXPERIMENTS

A. V. Tollestrup
Collider Detector
Fermilab National Accelerator Laboratory
Batavia, IL 60510

I. INTRODUCTION

The Standard Model requires the Top.

Quarks		
u	c	t
d	s	b

Leptons		
e	μ	τ
ν_e	ν_μ	ν_τ

The b was discovered in 1977, and speculation immediately began about whether or not it had a partner. A direct measurement of the weak isospin of the b is possible through the Z decay to $b\bar{b}$ at LEP. The following two diagrams interfere and give a forward-backward asymmetry to the decay.

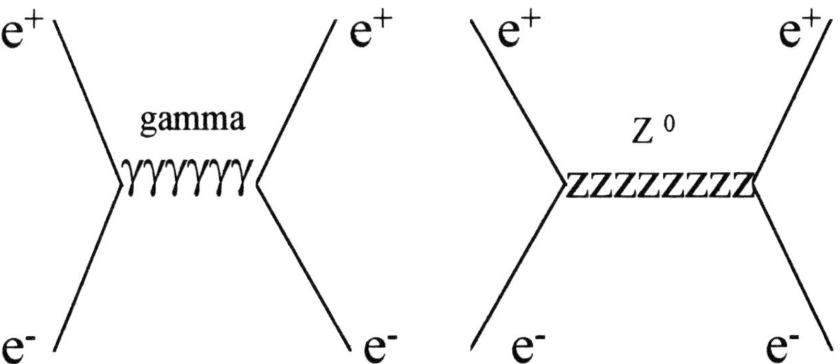

Figure 1. Two diagrams that interfere with each other in Z production .

Frontiers in Particle Physics: Cargèse 1994
Edited by M. Lévy *et al.*, Plenum Press, New York, 1995

The asymmetry is proportional to the coupling which is given by:

$$t_3 + 0.33 \sin^2 \theta_W = 0.07 \text{ if } t_3 = 0$$
$$= 0.43 \text{ if } t_3 = -1/2$$

Direct measurements at LEP have given the value for $t_3 = -0.504^{+0.18}_{-.011}$, indicating that the b is a weak isospin doublet. By definition the object with $t_3 = +1/2$ is the "top."

The mass of the top has been growing with time. The early searchers started at small multiples of the b mass, and a number of guesses were made at formulas that would relate the masses of the quarks and leptons to each other which were then extrapolated to predict the mass of the top. However, as higher energies became available, direct searches gave lower limits for the top mass that increased with time. The most exciting time came in 1983 when UA1 at CERN had evidence for a top with a mass in the range between 30 and 50 GeV, Ref. 1. This created great excitement in the community as it opened up the possibility that TRISTAN could make Toponium. However, it later turned out that the evidence at UA1 was a statistical fluctuation, and the limit for the mass of the top grew even higher.

LEP took up the search and came up with the direct limit of 46 GeV. Later in 1987 CDF set a limit that M_{top} was greater than 62 GeV from a measurement of the width of the W, Ref. 2. If the W can decay into top and b, then the width of the W is wider than if this decay cannot occur as is the case when the mass of the top is greater than the mass of the W. This particular test has an advantage that it would detect nonstandard decays (such as those involving a light Higgs) that a direct search might miss.

Assuming Standard Model top decays, CDF pushed the limit to 91 GeV in 1993, Ref. 3, and early in 1994, D0 increased this limit to 131 GeV, Ref. 4. These searches looked for the Standard Model decays of top to W + b, where the W could be either real or virtual.

Indirect effects from the existence of the top have allowed the LEP experiments to produce a set of mass predictions that have increased with time. The most recent prediction given at the Glasgow Conference was $M_{top} = 178 \pm 11^{+18}_{-19}$ GeV. An easy way to see how the top can show itself through an indirect effect is to look at the following pair of diagrams.

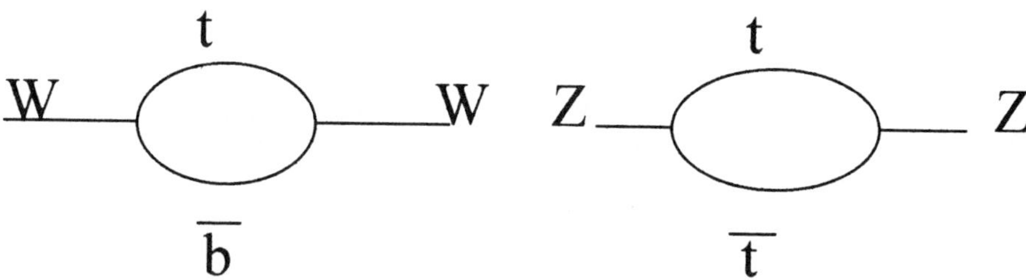

Figure 2. Diagrams that split the W and Z.

The virtual loop for the W contains a t and a $\bar{b}$, whereas the virtual loop for the Z contains a t and $\bar{t}$. The difference caused by these two loops splits the mass of the W and the Z. This splitting of mass is quadratic in the top mass and is logarithmically dependent upon the mass of the Higgs. Eventually a precise measurement of the top mass and the W mass will allow an indirect prediction of the mass of the Higgs. This is one of the simpler cases in which the result from a physical measurement is sensitive to virtual loops involving the top. There are many of these, and the LEP measurements have been analyzed carefully to give the prediction mentioned above. It is thus clear that we are now in the process of searching for an object that has a very high mass.

Dalitz, Ref. 5, shows the predicted lifetime for the top quark to decay as a function of its mass. When the mass is less than the mass of a W plus a b quark, the decay is through a virtual W, and the decay lifetime goes like the inverse mass of the top to the fifth power. When the mass becomes greater than this limit, the lifetime goes like the inverse mass cubed. For masses in the region indicated above, the width is of the order of 1 or more GeV. This makes the lifetime too short for Toponium to be observed and, in addition, the quark does not have time to clothe itself before it decays. Remember that the momentum transfers in a typical hadronization process for a quark are only of the order of 100 MeV, and thus these processes don't compete with the fundamental rapid decay of the top into a boson plus a quark. A very interesting observation in the future will be whether or not there is any non Standard Model interactions between and t and $\bar{t}$. We should be able to answer questions such as this within the next year.

Production and Decay of the Top

Let us now consider production of the top and its various decay channels that are useful for a search. Laenen et al., Ref. 6, have made the next-to-next leading order calculation for the production of the top. This is shown in Fig. 3. At masses

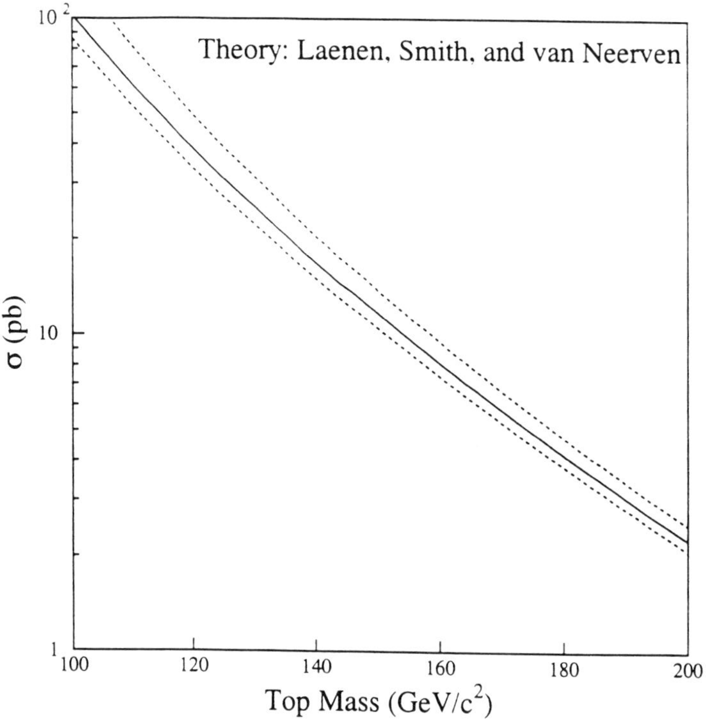

Figure 3: Top cross section according to ref 6. The dotted lines are the quoted estimates of the theoritical uncertainty.

STANDARD MODEL TOP DECAY

$$\bar{p}p \rightarrow t\bar{t} \rightarrow W^+ b\ W^- \bar{b}$$

W^+

	$\bar{e}$ ν	$\bar{\mu}$ ν	$\bar{\tau}$ ν	u $\bar{d}$	u $\bar{d}$	u $\bar{d}$	c $\bar{s}$	c $\bar{s}$	c $\bar{s}$
$e\bar{\nu}$	1	1	1	1	1	1	1	1	1
$\mu\bar{\nu}$	1	1	1	1	1	1	1	1	1
$\tau\bar{\nu}$	1	1	1	1	1	1	1	1	1
$\bar{u}$ d	1	1	1	1	1	1	1	1	1
$\bar{u}$ d	1	1	1	1	1	1	1	1	1
$\bar{u}$ d	1	1	1	1	1	1	1	1	1
$\bar{c}$ s	1	1	1	1	1	1	1	1	1
$\bar{c}$ s	1	1	1	1	1	1	1	1	1
$\bar{c}$ s	1	1	1	1	1	1	1	1	1

W^- labels the rows.

Figure 4: Different decay channels for the two W's

around 100 GeV, the diagrams involving gg collisions comprise about 30 percent of the cross section and $q\bar{q}$ going to $t\bar{t}$ comprise the rest. As M_t increases, the glue contribution decreases to only 7 percent at 200 GeV. The dotted lines shown on the graph reflect the uncertainty expected in the cross section due to structure function errors as well as diagrams that have been neglected. A top mass of 150 GeV has a cross section of about 10 picobarns. The experimental data that I am going to talk about in these lectures covers a running period in 1992-1993 of the Tevatron at Fermilab, and the integrated luminosity was about 20 inverse picobarns. This means that the experiments have to be sensitive to only a few hundred $t\bar{t}$ pairs, and the statistical fluctuations in the various production processes and backgrounds will dominate our discussion.

The search for top production is centered on identifying the products of the $t\bar{t}$ system when it decays. Since the primary decay process is dominated by t going to W + b, we can make the table shown in Fig. 4 for the various decay channels available. Each channel has a weight of 1, and the quarks are shown with their three color states. We see that there are a total of nine ways that a W can decay, and there are 81 ways that we can list for the two W's. The tau, since it decays into 2 neutrinos and a lepton, is not very useful. Hence, we will concentrate on only the electron and the muon. We see from the table that the branching ratio is 4 out of 81 to give us a dilepton mode where the dileptons are e's and μ's in any combination. There are 24 out of 81 combinations where we have a μ or an e plus jets, and there are 36 out of 81 combinations where the W's both decay hadronically.

Let's examine these various channels individually. In the case of the dilepton mode, we also have two neutrinos. Thus, we are looking for two leptons and two b jets plus a large amount of missing transverse energy which is carried away by the neutrinos. If both the b jets could be tagged by their decay, this would be a rather unique signature for this mode. However, we will see that the efficiency for tagging a b is only of the order of 20 to 30 percent, which when coupled with the small branching ratio of this mode makes these events rather rare. It is also obvious that we cannot reconstruct this mode uniquely because of the two neutrinos that are involved in the decays. However, it is true that given a large number of these events, one could obtain an estimate of the mass of the top by studying the momentum distribution of the leptons and the b's.

The next channel that we investigate involves one of the W's decaying hadronically, so that we have two jets from one of the W's plus two b jets, a lepton, and a neutrino. It turns out that this category of event can be reconstructed kinematically and, hence, an estimate of the top mass obtained. Also, the branching ratio of 24 out of 81 is 6 times larger than the dilepton signature. However, we will see that the background for this channel is higher than it is in the dilepton case, and it will require some additional information to separate it from the production of a W plus 4 QCD jets.

Finally, there is the case where both W's decay hadronically, and in this case one is looking at 6 jet events. Although the branching ratio of this channel, 36 out of 81, is high, it has an enormous background from the QCD production of 6 jet events. Kinematics can aid in separating out top decays, but it becomes imperative to also tag the b jets, if one is to study this channel. The b tag reduces the sensitivity of the search, and at present it looks possible but very difficult to identify $t\bar{t}$ production through this channel. Future success will require that the b jets be tagged with a high efficiency.

A summary of the experimental challenge is the following. We have a process with a very small cross section, and we are expecting to find a few events in 10^{12}. In order to establish that the top is really there, we must accomplish the following:

1. Establish a selection criteria for triggering the detector so that these events will be written to tape.

2. Measure the efficiency of the trigger.

3. Measure the efficiency of the offline event reconstruction program.

4. Measure the background:

   (a) Real processes that fake real events.

   (b) Mismeasurements due to detector errors that fake real events.

5. If the above process yields an excess of signal events over background events, then we must show that the events are characteristic of top decay. We must reconstruct the decay and show that it leads to a unique mass, and the ratio between the different channels should be consistent with that which we expect for the decay of the top.

Tevatron and Detectors

For the rest of these lectures, we will be concerned with experiments that have been done at the Tevatron at Fermilab. The Tevatron characteristics are shown in the following table:

TEVATRON CHARACTERISTICS

- Pbar P   900 x 900 GeV

- 6 bunches

- Bunch separation 3.5 micro sec

- Initial luminosity 1.2 x 10^{31}

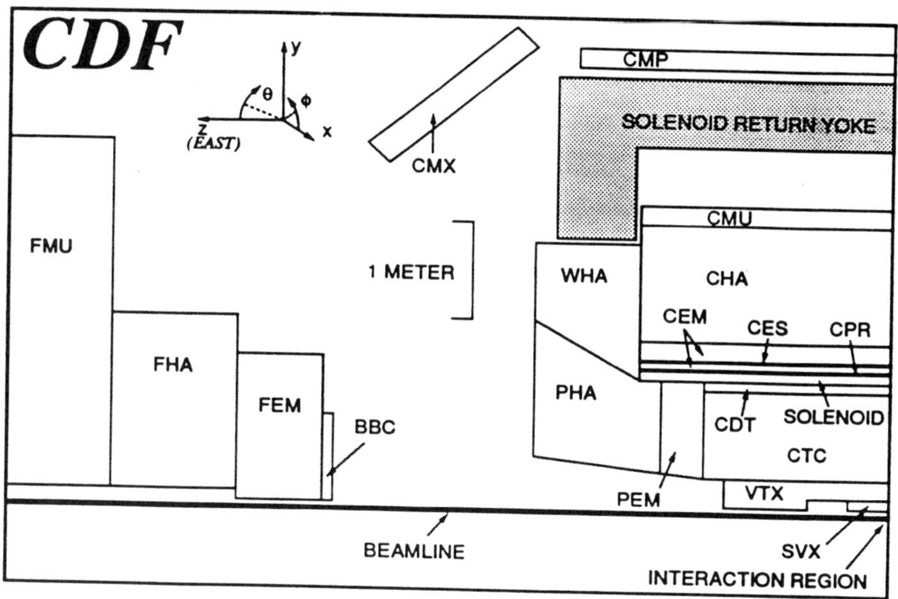

A side-view cross section of the CDF detector. The detector is forward-backward symmetric about the interaction region, which is at the lower-right corner of the figure. See text for detector component definitions.

System	η Range	Energy Resolution	Thickness
CEM	$\|\eta\| < 1.1$	$13.7\%/\sqrt{E_T} \oplus 2\%$	$18\ X_0$
PEM	$1.1 < \|\eta\| < 2.4$	$22\%/\sqrt{E} \oplus 2\%$	$18\text{-}21\ X_0$
FEM	$2.2 < \|\eta\| < 4.2$	$26\%/\sqrt{E} \oplus 2\%$	$25\ X_0$
CHA	$\|\eta\| < 0.9$	$50\%/\sqrt{E_T} \oplus 3\%$	$4.5\ \lambda_0$
WHA	$0.7 < \|\eta\| < 1.3$	$75\%/\sqrt{E} \oplus 4\%$	$4.5\ \lambda_0$
PHA	$1.3 < \|\eta\| < 2.4$	$106\%/\sqrt{E} \oplus 6\%$	$5.7\ \lambda_0$
FHA	$2.4 < \|\eta\| < 4.2$	$137\%/\sqrt{E} \oplus 3\%$	$7.7\ \lambda_0$

Summary of CDF calorimeter properties. The symbol $\oplus$ signifies that the constant term is added in quadrature in the resolution. Energy resolutions for the electromagnetic calorimeters are for incident electrons and photons, and for the hadronic calorimeters are for incident isolated pions. Energy is given in GeV. Thicknesses are given in radiation lengths (X_0) and interaction lengths (λ_0) for the electromagnetic and hadronic calorimeters, respectively.

Fig. 5

- Initial lifetime $\sim$ 12 hours increases to $\sim$ 20 hours

- About 2 interactions/crossing

- Beta* $\sim$ 3.5 meters

- Sigma X = Sigma Y = $\sim$ 60 microns

- Length of interaction region sigma Z $\sim$ 26 cm

- Protons/bunch 200 x 10^9 Pbar/bunch 60 x 10^9

- Pbar stacking rate 4 x 10^{10}/hour

- Integrated luminosity $\sim$ 30 pb^{-1} in 1994
  30 pb^{-1} in 1992
  9 pb^{-1} in 1988

Briefly there are 6 bunches of protons and 6 bunches of counter-rotating antiprotons in the machine. The bunches are spaced equally such that there is a collision every three and one-half microseconds at the two intersection regions ... B0 and D0, where large detectors are located. Electrostatic separators generate helical orbits for the bunches so that they only intersect at the detector location. This is necessary because of the large beam-beam turn shift that would result if the bunches crossed at 12 places. The energy is 1.8 TeV in the center of mass, and the initial luminosity of a store is 1.3×10^{31} with an initial lifetime of about 12 hours, and which increases to 20 hours as the luminosity decreases. At peak luminosity, there are about two interactions per crossing. This is achieved in a collision region that has a sigma of about 26 centimeters along the beam direction and has a circular cross section with a rms radius of about 60 microns. The integrated luminosity delivered to each of the experiments that are described here, was about 30 inverse picobarns in 1992. At present a new run is in progress where we have accumulated an additional 30 inverse picobarns, and we hope to have several times this amount of data before the end.

CDF Detector

The CDF Detector is described in detail in Ref. 7 and shown in Fig. 5. The features of it that are important for this discussion are the following:

1. A Silicon Vertex Detector located at few centimeters from the beam centerline which enables the impact parameter of a track to be measured with an accuracy of 15 to 20 microns (Ref. 8).

2. A large 3 meter diameter by 3 meter long central tracking chamber immersed in a 1.4 Tesla magnetic field that allows precise measurement of charged particle momenta.

PARTON ID SCHEMES

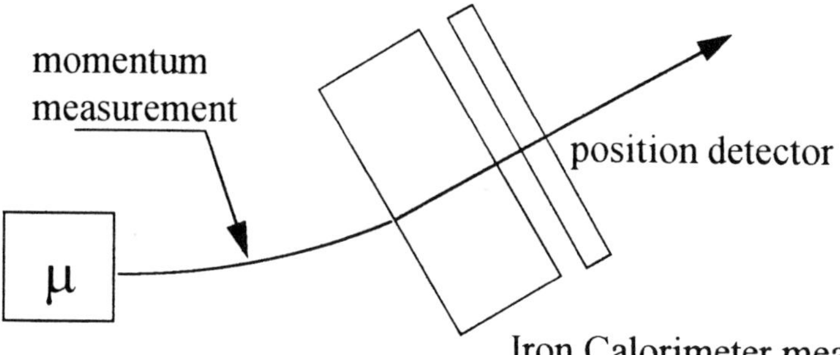

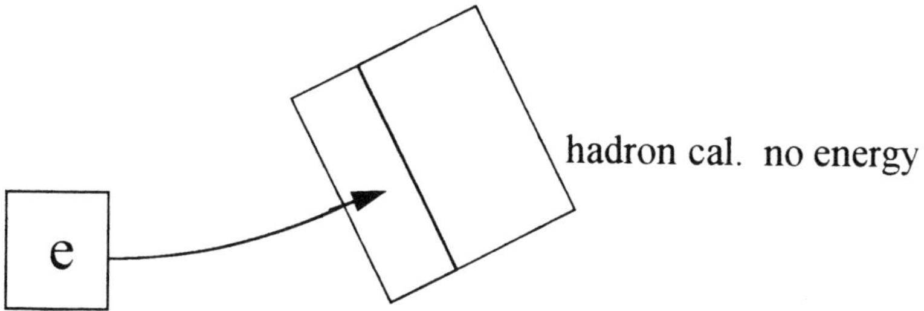

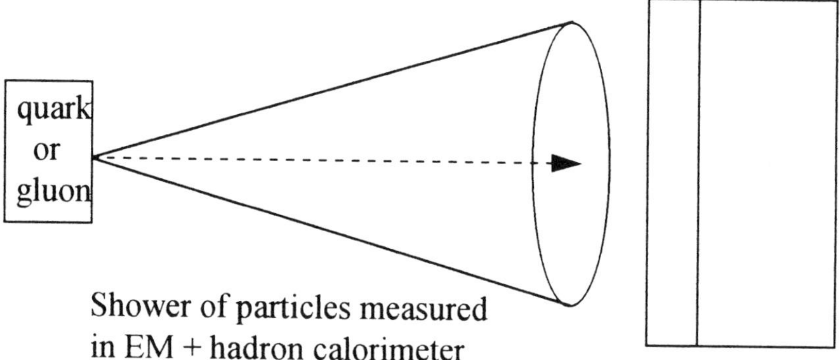

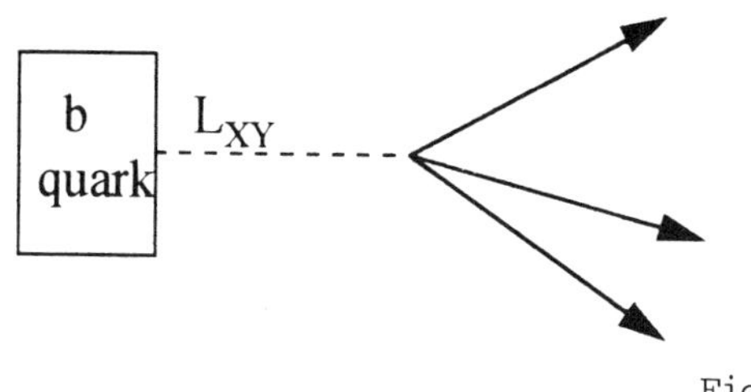

Fig. 6

3. Electromagnetic and hadronic calorimetry. In the central region this calorimeter uses scintillation plastic for a readout; in the forward and backward region, it uses proportional chambers.

4. A muon system that allows the identification and momentum measurement of muons when their transverse energy is over about 2 GeV.

Since we will be discussing the identification and detection of various kinds of particles in the rest of these lectures, we show in Fig. 6, in cartoon form, the technique for identifying electrons, muons, gluons, quarks, and b particles. These techniques are specific for CDF but are also widely used by all large modern particle detectors.

Muons are the easiest. Their momentum is measured in the central magnetic field and the tracking chamber with a precision of $\Delta p/p = 0.001p$. When a muon passes through the calorimeter, it deposits energy only through ionization loss, and hence leaves the signal of a minimum ionizing particle. Finally, it exits the calorimeter which has 5 or more absorption lengths in it, and its position and angle is detected by tracking chambers that surround the detector. The primary identification for the muon then comes from the minimum energy loss in the calorimeter plus the fact that it traversed an amount of absorber that would have removed a hadron through strong interactions, thus removing any track in the backup position detector.

Electrons are identified first of all by having their momentum measured in the central tracking chamber to the same precision as was given for the muons and by their total absorption in the lead absorber of the electromagnetic calorimeter. The criterion for an electron then is that the momentum measured in the magnetic field equals the energy loss in the electromagnetic calorimeter, and the fact that the position of the shower matches the entry point of the track.

Quarks and gluons, of course, are not measured directly as they hadronize and turn into a shower of particles. The size of the cone containing the energy of these particles is measured by its span in rapidity η and azimuth ϕ and is generally equal to a number between 0.7 and 1. At 90 degrees this amounts to an opening angle of between 40 and 50 degrees. A cone of this size does not completely contain the energy of the gluon, and corrections must be made for so-called out of cone losses. An additional correction must be made for the fact that the underlying event structure can also put energy into this cone which is not associated directly with the gluon or the quark under consideration. b-quarks are a special case in that during the hadronization process the quark will emit hadrons but also imbedded in the shower will be a B meson or a B hadron associated with the jet. The lifetime of these particles is generally of the order of 10^{-12} seconds and corresponds to a $c\tau$ of about 500 microns. Thus, if ones sees a shower and finds inside a detached vertex by means of using a silicon vertex detector, this shower can either be associated with the b or a c quark. Since a b quark has a mass of about 5 GeV, whereas the c quark is considerably lighter, the transverse energy of the decay helps distinguish these two

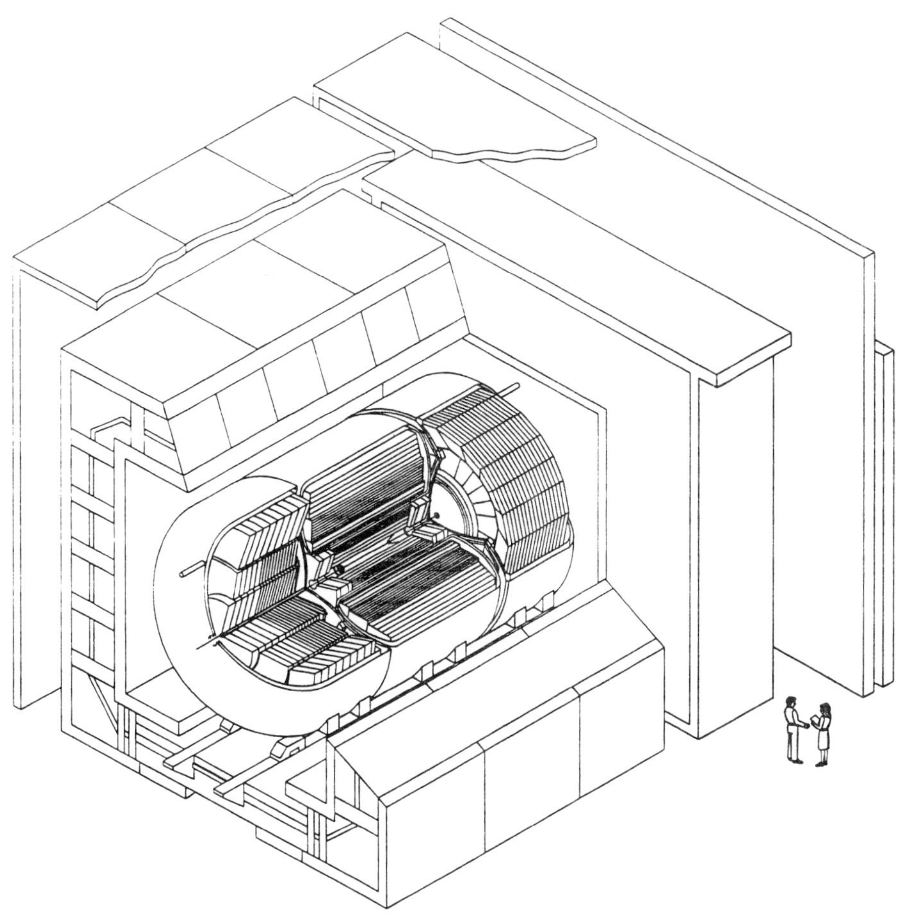

TRACKING

	Vertex Chamber	Central Drift Chamber	Forward Drift Chamber
Resolution in rϕ	60 μm	180 μm	200 μm

CALORIMETRY

Coverage $|\eta| < 4$ ($\vartheta > 2°$) Granularity $\delta\eta \times \delta\vartheta = 0.1° \times 0.1°$

Electron Energy Resolution $= 15\%/E^{.5}$ Hadron Energy Resolution $= 50\%/E^{.5}$

MUON SYSTEM

Coverage $\quad |\eta| < 3.3 \qquad (\vartheta > 5°)$

Resolution $\quad \delta P/P = [0.04 + (0.01P)^2]$

Figure 7 : D0 Detector

quarks from each other.

Finally, we come to the question of neutrinos. If there is a single neutrino associated with the event, it will reveal itself through the lack of the transverse momentum balance in the event. Since there is no transverse momentum in the initial state, the final state should sum to zero. This includes the momentum of all of the neutrinos plus all the charged particles and the leptons. Since there are errors associated with measuring the momentum of the quarks, there will be some error reflected in the measurement of missing E_T. The accuracy with which this variable can be measured is then determined by the resolution of the calorimetry plus the hermeticity. It is clear that any cracks or undetected energy that escapes the calorimetry will result in the missing E_T. Note also that p_z^ν is not measured.

This short summary of how various partons are identified is generic in nature, and the accuracy of the identification as well as the accuracy of the measurement depend upon the details of the detector. The numbers given above are typical of the CDF detector.

D0 Detector

A cross section of the D0 Detector is shown in Fig. 7 and described in Ref. 9. The main feature of the detector is the large uniform liquid argon calorimeter for measuring total particle energies. There is not a magnetic field in the central region, but the momentum of muons is measured in magnetized iron in a system that surrounds the liquid calorimeter. The very fined grained high resolution calorimetry provided by the liquid argon allows a better measurement of the missing energy in an event than is available in CDF. On the other hand at present there is not a silicon vertex detector nor a central field for measuring the momentum of the tracks. Thus the techniques used in the two detectors to search for $t\bar{t}$ events tend to be complimentary in nature.

The rest of these lectures will describe first the experiments that have taken place at CDF, and then we will continue on to describe the results from D0.

II. CDF EXPERIMENT

I am assuming that these notes are being read in conjunction with the papers that have been published by CDF and D0. CDF has published a complete paper, Ref. 10, on the experiment with an enormous amount of detail, and I consider that these notes are only a guide through that paper. The same applies to the D0 experiment, although only the notes given in the Glasgow 1994 Conference were available at the time of the School. Ref. 12 gives additional results that are more recent and includes additional information not available at the time of these lectures.

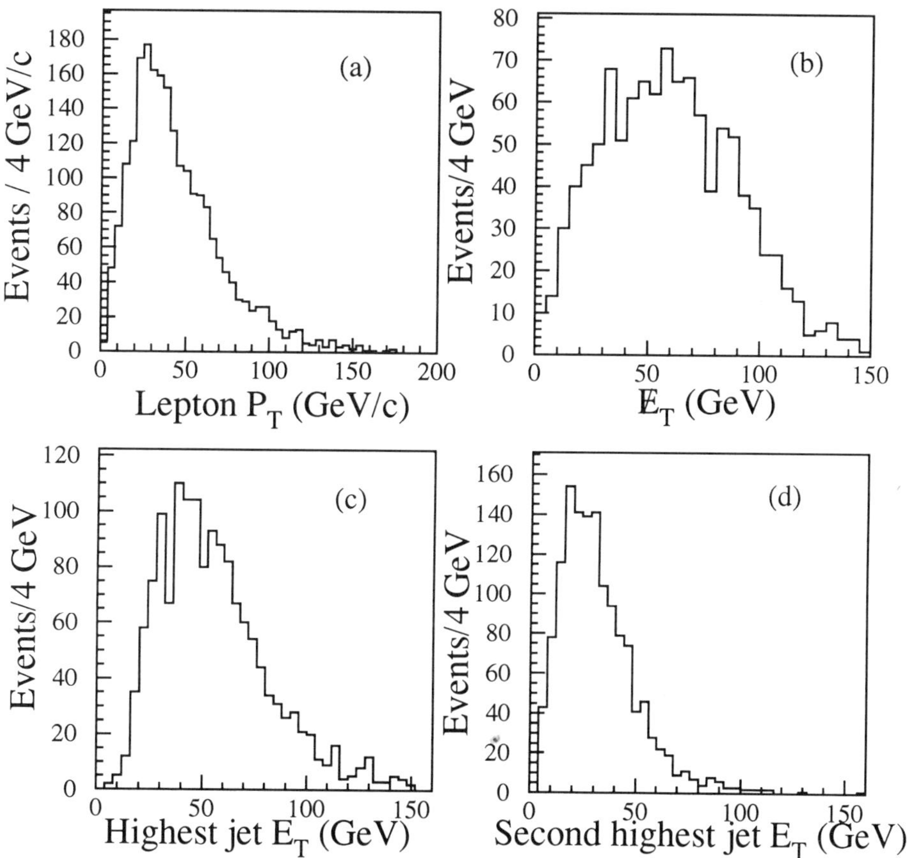

Figure 8. Monte Carlo distributions for $M_{top}=160$ GeV/c^2. a) Lepton P$_T$ spectrum from $t \rightarrow W \rightarrow \ell$. b) $\not{E}_T$ for events with two leptons with P$_T$ > 20 GeV/c. c) Leading-jet E$_T$ for dilepton events. d) Next-to-leading jet E$_T$ for dilepton events.

Cut	$e\mu$	ee	$\mu\mu$
P_T	8	702	588
Opposite-Charge	6	695	583
Isolation	5	685	571
Invariant Mass	5	58	62
$\not{E}_T$ magnitude	2	0	1
$\not{E}_T$ direction	2	0	0
Two-jet	2	0	0

Fig. 9 Number of data events surviving consecutive requirements.

		Without $\not{E}_T$ and two-jet cuts	Without two-jet cut	All cuts
$e\mu$	WW	1.1	0.74	0.10 ± 0.04
	$Z \to \tau\tau$	3.7	0.22	0.07 ± 0.02
	$b\bar{b}$	1.2	0.10	0.04 ± 0.03
	Fake	1.2	0.19	0.03 ± 0.03
	Total background	7.2	1.25	0.24 ± 0.06
	CDF data	5	2	2
$ee, \mu\mu$	WW	0.6	0.43	0.06 ± 0.02
	$Z \to \tau\tau$	3.0	0.20	0.06 ± 0.02
	$b\bar{b}$	1.6	0.12	0.05 ± 0.03
	Fake	1.7	0.25	0.04 ± 0.03
	Drell-Yan	113	0.28	$0.10^{+0.23}_{-0.08}$
	Total background	120	1.28	$0.31^{+0.24}_{-0.10}$
	CDF data	120	0	0

Fig. 10 Number of background events expected in 19.3 pb^{-1} and the number of events observed in the data.

High p_T Dilepton Search

We will now consider the dilepton channel. The first thing we must do is establish some kind of criterion for selecting the events. The variables that we have available are the p_T of the leptons, the missing E_T, and the energy of the jets associated with the event. The distributions of these variables are shown in Fig. 8. The lepton p_T is particularly useful as cuts on this variable can be implemented in a fairly fast fashion at the trigger level. Later in the analysis, considerably more sophisticated cuts can be made in the software analysis package. See Ref. 10 for details on the trigger.

After the events have been collected by either the inclusive electron or muon trigger, the additional cuts are implemented in the software. These cuts are as follows. Both of the leptons must have a p_T greater than 20 GeV and be of opposite charge. At least one of the tracks must have η less than 1.0 and be "isolated." The missing transverse energy E_T must be greater than 25 GeV. In addition, we will want to discuss the two b jets, and the cuts placed on these require that their transverse energies should be greater than 10 GeV, and their $|\eta|$ should be less than 2.4. These cuts were established after extensive work looking at the backgrounds from various processes and at the efficiency for finding top. Fig. 9 shows the number of data events surviving the consecutive requirements.

We will now show the results of this search in Fig. 10 and then come back and discuss the individual components. The rows labelled CDF data are the number of events surviving all of the cuts. In addition, the table shows in itemized fashion the backgrounds from various sources as well as the effect of the missing E_T and two jet cuts. The $e-\mu$ events are displayed separately from the ee and $\mu\mu$ events. The bottom line is that for all channels we observe two events with an expected background of $0.56^{+0.25}_{0.13}$.

Let us now examine these results in detail. First of all, it is necessary when considering the ee and $\mu\mu$ channels to make a cut on the invariant mass in order to eliminate the Z. The two additional jets can come from gluon radiation in the initial state of Z production thus faking the overall event. Therefore, all of the events with an invariant mass of the leptons between 75 and 105 GeV are removed. 80 percent of the dielectron and dimuon events from the $t\bar{t}$ are expected to pass this invariant mass cut. The effect of this cut is shown in Fig. 9 where we see that only 10 percent of the dilepton events are outside of this mass window and that the missing E_T cut removes essentially all of the rest.

Fig. 10 also lists other sources of background. For instance W pair production can lead to dilepton events where the two additional jets come from initial state radiation. This figure also shows the reason for the two-jet cut on the data. It is a cut that reduces the background by a factor of 4 or more, whereas the efficiency for top of 120 GeV is greater than 60 percent and grows with increasing M_T. The same effect of the two-jet cut can be seen in the rest of the channels as well.

A second source of background is $Z \rightarrow \tau\tau$. The missing E_T comes from the

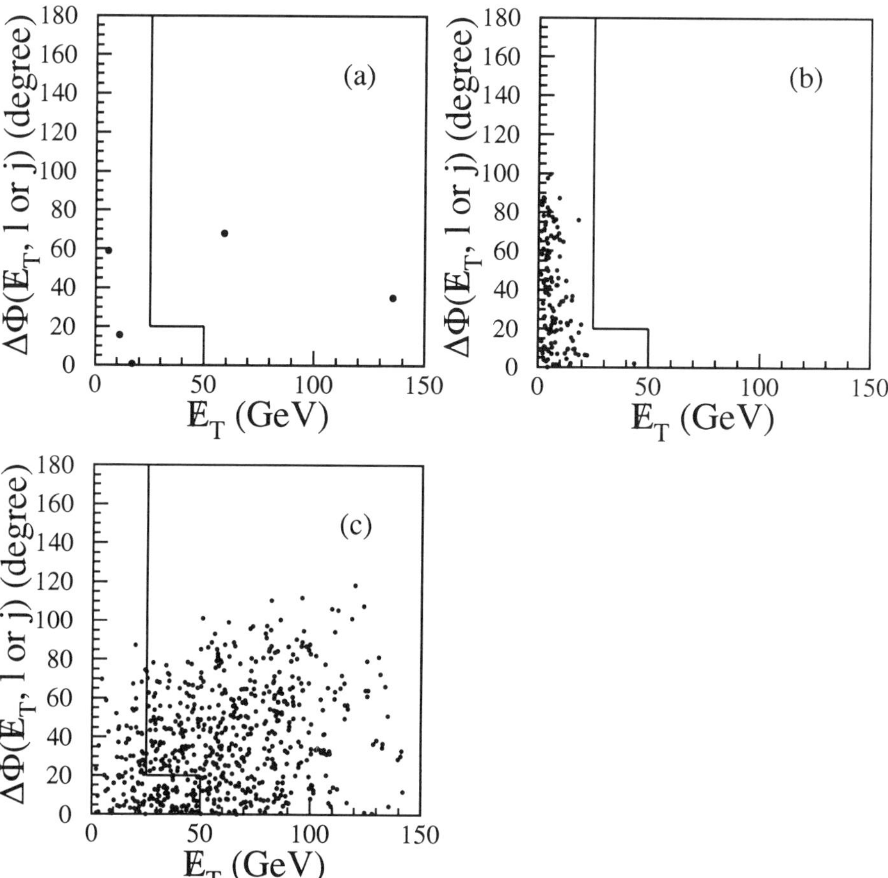

Figure 11 Distributions of the azimuthal angle between $\not{E}_T$ and the closest lepton or jet versus $\not{E}_T$. a) $e\mu$ data. b) Dielectron and dimuon data after the invariant mass cut. c) Monte Carlo events for $M_{top}=160$ GeV/c^2 (unnormalized). Events in the region to the left of the boundary in the figures are rejected by the $\not{E}_T$ cuts.

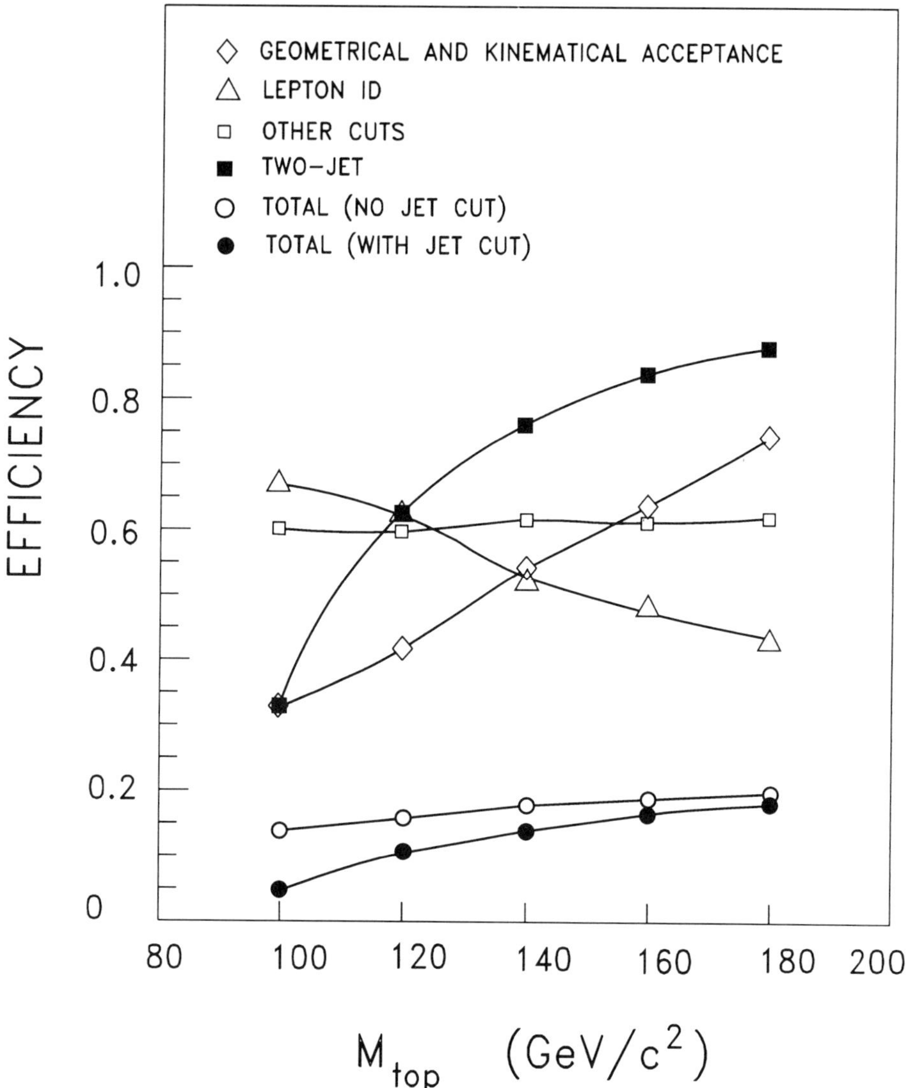

$$M_{top} \quad (GeV/c^2)$$

Figure **12.** Efficiencies of the dilepton selection as a function of M_{top}. 'Other cuts' corresponds to the combined efficiency for the isolation, topology (opposite-charge, mass, $\not{E}_T$) and trigger requirements.

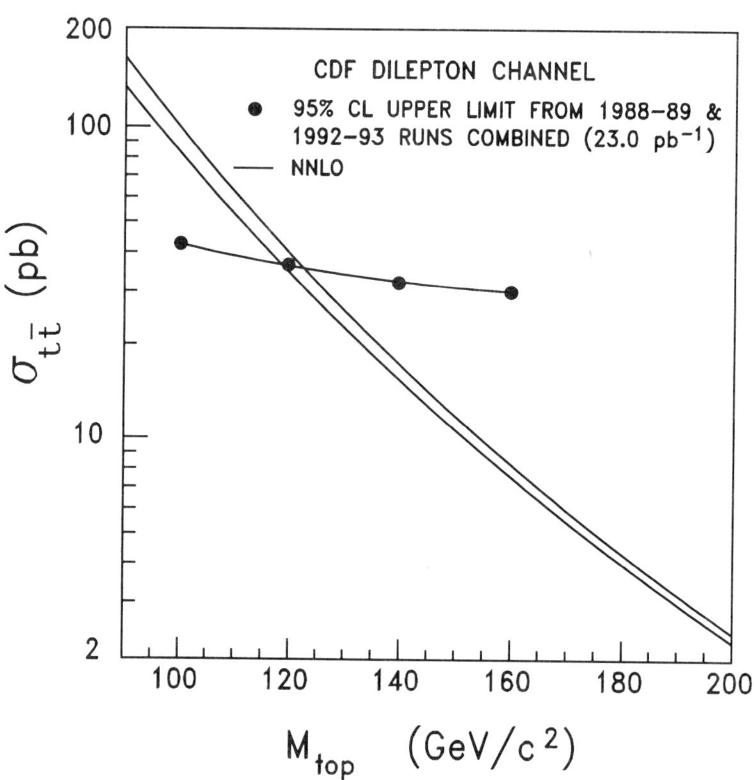

Figure 13: The upper limit at the 95%-C.L. on $\sigma_{t\bar{t}}$, overlaid with the theoretical lower bound and central value of a next to next to leading order (NNLO) calculation from Ref. [10]

neutrinos associated with the tau decay. However, the direction of this missing E_T would be expected to be closely collimated with the lepton direction as the τ is quite high energy. Thus a cut was instituted that increases the missing E_T required in the event to be greater than 50 GeV when the angle of the missing E_T with respect to either of the leptons is less than 20 degrees.

There are two other backgrounds listed in Fig. 10 labelled $b\bar{b}$ and fake. The $b\bar{b}$ production cross section is very large, and the background comes from the two b's decaying into leptons with associated QCD jets. The missing E_T comes from the neutrinos associated with the b decay or from a mismeasurement of the jet energies in the detector. Fakes come from QCD jet events in which the leptons are mimicked by rather rare jets that consist of only a single particle which in turn fakes a lepton. This is not a very probable process, but as the QCD jet cross section is very high a small background is generated. In this case the missing E_T comes from an incorrect measurement of the jet energies. Since an under measurement of a jet energy will lead to a missing E_T parallel in direction to the jet, a cut is made to decrease the probability of this process. If the missing E_T lines up within 20 degrees of the jet, the cut is increased from 25 GeV to 50 GeV. Fig. 11 is a a plot of the missing E_T versus the angle between the missing E_T and the closest lepton or the jet. Fig. 11a is for the $e - \mu$ case, and Fig. 11b is for the dielectron or dimuon data after the invariant mass cut. Fig. 11c shows the result that would be expected in the 160 GeV top Monte Carlo.

Fig. 12 shows a study carried out using a Monte Carlo for simulating top events and displays the efficiency of the various cuts versus the mass of the top. We note that the efficiency of the two-jet cut increases as the mass increases because of the higher energy given to the b jets for high mass top. The geometrical and kinematical acceptance also increases with energy as the events from high mass top tend to become more centrally located in the detector. The lepton I.D. efficiency falls with increasing mass because the events become more collimated and the chance increases of the leptons being covered up by other particles in the decay. Finally, we note that requiring two jets for top masses above 100 GeV is rather efficient.

As the mass of the top increases, it becomes easier to kinematically identify the products from the decay. Therefore, it is expedient to place a lower limit on what the top mass can be. D0 has set a limit of 130 GeV (Ref. 4), but for self-consistency of the analysis, CDF has used the dilepton events to set a lower limit on the mass of the top. This was done by simply looking at the dilepton themselves with the two jet requirement removed. This is necessary because if the mass is close to the W mass, the b jets have very low energy, and the efficiency for finding them is low. Thus, to set a limit one looks for simple $e - \mu$ events with a missing E_T cut greater than 25 GeV and compares this with the production expected for $t\bar{t}$. Fig. 13 shows the upper limit at the 95 percent confidence level on $\sigma_{t\bar{t}}$ for the combined 1988-1989 and 1992-1993 runs. The number obtained is that the top mass is greater than 118 GeV

T TBAR PRODUCTION AND DECAY

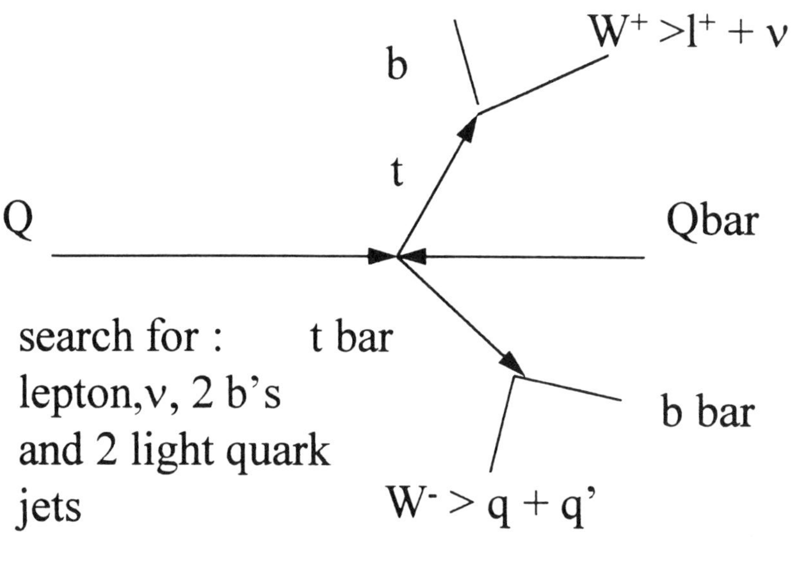

LEPTON PLUS JETS BACKGROUND

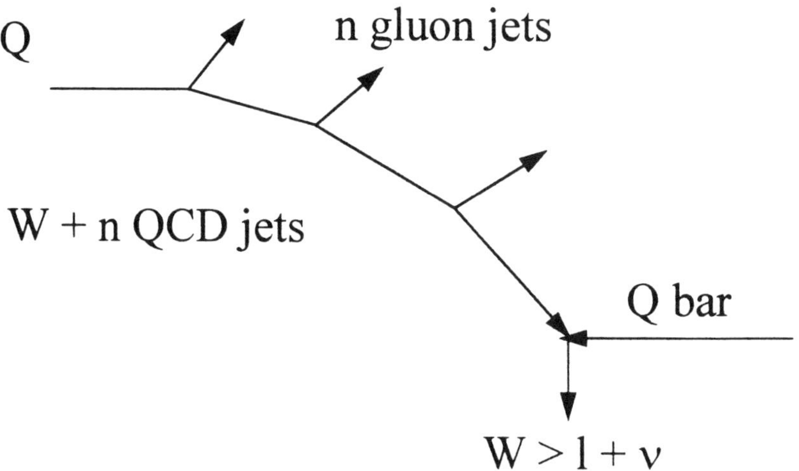

Fig. 14

Selection Criteria	Electrons	Muons
Good Lepton	28,522	17,994
Lepton Isolation Requirement	20,420	11,901
Z Removal	18,700	11,310
$\not{E}_T > 20$ GeV	13,657	8,724
Good Quality Run	12,797	8,272
Trigger Requirement	11,949	7,024

Fig.15. The number of events passing various consecutive selection criteria in data. The good lepton requirement includes all quality selection, fiducial requirements, E_T cuts, and conversion removal.

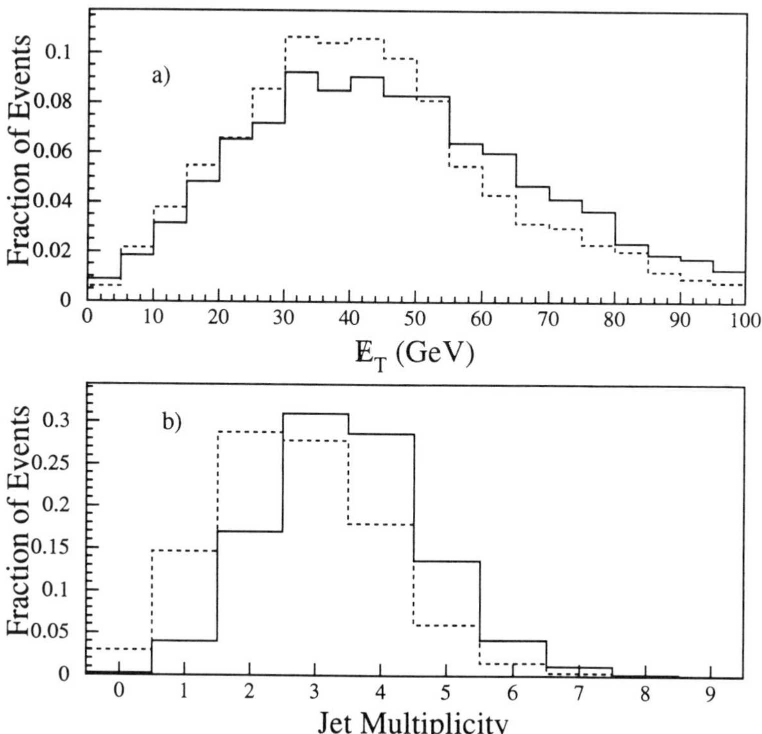

Fig.16. The $t\bar{t}$ Monte Carlo distribution of a) $\not{E}_T$ for events with a 20 GeV electron passing electron identification cuts, and b) the expected jet multiplicity distribution for events passing the W selection criteria. In both plots the dashed line is for $M_{top} = 120$ GeV/c^2 and the solid histogram is for $M_{top} = 180$ GeV/c^2.

with a 95 percent confidence level. For the rest of the experiment, the mass search concentrated on masses greater than 120 GeV.

Lepton Plus Jet Search

We will now consider the channel where only one of the W's decays leptonically, and the other one goes through the hadronic mode. Thus, the handles for this mode will be two b jets, two hadronic jets in the W, a missing E_T, and a lepton from the other W decay. This is shown schematically in Fig. 14. For a heavy top, the jets and leptons will be in the central region of the detector, and the event will be rather spherical in nature. There is a major background to this process; it is shown cartoon style in the same figure. It involves a W produced with initial state radiation in the form of four additional jets. The QCD radiation from the initial state tends to be along the forward and backward direction. However, since this is a strong QCD process, there is a probability that the tail of it can generate a W with high p_T jets that are in the central region of the detector. This will be a major background with which we must contend, and we will spend a considerable amount of time discussing it.

To select events for this mode, we use the following cuts: The electron has an E_T of greater than 20 GeV, muon p_T of greater than 20 GeV, missing E_T greater than 20 GeV, three or more jets with an E_T of greater than 15 GeV, and an η less than 2.0. The jet E_T is not corrected for detector effects and hence will tend to be associated with a parton whose energy is 20 GeV or more. The missing E_T is corrected for the muon only. Recall that the electron and muon modes are equal and together account for about 30 percent of the $t\bar{t}$ decays.

When we apply these cuts on the event sample, we find the results given in Fig. 15. The events are categorized by whether they are associated with an electron or a muon and then listed in terms of the number of jets associated with the event. The final sample of three of more jets contains 52 events total, and it is this set of events that we use for the top search. To get an idea of the efficiency of the cuts that we have made, Fig. 16 shows the spectrum of missing E_T expected for 120 GeV or 180 GeV top production. The lower part of the figure shows the jet multiplicity expected for these same mass tops. The cut on missing E_T greater than 20 GeV is seen to be highly efficient. The efficiency of the jet multiplicity cut is more dependent upon the mass of the top. The cut on the number of jets has been made at 3 or more, and approximately 75 percent of the $t\bar{t}$ events with a top mass of 160 GeV will pass this cut whereas less than one-half percent of all of the W events are retained. Cutting on Njet = 4 is not only less efficient, but also makes the cut highly sensitive to the top mass. The reason an intrinsic four-jet event can turn into less than four jets, is because some of the jets fall outside of the η cut or are such a low energy that they do not pass the E_T cut. In the 52 remaining events, one would expect to find a small number of $t\bar{t}$ events. It is thus clear that we require additional means for identifying

R= 7.9 cm
X₁
X₂
X₃
plane #4
55 μ
spacing
R=3.0 cm
plane #1
60 μ
spacing
Secondary B
decay vertex
L_XY
impact parameter d
primary vertex

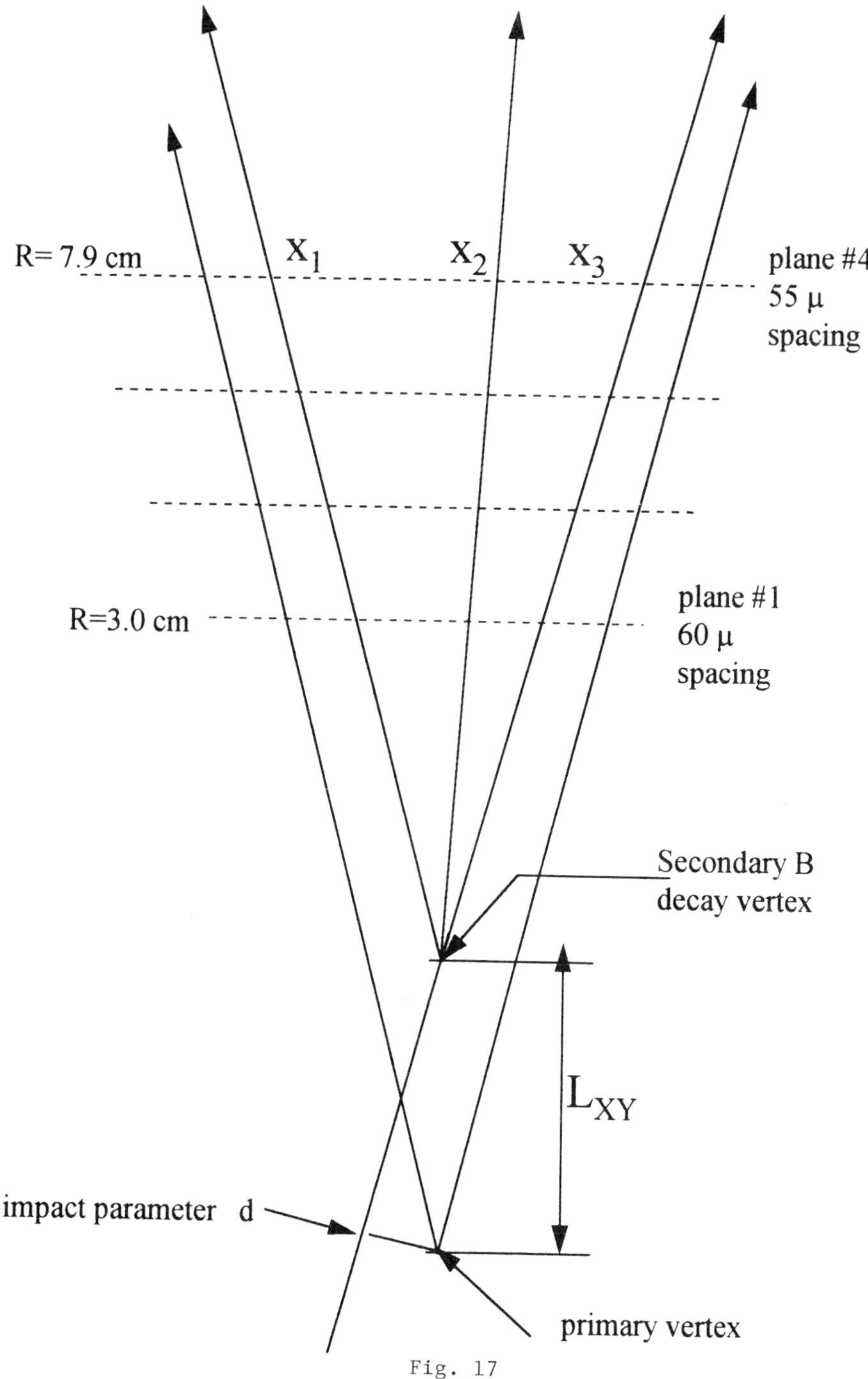

Fig. 17

the top and separating it out the W plus QCD jet production. We will discuss the techniques for doing this next.

Separating Top Candidates from Background

There are several ways to obtain the increased rejection that we need in order to find the $t\bar{t}$ signal in the W plus jet background. Since there are two b jets, it is possible to try to identify the b decays in the event. There are two ways of doing this. Since the $c\tau$ for a b is almost 500 microns, one can search for a secondary vertex. This technique requires a silicon vertex detector that can identify tracks coming within a few tens of microns of the primary vertex. A second technique is to look for the associated soft electron or muon accompanying a semileptonic decay. A generic B has a branching ratio of about 20 percent for semileptonic decay into an electron or muon. In this case one looks for either a muon or an electron in close association with a jet, and the cuts are designed to enhance the sensitivity of the measurement to the higher transverse mass of the b as compared to other quarks.

There are other techniques for discriminating between W plus QCD jets and $t\bar{t}$ production. These methods rely on the fact that for a heavy top the decay products will have a much more nearly spherical distribution in space than for the QCD production. For instance, the momentum distribution of the jets can be studied, and it is found that this provides a discriminant. Two variables are useful for this study. The first is the aplanarity of the event which measures it sphericity, and the second is the sum of the total transverse energy in the event which for a high mass intermediate state should increase as the mass increases. One can also examine the kinematics of the events and test whether the distribution in energy of the jets resemble that expected for $t\bar{t}$ production. In this case the Monte Carlo program called VECBOS is used to mimic the W plus QCD jet production, and ISAJET is used to simulate the $t\bar{t}$ production. The ultimate test, of course, is to reconstruct the mass of each event and look for a peak in the distribution corresponding to the top mass. We will investigate all of these avenues in turn.

It is worth pointing out, however, that these approaches are somewhat complimentary in nature, and that final identification of the top will rely on a combination of all of them. For instance, a set of events could have b's associated with them and yet not be $t\bar{t}$ production. Also a set of events could give a peak in the mass distribution and yet not have the kinematics of the individual events correspond with $t\bar{t}$ production. It is also not known what the correlations are among the various kinematic discriminants. Some studies are being done of the correlations and will be used in studying the larger data set from the present run.

Tagging with the SVX

A schematic diagram of the SVX (Ref. 8) is shown in Fig. 17. The beam pipe for CDF is made of beryllium and has a radius of 1.9 centimeters. Just outside of

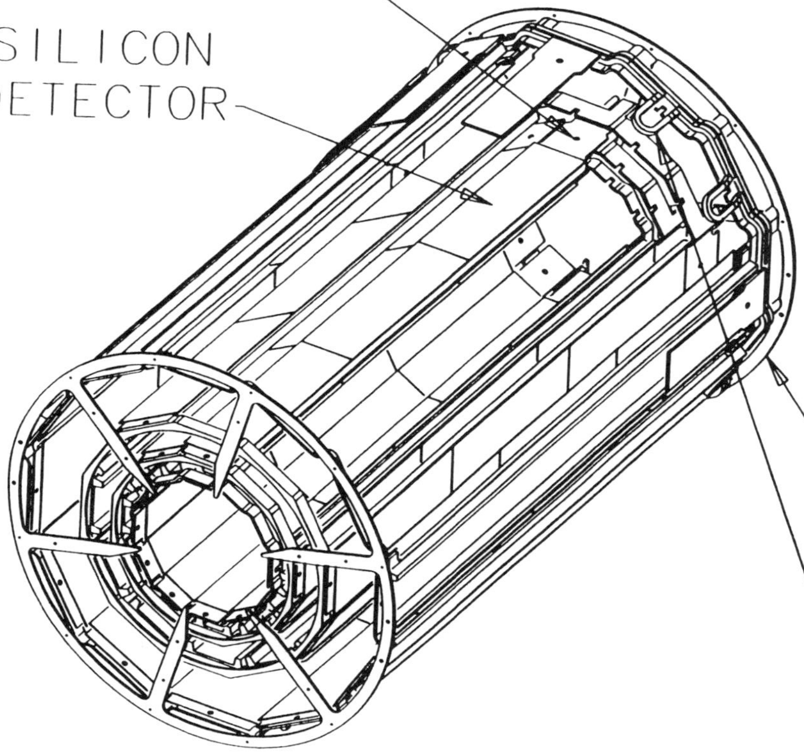

READOUT EAR
SILICON DETECTOR
READOUT END
BULKHEAD
COOLING TUBE

Fig. 18

this beam pipe is a four layer silicon microstrip vertex detector called the SVX. Since the interaction region has a length of about 50 centimeters, it is necessary to have a fairly long detector if it is to have a high efficiency. The SVX has a total length of 51 centimeters, but it is split into two sections at $Z = 0$. The microstrips are etched on 300 micron thick silicon wafers that are about 9 centimeters long. Three of these wafers have their microstrips connected in series in order to form the half-module. The flat silicon planes are configured in the form of a duodecagon around the axis of the beam. There are four layers located at 3.0, 4.2, 6.8, and 7.9 centimeters radius. The three innermost wafers have the strips etched with a 60 micron pitch, and the outer layer has the pitch reduced to 55 microns. There is no Z readout, and thus this detector gives an $r\phi$ view of the event, and the impact resolution in that plane at high momentum is measured to be 17 microns. The 1992-1993 run was the first time that a silicon detector had been operated in a hadron collider, and as a result it suffered a certain amount of radiation damage, resulting in some deterioration of the signal to noise ratio during the run. This detector has since been replaced with a radiation hard version of the electronics. Fig. 17 shows a cartoon of an event with a secondary decay vertex separated from the primary vertex and indicates how such a decay vertex can be reconstructed. The primary vertex is reconstructed in the same manner as the decay vertex and indeed the resolution of the SVX is high enough so that the distribution of the interactions in the $r\phi$ plane can be investigated. The beams have a radius of about 60 microns. Recall that the impact parameter resolution is of the order of 17 microns, and the decay distance $c\tau$ for a B is typically 450 microns.

A drawing of the SVX is shown in Fig. 18. It fits inside of a drift chamber, the VTX, that reconstructs the event in the rz plane. Both of these chambers fit inside of the CTC which has three-dimensional track reconstruction. The challenge of the tracking programs lies in attaching the tracks measured by the CTC which starts at a radius of about 27 centimeters to the measurements made in the SVX where the last plane is at 7.9 centimeters, and then associating these tracks with tracks in the VTX that give the Z position of the interaction.

In order to select the events, it is necessary to place some cuts on the significance of the tracks that are to be tested for association with a possible secondary vertex. The tracks must be associated with jets that have an E_T greater than or equal to 15 GeV and an η less than 2.0. An SVX track is said to be associated with the jet if the opening angle between the track direction and the jet direction is less than 35 degrees. The tracks must have a p_T greater than 2 GeV and must have an impact parameter significance of D/σ_D greater than 3. This sample of tracks is used to search for a secondary vertex as described in Ref. 10. If one is found, a cut is made on L_{XY}/σ_{XY} greater than 3. Fig. 19 shows the result of applying the jet vertex tagging algorithm to a sample of inclusive electron events. These events are heavily populated by b production. The histogram shows a Monte Carlo fit to the data using the world average b lifetime. It should also be noticed that there are a few events located at

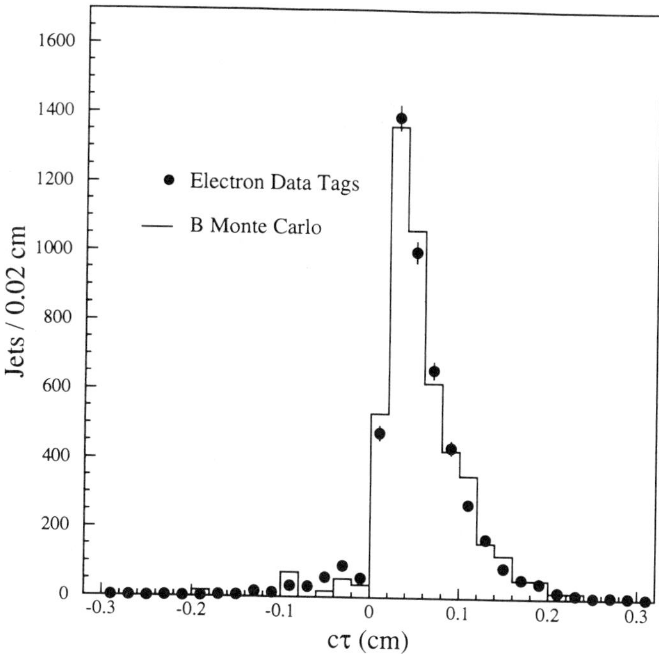

Fig. 19. The $c\tau_{\mathit{eff}}$ distribution for jets with a secondary vertex in the inclusive electron data (points with errors) compared to Monte Carlo simulation (histogram) with the world average B lifetime.

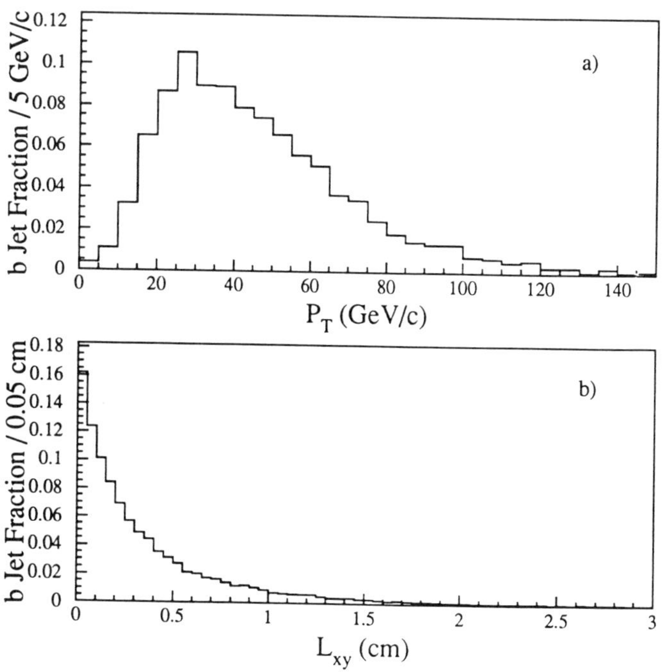

Fig. 20 a) The P_T spectrum for b hadrons from $t\bar{t}$ Monte Carlo events with M_{top} of 160 GeV/c². b) The transverse decay length distribution for the b hadrons, before detector resolution effects, in the same sample.

negative $c\tau$ that are due to tracking errors of one kind or another. Studies have shown that this type of background should be symmetric about zero, and this fact is used to estimate the number of erroneous events on the positive side of the origin. The p_T spectrum and the L_{XY} distribution expected for top production are shown in Fig. 20.

The results of applying the SVX tagging algorithms to the W plus jet sample is shown in Fig. 21. The observed number of tagged events as a function of the number of jets is shown in the last line. Of the 52 W plus three or more jets, six have observed tags. We must now consider two questions: First, what is the efficiency for tagging events, and, second, what is the background that one would expect in the tagged sample?

To measure the efficiency of tagging, it would be nice if we could place the SVX in a beam of b's and measure directly the efficiency for tagging the secondary vertex. This, of course, is not possible, but we can come close to that by performing the following experiment. We take a large sample of inclusive electron events selected by requiring an electron to be in the central region of the detector and have that a p_T greater than 10 GeV. It is known that this sample is rich in b decays. If we knew the fraction of b's in the sample of, then we could count events observed with the SVX and directly determine the efficiency for finding a secondary vertex. The fraction of semileptonic b's has been measured to be about 37 percent. This is determined by two methods. The first involves looking for an associated low p_T muon near the electron direction. A Monte Carlo is used to estimate how often the cascade decay of the b should give an observable μ. This method gives the fraction of b's in the inclusive electron sample $f_b = 37 \pm 8\%$. However, there is an alternative way that this fraction can be checked. This approach relies on kinematically reconstructing $D^0 \to K\pi$ decays. This directly tags the D associated with the semileptonic b decay and gives a number that is consistent with the previous described measurement.

Using these measured efficiencies and a Monte Carlo to describe the $t\bar{t}$ production and decay, a number for the efficiency for tagging a b in $t\bar{t}$ production can be obtained. Fig. 22 shows the efficiency for tagging one of the b's as a function of the top mass. The expected number of events obtained from using the theoretical cross section is also shown.

Next we must worry about the background associated with the tagging operation. This background can come from a number of sources which are listed in Fig. 21. The important components of this background come from the following considerations. First, in the W plus QCD jet production it is possible for one of the gluons to split into a $b\bar{b}$ pair. This would give an event with two real b's in it plus the W. Then, there is the possibility that the tag is an artifact of the tracking. This type of mistake is called a mistag. There are a few other small sources of background that are also listed.

In order to understand mistags a model for the SVX must be constructed that

	Source	$W + 1$ Jet	$W + 2$ Jets	$W+ \geq 3$ Jets
(1)	$Wb\bar{b}$, $Wc\bar{c}$ + Mistags, Method 1	12.7 ± 1.7	4.86 ± 0.63	1.99 ± 0.26
(2)	$Wb\bar{b}$, $Wc\bar{c}$ only, Method 2	2.7 ± 2.2	1.05 ± 0.85	0.37 ± 0.31
(3)	Mistags only, Method 2	4.8 ± 2.5	1.85 ± 0.98	0.76 ± 0.43
(4)	$Wb\bar{b}$, $Wc\bar{c}$ + Mistags, Method 2	7.5 ± 3.3	2.90 ± 1.30	1.13 ± 0.53
(5)	Wc	2.4 ± 0.8	0.66 ± 0.27	0.14 ± 0.07
(6)	$Z \to \tau\bar{\tau}$, WW, WZ	0.20 ± 0.10	0.19 ± 0.09	0.08 ± 0.04
(7)	Non-W, including $b\bar{b}$	0.50 ± 0.30	0.59 ± 0.44	0.09 ± 0.09
(8)	Total Method 1	15.8 ± 2.1	6.3 ± 0.8	2.30 ± 0.29
(9)	Total Method 2	10.6 ± 3.7	4.3 ± 1.4	1.44 ± 0.54
(10)	Events Before Tagging	1713	281	52
(11)	Observed Tagged Events	8	8	6

Fig. 21 Summary of Background and Observed Tags

M_{top} GeV/c^2	ϵ_{tag}	Expected # of Events
120	0.20 ± 0.05	7.7 ± 2.5
140	0.22 ± 0.06	4.8 ± 1.7
160	0.22 ± 0.06	2.7 ± 0.9
180	0.22 ± 0.06	1.4 ± 0.4

Fig. 22. Summary of SVX tagging efficiency (defined as the efficiency of tagging at least one jet in a $t\bar{t}$ event with three or more jets) and the expected number of SVX b-tagged $t\bar{t}$ events in the data sample.

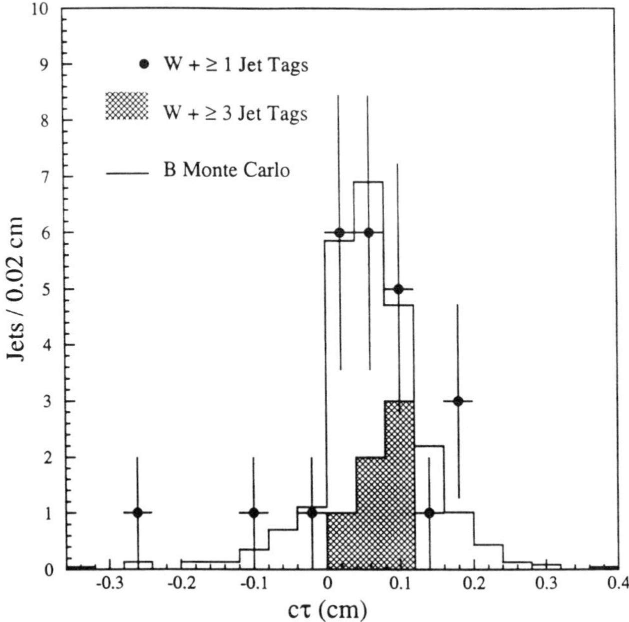

Fig.23 The $c\tau_{\textit{eff}}$ distribution for jets with a secondary vertex in the $W+jets$ data (points with errors) compared to b quark jets from Monte Carlo $t\bar{t}$ events (histogram normalized to data). The shaded histogram is the $W + \geq 3$ jets tags in the data. A $W+2$ jet event with a $c\tau_{\textit{eff}} = 1.2$ cm and a $W+1$ jet event with a $c\tau_{\textit{eff}} = -0.41$ cm are not shown.

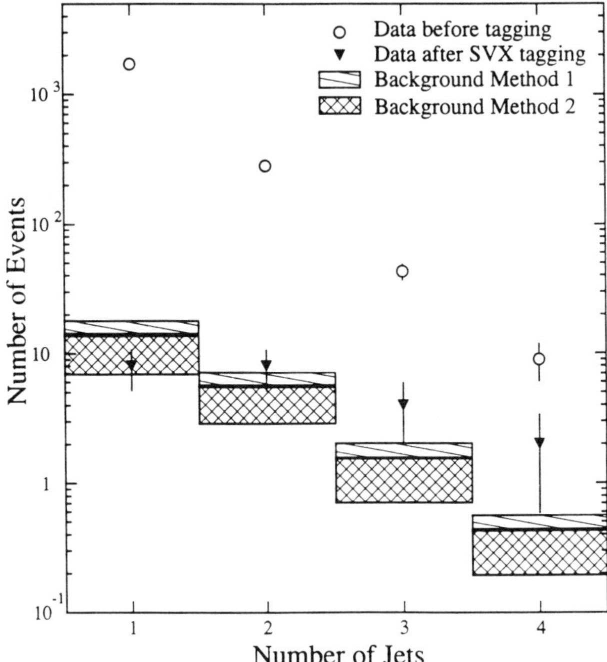

Fig.24 The W+jets distribution observed in the data. The open circles are before SVX tagging and the solid triangles are after SVX tagging. The cross-hatched boxes are the two after tagging background estimates. See text for description.

accurately predicts these mistakes in a variety of situations. The jets from W plus N jets can occur over a wide range of E_T, and with a large variation in the number of tracks associated with them. Thus we need a model for the SVX that accurately predicts it's behavior as a function of these variables. The model was constructed after studying a sample of 67,000 events that passed the 50 GeV jet trigger. These events containing 137,000 jets with E_T greater than 15 GeV were designated as generic jets in that they were not necessarily enriched in heavy flavor. The tagging rate, both positive and negative, was studied as a function of the jet E_T and the track multiplicity. The negative tag rate refers to the rate for a jet to produce a negative L_{XY}. For instance, jets with an E_T between about 20 and 120 GeV have a positive tag rate that varies between 2 and 3 percent and a negative tag rate of about 1 percent. Both rates are a function of the track multiplicity in the jet which can vary from a minimum of 2 to up to greater than 10 in the sample that was used. These empirical measurements were then used to construct a Monte Carlo model for the SVX that could predict both a negative and positive tagging rate for a generic jet. This model was checked against other samples obtained by means of different triggers. The agreement between the predictions and the measurements was excellent. See Ref. 10 for a complete description.

To predict the number of the background events in W plus N jet, we will make the assumption that the tagging rate for W plus N jet is the same as it would be for generic jets. This assumption will be an overestimate because the generic jets contain some direct $b\bar{b}$ production in addition to gluon splitting whereas W production only contains gluon splitting. Thus the model that we have constructed gives a conservative estimate of the b content in W plus N jets and is called method 1.

A second approach is possible. The mistag rate should be correct as it comes from a prediction of the negative L_{XY} tags. It is possible to use theory to directly calculate the expected $Wb\bar{b}$ cross section. Combining these two numbers should give the background actually expected. It has the weakness of having to rely on theory for a calculation of an important contribution to the background. We call this method 2.

The first line of Fig. 21 assumes that the generic jets model the b content accurately, and this conservative number has been used in order to estimate the background. A comparison of method 1 and 2 is shown in Lines 8 and 9. Thus, of the 6 tagged events, we conservatively predict a background of 2.3 $\pm$ .29. A summary of these results is shown in Fig. 23 for the $c\tau$ distribution of all of the W plus jet sample. There are four negative tags, but predominately the tags are consistent with b production. The predicted tags are shown as a histogram and compare well with the measurements. The shaded region show the tags for events with three or more jets. Fig. 24 shows this data in yet another form. It plots the number of events versus the number of jets, both for the tag and untagged data as well as the background from method 1 and method 2.

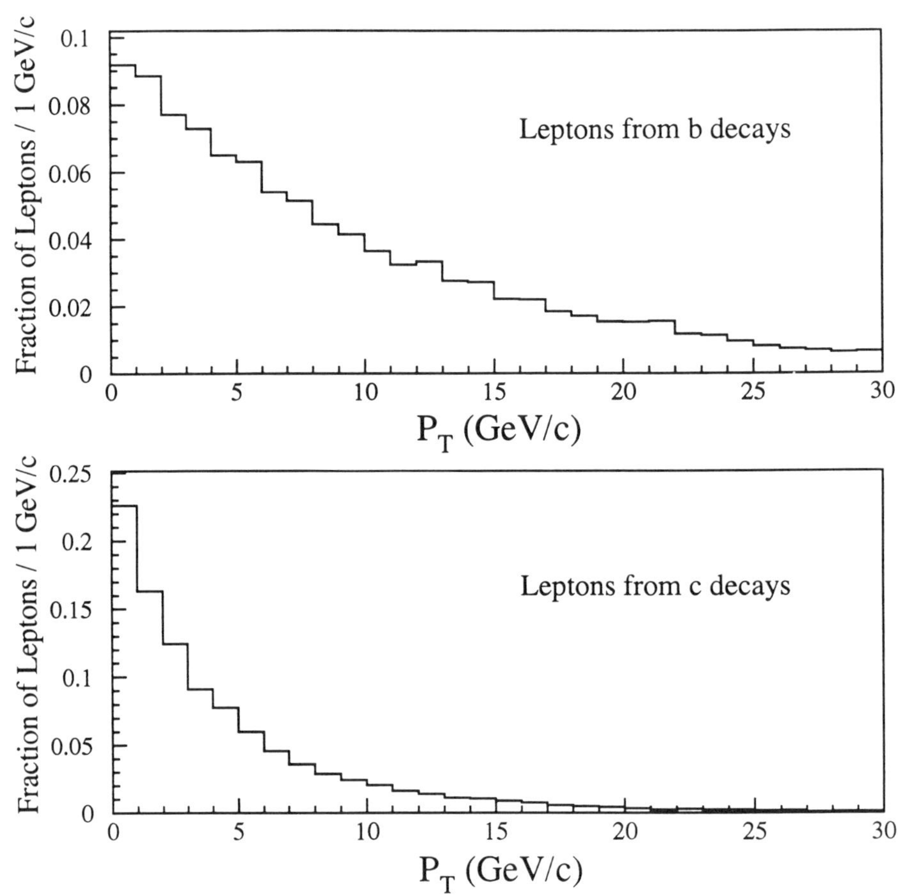

Fig. 25. P_T spectra of leptons from the decay of b and c quarks in top Monte Carlo events (M_{top}=160 GeV/c^2).

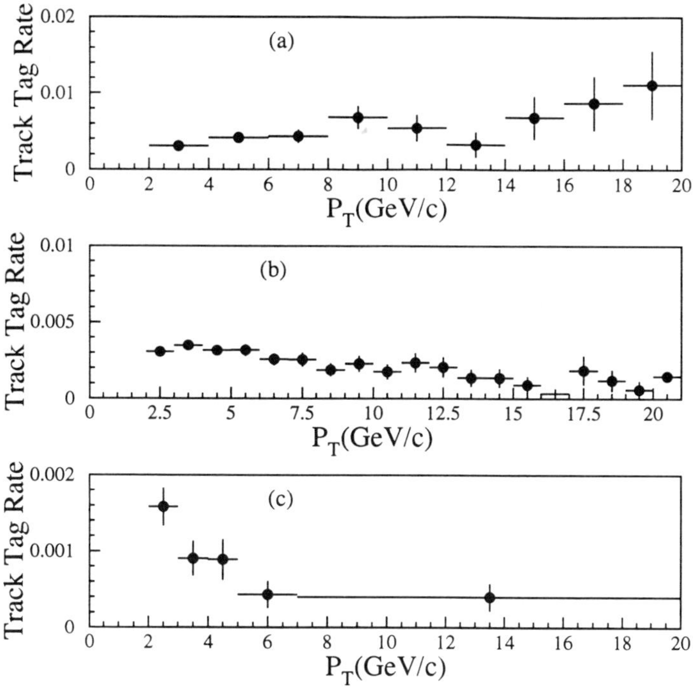

Fig. 26 Track-tag rates for the electron search for tracks satisfying (a) $\frac{\sum p}{p} < 0.2$, where $\sum p$ is the scalar sum of the momenta of all other tracks within a cone of 0.2, and p is the momentum of the track. (b) $0.2 < \frac{\sum p}{p} < 5$. (c) $\frac{\sum p}{p} > 5$.

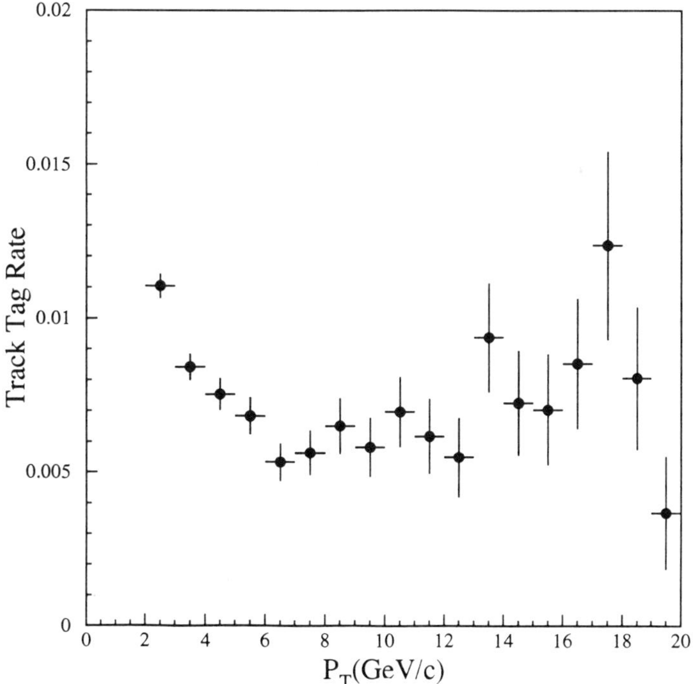

Fig. 27. Track-tag rate for muons in generic jets.

Sample	Number of Electrons		Number of Muons	
	Predicted	Observed	Predicted	Observed
100 GeV jet trigger	598	531	487	471
70 GeV jet trigger	621	631	511	546
50 GeV jet trigger	502	531	374	375
20 GeV jet trigger	757	785	556	557
16 GeV photon sample	30	37	129	128
Six jet sample	65	60	143	144
$\sum E_T$	259	203	762	682
Minimum Bias	25	21	50	47
Z + jets	1.4	2	2.7	4

Fig. 28. A comparison of the observed number of lepton candidates in different samples with the prediction from the track-tag rate parametrizations. The track-tag rate parametrizations were obtained from a mixture of the 20, 50, 70, and 100 GeV inclusive-jet triggers. A trigger bias is present in the muon yields for the inclusive-jet triggers because the energies of jets containing hadrons that do not interact in the calorimeter are measured systematically low. For this reason, only tracks well separated from a trigger-jet are considered in the muon analysis. The statistical uncertainties on the predictions are negligible.

Source		$W + 1$ Jet	$W + 2$ Jets	$W + \geq 3$ Jets
Fakes$+Wb\bar{b}+Wc\bar{c}$	e tags	9.9 ± 1.5	2.9 ± 0.4	0.88 ± 0.13
	μ tags	19.2 ± 1.9	5.9 ± 0.6	1.82 ± 0.18
	e $+$ μ tags	**29.1 ± 2.9**	**8.8 ± 0.9**	**2.70 ± 0.27**
$b\bar{b}$	e tags	0.8 ± 0.6	0.14 ± 0.10	0.03 ± 0.02
	μ tags	0.9 ± 0.6	0.14 ± 0.10	0.03 ± 0.02
	e $+$ μ tags	**1.7 ± 1.2**	**0.28 ± 0.20**	**0.05 ± 0.03**
Diboson	e tags	0.25 ± 0.12	0.11 ± 0.06	0.03 ± 0.02
	μ tags	0.28 ± 0.13	0.03 ± 0.02	0.01 ± 0.01
	e $+$ μ tags	**0.53 ± 0.25**	**0.14 ± 0.08**	**0.04 ± 0.03**
$Z \to \tau\tau$	e tags	0.37 ± 0.13	0.11 ± 0.05	0.08 ± 0.03
	μ tags	0.30 ± 0.11	0.07 ± 0.04	0.06 ± 0.03
	e $+$ μ tags	**0.67 ± 0.24**	**0.18 ± 0.09**	**0.14 ± 0.06**
Drell-Yan	e tags	0.15 ± 0.10	0.03 ± 0.03	0.03 ± 0.03
	μ tags	0.15 ± 0.10	0.03 ± 0.03	0.03 ± 0.03
	e $+$ μ tags	**0.30 ± 0.20**	**0.05 ± 0.05**	**0.05 ± 0.05**
$W + c$	e tags	0.4 ± 0.1	0.10 ± 0.03	0.02 ± 0.01
	μ tags	1.4 ± 0.5	0.32 ± 0.08	0.06 ± 0.02
	e $+$ μ tags	**1.8 ± 0.6**	**0.42 ± 0.11**	**0.08 ± 0.03**
Total	e tags	11.9 ± 1.6	3.4 ± 0.4	1.1 ± 0.2
	μ tags	22.2 ± 2.1	6.5 ± 0.6	2.0 ± 0.2
	e $+$ μ tags	**34.1 ± 3.3**	**9.9 ± 1.0**	**3.1 ± 0.3**
Events Before Tagging		1713	281	52
Events After Tagging	e tags	17	2	4
	μ tags	16	10	3
	e $+$ μ tags	**33**	**12**	**7**

Fig. 29. Summary of SLT backgrounds as a function of jet multiplicity.

Tagging the b with Soft Leptons

As mentioned earlier, we can tag the b's by looking for their semileptonic decay: $b \rightarrow e\nu X$ or $b \rightarrow \mu\nu X$. Calculations indicate that there is about .8 of an e or μ for each $t\bar{t}$ event. As before we have two questions that have to be answered. One is the efficiency for tagging an event which gives us the signal, and the second is the mistag rate which gives the background.

The probability of finding the e or the μ depends upon the momentum spectrum in the decay. Fig. 25 shows the P_T spectrum of the leptons from b decays as well as the lepton spectrum from c decays that are the secondary of b decays. The hardness of the spectrum, of course, depends upon the mass of the top, and that has been chosen to be 160 GeV for Fig. 25. It is necessary to make a low momentum cut on either the electron or the muon in order to eliminate a large amount of background that would come in from extraneous processes. In the case of the muon this low momentum cut must be higher than 2 GeV because that is the energy required for a muon to traverse the hadron calorimeter and be detected in the chambers just to the rear. A study of the electron backgrounds indicated that this was also a sensible place to make the cut for electrons. The efficiency of these cuts is seen to be very high.

The background in both cases is associated with the probability that a track will fake a lepton. For instance, a muon can be faked by a pion decay in flight or an electron can be faked by a pion giving a big interaction in the electromagnetic calorimeter. To calculate the background then requires a detailed study of these probabilities which can depend on the track momentum as well as a number of other cuts that are made in the calorimetry. Details of these are given in Ref. 10. Fig. 26 and 27 show the tag rate per track for electrons and muons in generic jets. It is seen that this tag rate in both cases is less than 1 percent. The background then for the tagging algorithm consists of folding this information about the fake track tagging rate into the distribution of tracks expected from the jets that are being studied. Again, as in the case of the SVX, a number of independent sources of jets were examined to see how well the predicted and observed number of tracks agreed with each other. Fig. 28 shows a summary of this information. And it can be seen that the predicted numbers agree quite well with those actually observed. The deviation between and predicted numbers and the observed numbers is used to estimate the systematic error on this procedure.

Fig. 29 shows a summary of the backgrounds as well as the tagging rate for SLT events. Again, as in the case of the SVX, we assume that a generic jet has the same b content as the W + jets and again, we understand that this is a conservative assumption as it is probably an overestimate of the $Wb\bar{b}$ contribution. The summary is given in the bottom line where we observe that the 52 W plus more than three jets events have seven tags and an estimated background 3.1 ± 0.3 events.

Channel:	SVX	SLT	Dilepton
Expected # events $M_{top} = 120$ GeV/c^2	7.7 ± 2.5	6.3 ± 1.3	3.7 ± 0.6
Expected # events $M_{top} = 140$ GeV/c^2	4.8 ± 1.7	3.5 ± 0.7	2.2 ± 0.2
Expected # events $M_{top} = 160$ GeV/c^2	2.7 ± 0.9	1.9 ± 0.3	1.3 ± 0.1
Expected # events $M_{top} = 180$ GeV/c^2	1.4 ± 0.4	1.1 ± 0.2	0.68 ± 0.06
Expected Bkg.	2.3 ± 0.3	3.1 ± 0.3	$0.56^{+0.25}_{-0.13}$
Observed Events	6	7	2

Fig. 30 Numbers of $t\bar{t}$ events expected, assuming the theoretical production cross sections shown in Table 32, and the numbers of candidate events observed with expected backgrounds.

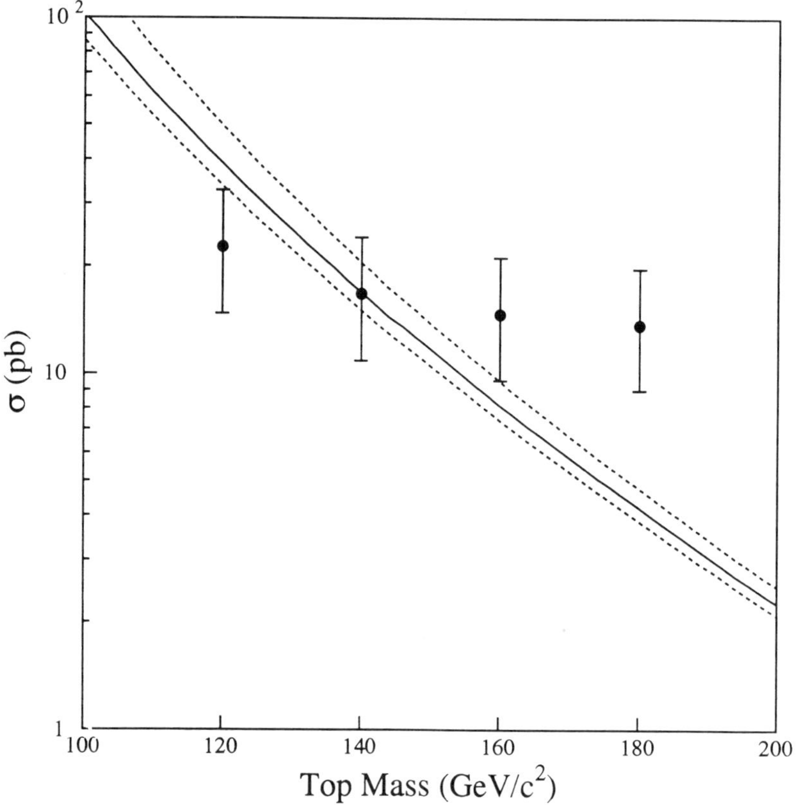

Fig. 31. Combined $t\bar{t}$ production cross section vs. M_{top} from data (points) and theory [10]. The dashed lines are estimates of the theoretical uncertainty quoted in Reference [10].

Statistical Significance of the Counting Experiments

We are now in a position to evaluate the statistical significance of the results from the counting experiments. We have three channels each showing an excess. The numbers are shown in Fig. 30. The bottom line gives the observed number of events in each channel and just above, is the expected background. For comparison, the first four lines of the table give the number of events predicted from using the theoretical value of the cross section for four different top masses. If we treat the channels independently, we can calculate the probability that the estimated background has fluctuated up to a number greater than or equal to the number of events seen. We find P_{DIL} is equal to 12 percent, P_{SVX} is equal to 3.2 percent, and P_{SLT} is equal to 3.8 percent. However, one can make a stronger statement by calculating a combined probability for the three results. Recall that there are two dilepton tags (one event has both an SLT and SVX tag). There are 6 SVX tags and 7 SLT tags. However, 3 of the SVX events overlap 3 of the SLT events. The question of how to combine this data was investigated at length, and the following ansatz was finally used. Instead of using tagged events, the number of tags in the sample was taken as the variable except in the dilepton case, where events were used. Thus, there are 15 "counts;" the 2 dilepton events, the 6 SVX tags, and the 7 SLT tags. This procedure gives extra weight to the double-tagged events which are more likely to be real than false and, therefore, have a considerably smaller background than single tagged events. However, there are still correlations among the experiments that must be properly understood in order to calculate correctly the combined probability. A Monte Carlo program was used which generated many samples of the 52 events with the background such as W $+ b\bar{b}$, etc. fluctuating around their mean value. The procedure is described in great length in Ref. 10 and leads to the result that $P_{combined}$ is equal to 0.26 percent which, if it were a Gaussian probability, would be a $2.8\,\sigma$ excess.

Assuming that these excess events come from $t\bar{t}$ production, one can calculate the cross section as a function of M_T. The dependence on M_T enters because the acceptance of the experiment is slightly dependent upon the top mass. The results are shown in Fig. 31. The next task is to estimate the mass from the kinematics of the events.

Checks on the Counting Experiments

Before we study the behavior of the kinematic variables, we will describe briefly some of the checks that are made on the counting experiment. An obvious place to test the validity of the procedure would be to study the corresponding situation in Z + jets. In this case, no top signal is expected, however, the smaller number of events in which the Z is identified through its e^+e^- decay mode will make these checks statistically rather limited. In order to compare W + jets with Z + jets, we subtract the top signal from the W + jet sample. This is possible because we know

Jet Multiplicity	Data	Top	Other backgrounds	QCD W + jets
1 Jet	1713	$1.1^{+0.5}_{-0.4}$	284 ± 89	1428 ± 98
2 Jets	281	$5.0^{+2.3}_{-1.7}$	54 ± 15	222 ± 23
3 Jets	43	$10.0^{+4.8}_{-3.5}$	8.9 ± 2.5	$24.1^{+8.0}_{-8.5}$
$\geq$ 3 Jets	52	$21.6^{+9.8}_{-7.6}$	10.8 ± 3.1	$19.6^{+10.9}_{-12.6}$
$\geq$ 4 Jets	9	$11.6^{+5.6}_{-4.5}$	1.9 ± 0.6	$0^{+3.5}_{-0.0}$

Fig. 32. Number of events in the data, number of expected top events, assuming the top cross section measurement from Section 7.1, and number of background events. The number of QCD $W+$ jets events is obtained by subtracting from the data the top and non-W background contributions. For $W+$ 4 or more jets, this subtraction yields the unphysical value $-4.5^{+5.4}_{-6.4}$. The value $0^{+3.5}_{-0.0}$ given in the Table is obtained by imposing the constraint that the number of QCD $W+$ 4 or more jets should be ≥ 0.

Jet Multiplicity	QCD W + jets	VECBOS ($Q^2 =< P_T >^2$)
1 Jet	1428 ± 98	$1571 \pm 82^{+267}_{-204} \pm 55$
2 Jets	222 ± 23	$267 \pm 20^{+77}_{-53} \pm 9$
3 Jets	$24.1^{+8.0}_{-8.5}$	$39 \pm 3^{+11}_{-9} \pm 2$
$\geq$ 4 Jets	$0^{+3.5}_{-0.0}$	$7 \pm 1^{+3}_{-2} \pm 0.2$

Fig. 33. Comparison of QCD $W+$jet yields from Table 36 with expectations from the VECBOS Monte Carlo. The first uncertainty on the VECBOS prediction is due to Monte Carlo statistics, the second to the jet energy scale and lepton identification efficiency uncertainties, and the third to the uncertainty on the luminosity normalization. The additional uncertainty related to the choice of the Q^2 scale in the VECBOS Monte Carlo program is discussed in the text. The VECBOS predictions include the $W \rightarrow \tau\nu$ contribution.

Jet Multiplicity	W + jets	Z + jets	R_{wz}
1 Jet	1428 ± 98	176	8.1 ± 0.9
2 Jets	222 ± 23	21	10.6 ± 2.6
3 Jets	$24.1^{+8.0}_{-8.5}$	3	$8.0^{+4.3}_{-4.7}$
$\geq$ 3 Jets	$19.6^{+10.9}_{-12.6}$	5	$3.9^{+2.4}_{-2.6}$
$\geq$ 4 Jets	$0^{+3.5}_{-0.0}$	2	$0^{+1.5}_{-0.0}$

Fig. 34. $W+$ jets and $Z+$ jets event rates from Tables 36 and 26 as a function of jet multiplicity. R_{wz} is the ratio of the number of W and Z events.

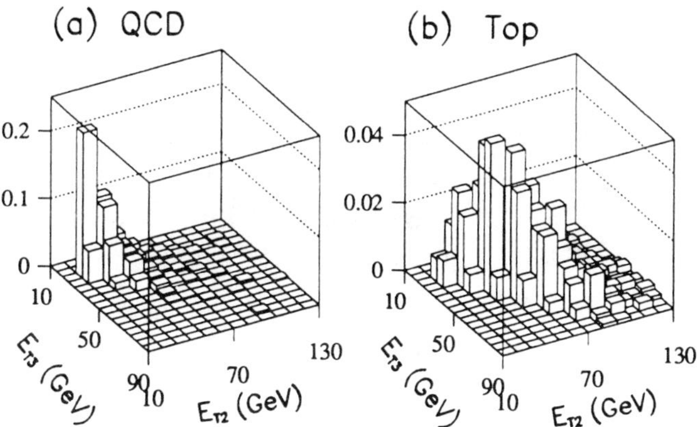

Fig. 35 $d\sigma^2 / dE_{T2}dE_{T3}$ for (a) QCD $W+3$ jet and (b) top ($M_{top} = 170$ GeV/c^2) Monte Carlo events. The vertical scale is in arbitrary units.

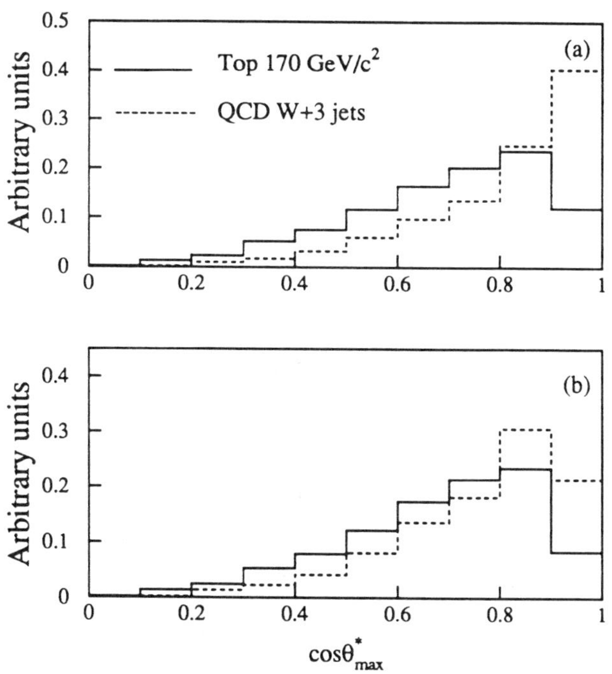

Fig. 36 $\cos\theta^*_{max}$ for HERWIG top ($M_{top}=170$ GeV/c^2) and VECBOS $W+3$ jet events. (a) inclusive distribution, (b) after applying a cut on $|\eta(\text{jets})| < 2$. The distributions are normalized to unit area.

the efficiency for tagging a top event, and we also know from Monte Carlo studies the population of the top events in the W + N JETs sample. It is true that there is a small variation of tagging efficiency with mass, but this variation is less than 10 percent for the SVX and less than 5 percent for the SLT over a top mass range from 120 to 180 GeV. Fig. 32 then shows the corrected number of W + QCD jets that are observed. Notice that the contribution from the top is so large that it completely accounts for all of the events observed in W + four or more jets. To see if this is reasonable, we compare these numbers with a VECBOS calculation in Fig. 33, and there seems to be a deficit in the W + four jet events. However, the uncertainty on the VECBOS predictions due to the choice of the Q^2 scale dependence makes the uncertainties hard to quantify.

The numbers from Fig. 33 are shown in Fig. 34 along with the experimental numbers from a study of Z + N jets. The last column shows the ratio between the W and the Z columns. Again, in the case of three of more jets, there seems to be a deficit of events in the W + N jet case, but the statistics is unfortunately rather limited. An additional feature of the Z events is that there are two b-tagged Z events with greater than or equal to 3 jets where only .64 is expected. The resolution of these questions will have to await additional experimental data.

The Analysis of the Event Structure

So far we have been considering the search for the top as a counting experiment, that is to say, was there an excess number of W + 3 or more jets in the data, or was there an excess of dilepton events. The question of whether the kinematics of the event describes a $t\bar{t}$ production and decay has arisen only indirectly in calculating the detector acceptance. However, it is clear that a study of the event variables may be able to distinguish between QCD processes and $t\bar{t}$ production. We investigate that question now.

Fig. 35 shows a lego plot of E_{t2} versus E_{t3} for W + 3 or more jets where the VECBOS calculation has been used for the QCD background and ISAJET has been used for the $t\bar{t}$ case. A top mass of 170 GeV has been assumed. This figure graphically illustrates the fact that a heavy mass top tends to populate the central regions of the detector with rather high jets. The fourth jet would also show this effect. However, in the interest of maximizing the signal and minimizing the systematic errors at low jet energy, we initially exclude consideration of the fourth jet.

Fig. 36 shows the $\cos\theta^*_{\max}$ predicted by Herwig for top production, and by VEC-BOS for W + 3 jet events. The upper figure shows the inclusive distribution, and the lower figure shows the distribution after applying a rapidity cut to the jets which requires them to be in the central region of the detector. $\cos\theta^*_{\max}$ is the maximum $\cos\theta$ of the three jets. The curves have been normalized to the same area for comparison. If one cuts on $|\cos\theta^*_{\max}|$ then the region greater than 0.7 will contain an enhanced

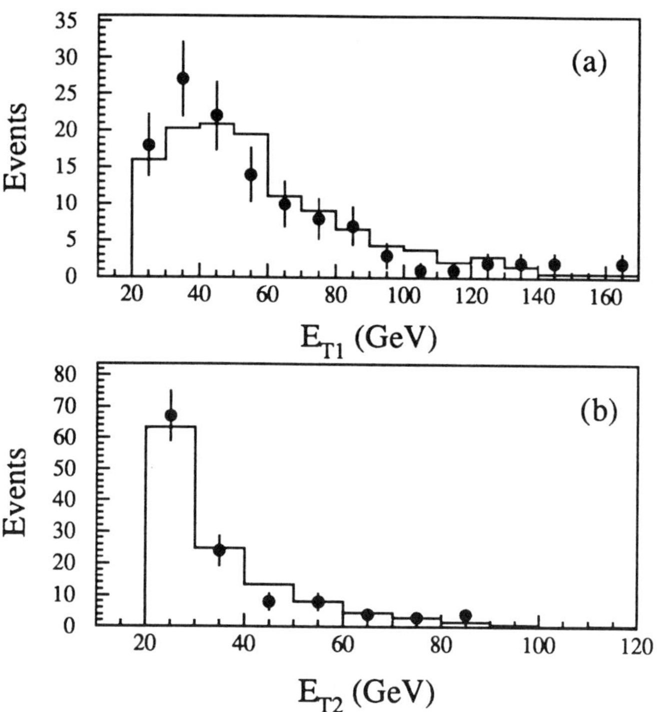

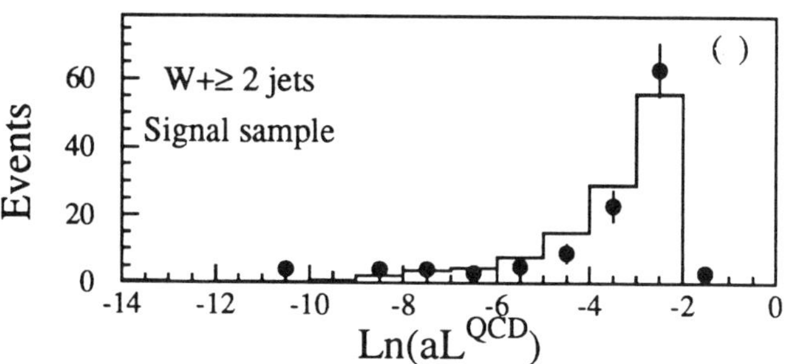

Figure 37: E_T distributions for W + 23 or more jets data (points) and the VECBOS predictions for W + 2 jets (histogram). (a) leading jet, (b) second jet. (c) shows the $Ln(aL^{QCD})$ for Signal sample.

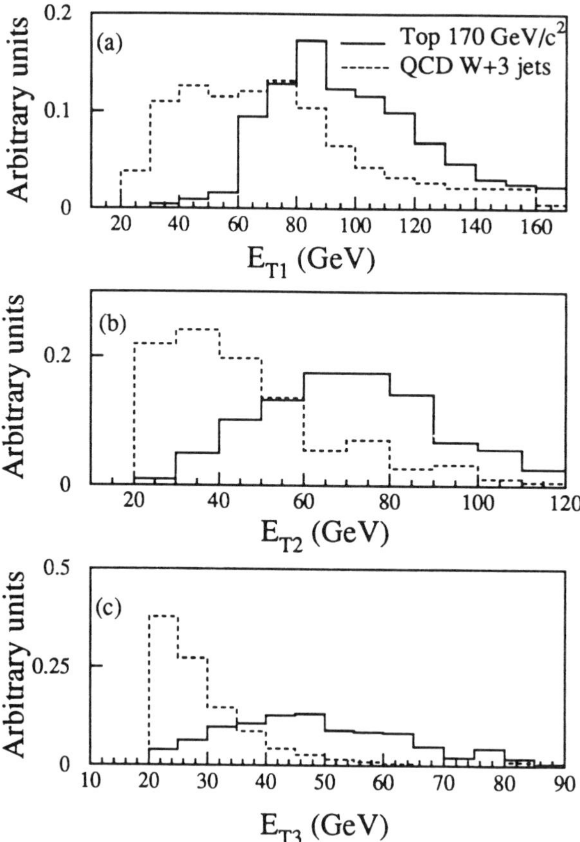

Fig. 38 **Jet energy distributions for HERWIG top (solid line) and VECBOS $W + 3$ jet events (dashed line) passing the signal sample selection cuts. Each distribution is normalized to unit area.**

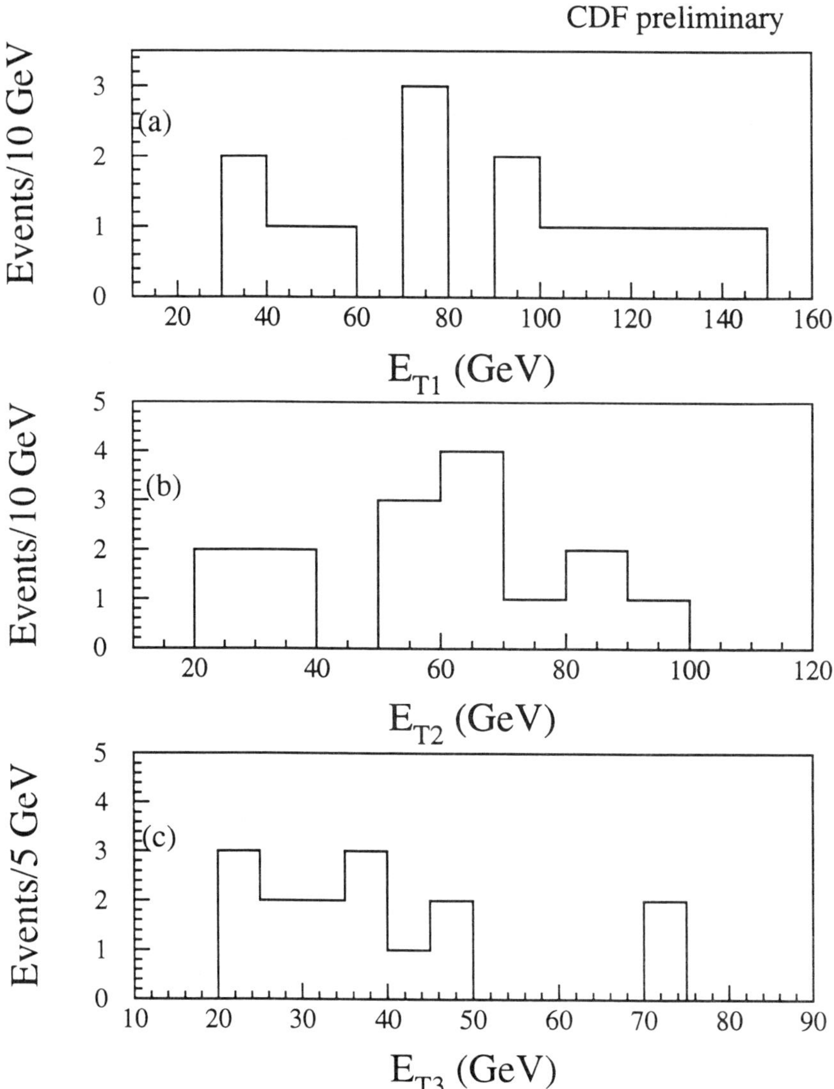

Fig. 39 **Jet energy distributions for the three leading jets in the 14 events passing the signal sample selection cuts. There is one overflow in E_{T1} at $E_{T1} = 224$ GeV.**

background. The number of top events in the two samples should be about equal, but in the latter sample the background should be three times higher. The region of large $\cos\theta$ is called the background region, and the region less than 0.7 is called the signal region in the following discussion.

Since we will be comparing top decay with $W + QCD$ jets, it is imperative that we have the good model for the QCD process. The model used here is VECBOS. However, in using VECBOS, it is necessary to define the Q^2 scale for α_s. The VECBOS program allows generation of W events with $N = 0, 1, 2, 3,$ and 4 partons. We require the p_T of the parton to be greater than 10 GeV, and the η of the parton to be less than 3.5 as well as the ΔR separation of 2 partons to be greater than .4 in order to avoid infrared divergences. The partons have been fragmented using Herwig as well as Field-Feynman. The results are not sensitive to this feature. However, they are somewhat sensitive to the Q^2 scale that is chosen. In this study $Q^2 = M_W^2$ has been used as it yields the hardest distribution for the jet partons. Two checks of this model are possible. The E_T distribution for the jets in the $W + 2$ or more jets sample can be studied as well as the complementary reaction with the Z. In both cases, reasonable agreement with the model is found.

To display this data, we define an absolute likelihood as follows:

$$ aL \;=\; \left(\frac{1}{\sigma} \frac{d\sigma}{d\,E_{T1}} \right) \times \left(\frac{1}{\sigma} \frac{d\sigma}{d\,E_{T2}} \right) $$

E_{T1} and E_{T2} are the energies of the highest two jets in the $W + 2$ or more jet sample. The distributions in E_{T1} and E_{T2} are shown in Fig. 37, and the distribution in absolute likelihood as defined above, is shown in the lower histogram. It is seen that the agreement between the model and the experimental data is quite good, although the data may be slightly softer than the model.

We now proceed to the $W + 3$ or more jet events, and we now expect both QCD background plus real top to be present. As described above, we can enhance the signal by making a cut $|\cos\theta^*_{max}| < 0.7$. The distributions expected from $t\bar{t}$ events and from VECBOS plus 3 or more jet events in shown in Fig. 38. The curves have been normalized to unity for reasons that will become apparent shortly. The top curves have been drawn for 170 GeV top, and it is apparent that the E_{T1} and E_{T2} and E_{T3} spectra are considerably harder than would be expected for the QCD events. The experimental data are shown in Fig. 39.

We now need a way to test whether an event is more like the QCD case or more like the top case in its characteristics. We define an absolute likelihood in analogy with the 2-jet case but use E_{T2} and E_{T3}. We note that given an event with an E_{T2} and a E_{T3}, we could use either of the distributions shown in Fig. 38 to calculate an absolute likelihood. That is, we could use the QCD distribution to measure a likelihood that it is similar to a QCD event or we could use the distribution from the top Monte Carlo to measure the probability that it resembles the top. A convenient way

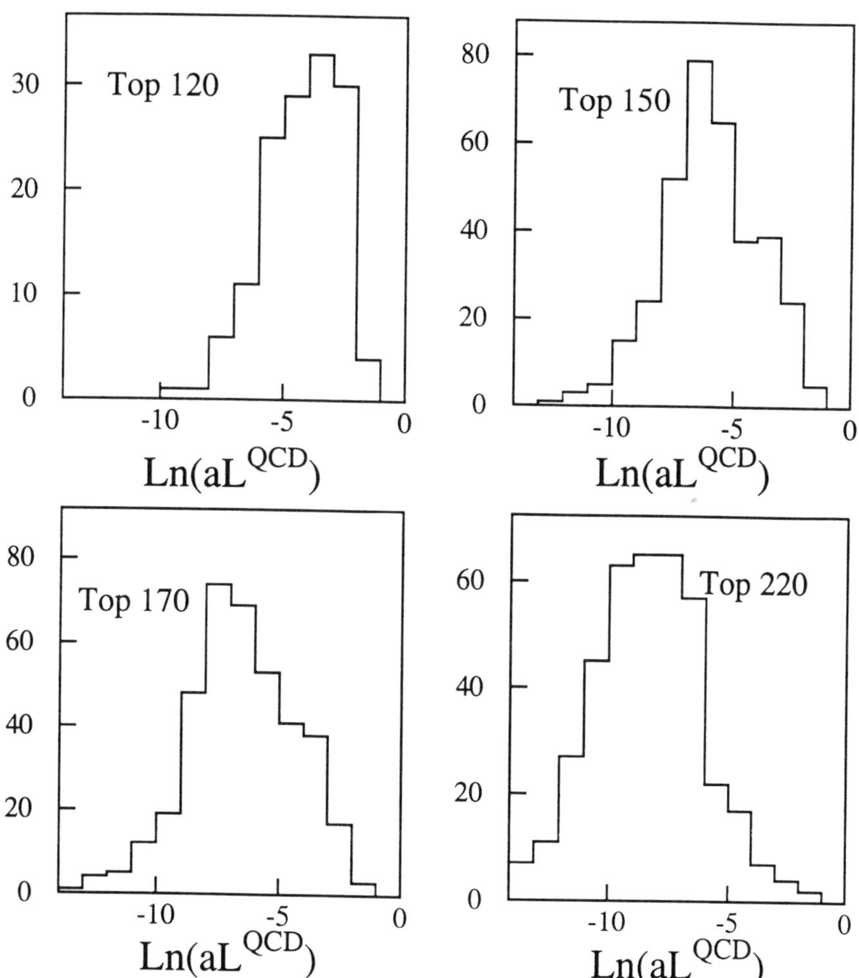

Figure 40: This figure shows the expected distributions of aL^{QCD} for different sets of Monte Carlo t tbar events where the top mass is varied from 130 GeV to 220 GeV.

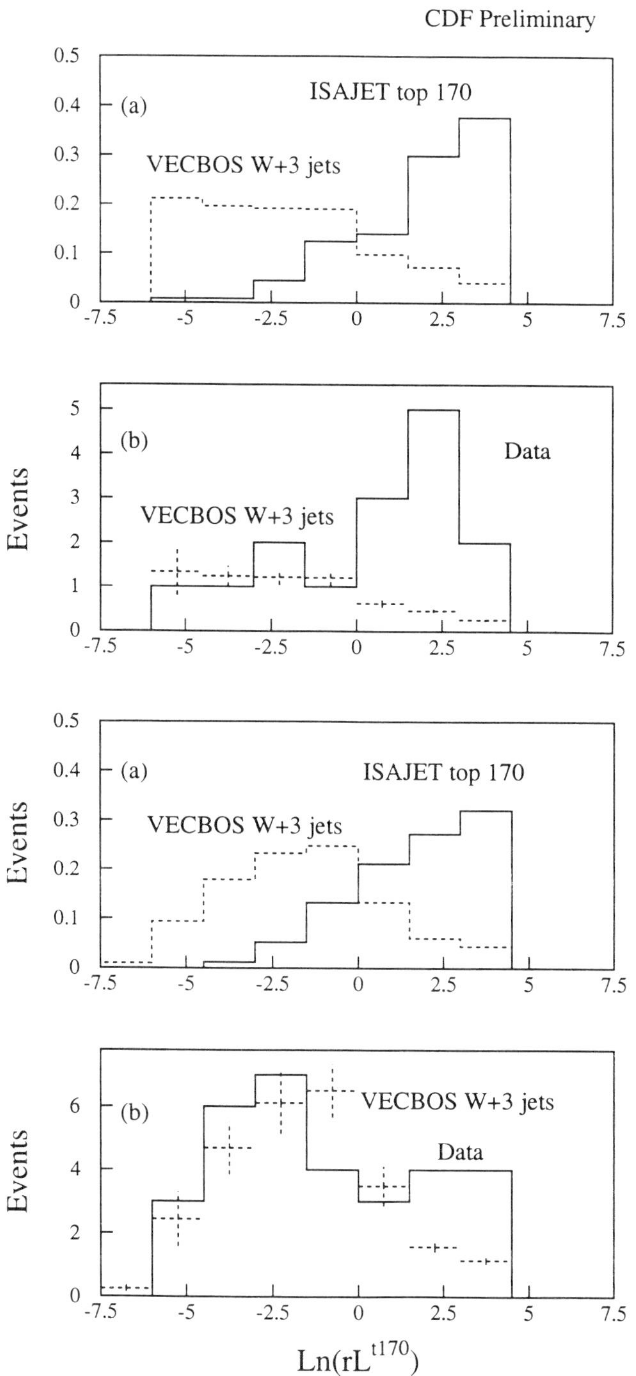

Figure 41: Ln(rL^{t170}) for QCD VECBOS, top Isajet and data events for W plus 3 or more jets. (a) and (c) have had their histograms normalized to 1.0. (a) and (b) are for events in the signal and (c) and (d) are for events in the control region. For (b) and (d) VECBOS MC has been normalised to the data for ln(rL) < 0.

272

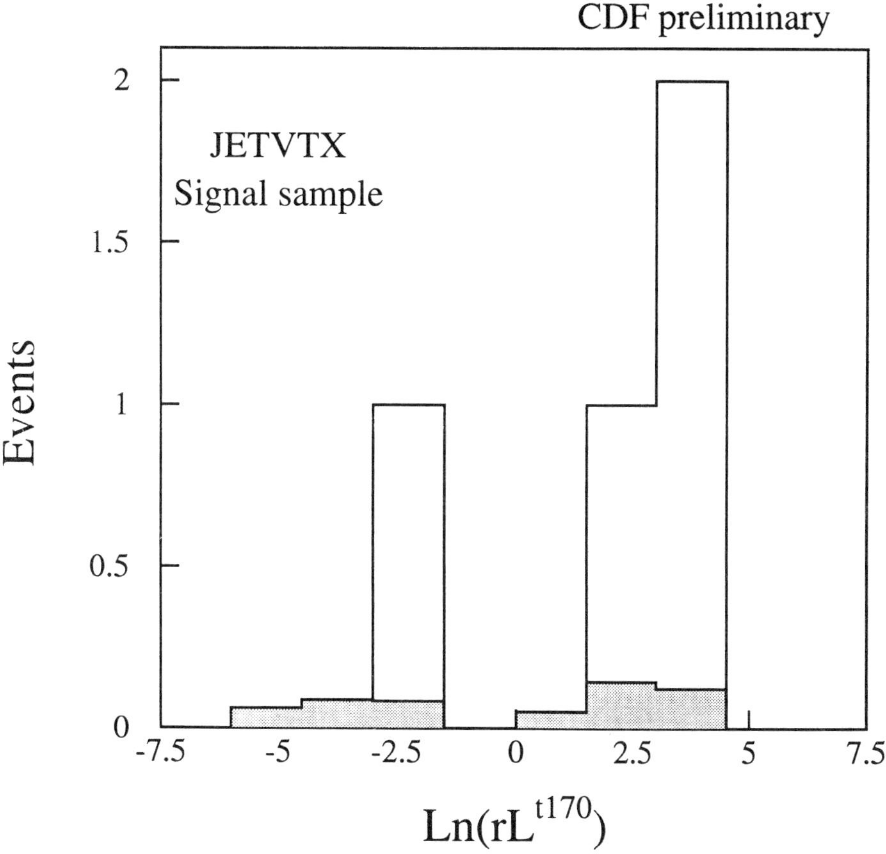

Figure 42: Distribution in ln(rL) of the 4 events of the signal sample tagged by the JETVTX algorithm. The expected fakes are shown as a shaded histogram.

to display this data then is to define a relative likelihood by the following formula. This relative likelihood is the ratio between the absolute likelihood that the event is a top and the absolute likelihood that it is a QCD event.

$$rL \; = \; aL^{top}/aL^{QCD}$$

The normalized histograms of Fig. 38 are used for calculating this ratio.

Large values of relative likelihood will indicate a top-like candidate. And small values of relative likelihood will indicate a QCD-like event. Note that this whole process is sensitive to the parent distributions which involve the Q^2 scale for VECBOS and also the mass used for the top in the $t\bar{t}$ distributions. The dependence on the mass assumed for the top is shown in Fig. 40.

The distributions predicted by a Monte Carlo calculation of rL are shown in Fig. 41a for the signal region and 41c from the control region. The solid curve is from top production and the dotted from the VECBOS Monte Carlo. The curves have all been normalized to unity.

The distribution of the data in the two regions is displayed in Fig. 41b and Fig. 41c. The data is shown as a solid line and the VECBOS predictions as crosses. The VECBOS points have been normalized to the region $\ell n(rL) < 0$. It is seen that there is an indication of a top-like signal in the data.

We have one more test of the nature of these events in that we can look at the b-tags in the SVX and SLT. There are 14 events in the signal sample, and four of these events have an SVX tag. The distribution of the tagged events is shown in Fig. 42. The shaded region is an estimation of the tags that would be expected from background processes. The method for estimating this background is similar to that described in the SVX and SLT search. There is one event in the background region, and three events in the top region where the expected background is $0.58^{+.12}_{-.09}$. The probability that the observed number of tags is due to a statistical fluctuation of the background is 0.4 percent. Four of the 14 events include a soft lepton tag, and the expected background in this case is $1.2 \pm .3$ events with a probability of the background fluctuating up to 4 or more events being 4 percent. In the control sample there is one SVX tag and one soft lepton tag, and the expected number of tags is of the order of 2.

Thus, within the limited statistics that are available, the kinematic structure shows a top-like signal. In the future when a large sample of events is available, this will become an important technique for demonstrating that the events have the distributions in E_{T2} and E_{T3} corresponding to that expected for a top. We now proceed to the reconstruction of mass and note that it would be possible to have events reconstruct to a top mass without having the distribution of the kinematic variables fit the $t\bar{t}$ hypothesis. Thus, the event structure analysis gives independent evidence as to the nature of the events.

Mass Reconstruction

If it is assumed that the excess of b-tagged events described in the preceding sections comes from $t\bar{t}$ production, then it should be possible to determine the mass of the top directly by reconstruction. In order to do this, it is necessary to have access to all four jets. For this reason, we will change the cuts slightly to increase the acceptance for a fourth jet which will now be included if it has an uncorrected E_T greater than 8 GeV and an η less than 2.4. Monte Carlo studies show that for 170 GeV mass, 60 percent of the events having three jets will also have a fourth jet passing the standard criteria, while 86 percent will have a fourth jet passing passing these relaxed criteria. Of the 10 b-tagged events, 7 pass the relaxed criterion for having a fourth jet.

For the purposes of making a constrained fit, we assume that the production and decay process goes through the following steps.

1. $\bar{p}p \rightarrow t_1 + t_2 + x$

2. $t_1 \rightarrow b_1 + W_1$

3. $t_2 \rightarrow b_2 + W_2$

4. $W_1 \rightarrow \ell + \nu$

5. $W_2 \rightarrow j_1 + j_2$

This is a five vertex system in which we make measurements of the jet energies, the lepton energy, and the missing E_T. It is assumed that the initial state transverse momentum is zero. The overall kinematic fit has two degrees of freedom. There are 20 equations and 18 unknowns. However, the association between the jets and the partons is not unique. If both of the b jets were correctly tagged, there would still be multiple solutions. First, there are two solutions for the p_Z for the neutrino, and there would be an additional two combinations in the association of the b with the correct top. However, we only have one b-jet tag, and hence there are 12 different configurations that we must choose between. If none of the b jets are tagged, then there are 24 possible configurations. To chose among the different configurations, we calculate χ^2 and demand that $\chi^2 < 10$. We will discuss the efficiency of this method shortly. The calculation is also complicated by the possibility that one of the jets may come from initial state radiation and is not even associated with the t or $\bar{t}$ decay.

The outline of the solution above requires that we know the parton momenta. However, the detector measures jet energies. In order to do the reconstruction, we need to relate the jet energy to the parton energy. Furthermore, in order to calculate a χ^2, we need to estimate the error on the parton energy that arises because of

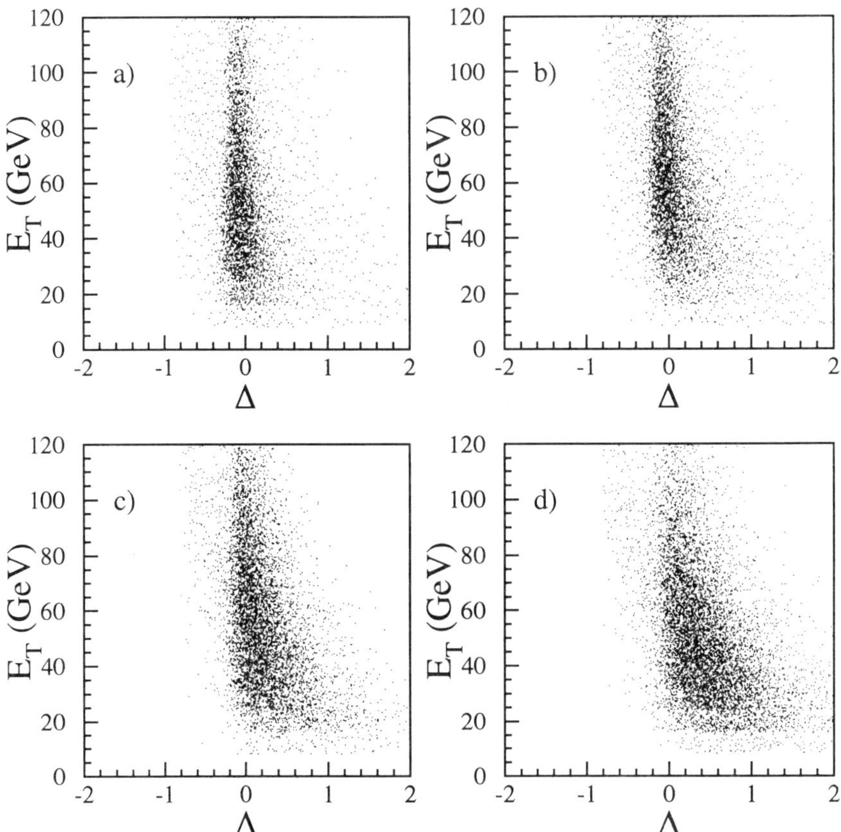

Fig. 43. Scatter plots of jet E_T corrected with standard jet corrections vs. Δ for the four jet types: a) light quarks b) generic b jets c) $b \to e\nu$ X jets d) $b \to \mu\nu X$. $\Delta = (P_T(\text{parton}) - E_T(\text{jet}))/E_T(\text{jet})$. The Monte Carlo events have been generated with HERWIG at $M_{top} = 170$ GeV/c^2.

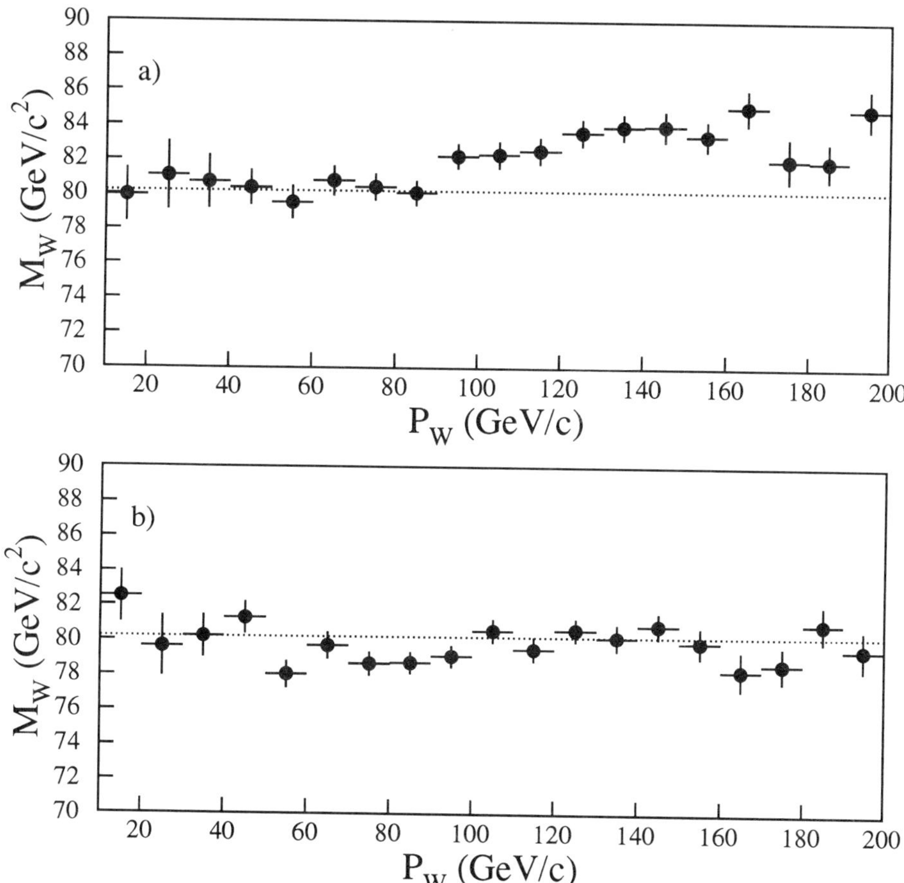

Fig. 44. Mass of the W calculated using the 4-momenta of the jets as a function of the W momentum. The jet momenta are corrected with standard jet corrections (top plot) and with the jet corrections used in the mass analysis (bottom plot). The events plotted are generated with HERWIG at $M_{top} = 170$ GeV/c^2.

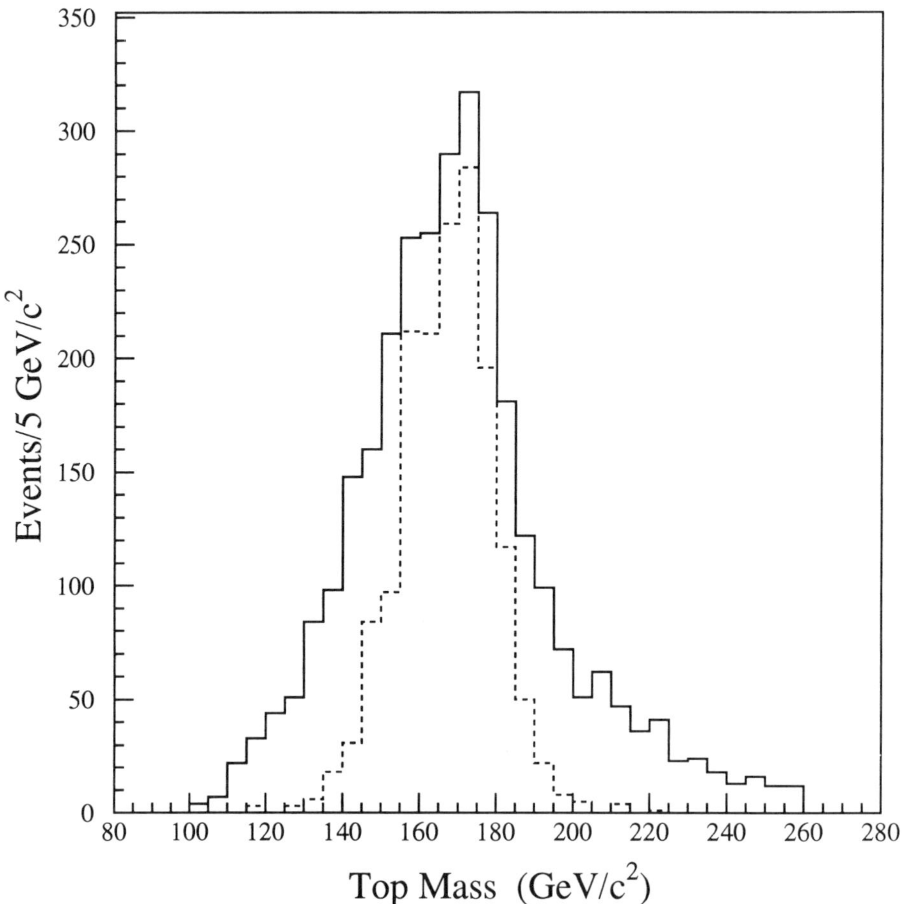

Fig. 45. Reconstructed top mass distribution for Monte Carlo events generated with $M_{top} = 170$ GeV/c^2. The full histogram corresponds to the best fit obtained by the fitting program when requiring that one of the b jets is a b in the fit. The dashed histogram refers to the fit with the correct assignment for each of the jets.

278

uncertainties in the jet measurement. It should be remembered that the major uncertainties in this process come from the jet measurements as the lepton is measured quite accurately.

Fig. 43 illustrates the problem of associating parton energy with jet energy. A large sample of Monte Carlo events was generated with Herwig using a top mass of 170 GeV. The jets generated by this process can be associated with the b jet or with light quark jets coming the W decay. In addition the b jets can be categorized as generic b jets or as b jets that decay semileptonically with an electron or with a muon. The jets from Herwig have been run through the CDF detector simulation, and the horizontal axis is the difference between the parton energy and the reconstructed jet energy using the standard calorimetry codes. Fig. 43a shows the spread in reconstructed energies versus the E_T of the jet. The spread is reasonably Gaussian and is determined by the statistical processes that take place in the calorimetry. Fig. 43b shows generic b jets, and it can be seen that the neutrino is making a non Gaussian tail due to the fact that it has taken away a fair amount of energy from the jet. Fig. 43c and 43d further elucidates this feature for the case of semileptonic decays involving an electron and a muon. Since the electron is well measured by the calorimeter, this skewing in c is less than that in d where the muon only deposits a minimum amount of energy calorimetrically.

As a result of this study, a new correction code for jets was generated. This algorithm attempted to relate directly the parton energy to the observed jet energy, and by studying the deviations shown in Fig. 43, the uncertainty in the parton energy from the jet measurement was evaluated. Fig. 44 shows an interesting example of the effect of this correction. The top plot shows the mass of the W calculated using jets with only standard corrections, and the bottom plot shows the mass using the new algorithm. The horizontal axis is the momentum of the W. Note, that in the future, when one has a large sample of $t\bar{t}$ decays to study, it will be possible for the first time to study the accuracy of reconstruction of events using calorimetric data. The check on the process will come from measuring how well the W mass can be resolved.

A number of systematic effects in this model were studied. One of the most important tests verified that the reconstructed top mass coincided with the input mass for the top that was used in the Monte Carlo generator over the range between 120 and 200 GeV. The jet energy scale of the calorimeter is also an important number in determining the mass. Fortunately, uncertainty in the scale of 10 percent results in a top mass uncertainty of the order of 5 percent because the lepton energy is very well measured, and also because there are additional constraints on the W mass in the fitting procedure.

Fig. 45 shows a reconstructed top mass distribution for Monte Carlo generated events with $M_{top} = 170$ GeV. The full histogram corresponds to the best fit obtained by the program when requiring that one of the b jets is a b in the fit. The dashed histogram refers to the fit with a correct assignment for each jet. The χ^2 assignments

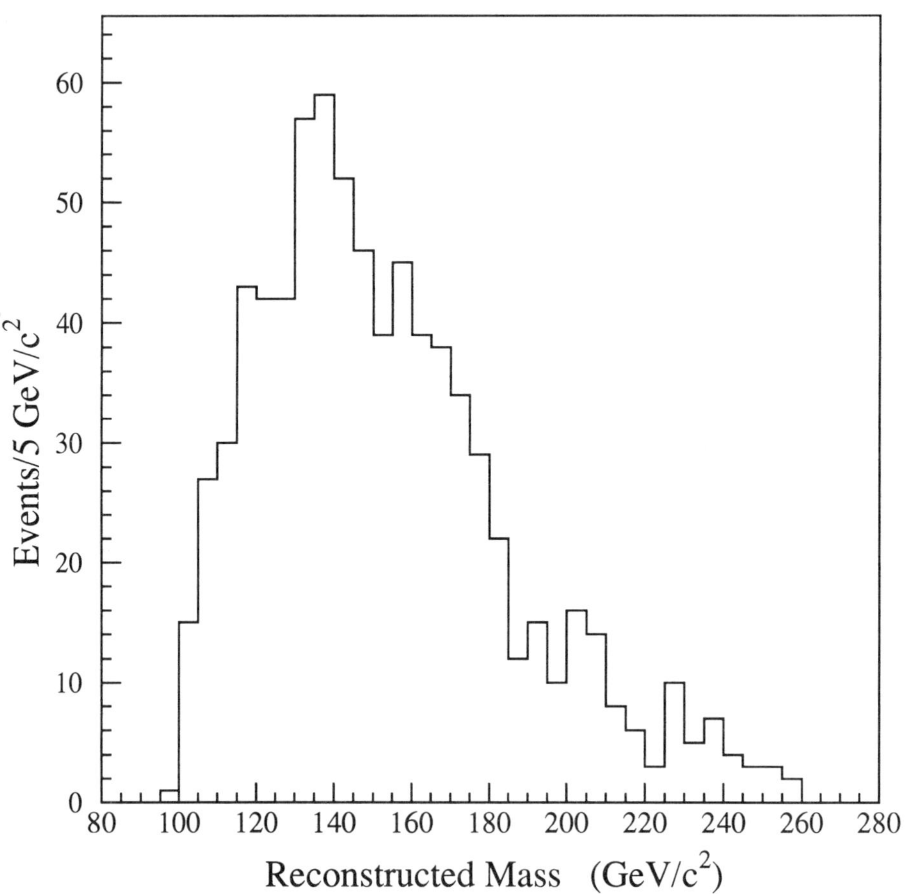

Fig. 46. Reconstructed mass distribution for W + multijet Monte Carlo events.

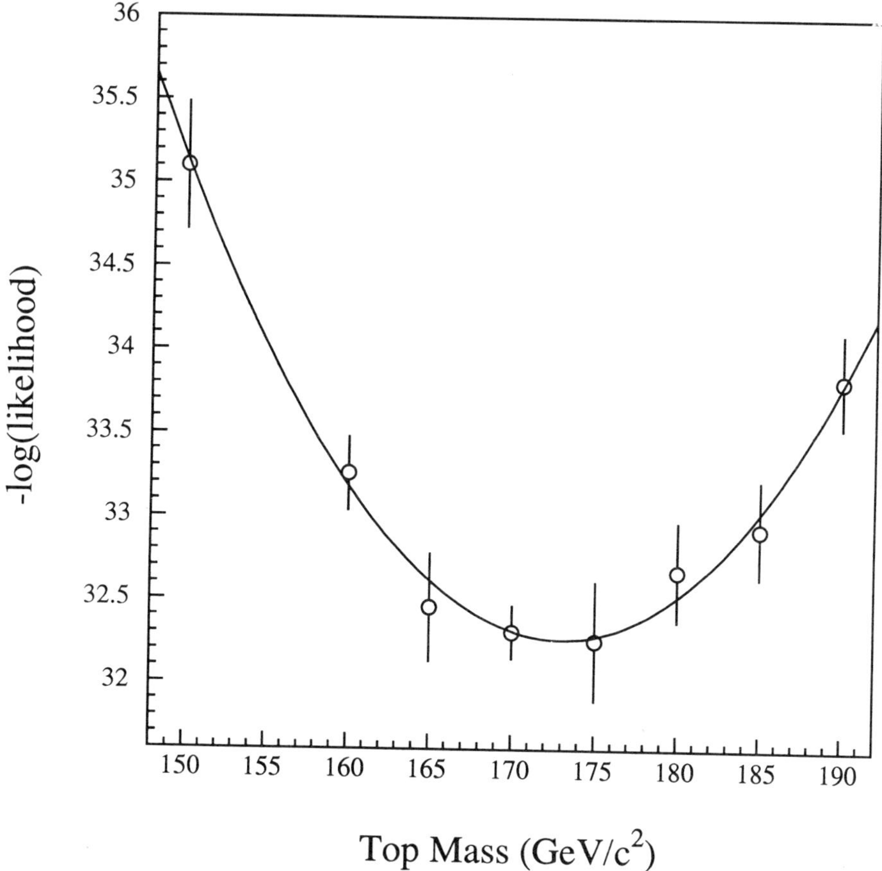

Fig. 47.   Likelihood fit of the top mass.

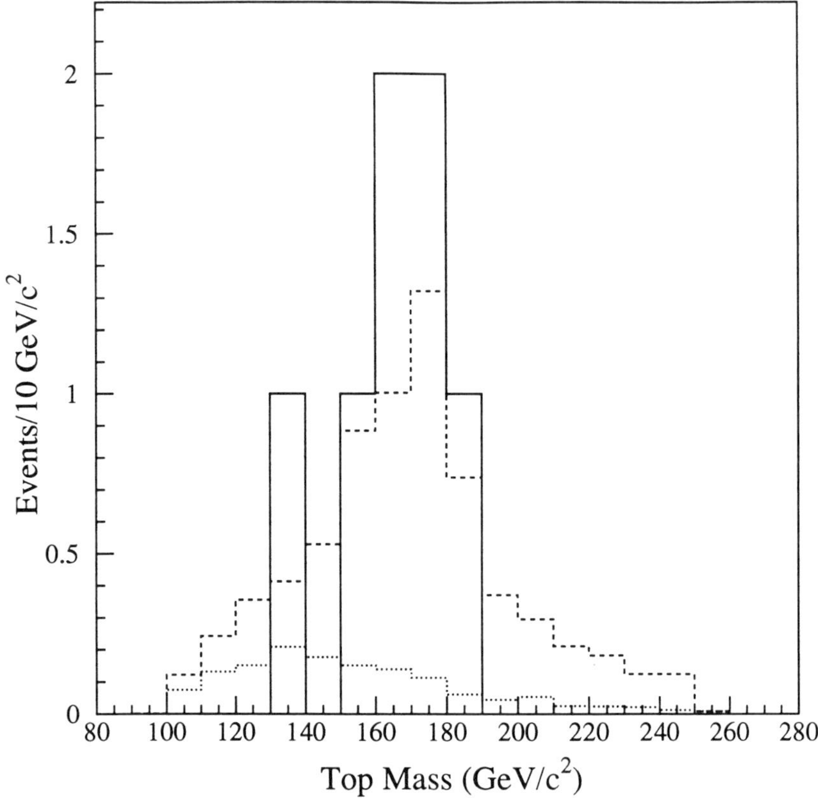

Fig. 48. Top mass distribution for the data (solid histogram) and the background of 1.4 events (dots) obtained from the $W+$ multijets VECBOS events. The dashed histogram represents the sum of 5.6 $t\bar{t}$ Monte Carlo events (from the $M_{top}=175$ GeV/c^2 distribution) plus 1.4 background events.

	Systematic uncertainties	(%)
a.	Jet Energy Scale (detector effects)	1.8
b.	Gluon radiation effects on parton energy	4.4
c.	Different backgrounds	$^{+5.3}_{-4.4}$
d.	Effects due to tagging algorithms	1.4
e.	Different likelihood fits	1.1

Fig. 49. Systematic uncertainties in the top mass measurement

of the jets only lead to a correct assignment in 31 percent of the time, and the long tail on the mass distribution is due to an incorrect assignment of the jets to the partons. It is interesting to note that even if no b tagging is used, one still obtains a peak at the correct mass but with somewhat worse tails. Picking the event with the best χ^2 is fairly effective at generating the correct mass.

W plus multijets were generated by VECBOS and studied, and it is found that 83 percent of the events that pass our selection criterion can be fit with the $t\bar{t}$ hypothesis. The mass spectrum from these events is shown in Fig. 46, and peaks at about 140 GeV.

We now consider the sample of 7 tagged events and estimate the background in this sample to be $1.4^{+2}_{-1.1}$ events. This estimate corresponds essentially to method 2, since in this case we are not doing a counting experiment, we will not use the most conservative estimate for the background but rather our best estimate of what it should be. A likelihood function is constructed which includes the number of background events and the number of signal events, the sum of which is constrained to be 7. The likelihood fit is shown in Fig. 47 and has a minimum at 174 GeV. The best estimate for the background fraction is $0.16^{0.16}_{-0.14}$ compared with the estimated value of .20. If one imposes the constraint that the number of top events is 0, the hypothesis that W + jet background spectrum fits the observed spectrum is 2.3 standard deviations away from the top + background hypothesis. Fig. 48 shows the top mass distribution as a solid histogram on the expected background of 1.4 events. The dashed histogram represents the sum of 5.6 top events and 1.4 background events as calculated from Monte Carlos.

The systematic errors on the mass measurement are given in Fig. 49. They come from the absolute energy scale of the calorimeter, the uncertainty due to gluon radiation effects being modelled correctly in the Monte Carlo, and an uncertainty in the shape of this background that is modelled by using VECBOS. Using a different scale for Q^2 and different fragmentations can change the shape of the background slightly. These uncertainties combined in a quadrature manner yield the final value for the top mass $M_{top} = 174 \pm 10^{+13}_{-12}\,\mathrm{GeV/c^2}$. Using the acceptance for the top mass of 174 GeV gives a $\sigma t\bar{t}(174) = 13.9^{+6.1}_{-4.8}$ pb.

Summary of the Collider Detector Experiment

In summary, the CDF experiment has some strong evidence for the top, but there are some observations that do not support this conclusion.

In support of the hypothesis, we observe two dilepton events with a background of 0.56. In addition we observe 6 lepton + jet events with b tagging information from the SVX on a expected background of 2.3, and 7 events on a background of 3.1 using the soft lepton from the semileptonic b decay to identify the b. The background has been estimated in a conservative manner from the data. In addition, one of the dilepton

TABLE

m_t [GeV/c^2]		$e\mu$	ee	$\mu\mu$	e + jets	μ + jets	e + jets(μ)	ALL
140	$\varepsilon \times B(\%)$	$.32 \pm .06$	$.18 \pm .02$	$.11 \pm .02$	1.2 ± 0.3	$.8 \pm 0.2$	0.6 ± 0.2	
	$\langle N \rangle$	$.72 \pm .16$	$.41 \pm .07$	$.24 \pm .05$	2.8 ± 0.7	1.3 ± 0.4	1.3 ± 0.4	6.7 ± 1.2
160	$\varepsilon \times B(\%)$	$.36 \pm .07$	$.20 \pm .03$	$.11 \pm .01$	1.6 ± 0.4	1.1 ± 0.3	0.9 ± 0.2	
	$\langle N \rangle$	$.40 \pm .09$	$.22 \pm .04$	$.12 \pm .02$	1.8 ± 0.5	0.9 ± 0.3	1.0 ± 0.2	4.4 ± 0.7
180	$\varepsilon \times B(\%)$	$.41 \pm .07$	$.21 \pm .03$	$.11 \pm .01$	1.7 ± 0.4	1.2 ± 0.3	1.1 ± 0.2	
	$\langle N \rangle$	$.23 \pm .05$	$.12 \pm .02$	$.06 \pm .01$	1.0 ± 0.2	0.5 ± 0.2	0.6 ± 0.2	2.5 ± 0.4
Background		$.27 \pm .09$	$.16 \pm .07$	$.33 \pm .06$	1.2 ± 0.7	0.6 ± 0.5	0.6 ± 0.2	3.2 ± 1.1
$\int \mathcal{L}dt$ [pb^{-1}]		13.5 ± 1.6	13.5 ± 1.6	9.8 ± 1.2	13.5 ± 1.6	9.8 ± 1.2	13.5 ± 1.6	
Data		1	0	0	2	2	2	7

Fig. 50. Efficiency $\times$ branching fraction ($\varepsilon \times B$), expected number of events ($\langle N \rangle$) for signal and background sources for the observed integrated luminosity ($\int \mathcal{L}dt$), and number of events observed in the data.

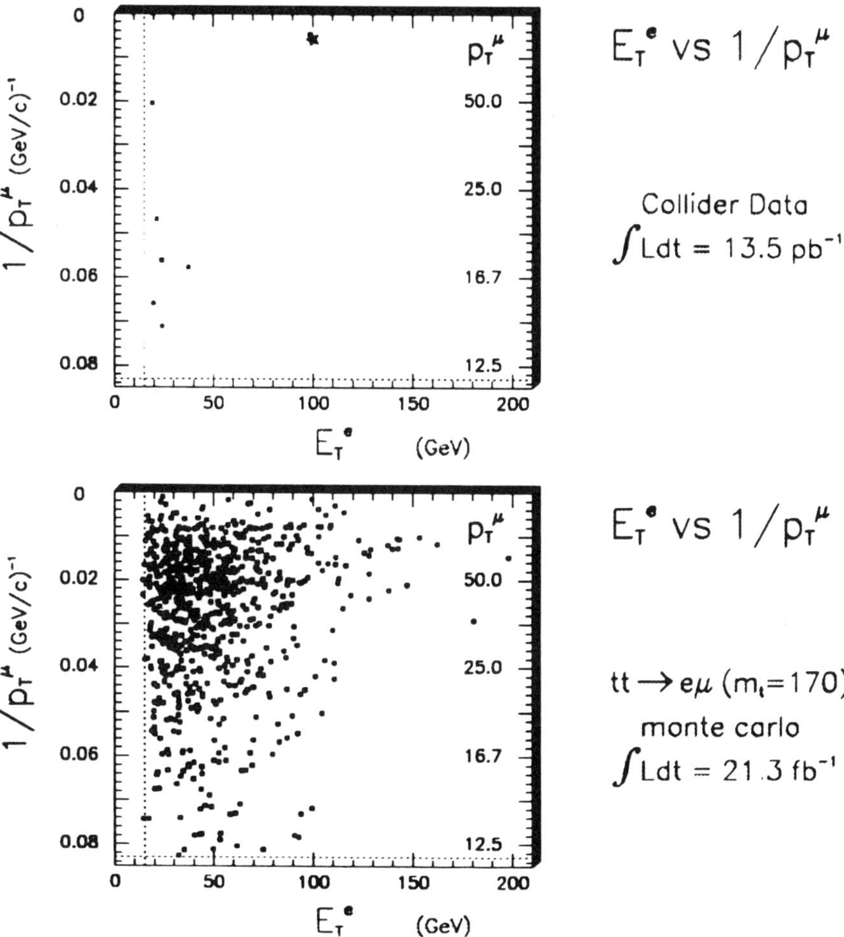

Figure 51: Distribution of events in E_T^e and $1/p_T^\mu$ for the data (before application of the final cut requiring two jets) and for Monte Carlo t tbar events with $M_T = 170$ GeV/c^2 The Monte Carlo corresponds to about 1600 times the luminosity shown for the data.

candidates is tagged by both b tagging techniques. This, together with the observed excess of lepton + jet events, gives evidence for both $Wb\bar{b}$ and $WWb\bar{b}$ production as would be expected in $t\bar{t}$ decays. There is evidence in the lepton + jet events that the kinematics of the decays are consistent with the $t\bar{t}$ hypothesis, and in fact a kinematic reconstruction of the events yields a mass of 174 GeV. This mass also agrees with the mass inferred in precision electroweak measurements.

On the other hand some features of the data do not support this hypothesis. Z + multijet events have been studied, and 2 tagged events are seen in the Z + 3 or more jets where only 0.64 would be expected. In addition, the $t\bar{t}$ cross section that we find is large enough so that it absorbs all of the rate for W + multijet production that should be seen in the W + 4 jet events. It is imperative to have more data to answer some of the questions that have been raised by this analysis. At present the machine is running again, and there is already additional data equal to the amount presented in this analysis.

III. SEARCH FOR THE TOP QUARK AT THE D0 DETECTOR

We now discuss the results found by the D0 Collaboration. The most complete reference at this point is the report from the Glasgow Conference, Ref. 11. And, as in the case of the CDF experiment, this report should be consulted, along with these lecture notes. An additional paper is now available, Ref. 12, which includes additional results from this experiment which were not available at the time of these lectures.

A cross section of the D0 detector is shown in Fig. 7. The main feature of the detector is the large uniform liquid Argon calorimeter for measuring total particle energies. There is not a magnetic field in the central region, but the momentum of muons is measured in magnetized iron in a system that surrounds the liquid calorimeter. The very fine grained, high resolution calorimetry provided by the liquid Argon allows a better measurement of the missing E_T in an event than is available in CDF. On the other hand, at present there is not silicon vertex detector. Thus, the techniques used in the two detectors to search for $t\bar{t}$ events tend to be complimentary in nature.

Dilepton Search

The dilepton analysis is reported in Ref. 4 and was updated at Glasgow. It requires the presence of 2 high p_T leptons, a large missing E_T, and 2 jets with E_T jet greater than 15 GeV. The results are shown in Fig. 50, along with the expected top signal and backgrounds. Fig. 51 shows these events plotted in the E_T, p_T plane for the data before the final cuts requiring 2 jets. The Monte Carlo prediction for a top mass equal to 170 GeV is also shown. There is one event observed in this data, and a likelihood analysis of the kinematics would indicate a value for the mass in the vicinity of 150 GeV.

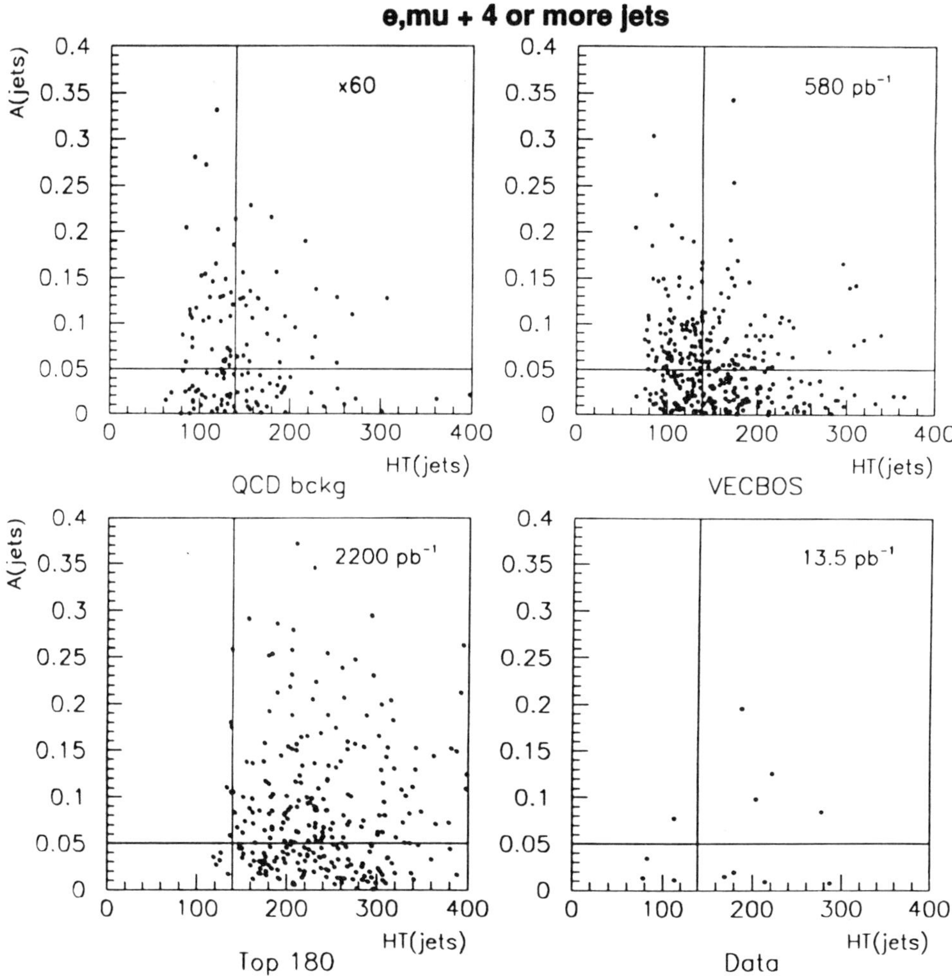

Figure 52: Event distributions vs aplanarity and H_T for QCD multijets (upper left). W + jets (upper right), a t tbar Monte Carlo with $M_T = 180$ GeV/c² (bottom left) and for data (lower right).

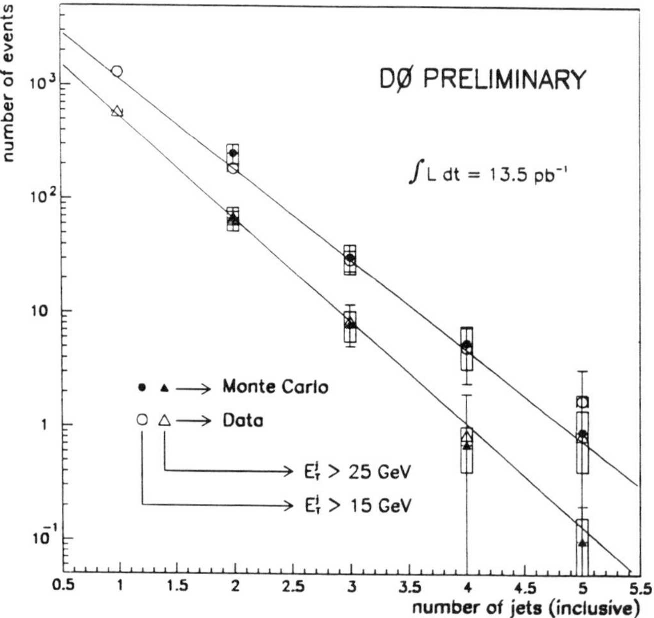

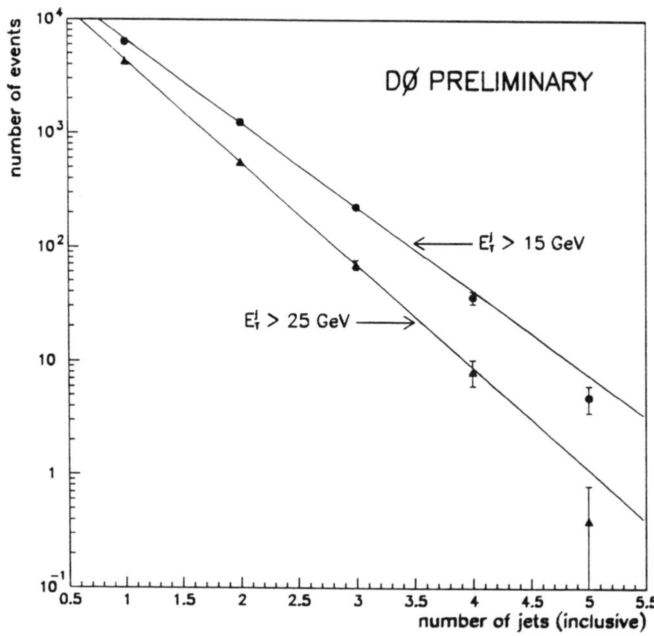

Figure 53: (a) Number of W + jets events (W > e.nu) vs. the inclusive number of jets for $E_T > 15$ GeV (upper points) abd $E_T > 25$ GeV (lower points). The open symbols denote data and the solid show the Monte Carlo. The lines are fits to the data for $1 < N_{jet} < 3$. (b) Number of multijet events for E_T 15 GeV (upper points) and $E_T > 25$ GeV (lower points). The filled symbols denote data and the lines are fits to the data for $1 < N_{jet} < 4$.

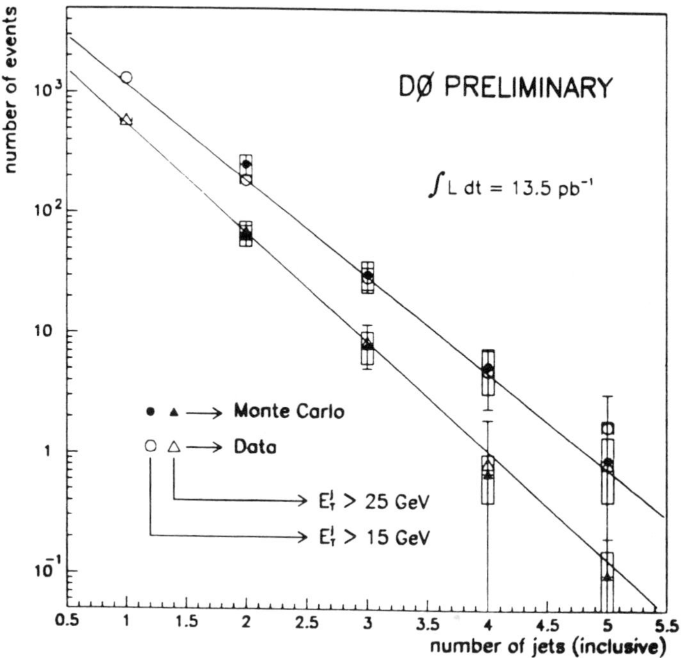

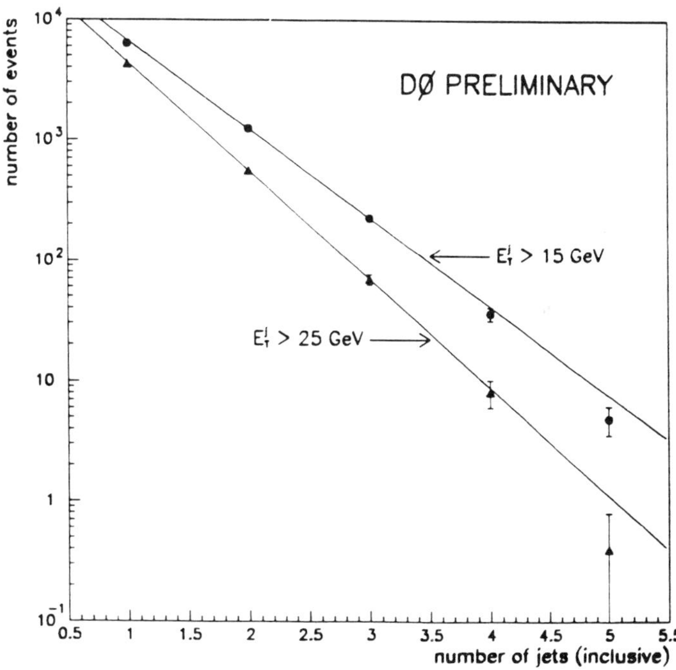

Figure 53: (a) Number of W + jets events (W > e.nu) vs. the inclusive number of jets for E_T > 15 GeV (upper points) abd E_T > 25 GeV (lower points). The open symbols denote data and the solid show the Monte Carlo. The lines are fits to the data for $1 < N_{jet} < 3$. (b) Number of multijet events for E_T 15 GeV (upper points) and E_T > 25 GeV (lower points). The filled symbols denote data and the lines are fits to the data for $1 < N_{jet} < 4$.

Search in Electron + Jet Mode

We have already studied in detail the difficulty of isolating $t\bar{t}$ events from W + QCD events in the mode of lepton plus missing E_T plus jets. Some additional discrimination is needed in the W + 4 jets in order to isolate the top production. The D0 experiment has two techniques for dealing with this. The first is use the kinematic differences between the $t\bar{t}$ production and the W + QCD jets to isolate the top events. They also have developed a way to search for a soft secondary muon which would identify a b jet in the event. This is similar to the muon SLT search described in the CDF experiment. We will first of all consider the kinematic technique.

The two variables that are chosen are the aplanarity A of the event which is defined to be 1.5 times the smallest eigenvalue of the normalized momentum tensor constructed in the overall $p\bar{p}$ frame from jets with η less than 2. The second variable called H_T is defined as the sum of the scalar transverse momentum of all final state jets. Large A and large H_T correspond to decay of high mass states. The cuts placed on the events to select them are as follows: Either the electron E_T^e is greater than 20 GeV and $|\eta_e| < 2$ or $p_T^\mu > 15$ GeV and η_μ is less than 1.7. Missing E_T must be greater than 25 GeV for the electronic mode and greater than 20 GeV for the muonic mode. And finally there must be least four jets with E_T greater than 15 GeV in the region of $|\eta| < 2$. Furthermore events with a soft muon are eliminated to keep this search statistically independent of the one that we will describe shortly. Fig. 52 displays a Monte Carlo study of how these variables distinguish events. The upper left hand plot shows A versus H_T for QCD multijets, and the right-hand side shows W + jets. A $t\bar{t}$ Monte Carlo is shown in the lower left-hand corner, and data from the experiment is shown in the lower right.

The event distributions shown in Fig. 52 can be used to directly estimate the fraction of events for each of the two processes which fall in each of the four quadrants of the A − HT space. Using these fractions, one can then fit the data in the lower right-hand corner directly to obtain the background and the signal. There are a total of four events in the signal region, and the background is estimated to be $1.7 \pm 0.8 \pm 0.4$ events.

A second method of obtaining the background after the topological cuts is to study the behavior of the background W + QCD jets as a function of the number of jets. This information is then used to predict the QCD background in the lepton + four jet category. Once this number is obtained the cuts shown in Fig. 52b give the fraction of these events that will wind up in the signal region.

Fig. 53 summarizes the result of the study. The top curve shows the number of W + jet events versus the inclusive number of jets for E_T greater than 15 GeV and E_T greater than 25 GeV. The open symbols are the data, and the filled symbols show the prediction of the Monte Carlo, and the lines are fit to the data for the interval between 1 and 3 jets. A similar study carried out for the case of QCD multijets is

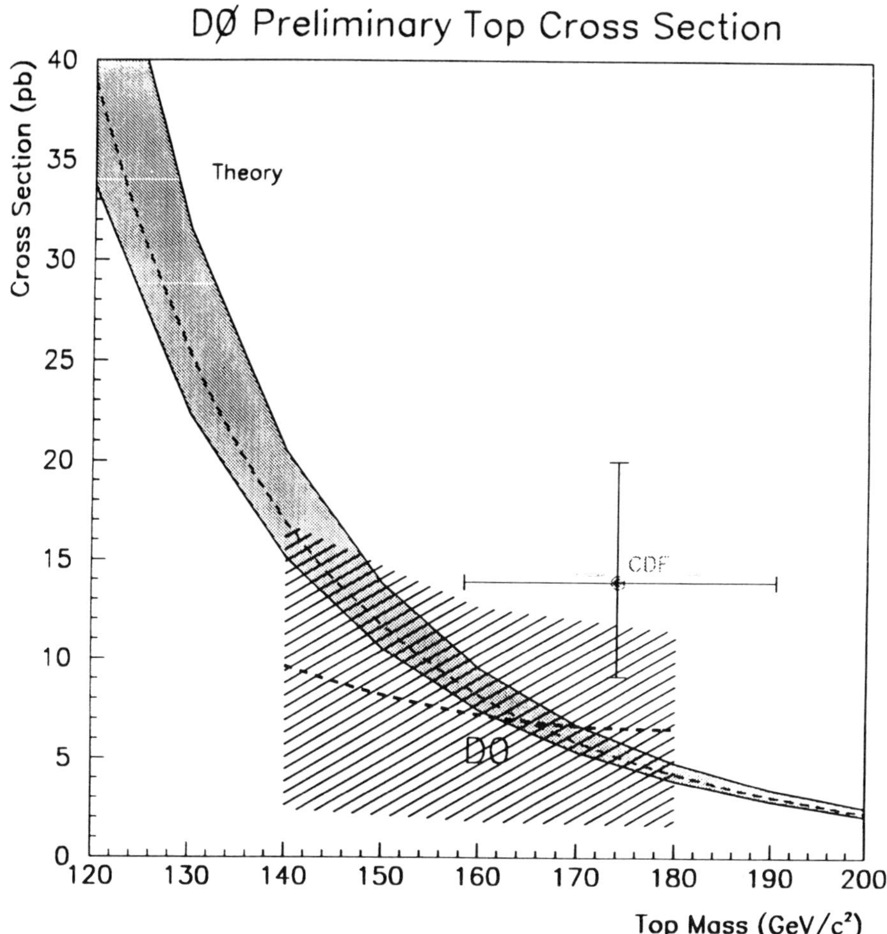

Figure 54. Cross-section vs. M_T. The dotted line and the cross-hatched area give the D0 preliminary result for the range of possible quark masses. The band is the theory curve shown in Figure 3. The cross is the CDF result.

shown in Fig. 53b. In this case, the selection of the sample is made from multijets, where one of the jets fakes an electron, and where there is also a missing E_T that is less than 25 GeV. This sample should contain no signal from the top. The slope is similar to the slope shown in Fig. 53a, and again the scaling hypothesis seems to work rather well. Therefore, the extrapolation of the curves to N = 4 jets is considered a reliable way to estimate the background. The number of predicted background events is then decreased by the fraction that would fall in the signal region of the AH_T space. The background predicted by this technique is $1.8 \pm 0.8 \pm 0.4$, agreeing well with the direct fitting procedure described above.

Muon Tagging

The muon discrimination in the D0 detector is very good, and hence they can use this to look for a secondary muon associated with a b jet in order to tag it. This search is performed on the e + multijet sample. The results of this search are presented in Fig. 50 along with the other channels. The bottom line gives the data for the various channels, and the line just above the estimated background. The overall search finds 7 events on an expected background of $3.2 \pm .1$. The probability that the background alone could fluctuate and give the 7 events < 7.2 percent or about 1.5 standard deviations in a Gaussian approximation. If this result is combined with the acceptance of the detector which varies with top mass, then the D0 results can be presented as shown in Fig. 54. The CDF result is shown as a cross. See Ref. 12 for more complete D0 results.

IV. SUMMARY AND CONCLUSIONS

It is still early to make a comparison in detail of the two experiments. However, Fig. 55 shows the acceptances and the background rate of the two detectors in the various channels as reported at the Glasgow Conference. It is seen that the acceptance of the two experiments is comparable. The major difference comes in the way that lepton + jet events are treated. In the case of CDF, these events are analyzed with the SVX and with the SLT technique to identify b jets. The advantage of a secondary vertex detector is that it enables the systematics of the tagging process to be investigated in much greater detail. The topology is then used as an independent check of the likelihood that the events are top. D0 uses the topology to select the events except in the case where they use a secondary muon tag. Shortly D0 will have the soft muon tag working for the muon + jet events.

Within the next year or so, there should be several times the amount of data available that has been presented in these lectures. An identification of the top will have decays seen in all the channels. One would hope to see lepton + jet events and dilepton events, where both of the b jets are tagged although this will be rare. The properties of the B may even give more information that will help associate it with

	CDF Dilepton search	D0 Dilepton search	D0 e+jets Topology	D0 μ+jets Topology	D0 e+μ tag soft lepton	CDF e,μ + jets SVX	CDF e,μ + jets SLT
Acceptance	.78%	.67%	1.6%	1.1%	0.9%	1.69%	1.2%
Background per pb^{-1}	.028	.066	.089	.061	.044	.119	.16

Figure 55. Table showing the published acceptances of the CDF and D0 experiments. The last line shows the background events per pb^{-1} for each channel.

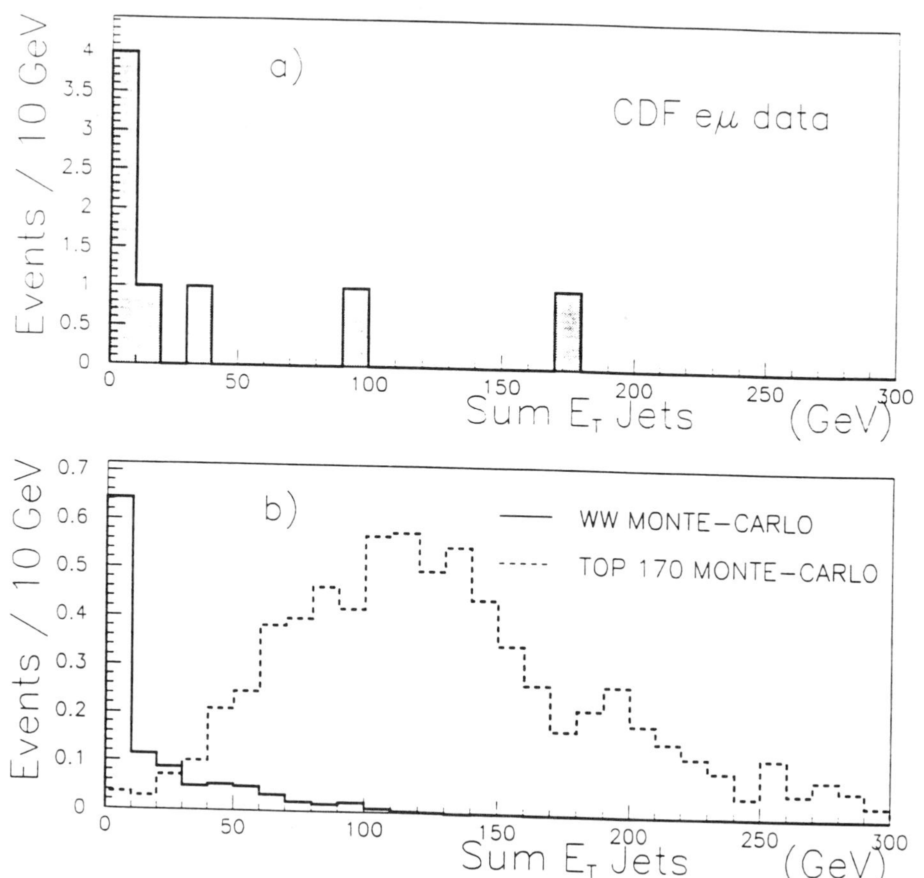

Figure 56: (a) The sum of ET(jet) for the 8 e mu events passing the pt > 20 GeV/c requirement on each lepton. Only jets with ET > 10 GeV and abs(eta) < 2.4 are included in the sum. The two events in the signal region of the dilepton analysis are the two events with the highest sum ET(jets). The 6 events at low sum ET fail both the two-jet cut and the MET cut. (b) Monte Carlo sum ET(jets) for t tbar, and for electroweak WW production, which is one of the backgrounds to the top search. The WW histogram is normalized to 19.3 pb^{-1}, while the t tbar is shown for 150 pb^{-1}. Note that the six events at low sum ET in (a) are unlikely to be mostly WW since they have low MET.

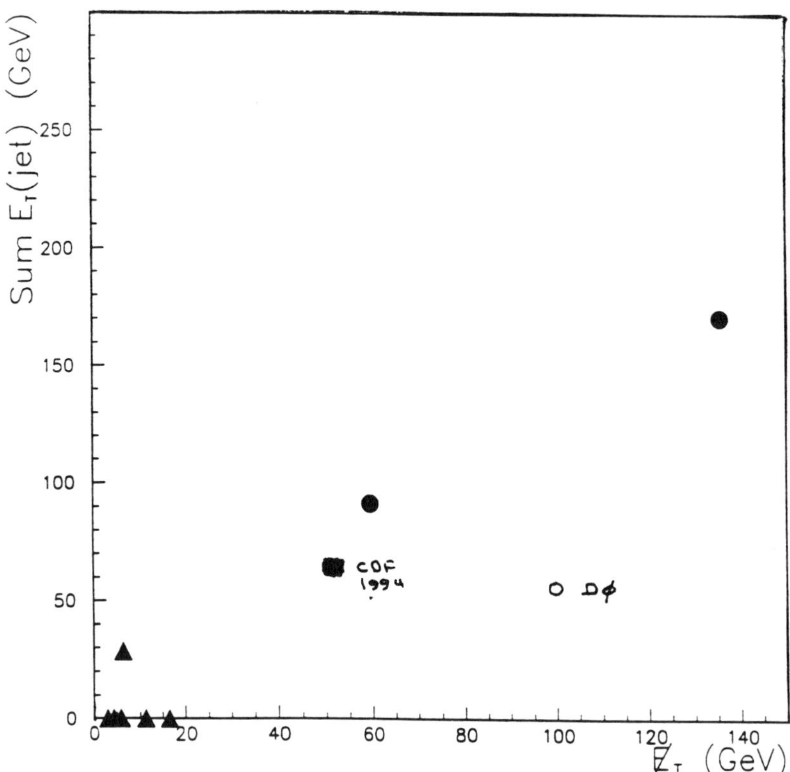

Figure 57.   The published dilepton events shown in the Sum E_T (jets) vs the missing E_T plane. The solid symbols are for the events shown in Figure 56 from CDF and the open symbol is the D0 event. The square is an event from CDF in the 1994 run.

the correct W. However, it is clear at this point that the dilepton events themselves present a very strong case for a new class of events. As a group they are remarkable! Fig. 56 is a histogram of the sum E_T of the jets in the CDF dilepton events. The upper figure displays the data of the 8 $e\mu$ events from Fig. 9. The lower figure shows the histogram from a Monte Carlo study of WW events compared to that expected for top production. It serves as an example of how $t\bar{t}$ production compares to a typical background. In Fig. 57, I show a different plot of the dilepton events. The vertical axis in the missing E_T and the horizontal axis is the jet sum E_T as in Fig. 56. I have included the D0 event as reported at Glagow and Ref. 4, as well as an additional CDF event from early in the 1994 run. Although one cannot conclude from this meager sample that the events are top, it is clear that they are unique events!

The future is exciting. Shortly there will be enough new data available to answer all of the unanswered questions raised in these lectures. We will be able to actually study how accurately jet spectroscopy is able to measure the mass of the top. There will be internal consistency checks within the reconstruction due to the hadronic decay of one of the W's. The study of the interaction between the t and $\bar{t}$ could lead to exciting new physics. There will be information from the spin correlations that will help check our understanding of the production and decay. Finally, combining an accurate measurement of the top mass with the precision measurement of the W that will be available from CDF and D0 will give the first solid prediction for the mass of the elusive Higgs. There is still some fun left!

I would like to thank my many colleagues in both CDF and D0 for help in assembling this information for these notes, especially Carol Picciolo for transcribing these notes.

REFERENCES

1. G. Arnison et al., UA1 Collaboration, Associated Production of an Isolated, Large Transverse Momentum Lepton (Electron or Muon), and Two Jets at the CERN $\bar{p}p$ Collider, *Phys. Lett.* 147b, 493 (1984).

2. F. Abe et al., The CDF Collaboration, Top-Quark Search in the Electron + Jets Channel in Proton-Antiproton Sollicions at $\sqrt{s} = 1.8$ TeV," *Phys. Rev. D* 43, 664 (1991).

3. F. Abe et al., The CDF Collaboration, Lower Limit on the Top-Quark Mass from Events with Two Leptons in $p\bar{p}$ Collisions at $\sqrt{s} = 1.8$ TeV," *Phys. Rev. Lett.* 68:447 (1992).

4. S. Abachi et al., D0 Collaboration, Search for the Top Quark in $\bar{p}p$ Collisions at $\sqrt{s} = 1.8$ TeV, *Phys. Rev. Lett.* 72:2138 (1994).

5. R. H. Dalitz et al., Where is the Top?, *Int. J. of Modern Phys.* 9A:635 (1994).

6. E. Laenen, J. Smith, W. Van Neerven, Top Quark Production Cross Section, *Phys. Lett.* 321B:254 (1994).

7. F. Abe et al., The CDF Collaboration, The CDF Detector: An Overview, *Nucl. Instrum. Methods Phys. Rev., Sect. A* 271:387 (1988).

8. D. Amidei et al., The Silicon Vertex Detector of the Collider Detector at Fermilab, Nucl. Instrum. Methods Sect. A 350:73 (1994).

9. S. Abachi et al., D0 Collaboration, The D0 Detector, *Nucl. Instrum. Methods A* 338:185 (1994).

10. F. Abe et al., The CDF Collaboration, Evidence for Top Quark Production in $\bar{\text{p}}$p Collisions at $\sqrt{s} = 1.8$ TeV," *Phys. Rev. D* 50:2966 (1994).

11. 27th International Conference on High Energy Physics, University of Glasgow, Glasgow, Scotland, July 20-27, 1994. See contributions by:

    (a) P. Grannis

    (b) S. Protopopescu

    (c) S. J. Wimpenny

    (d) R. Raja

12. S. Abachi et al., D0 Collaboration, Search for High Mass Top Quark in p$\bar{\text{p}}$ Collisions at $\sqrt{s} = 1.8$ TeV, submitted to *Phys. Rev. Lett.* November 1994. FERMILAB-PUB-94/354-E.

SUPERSYMMETRIC GRAND UNIFIED THEORIES AND YUKAWA UNIFICATION

B. C. Allanach

Physics Department
University of Southampton
Southampton
SO9 5NH
UK

INTRODUCTION

In this paper, I intend to motivate supersymmetric grand unified theories (SUSY GUTs), briefly explain an extension of the standard model based on them and present a calculation performed using certain properties of some SUSY GUTs to constrain the available parameter space.

Why GUTs?

Much work has been done on the running of the gauge couplings in the standard model, as prescribed by the renormalisation group. Amazingly, when the couplings α_1, α_2 and α_3 were run up to fantastically high energies $\sim \mathcal{O}(10^{14})$ GeV, they seemed to be converging[1] to one value. This is a feature naturally explained by many GUTs such as $SU(5)$[2,3] and reflects the fact that the strong, weak and electromagnetic forces seen today are different parts of the same grand unified force. It was realised that GUTs could also provide relations between the masses of the observed fermions, the structure and hierarchy of which are as yet unexplained. Despite these attractive features, several problems arose which detracted from the idea.

Unfortunately, the three couplings do not quite converge by $\sim \mathcal{O}(7\sigma)$, and many GUTs, notably $SU(5)$, predict proton decay much faster than the lower experimental bounds. Also incredible fine tuning is required for the so-called 'hierarchy problem'. This stems from the fact that M_W changes through radiative corrections of order the new physics scale (Fig.1) , say the Planck mass $\sim 10^{19}$ GeV, if there is no new physics at smaller energies. M_W is therefore unstable to the corrections and vast cancellations in the couplings are required to motivate the correct phenomenology.

Frontiers in Particle Physics: Cargèse 1994
Edited by M. Lévy *et al.*, Plenum Press, New York, 1995

Figure 1: One loop corrections to m_h^2. The first diagram gives a $\sim \mathcal{O}(M_{Pl}^2)$ contribution.

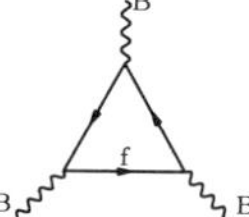

Figure 2: One loop triple hypercharge boson anomally.

Why SUSY?

Supersymmetry is an extra symmetry relating fermions and bosons, and so provides some explanation of how particles of differing spin should be related to one another. In an unbroken supersymmetric theory, each fermion has a degenerate bosonic partner. Of course, these so called superpartners are not observed, so that if supersymmetry was ever the correct theory, it must have been broken. However, with the introduction of superpartners at some rough energy scale M_{SUSY}, the renormalisation group running of the gauge couplings changes. The coupling constants are now seen to meet at a scale $\mathcal{O}(10^{16})$ GeV, as reflected by the correct $\sin^2 \theta_w$ prediction[4]. M_W becomes stabilised because supersymmetry induces cancellations between the bosonic and fermionic loop corrections to the mass. The quadratic divergences induced by the loop corrections now add to zero and one is left with merely logarithmic divergences.

THE MINIMAL SUPERSYMMETRIC STANDARD MODEL (MSSM)

The MSSM is a minimal extension of the standard model into supersymmetry. In the model, every particle of the standard model has a superpartner associated with it that transforms identically under the standard model gauge group but have spin different by $\frac{1}{2}$. So for example, each quark has a scalar "squark" superpartner, the gluons have "gluinos" etc. At first sight however, the model has a $U(1)_Y^3$ gauge anomaly. This originates from the diagram with three B gauge bosons connected to an internal loop through which any fermions may run (cf Fig.2) and the counter term to it would destroy gauge invariance. The diagram is proportional to $\sum_i (Y_i/2)^3$ where i runs over all active fermions. Through the hypercharge assignments, this cancels in the standard model but in the MSSM the superpartner of the Higgs called the Higgsino with $Y = 1$ may run around the loop. To cancel this effect, a second Higgs H_2 must be introduced which transforms in the same way to H_1 except for having $Y = -1$.

The new Higgs must also develop a vev v_2 to give masses to up quarks and the two vevs are related by

$$\tan \beta = \frac{v_2}{v_1} \tag{1}$$

where $v_1^2 + v_2^2 = v^2$ and $v = 246$ GeV, the measured vev of the standard model.

In chiral superfield form, the superpotential looks like

$$W_{MSSM} = UQH_2u^c + DQH_1d^c + ELH_1e^c + \mu H_1 H_2 \tag{2}$$

where U, D and E are the up, down and charged lepton Yukawa matrices respectively and all gauge and family indices have been suppressed.

One possible problem with this superpotential is the dimensionful parameter μ. μ needs to be $\mathcal{O}(M_Z)$ to give the right electroweak symmetry breaking behaviour whereas one would expect it to be of order of the new physics scale M_{GUT}. One solution to this problem is described in the Next to Minimal Supersymmetric Standard Model (NMSSM).

The NMSSM

The μ term in Eq.2 is replaced by $\lambda N H_1 H_2$ where N is a gauge singlet and therefore doesn't affect the coupling constant unification. In certain supergravity models, N develops a vev naturally of order M_Z and so the μ term is generated without having to put μ in "by hand." The superpotential now has a discrete Peccei-Quinn symmetry which leads to phenomenologically unacceptable low energy axions and so a term $-\frac{k}{3}N^3$ is added which breaks it. [1]

GUTS WITH YUKAWA UNIFICATION

GUTs can quite naturally provide Yukawa unification relations between the quarks and/or leptons. For example in SU(5), the right handed down quarks and conjugated lepton doublet lie in a $\underline{5}$ representation. When a mass term $\sim 5^i 5_i$ is formed, the Yukawa relation

$$\lambda_b(M_{GUT}) = \lambda_\tau(M_{GUT}) \tag{3}$$

applies. Also in SO(10), the whole of one family and a right handed neutrino is contained in one $\underline{16}$ representation, leading to triple Yukawa unification, where the top, bottom and charged lepton Yukawa couplings are equal at the GUT scale.

These relations can be used to constrain the parameter space of m_t and $\tan \beta$, which has been done for the MSSM[5]. Our idea was to repeat this calculation for the NMSSM, to see how much the viable parameter space changes in the model.

THE CALCULATION

The basic idea is to choose some $\tan \beta$ and m_t and run λ_b and λ_τ up to $M_{GUT} \sim 10^{16}$ GeV. Then, to some arbitrary accuracy, one can determine whether the GUT relation Eq.3 holds. If it does, then SU(5) and the other Yukawa unifying extensions of the standard model are possible on this point in parameter space. The procedure is iterated over all reasonable values of $\tan \beta$ and m_t. The calculation is presented in more detail in Ref.6.

---

[1] λ and k are merely coupling constants

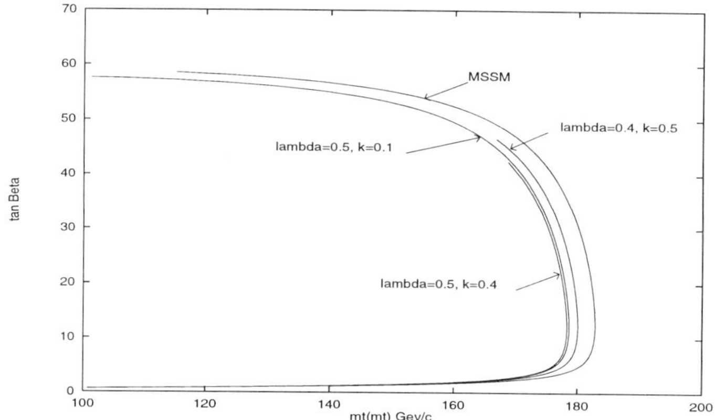

Figure 3: Viable Range of Parameter Space For $\alpha_S(M_Z) = 0.11$, $m_b = 4.25$ GeV. λ and k values are quoted at m_t.

Starting Point M_Z.

We use the definitions of the gauge couplings at M_Z: $\alpha_1^{-1}(M_Z) = 58.89$, $\alpha_2^{-1}(M_Z) = 29.75$ and $\alpha_3^{-1}(M_Z) = 0.11 \pm 0.01$. The first two gauge couplings are determined accurately enough for our purposes whereas the third needs to be used as a parameter, on account of its large uncertainty.

In order to convert masses of quarks to Yukawa couplings, we simply need to read them off the potential Eq.2 at some energy scale (taken here to be m_t):

$$\lambda_t(m_t) = \frac{\sqrt{2}m_t(m_t)}{v \sin\beta} \tag{4}$$

$$\lambda_b(m_t) = \frac{\sqrt{2}m_b(m_b)}{\eta_b v \cos\beta} \tag{5}$$

$$\lambda_\tau(m_t) = \frac{\sqrt{2}m_\tau(m_\tau)}{\eta_\tau v \cos\beta}. \tag{6}$$

where

$$\eta_f = \frac{m_f(m_f)}{m_f(m_t)}. \tag{7}$$

Note that whereas the m_t referred to here is always the running one, it can be related to the physical mass by[5]

$$m_t^{phys} = m_t(m_t)\left[1 + \frac{4}{3\pi}\alpha_3(m_t) + O\left(\alpha_3^2\right)\right]. \tag{8}$$

To determine η_b and η_τ, the masses are run up from the on shell mass to m_t using effective 3 loop QCD $\otimes$ 1 loop QED [7,8,9,10]. Note that these factors will depend of $m_b = 4.25 \pm 0.15$ GeV and $\alpha_3(M_Z)$. m_t is assumed to be the rough energy scale when the whole supersymmetric spectrum kicks in. While being unrealistic, trials with $M_{SUSY} = 1$ TeV show only a few percent deviation from the predictions with $M_{SUSY} = m_t$. So, having determined the gauge and relevant Yukawa couplings at m_t,

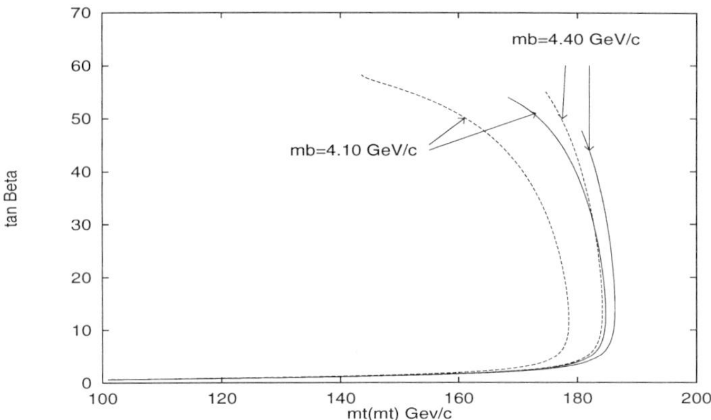

Figure 4: Viable Range of Parameter Space For $\alpha_S(M_Z) = 0.12$ and experimental bounds of $m_b = 4.1\text{–}4.4$ GeV. The left most lines are for $\lambda_b = 0.9\lambda_\tau$.

we need RG equations to run them up to M_{GUT} in the NMSSM. To derive these, we used results from a general superpotential[11] to obtain

$$
\begin{aligned}
16\pi^2 \frac{\partial \lambda_t}{\partial t} &= \lambda_t \left[6\lambda_t^2 + \lambda_b^2 + \lambda^2 - \left(\frac{13}{15}g_1^2 + 3g_2^2 + \frac{16}{3}g_3^2 \right) \right] \\
16\pi^2 \frac{\partial \lambda_b}{\partial t} &= \lambda_b \left[6\lambda_b^2 + \lambda_\tau^2 + \lambda_t^2 + \lambda^2 - \left(\frac{7}{15}g_1^2 + 3g_2^2 + \frac{16}{3}g_3^2 \right) \right] \\
16\pi^2 \frac{\partial \lambda_\tau}{\partial t} &= \lambda_\tau \left[\lambda_\tau^2 + 3\lambda_b^2 + \lambda^2 - \left(\frac{9}{5}g_1^2 + 3g_2^2 \right) \right] \\
16\pi^2 \frac{\partial \lambda}{\partial t} &= \lambda \left[4\lambda^2 + 2k^2 + 3\lambda_\tau^2 + 3\lambda_b^2 + 3\lambda_t^2 - \left(\frac{3}{5}g_1^2 + 3g_2^2 \right) \right] \\
16\pi^2 \frac{\partial k}{\partial t} &= 6k \left[\lambda^2 + k^2 \right]
\end{aligned}
\tag{9}
$$

in the limit that the lighter two families have negligible contributions (a very good approximation).

The Yukawa couplings can now be run from m_t to 10^{16} GeV using numerical techniques. The parameters λ and k particular to the NMSSM are unconstrained at m_t so they are merely varied for different curves.

Our results are displayed in Fig. 3 as contours in the $\tan\beta - m_t$ plane consistent with Eq.3. We take $\alpha_3(M_Z) = 0.11$, $m_b = 4.25 GeV$ and the NMSSM parameters $\lambda(m_t)$ and $k(m_t)$ as indicated. The MSSM contour is shown for comparison and is indistinguishable from the NMSSM contour with $\lambda(m_t) = 0.1$ and $k(m_t) = 0.5$. In fact our plot for the MSSM based on 1-loop RG equations is very similar to the 2-loop result in ref.5. The deviation of the NMSSM contours from the MSSM contour depends most sensitively on $\lambda(m_t)$ rather than $k(m_t)$. Two of the contours are shortened due to either λ or k blowing up at the GUT scale. For $\lambda(m_t) = 0.5$, $k(m_t) = 0.5$, no points in the $m_t - \tan\beta$ plane are consistent with Eq.3 Yukawa unification, while for $\lambda(m_t) = 0.1$, $k(m_t) = 0.1 - 0.5$ the contours are virtually indistinguishable from the MSSM contour. In general we find that for any of the current experimental limits on α_3 and m_b, the maximum value of $\lambda(m_t)$ or $k(m_t)$ is ~ 0.7 for a perturbative solution to Eq.3.

Fig.4 shows the effects of particle thresholds, which can modify Eq.3 to $\lambda_b = 0.9\lambda_\tau$. Our treatment does not treat supersymmetric or heavy thresholds exactly and so some sort of corrections like those shown are expected. The curves are at $\alpha_S(M_Z) = 0.12$ and $m_b = 4.1\text{--}4.4$ GeV to illustrate that uncertainties in these quantities make a large difference to the parameter space. These uncertainties are much bigger than those associated with the NMSSM, and so the MSSM and NMSSM would be practically indistinguishable given the parameters m_t and $\tan\beta$.

Other Yukawa Parameters

The next useful step is to notice that Eqs.9 are all of the form

$$16\pi^2 \frac{\partial \lambda_a}{\partial t} = \lambda_a \left[\sum_i M_i^a \lambda_i^2 - \sum_{j=1}^{3} c_j^a g_j^2 \right], \tag{10}$$

where M_i^a and c_j^a are constants supplied by the relevant RG equation. When the β function

$$\frac{dg_i}{dt} = \frac{b_i g_i^3}{16\pi^2} \tag{11}$$

is inserted, and the RG equations are reparameterised in terms of the flow and not the trajectory of the solutions, we obtain

$$\frac{\lambda_a(M_{SUSY})}{\lambda_a(M_{GUT})} = \xi^a \exp(-\sum_i M_i^a I_i), \tag{12}$$

where

$$\xi^a = \prod_{i=1}^{3} \left(\frac{\alpha(M_{GUT})}{\alpha_i(M_{SUSY})} \right)^{\frac{c_i^a}{2b_i}} \tag{13}$$

contains all the information about the gauge couplings and

$$I_i = \frac{1}{16\pi^2} \int_{\ln(M_{SUSY})}^{\ln(M_{GUT})} \lambda_i^2 dt \tag{14}$$

concerns the Yukawa couplings.

With this formulation, the running of the physically relevant Yukawa eigenvalues and mixing angles can be expressed in simple terms as shown below,

$$
\begin{aligned}
\left(\frac{\lambda_{u,c}}{\lambda_t} \right)_{M_{SUSY}} &= \left(\frac{\lambda_{u,c}}{\lambda_t} \right)_{M_{GUT}} e^{3I_t + I_b} \\
\left(\frac{\lambda_{d,s}}{\lambda_b} \right)_{M_{SUSY}} &= \left(\frac{\lambda_{d,s}}{\lambda_b} \right)_{M_{GUT}} e^{3I_b + I_t} \\
\left(\frac{\lambda_{e,\mu}}{\lambda_\tau} \right)_{M_{SUSY}} &= \left(\frac{\lambda_{e,\mu}}{\lambda_\tau} \right)_{M_{GUT}} e^{3I_\tau} \\
\frac{|V_{cb}|_{M_{GUT}}}{|V_{cb}|_{M_{SUSY}}} &= e^{I_b + I_t},
\end{aligned} \tag{15}
$$

with identical scaling behaviour to V_{cb} of V_{ub}, V_{ts}, V_{td}. To a consistent level of approximation V_{us}, V_{ud}, V_{cs}, V_{cd}, V_{tb}, λ_u/λ_c, λ_d/λ_s and λ_e/λ_μ are RG invariant. The CP violating quantity J scales as V_{cb}^2. Eqs. 15, 14 also apply to the NMSSM since the extra λ and k parameters cancel out of the RG equations in a similar way to the gauge contributions as can easily be seen from Eq.9. The only difference to these physically relevant quantities is therefore contained in I_τ, I_b and I_t.

302

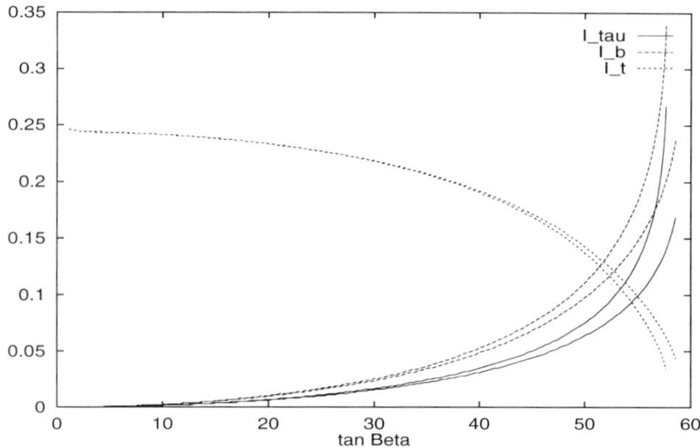

Figure 5: I_i integrals for $\alpha_S(M_Z) = 0.11$, $m_b = 4.25$ GeV.

These I_i integrals are shown in Fig.5 and the NMSSM results are the upper lines of each pair, and it is clear that the deviation between the two models is small again.

We emphasise that the results of the I_i integrals shown in Fig.5 play a key role in determining the entire fermion mass spectrum via the scaling relations of Eq.15. The small deviation between the NMSSM and the MSSM results compared to the experimental uncertainties means that the recent GUT scale texture analyses of the quark mass matrices which were performed for the MSSM are equally applicable to the NMSSM. For example, the recent Ramond, Roberts and Ross (RRR)[12] texture analysis is also based upon Eq.3 and assumes a Georgi-Jarlskog (GJ)[13,14] ansatze for the charged lepton Yukawa matrices, although their results in the quark sector are insensitive to the lepton sector. It is clear that all the RRR results are immediately applicable to the NMSSM since the only difference between the two models enters through the scaling integrals I_i whose deviation we have shown to be negligible compared to the experimental errors.

CONCLUSIONS

We have discussed the unification of the bottom quark and tau lepton Yukawa couplings within the framework of the NMSSM. By comparing the allowed regions of the m_t-$\tan\beta$ plane to those in the MSSM we find that over much of the parameter space the deviation between the predictions of the two models which is controlled by the parameter λ is small, and always much less than the effect of current theoretical and experimental uncertainties in the bottom quark mass and the strong coupling constant. We have also discussed the scaling of the light fermion masses and mixing angles, and shown that to within current uncertainties, the results of recent quark texture analyses[12] performed for the minimal model also apply to the next-to-minimal model. There are however two distinguishing features of the NMSSM. Firstly, the scaling of the charged lepton masses will be somewhat different, depending on λ and k. Although this will not affect the quark texture analysis of RRR, it may affect the success of the GJ ansatze[13,14] for example. Secondly, the larger $\tan\beta$ regions may not be accessible in the NMSSM for large values of λ and k, so that full Yukawa unification may not be possible in this case.

REFERENCES

[1] V. Barger and R. J. N. Phillips, Preprint MAD/PH/752 (1993).

[2] M. Chanowitz, J. Ellis, and M. K. Gaillard, Nuclear Physics B128, 506 (1977).

[3] A. J. Buras, J. Ellis, M. K. Gaillard, and D. V. Nanopoulos, Nucl. Phys. B135, 66 (1978).

[4] H. E. Haber, Preprint SCIPP 92/33 (1993).

[5] V.Barger, M.S.Berger, and P.Ohmann, Phys. Rev. D47, 1093 (1993).

[6] B. C. Allanach and S. F. King Phys. Lett. B328, 360 (1994).

[7] S.G.Gorishny, A.L.Kataev, S.A.Larin, and L.R.Surgaladze, Mod. Phys. Lett. A5, 2703 (1990).

[8] O.V.Tarasov, A.A.Vladimirov, and A.Yu.Zharkov, Phys. Lett. B93, 429 (1980).

[9] S.G.Korishny, A.L.Kataev, S.A.Larin, and P. Lett., Phys. Lett. B135, 457 (1984).

[10] L.Hall, Nucl. Phys. B75 (1981).

[11] S.P.Martin and M.T.Vaughn, NUB-3081-93TH hep-ph 9311340 (1993).

[12] P. Ramond, R. Roberts, and G. Ross, RAL-93-010 UFIFT-93-06 (1993).

[13] H. Georgi and C. Jarlskog, Phys. Lett. B86, 297 (1979).

[14] S. Dimopoulos, L. Hall, and S. Raby, Phys. Rev. D45, 4192 (1992).

CHIRAL SYMMETRY BREAKING
FOR FUNDAMENTAL FERMIONS

A. Bashir

Centre for Particle Theory,
University of Durham,
Durham DH1 3LE, U.K.

INTRODUCTION

Massive fermions have long been a problem in gauge theories. Unification of electromagnetic and weak forces was once hindered by the fact that the introduction of mass terms broke the gauge invariance of the theory. This problem was solved by the introduction of the Higgs field. Spontaneous breakdown of the SU(2) × U(1) symmetry then takes place. The gauge bosons gain mass and the masses for the fermions are generated through their Yukawa interaction with this Higgs field. However, there has been a widespread dissatisfaction with this mechanism since the masses are not predictable. Rather, they must be fixed by experiment. Studying the non-perturbative behaviour of gauge theories provides an alternative. If the interactions are strong enough, they are capable of generating masses for the particles dynamically even if they start with zero bare mass. Moreover, experiment tells us that the top quark is very heavy and so the Yukawa coupling g_t for top-Higgs interaction is O(1). Then one naturally expects that non-perturbative effects become important. Indeed, it has been suggested [1] that the top quark may acquire mass non-perturbatively through four-fermion interactions, and the Higgs can then be viewed as the condensate of the top and the antitop. However, in an attempt to include the effects of gauge boson exchange term, one loses gauge invariance of the physical quantities. Of course, physical quantities must be gauge independent. This motivates the study of how to achieve this in non-perturbative calculations. Quenched QED provides a toy model in which to study this problem, as we discuss.

DYSON-SCHWINGER EQUATIONS

Our starting point is the set of Dyson-Schwinger equations. These are an infinite system of coupled equations for all the Green's functions, which are non-perturbative in nature. Their structure is such that the 1-point function is related to the 2-point function, the 2-point function is related to the 3-point function, etc. ad infinitum. As it is impossible to solve the complete set of equations, one has to truncate this infinite tower in a physically acceptable way to reduce them to something that is soluble. A

Frontiers in Particle Physics: Cargèse 1994
Edited by M. Lévy *et al.*, Plenum Press, New York, 1995

familiar way to do this is perturbation theory. However, if one wishes to generate masses for particles, a non-perturbative way has to be sought.

To see how to do this consider two of the Dyson-Schwinger equations, one for the fermion propagator, and the other for the photon propagator. These are shown below diagrammatically together with their corresponding mathematical expressions:

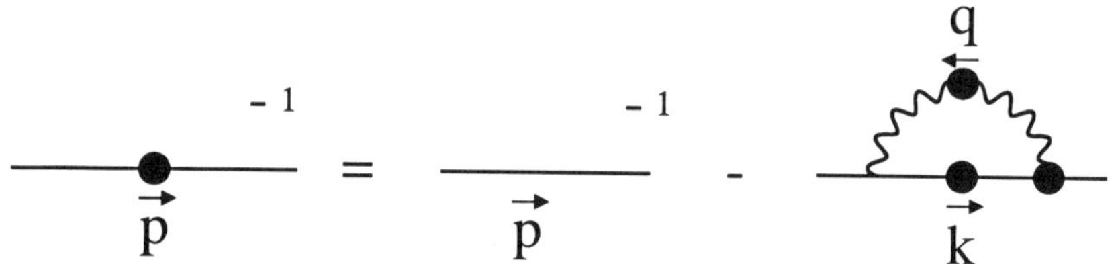

FIG. 1. Dyson-Schwinger equation for fermion propagator.

$$i S_F^{-1}(p) = i S_F^{0\,-1}(p) - e^2 \int \frac{d^4 k}{(2\pi)^4} \, \gamma^\mu \, S_F(k) \, \Gamma^\nu(k,p) \, \Delta_{\mu\nu}(q) \quad , \tag{1}$$

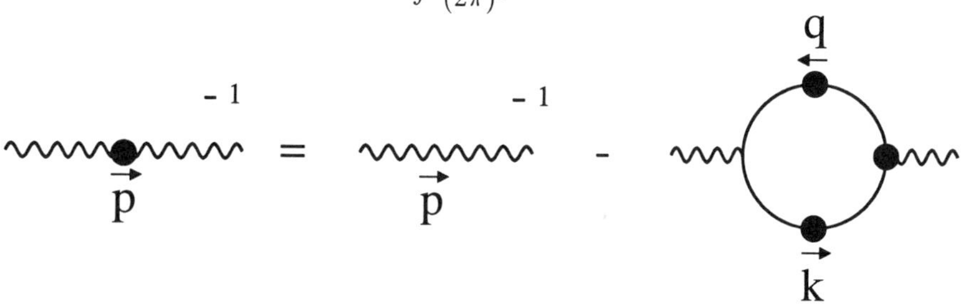

FIG. 2. Dyson-Schwinger equation for photon propagator.

$$i \Delta_{\mu\nu}^{-1}(p) = i \Delta_{\mu\nu}^{0\,-1}(p) - e^2 N_f \int \frac{d^4 k}{(2\pi)^4} \, \gamma^\mu \, S_F(k) \, \Gamma^\nu(k,p) \, S_F(q) \quad , \tag{2}$$

where the quantities with the superscript '0' are bare quantities, and the others are full ones. Quenched QED corresponds to making the assumption that the full photon propagator can be replaced by its bare counterpart. This limit is achieved by regarding N_f as a mathematical parameter, which is set equal to zero. As an example, to begin with, we make a further simplification by replacing the full vertex by the bare one. Eq. (1) then reduces to:

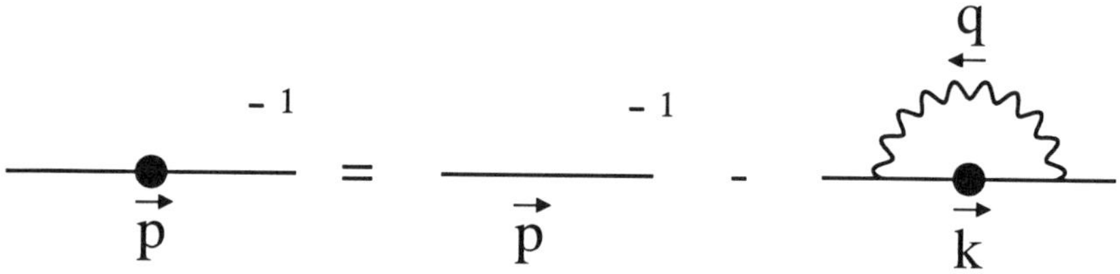

FIG. 3. Rainbow approximation.

$$i S_F^{-1}(p) = i S_F^{0\,-1}(p) - e^2 \int \frac{d^4 k}{(2\pi)^4} \, \gamma^\mu \, S_F(k) \, \gamma^\nu(k,p) \, \Delta_{\mu\nu}^0(q) \quad , \tag{3}$$

in what is known as the rainbow approximation, where

$$S_F(k) = \frac{F(k^2)}{\slashed{k} - \mathcal{M}(k^2)} \quad ,$$

$$S_F^0(k) = \frac{1}{\not{k} - m_0} \quad,$$

$$\Delta_{\mu\nu}^0(q) = \frac{1}{q^2}\left(g_{\mu\nu} + (\xi - 1)\frac{q_\mu q_\nu}{q^2}\right) \quad.$$

Eq. (3) is a matrix equation which corresponds to two equations in $\mathcal{M}$ and F. We can project out equations for these by taking the trace of Eq. (3) having multiplied by $\not{p}$ and 1 in turn to obtain:

$$\frac{1}{F(p^2)} = 1 - \frac{\alpha}{4\pi^3}\frac{1}{p^2}\int d^4k \frac{F(k^2)}{k^2 + \mathcal{M}^2(k^2)}\frac{1}{q^2} \cdot$$
$$\left\{-2k \cdot p - \frac{(\xi - 1)}{q^2}\left[2k^2 p^2 - (k^2 + p^2)k \cdot p\right]\right\},$$

$$\frac{\mathcal{M}(p^2)}{F(p^2)} = m_0 - \frac{\alpha}{4\pi^3}\int d^4k \frac{F(k^2)\mathcal{M}(k^2)}{k^2 + \mathcal{M}^2(k^2)}\frac{1}{q^2}(3 + \xi) \cdot$$

where as usual $\alpha = e^2/4\pi$. On carrying out the angular integrations, and putting the bare mass equal to zero, we have

$$\frac{1}{F(p^2)} = 1 + \frac{\alpha\xi}{4\pi}\int_0^{\Lambda^2} dk^2 \frac{F(k^2)}{k^2 + \mathcal{M}^2(k^2)}\left[\frac{k^4}{p^4}\theta(p^2 - k^2) + \theta(k^2 - p^2)\right] \quad, \quad (4)$$

$$\frac{\mathcal{M}(p^2)}{F(p^2)} = \frac{\alpha(3 + \xi)}{4\pi}\int_0^{\Lambda^2} dk^2 \frac{\mathcal{M}(k^2)F(k^2)}{k^2 + \mathcal{M}^2(k^2)}\left[\frac{k^2}{p^2}\theta(p^2 - k^2) + \theta(k^2 - p^2)\right] \quad, \quad (5)$$

where Λ is the ultra-violet momentum cutoff. It is easiest to solve these equations in the Landau gauge where they decouple. $F(p^2)$ is obviously 1. Moreover, there is a non-trivial solution [2] for the mass function $\mathcal{M}$ for the coupling larger than a critical value of $\alpha_c = \pi/3$. This is best illustrated by plotting the Euclidean mass $M = \mathcal{M}(M^2)$ as a function of α, as found by Curtis and Pennington [3]:

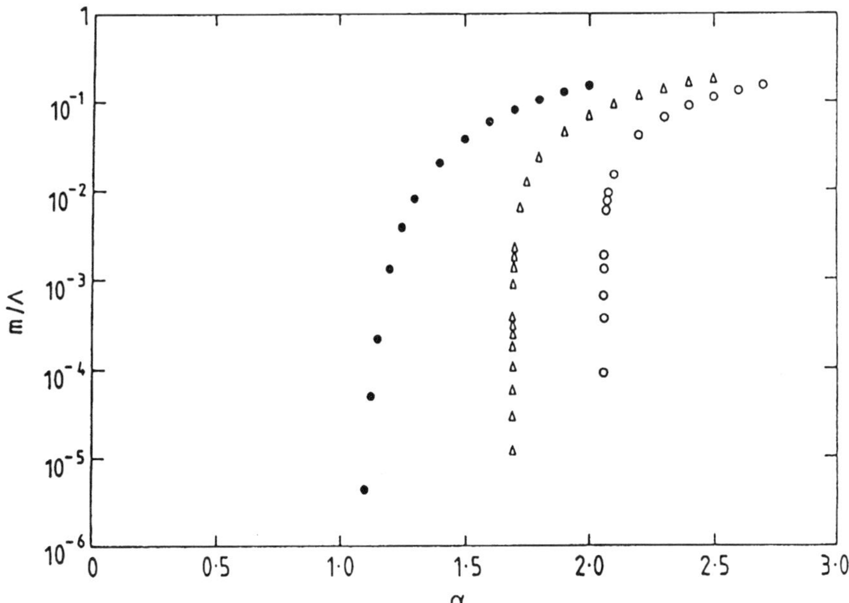

FIG. 4. Euclidean mass, $M = \mathcal{M}(M^2)$ dynamically generated in the rainbow approximation as a function of the coupling α in three different gauges: Landau ($\xi = 0$) •, Feynman ($\xi = 1$) ∆, and Yennie ($\xi = 3$) ∘ gauges.

Note that $\mathcal{M} = 0$ is always a solution to Eq. (5). However, beyond the critical value of the coupling, the non-zero solution bifurcates away from the trivial solution. This is in complete contrast with the perturbation theory, where, even if we perform an all orders resummation using the Renormalization Group Equation, we end up with a result of the following form,

$$
\begin{aligned}
\mathcal{M}(p^2) &= m_0 X(p^2) \\
X(p^2) &= \sum_n^{\infty} \sum_m^{n} \alpha^n A_n B_{m,n} \ln^m(p^2/\Lambda^2)
\end{aligned}
$$

and the field remains massless to all orders if we start with a zero bare mass, $m_0 = 0$.

In contrast, non-perturbative dynamics is able to generate masses for particles even if they have zero bare mass. However, there are problems. As the critical coupling corresponds to a change of phase, we expect it to be independent of the gauge parameter. But when one solves the Eqs. (4) and (5) for different gauges, one finds that this is not the case, as depicted in Fig. 4. However, it is not difficult to trace the root of this problem. The full vertex of Eq. (1) has to satisfy the Ward-Takahashi identity for the fermion propagator to ensure its gauge covariance. However, the bare vertex that was used in Eq. (3) does not obey this identity. Therefore, one should not expect physical outputs to be gauge independent when the input is not.

THE VERTEX

We expect that any reasonable *ansatz* for the vertex should fulfill the following requirements:

- It must satisfy the Ward-Takahashi Identity in all gauges.

$$
q^\mu \Gamma_\mu = S_F^{-1}(k) - S_F^{-1}(p)
$$

- It must ensure that the fermion propagator of Eq. (1) is multiplicatively renormalizable.

- It must result in a critical coupling, at which mass is generated dynamically, that is gauge independent.

- It must be free of any kinematic singularities, i.e. it should have a unique limit when $k^2 \to p^2$.

- It must have the same transformation properties as the bare vertex γ^μ under C and P.

Keeping in mind the form of the Ward-Takahashi identity, one can split the full vertex into two components, longitudinal and transverse:

$$
\Gamma^\mu(k,p) = \Gamma_L^\mu(k,p) + \Gamma_T^\mu(k,p) , \tag{6}
$$

where, the transverse part of the vertex is defined by:

$$
q_\mu \Gamma_T^\mu(k,p) = 0 . \tag{7}
$$

The Ward-Takahashi identity uniquely fixes the longitudinal part of the vertex, as shown by Ball and Chiu [4], to be

$$
\begin{aligned}
\Gamma_L^\mu(k,p) &= a(k^2,p^2)\gamma^\mu + b(k^2,p^2)(\slashed{k} + \slashed{p})(k+p)^\mu \\
&\quad - c(k^2,p^2)(k+p)^\mu ,
\end{aligned} \tag{8}
$$

where

$$a(k^2,p^2) = \frac{1}{2}\left(\frac{1}{F(k^2)} + \frac{1}{F(p^2)}\right) \quad,$$

$$b(k^2,p^2) = \frac{1}{2}\left(\frac{1}{F(k^2)} - \frac{1}{F(p^2)}\right)\frac{1}{k^2-p^2} \quad,$$

$$c(k^2,p^2) = \left(\frac{\mathcal{M}(k^2)}{F(k^2)} - \frac{\mathcal{M}(p^2)}{F(p^2)}\right)\frac{1}{k^2-p^2} \quad.$$

However, the transverse part remains arbitrary. Ball and Chiu [4] enumerated a basis of eight independent tensors in terms of which the most general form for the transverse part of the vertex can be written:

$$\Gamma_T^\mu(k,p) = \sum_{i=1}^{8} \tau_i(k^2,p^2,q^2)T_i^\mu(k,p) . \tag{9}$$

Following is the list of all the eight tensors:

$$
\begin{aligned}
T_1^\mu(k,p) &= p^\mu(k.q) - k^\mu(p.q) \\
T_2^\mu(k,p) &= (p^\mu(k.q) - k^\mu(p.q))(\slashed{k}+\slashed{p}) \\
T_3^\mu(k,p) &= q^2\gamma^\mu - q^\mu\slashed{q} \\
T_4^\mu(k,p) &= T_1^\mu p^\nu k^\rho \sigma_{\nu\rho} \\
T_5^\mu(k,p) &= \sigma^{\mu\nu}q_\nu \\
T_6^\mu(k,p) &= \gamma^\mu(k^2-p^2) - (k+p)^\mu(\slashed{k}-\slashed{p}) \\
T_7^\mu(k,p) &= \frac{1}{2}(k^2-p^2)[\gamma^\mu(\slashed{k}+\slashed{p}) - p^\mu - k^\mu] + (k+p)^\mu p^\nu k^\rho \sigma_{\nu\rho} \\
T_8^\mu(k,p) &= -\gamma^\mu p^\nu k^\rho \sigma_{\nu\rho} + p^\mu\slashed{k} - k^\mu\slashed{p} ,
\end{aligned}
\tag{10}
$$

with

$$\sigma_{\mu\nu} = \frac{1}{2}[\gamma_\mu,\gamma_\nu].$$

The simplest choice is to take the transverse part to be zero. But Curtis and Pennington [5] showed that if we take the transverse part of the vertex to be zero, the fermion propagator is no longer multiplicatively renormalizable. They suggested the following transverse part of the vertex satisfying this requirement.

$$\Gamma_T^\mu(k,p) = \frac{1}{2}\left(\frac{1}{F(k^2)} - \frac{1}{F(p^2)}\right)\frac{1}{d(k^2,p^2)}T_6^\mu(k,p) , \tag{11}$$

where, $d(k^2,p^2) = k^2$ for $k^2 \gg p^2$. $d(k^2,p^2)$ must be symmetric in k and p and free of kinematic singularities leading to the proposal:

$$d(k^2,p^2) = \frac{(k^2-p^2)^2 + [\mathcal{M}^2(k^2) + \mathcal{M}^2(p^2)]^2}{k^2+p^2} \quad. \tag{12}$$

The vertex specified by Eqs. (8-12) will be referred to as the CP-vertex [5]. Curtis and Pennington solved the coupled equations for F and $\mathcal{M}$ from Eq. (1), using this *ansatz*. They found that the gauge-dependence of the critical coupling at which the non-perturbative behaviour bifurcates away from the perturbative one reduces considerably, as seen by comparing Figs. 4 and 5.

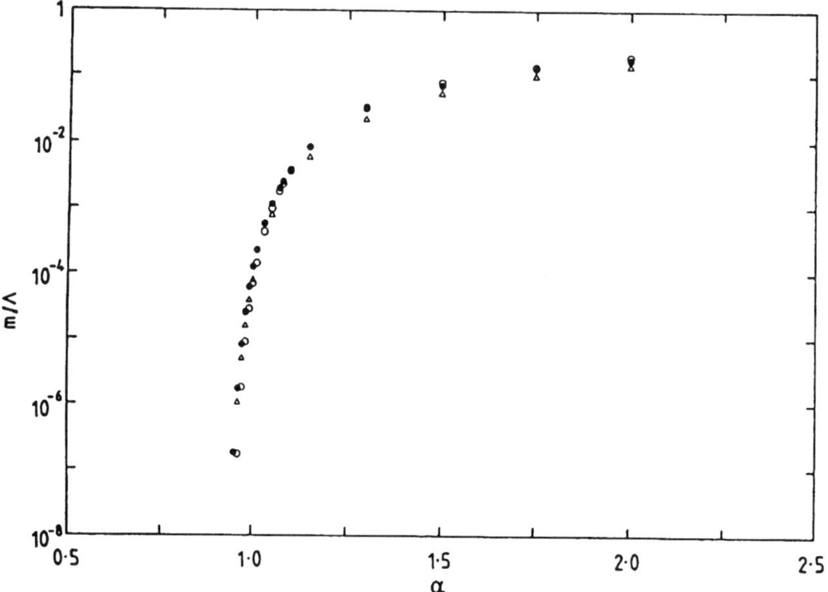

FIG. 5. Euclidean mass, $M = \mathcal{M}(M^2)$ dynamically generated using the CP-vertex as a function of the coupling α in three different gauges: Landau ($\xi = 0$) •, Feynman ($\xi = 1$) ▵, and Yennie ($\xi = 3$) ○ gauges. This plot is to be compared with the rainbow approximation results of Fig. 4.

BIFURCATION ANALYSIS

To see this, Atkinson et al. [6] recently suggested a bifurcation analysis to study the phase change near the critical coupling. This is a precise way to locate the critical coupling as compared to the previous methods which rely on numerical calculations. This method amounts, in practice, simply to throwing away all terms that are quadratic or higher in the mass-function $\mathcal{M}$. Employing this procedure, and using the fact that at the critical coupling, $\mathcal{M}(p^2) \sim (p^2)^{-s}$ and $F(p^2) \sim (p^2)^{\nu}$ in Eq. (1), one arrives at the following equation in an arbitrary gauge:

$$\nu = \frac{\alpha\xi}{4\pi}$$

$$\xi = \frac{3\nu(\nu - s + 1)}{2(1 - s)}\left[3 - \pi \cot \pi(\nu - s) + 2\pi \cot \pi s - \pi \cot \pi\nu\right.$$

$$\left. + \frac{1}{\nu} + \frac{1}{\nu + 1} + \frac{1}{\nu} + \frac{2}{1 - s} + \frac{3}{s - \nu} + \frac{1}{s - \nu - 1}\right] .$$

There are two roots of this latter equation for s between 0 and 1. Bifurcation occurs when the two roots for s merge at a point specified by $\partial\xi/\partial s = 0$. The bifurcation point defines the critical coupling, α_c. Numerically, $\alpha_c = 0.933667$ in the Landau gauge. For each value of the gauge parameter, these equations can be solved for ν, s_c and α_c. The solution found by Atkinson et al [6] is displayed in Fig. 6. For comparison, the points for the bare vertex have also been shown. One can see that the gauge dependence has considerably been reduced, as was seen earlier. However weak this variation, any gauge dependence shows that the CP vertex cannot be the exact choice.

310

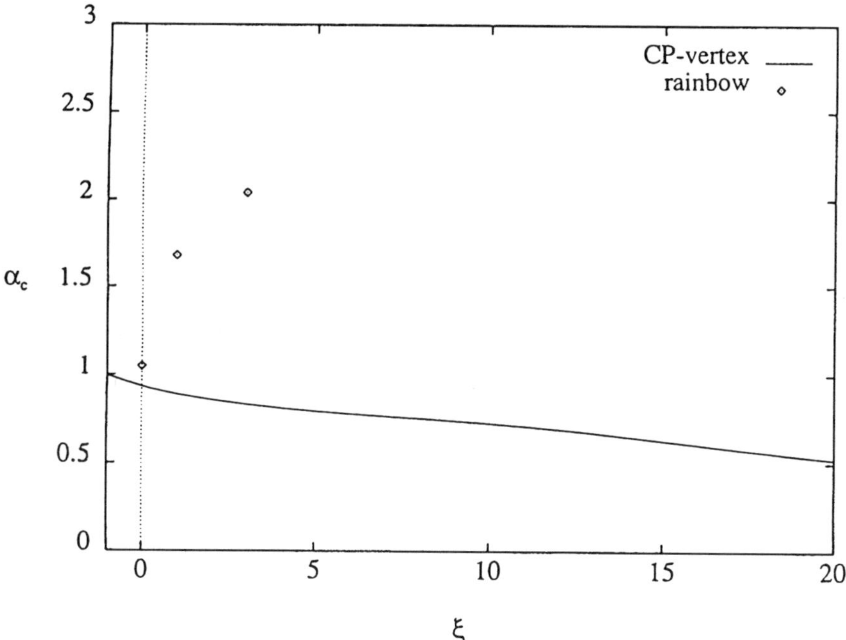

FIG. 6. Critical coupling, α_c, as a function of the gauge parameter, ξ (solid line). The corresponding values for the rainbow approximation have also been shown $\diamond$.

CONSTRAINTS OF MULTIPLICATIVE RENORMALIZABILITY

To find a vertex that ensures the gauge independence of the critical coupling, we start off by making three assumptions. Firstly, we demand that a chirally-symmetric solution should be possible when the bare mass is zero, just as in perturbation theory. This is most easily accomplished if the sum in Eq. (9) involves just $i = 2, 3, 6$ and 8. The second assumption is that the functions, τ_i, multiplying the transverse vectors, Eq. (9), only depend on k^2 and p^2, but **not** q^2. The third assumption is that the transverse part of the vertex vanishes in the Landau gauge. The motivation for this comes from the lowest order perturbative calculation for the transverse vertex, satisfied by Eq. (11). These conditions fix the τ_i of Eq. (9). Multiplicative renormalizability of the wavefunction renormalization $F(p^2)$ enables us to write τ_6 and $\overline{\tau}$ in terms of one function $W_1(x)$ [7] :

$$\overline{\tau}(k^2,p^2) = \frac{1}{4}\frac{1}{k^2-p^2}\frac{1}{s_1(k^2,p^2)}\left[W_1\left(\frac{k^2}{p^2}\right) - W_1\left(\frac{p^2}{k^2}\right)\right], \tag{13}$$

$$\tau_6(k^2,p^2) = -\frac{1}{2}\frac{k^2+p^2}{(k^2-p^2)^2}\left(\frac{1}{F(k^2)} - \frac{1}{F(p^2)}\right) + \frac{1}{3}\frac{k^2+p^2}{k^2-p^2}\overline{\tau}(k^2,p^2)$$
$$+\frac{1}{6}\frac{1}{k^2-p^2}\frac{1}{s_1(k^2,p^2)}\left[W_1\left(\frac{k^2}{p^2}\right) + W_1\left(\frac{p^2}{k^2}\right)\right], \tag{14}$$

where $\overline{\tau}$ is the combination of τ_i given by:

$$\overline{\tau}(k^2,p^2) = \tau_3(k^2,p^2) + \tau_8(k^2,p^2) - \frac{1}{2}\left(k^2+p^2\right)\tau_2(k^2,p^2),$$

and

$$s_1(k^2,p^2) = \frac{k^2}{p^2}F(k^2) + \frac{p^2}{k^2}F(p^2).$$

The condition of multiplicative renormalizability, i.e, $F(p^2) \sim (p^2)^\nu$, constrains the otherwise arbitrary function W_1 as follows:

$$\int_0^1 dx \, W_1(x) = 0 \ .$$

It should be noted that, with the simplest choice $W_1 = 0$, the massless CP-vertex, Eqs. (11,12) emerges.

CONSTRAINTS OF GAUGE INVARIANCE

At the bifurcation point, as stated before, multiplicative renormalizability forces a simple power behaviour for the mass function as well as for the wavefunction renormalization. Such a multiplicatively renormalizable mass function must exist in all gauges. Consequently, the exponent, s_c, must be gauge independent. Moreover, dynamical mass generation marks a physical phase change and so the critical coupling, α_c, must also be gauge independent. Thus the critical values, α_c, s_c, found in the Landau gauge must hold in all gauges. Using this physically motivated argument, the equation for the mass function gives τ_2, τ_3 and τ_8 in terms of a function $W_2(x)$ [8]:

$$
\begin{aligned}
\tau_2(k^2, p^2) \;=\; & \frac{2\xi}{(k^2 - p^2)^2} \frac{q_2(k^2, p^2)}{s_2(k^2, p^2)} - 6 \frac{\tau_6(k^2, p^2)}{(k^2 - p^2)} \\
& - \frac{1}{(k^2 - p^2)^2} \frac{1}{s_2(k^2, p^2)} \left[W_2\left(\frac{k^2}{p^2}\right) + W_2\left(\frac{p^2}{k^2}\right) \right] \\
& - \frac{k^2 + p^2}{(k^2 - p^2)^3} \frac{1}{s_2(k^2, p^2)} \left[W_2\left(\frac{k^2}{p^2}\right) - W_2\left(\frac{p^2}{k^2}\right) \right] \ , \\[2ex]
\tau_3(k^2, p^2) \;=\; & -\frac{k^2 + p^2}{k^2 - p^2} \tau_6(k^2, p^2) \\
& + \frac{1}{k^2 - p^2} \frac{1}{s_2(k^2, p^2)} \left[\frac{1}{2} r_2\left(\frac{k^2}{p^2}\right) - \frac{\xi}{3} q_3(k^2, p^2) \right] \\
& - \frac{1}{6} \frac{k^2 + p^2}{(k^2 - p^2)^2} \frac{1}{s_2(k^2, p^2)} \left[W_2\left(\frac{k^2}{p^2}\right) + W_2\left(\frac{p^2}{k^2}\right) \right] \\
& + \frac{1}{6} \frac{k^4 + p^4 - 6k^2 p^2}{(k^2 - p^2)^3} \frac{1}{s_2(k^2, p^2)} \left[W_2\left(\frac{k^2}{p^2}\right) - W_2\left(\frac{p^2}{k^2}\right) \right] \ , \\[2ex]
\tau_8(k^2, p^2) \;=\; & -2 \frac{k^2 + p^2}{k^2 - p^2} \tau_6(k^2, p^2) + \overline{\tau}(k^2, p^2) \\
& - \frac{1}{k^2 - p^2} \frac{1}{s_2(k^2, p^2)} \left[\frac{1}{2} r_2\left(\frac{k^2}{p^2}\right) - \frac{\xi}{3} q_8(k^2, p^2) \right] \\
& - \frac{1}{3} \frac{k^2 + p^2}{(k^2 - p^2)^2} \frac{1}{s_2(k^2, p^2)} \left[W_2\left(\frac{k^2}{p^2}\right) + W_2\left(\frac{p^2}{k^2}\right) \right] \\
& - \frac{2}{3} \frac{k^4 + p^4}{(k^2 - p^2)^3} \frac{1}{s_2(k^2, p^2)} \left[W_2\left(\frac{k^2}{p^2}\right) - W_2\left(\frac{p^2}{k^2}\right) \right] \ ,
\end{aligned}
$$

where

$$
r_1\left(\frac{k^2}{p^2}\right) \;=\; \left(\frac{k^2}{p^2}\right) \left[1 - \left(\frac{k^2}{p^2}\right)^\nu \right] - \left(\frac{k^2}{p^2}\right)^{-1} \left[1 - \left(\frac{k^2}{p^2}\right)^{-\nu} \right] \ ,
$$

$$r_2\left(\frac{k^2}{p^2}\right) = \left(\frac{k^2}{p^2}\right)^{\frac{1}{2}-s_c}\left[1-\left(\frac{k^2}{p^2}\right)^{\nu}\right] - \left(\frac{k^2}{p^2}\right)^{s_c-\frac{1}{2}}\left[1-\left(\frac{k^2}{p^2}\right)^{-\nu}\right] \ ,$$

$$s_1(k^2,p^2) = \frac{k^2}{p^2}\,F(k^2) + \frac{p^2}{k^2}\,F(p^2) \ ,$$

$$s_2(k^2,p^2) = \frac{k}{p}\frac{\mathcal{M}(k^2)}{\mathcal{M}(p^2)}F(k^2) + \frac{p}{k}\frac{\mathcal{M}(p^2)}{\mathcal{M}(k^2)}F(p^2) \ ,$$

$$q_2(k^2,p^2) = \frac{1}{k^2-p^2}\left[\frac{k^3}{p}\frac{\mathcal{M}(k^2)F(k^2)}{\mathcal{M}(p^2)F(p^2)} - \frac{p^3}{k}\frac{\mathcal{M}(p^2)F(p^2)}{\mathcal{M}(k^2)F(k^2)}\right] \ ,$$

$$q_3(k^2,p^2) = \frac{kp}{(k^2-p^2)^2}\left[(p^2-3k^2)\frac{\mathcal{M}(k^2)F(k^2)}{\mathcal{M}(p^2)F(p^2)} - (k^2-3p^2)\frac{\mathcal{M}(k^2)F(k^2)}{\mathcal{M}(p^2)F(p^2)}\right] \ ,$$

$$q_8(k^2,p^2) = \frac{1}{(k^2-p^2)^2}\left[\frac{k}{p}(3k^4+p^4)\frac{\mathcal{M}(k^2)F(k^2)}{\mathcal{M}(p^2)F(p^2)} - \frac{p}{k}(k^4+3p^4)\frac{\mathcal{M}(k^2)F(k^2)}{\mathcal{M}(p^2)F(p^2)}\right] \ .$$

and the function W_2 is constrained, by the gauge invariance of the mass function and the critical coupling, to obey the following integral equation,

$$\int_0^1 \frac{dx}{\sqrt{x}}\,W_2(x) = 0 \ ,$$

at the critical coupling $\alpha = \alpha_c$. In order to make sure that none of the functions τ_i has kinematic singularities as $k^2 \to p^2$, W_1 and W_2 should also satisfy the following conditions:

$$
\begin{aligned}
W_1(1) + W_1'(1) &= -6\nu \ , \\
W_2(1) + 2W_2'(1) &= 2\xi(\nu - s - 1) \ .
\end{aligned}
$$

This defines the construction of the full vertex via Eqs. (6-10) [8].

CONCLUSIONS

Above we have presented a truncation of the fermion Schwinger-Dyson equation for the quenched QED, which respects the key properties of the theory. We have constructed a non-perturbative vertex in terms of the constrained functions $W_i(x)(i = 1, 2)$. It satisfies the Ward-Takahashi identity, ensures the fermion propagator is multiplicatively renormalizable, agrees with one loop perturbation theory for large momenta and enforces a gauge independent chiral symmetry breaking phase transition. This study motivates the need for a realistic investigation of $t\bar{t}$ condensates as the source of the electroweak symmetry breaking. Including the four fermion interaction, the Dyson-Schwinger equation for the fermion propagator becomes:

FIG. 7. Dyson-Schwinger equation for the fermion propagator, including the four fermion interaction term.

We need to solve this equation in a gauge invariant way. The study of quenched QED presented here suggests that a proper choice of the vertex can guarantee the gauge independence of the physical observables. However, a realistic calculation, of course, requires the unquenching of the theory which complicates the problem significantly. The fermion-boson vertex (in particular its transverse part) will intimately depend on the photon renormalization function in a non-perturbative way not yet understood. The discussion for quenched QED presented here provides the starting point for such an investigation of full QED.

ACKNOWLEDGEMENTS

This work was performed in collaboration with M.R. Pennington. I wish to thank the Government of Pakistan for a research studentship and the University of Durham and Institut d'Etudes Scientfiques de Cargèse for providing me with the funds to attend the School.

REFERENCES

[1] W.A. Bardeen, C.T. Hill and M. Lindner, Phys. Rev. **D41** 1647 (1990).

[2] V.A. Miransky, Nuovo Cim. **90A** 149 (1985) ;
Sov. Phys. JETP **61** 905 (1985) ;
P.I. Fomin, V.P. Gusynin, V.A. Miransky and Yu.A. Sitenko,
La rivista del Nuovo Cim. **6**, numero 5, 1 (1983).

[3] D.C. Curtis and M.R. Pennington, Phys. Rev. **D48** 4933 (1993).

[4] J.S. Ball and T.W. Chiu, Phys. Rev. **D22** 2542 (1980).

[5] D.C. Curtis and M.R. Pennington, Phys. Rev. **D42** 4165 (1990).

[6] D. Atkinson, J.C.R. Bloch, V.P. Gusynin, M.R. Pennington and
M. Reenders, Phys. Lett. **B329** 117 (1994).

[7] Z. Dong, H.J. Munczek and C.D. Roberts, preprint
ANL-PHY-7711-94.

[8] A. Bashir and M.Pennington, " Gauge Independent Chiral Symmetry
Breaking in Quenched QED ", University of Durham preprint DTP-94/48
(June, 94) Phys. Rev. (to be published).

DARK MATTER, A CHALLENGE FOR PARTICLE ASTROPHYSICS

Bernard Sadoulet

Center for Particle Astrophysics
Lawrence Berkeley Laboratory
and Physics Department, University of California, Berkeley

INTRODUCTION

There is mounting evidence that at least 90% of the mass in the universe is "dark." By dark we mean that it is does not emit nor absorb any kind of electromagnetic radiation and is only seen by its gravitational effect on visible objects. We do not yet know the exact amount, nor the nature, of this obviously major component of the physical universe. This fundamental puzzle constitutes the "dark matter problem" which dates back to Zwicky,[1] and has been often reviewed in the past[2]. Its solution touches central issues in cosmology and astrophysics, and probably also involves particle physics.

EVIDENCE FOR DARK MATTER

Although Zwicky's initial suggestion of the presence of dark matter remained controversial for a long time, there is a growing consensus in the astronomy community that dark matter indeed exists. Four types of evidence have been found.

Rotation Curves

In individual spiral galaxies, the measured velocities of objects (typically isolated stars, gas clouds or globular clusters) that are apparently bound to the galaxies, allow us to estimate the centripetal gravitational force which has to balance the centrifugal force:

$$\frac{GM(r)}{r^2} = \frac{v^2}{r}$$

Figure 1 shows the famous measurements made by V. Rubin and collaborators[3]. The needed centripetal force is much larger than the gravitational force generated by the stars we can see in the galaxies, and as the velocity appears constant at high radii, the mass $M(r)$ enclosed in the orbit has to increase as r, in a region where practically no more stars are observed. Similarly it is often possible to measure[4] HI rotation curves out to many times the scale

Frontiers in Particle Physics: Cargèse 1994
Edited by M. Lévy *et al.*, Plenum Press, New York, 1995

length characterizing the exponential decrease of the surface brightness away from the galactic center. The dark halo clearly dominates the dynamics and typical mass to light ratios $M(r)/L \approx$ 3-5 M_{sun}/L_{sun} are obtained, increasing with radius even when no light is seen.

Elliptical galaxies also contain large amounts of dark matter. While velocity dispersion measurement of stars probes a region where the dark matter is not dominant, the studies of globular clusters[5] and planetary nebulae typically show an increase of M/L from 3 in the inner part to 15 in outer parts. The extended x-ray emission[6] detected by x-ray satellites implies even larger values (>70).

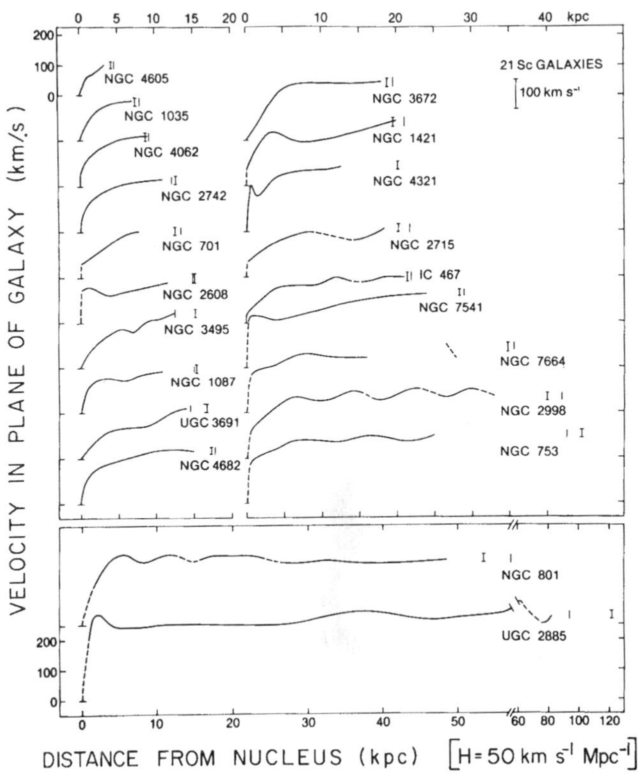

Figure 1. Rotation curves for many spiral galaxies obtained from optical measurement (From V.C. Rubin *et al.,* 1980).

Velocity Dispersion in Cluster of Galaxies

In a cluster of galaxies, each galaxy has a finite "peculiar" velocity, oscillating back and forth in the potential well created by the galaxy concentration. As these systems are believed to be bound and stationary, the dispersion of these velocities can be related to the depth of the potential well through the virial theorem

$$\langle\text{kinetic energy}\rangle = -1/2 \; \langle\text{potential energy}\rangle.$$

Figure 2 shows the observations for the Coma cluster. The large observed dispersion velocities of some 1000 km/*s* to 1500 km/*s* implies a mass to light ratio of 400*h* times that of the sun, where *h* is the Hubble expansion parameter measured in the usual units of 100 km/s/Mpc (experimentally $1/2 \leq h \leq 1$).

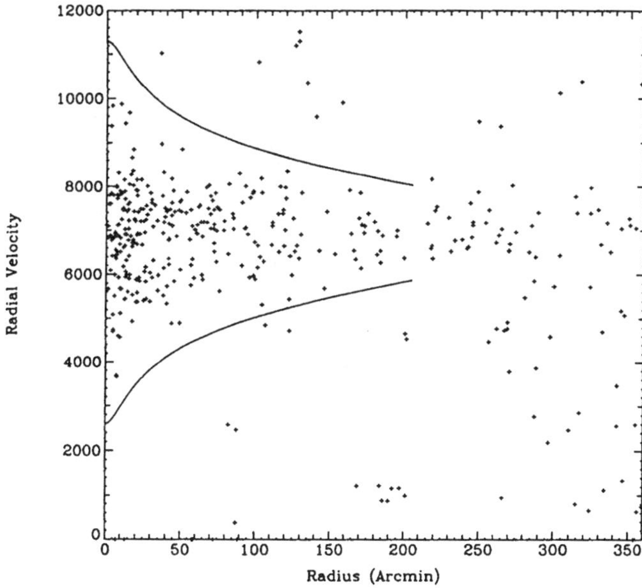

Figure 2. Line-of-sight velocities of galaxies in the Coma cluster (in kms^{-1}) as a function of distance from the cluster center in minutes of arc. The curves mark the authors' estimate of the boundary between cluster members and interlopers. At the distance of Coma 1 arcmin = 20h^{-1} kpc. (From Kent and Gunn, 1982.)

X-ray Emission by Galaxy Clusters

Similar velocity dispersions can be inferred from the temperature of the x-rays emitted by the intergalactic gas which appears[7] to be present in clusters of galaxies (Figure 3). If the gas has the same spatial distribution as the galaxies, we expect that

$$\frac{3}{2}kT_{gas} = \frac{1}{2}m\sigma^2,$$

where σ is the dispersion velocity of gas molecules. A typical 5 keV x-ray temperature corresponds to $\sigma \approx 1200$ km/s, a value analogous to that of galaxies. A detailed analysis is impeded by the lack of precise measurement of the temperature profile[8], but the results are in general agreement with those derived from the galaxy velocity dispersion.

Gravitational Lensing by Galaxy Clusters

New independent evidence on the depth of potential wells in clusters of galaxies has been obtained by Tyson and colleagues[9] who have studied the gravitational lensing of distant galaxies by foreground clusters. A round object located far behind the cluster will appear elliptical after lensing, with its major axis tangential to the mass distribution contours. In extreme cases, arcs and arclets should be seen; this indeed is what is observed. In Figure 4, most of the galaxies of the foreground cluster Abell 1689, which is roughly in the center and provides the gravitational lens, have been subtracted (using their reddish color) and a definite trend towards tangential structures is evidenced. From these many lensed objects it is possible to reconstruct the mass distribution of dark matter (Figure 5) and we can "see" the accumulation of mass in the center of the cluster, in a way similar to the distribution of gas. A high velocity dispersion of some 1200 km/s is also inferred, confirming the independent analysis of velocity dispersion and x-rays. This method promises to allow a detailed mapping of dark matter in large structures.

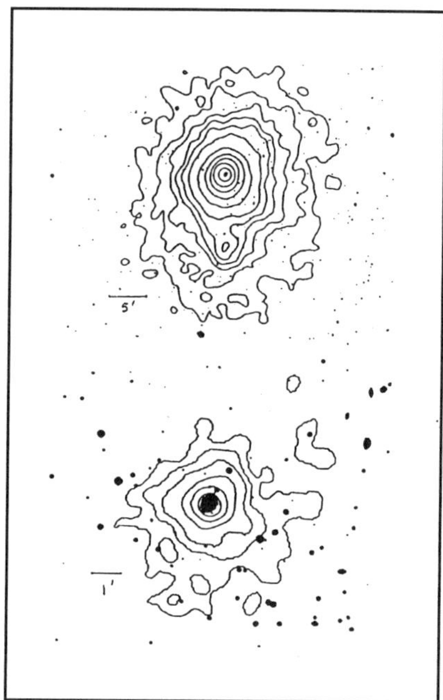

Figure 3. X-ray isodensity contours for cluster Abell 85, at two different resolutions (top with the Einstein Satellite IPC, bottom with the HRS). (From Forman and Jones, 1982.)

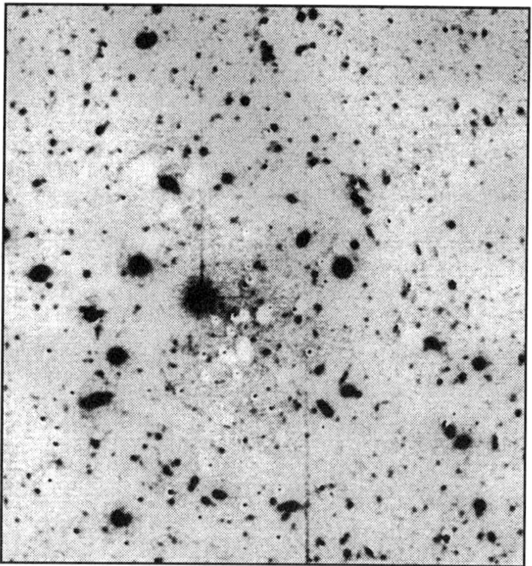

Figure 4. Image of background galaxies (in the blue) lensed by cluster A 1689. The cluster light has been eliminated by subtraction of a scaled red image. Several nearby arcs and systematic distortion along circles about the lens center can be seen.

Potential Loopholes

Many authors have outlined the potential loopholes in the above arguments. In particular, our kinematic arguments have implicitly assumed spherical symmetry, and it is indeed possible to explain individual rotation curves by peculiar, highly non-spherical, matter distributions without introducing dark matter. However, for each type of observation and at each scale, an ad hoc argument has to be devised. In such an approach, the compatibility observed in clusters between results of the virial theorem, the x-ray temperature and gravitational lensing would appear accidental. In spite of its far-reaching consequences, the dark matter explanation is much more natural.

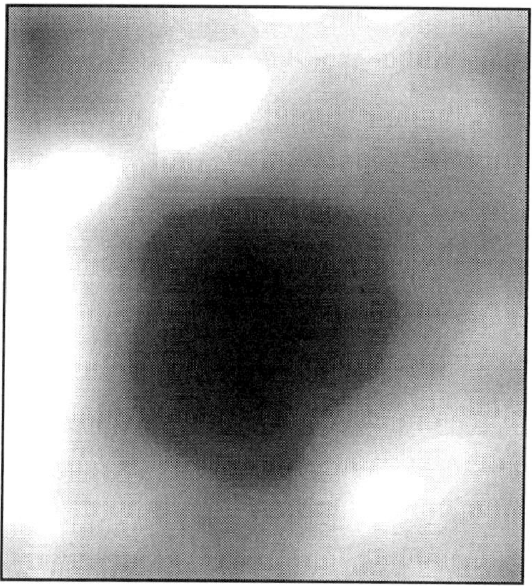

Figure 5. Distribution of dark matter in cluster A 1689 as inferred from Fig. 4.

It has also been proposed that Newtonian gravity could be modified[10] on scales larger than the solar system or the binary pulsar, where General Relativity has been extensively tested. So far, it appears difficult to construct such alternative gravity models in a way consistent with General Relativity at smaller scales, they fail to explain the magnitude of the gravitational lensing effects[11], and they may be incompatible with the compact dark halos observed around dwarf galaxies[12].

THE VALUE OF Ω

Arguments are therefore strong for the existence of dark matter. However, its amount is still uncertain by about a factor 5. As it dominates the mass density in the universe, measuring the amount of dark matter is essentially equivalent to the determination of Ω, the average density in units of the critical density

$$\Omega = \frac{\rho}{\rho_c} \text{ with } \rho_c = \frac{3H^2}{8\pi G},$$

where H is the Hubble expansion parameter. As is well known, Ω is related to the curvature of space and to the ultimate fate of the universe[13]. $\Omega > 1$ corresponds to a closed (i.e., spherical) universe which will eventually recollapse, while $\Omega < 1$ corresponds to an open (i.e., hyperbolic) universe which expands forever. For $\Omega = 1$, space is flat, and the expansion will stall at infinite time.

We examine in turn the three types of methods which astrophysicists attempt to use to determine the universe curvature and a potential cosmological constant.

Direct Summation

We may first attempt to sum the mass observed with the virial theorem in various systems. Usually it is done with the mass to light ratio *M/L*, which allows us to compute the mass density from the luminosity distribution of the considered objects:

$$\rho_m = \int \left(\frac{M}{L}\right) L \left(\frac{d^2 N}{dV dL}\right) dL$$

where *L* is the absolute luminosity of the objects. Typical values are significantly smaller than 1. We would need a *M/L* of $1600h$ (where h is the Hubble parameter in usual units of 100 km/s/Mpc), in order to have $\Omega = 1$. None of the known objects have such high mass to light ratio. However, it should be noted that the virial theorem is only sensitive to inhomogeneities of the mass distribution, and that such methods can only give a lower limit of Ω. Another problem is that it is difficult to know where the system stops. At large distances from the center of spiral galaxies we are running out of objects to measure and galaxies far from the cluster core fade in the field.

Dynamical Methods

Attempts to apply the virial theorem on even larger scale structure are prevented by the fact that the systems are not stationary, with structure still clearly forming. Not only do we observe on the large scale a complex distribution of galaxies[14] (Figure 6) with voids, filaments, bubbles and the like, not typical of a virialised system, but large scale velocity coherent flows can be deduced from the comparison of the distance and redshift of (relatively close-by) galaxies[15], or reconstructed[16] from the density of galaxies detected in the infrared by the IRAS satellite (Figure 7).

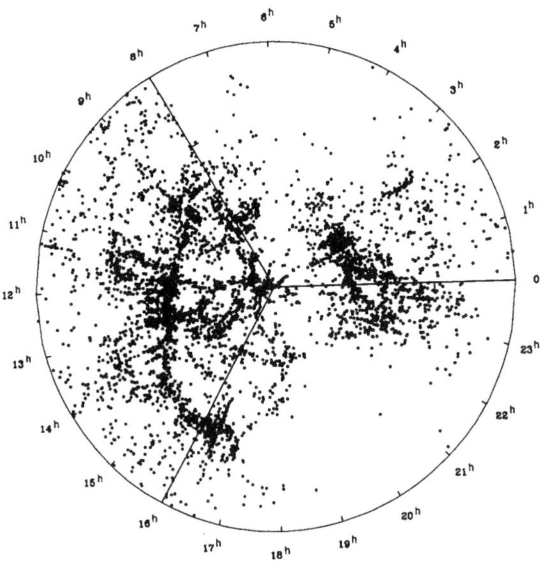

Figure 6. Distribution of galaxies as a function of redshift. We are at the center. Each galaxy is plotted at a radius proportional to its recession velocity and an azimuth equal to its right ascension. This is a slice in declination between 10° and 40°. The magnitude limit of the sample is 14.5. (From Huchra and Geller, 1989.)

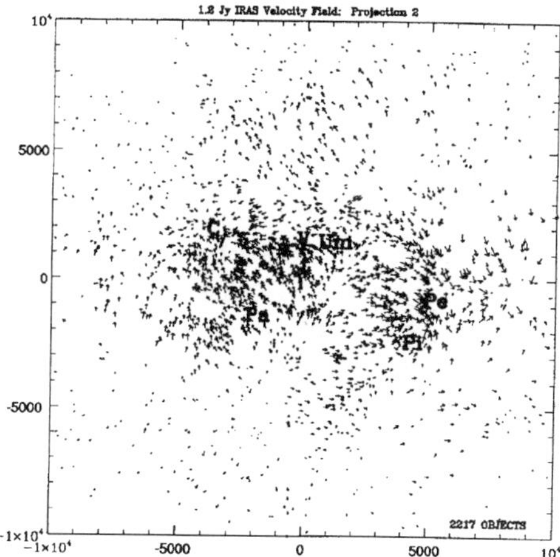

Figure 7. Velocity flows reconstructed from a redshift survey of galaxies selected with the IRAS catalog. The Great Attractor is seen on the left and the Perseus Pisces cluster is identifiable at the right. (From Strauss and Davis, 1987.)

As the time that structure takes to form is sensitive to the underlying averaged density, one can attempt to use the large scale distribution of galaxies and the associated velocity flows to estimate Ω. Basically, one writes the equations

observed velocities = acceleration (due to density inhomogeneities) $\times$ time of formation,

or

observed inhomogeneities = rate of variation (due to peculiar velocities) $\times$ time.

In the linear regime[17], these equations become

$$\vec{v} = G\rho_b \vec{\nabla} \int d^3x' \frac{\dfrac{\delta\rho}{\rho_b}}{|x-x'|} \times \frac{2f}{3H\Omega} = \frac{Hf}{4\pi} \vec{\nabla} \int d^3x' \frac{\dfrac{\delta\rho}{\rho_b}}{|x-x'|}$$

$$\text{and} \quad \frac{\delta\rho}{\rho_b} = -\vec{\nabla}.\vec{v} \times \frac{1}{Hf} \quad \text{with} \quad f \approx \Omega^{0.6}.$$

We can either compare the measured value of our peculiar velocity with respect to the cosmic microwave background to that obtained from summing the acceleration due to observed density inhomogeneities, or compare the observed density fluctuations to those predicted by the divergence of the observed velocity field. The second method is very recent, and became possible through the realization[18] that the three dimensional velocity field can be deduced from the radial components which are the only measured quantities, through the natural assumption of zero vorticity ($\vec{\nabla} \times \vec{v} = 0$), as any initial vorticity would be erased by the Hubble expansion. This procedure, which is called "POTENT," is demonstrated in Figure 8, which compares the density contrast observed in the IRAS catalog[16] to that reconstructed through POTENT from measured peculiar velocities[15]. Density enhancements in the region of the Great Attractor (left) and in the Perseus-Pisces region (right) are seen in the two plots.

However, there is a fundamental difficulty in applying these methods, as we cannot measure directly the background density contrast $\delta\rho/\rho_b$, but only the density contrast $\delta n/n$ in the number of galaxies, and we have numerous indications that light does not trace mass. It is usually assumed that the two densities are related by

$$\frac{\delta n}{n} = b \frac{\delta\rho}{\rho_b},$$

321

where b is the "biasing parameter" which is usually taken as constant (but does not need to be!). The POTENT method can then only give the value of $\beta = \Omega^{0.6}/b$. From studies such as exemplified in Figure 8 it can be deduced[19] that typically

$$\beta = \Omega^{0.6}/b = 0.6 - 1.2 \, .$$

As b is believed not to be much smaller than 1, this result points to a large value for Ω. Taking into account non-linear effects, it may be possible to extract Ω and the biasing parameter b separately. Although these procedures are very uncertain, both quantities appear to be greater that 0.5. In order to be put on firmer ground such a method would require the measurement of the peculiar velocities of a much larger sample of galaxies. This would certainly be one of the exciting products of the Million Galaxy Redshift Survey that Chicago, Princeton and Fermilab are starting.

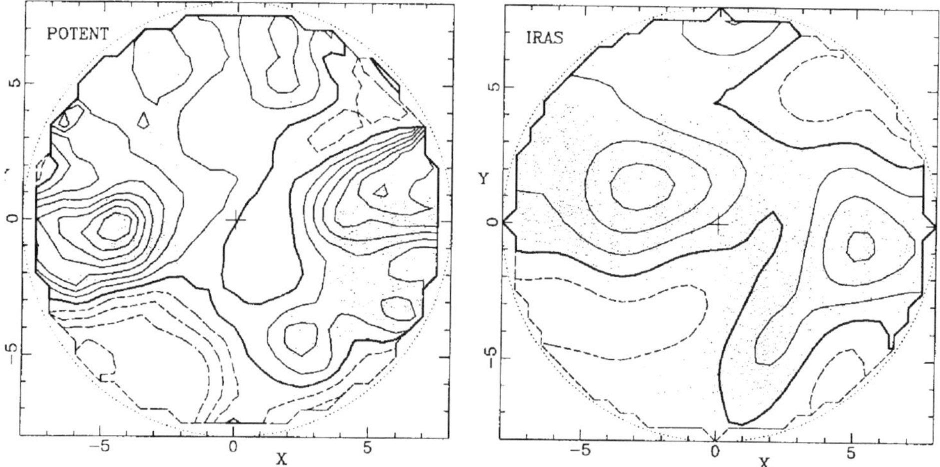

Figure 8. *Right:* IRAS galaxy density contrast $\delta n/n$. *Left:* The underlying mass-density $\delta\rho/\rho$ contrast reconstructed by POTENT from the Mark III peculiar velocities. The effective smoothing in both cases has a radius of $12h^{-1}$Mpc. Spacing between contours is 0.2. (From A. Dekel, 1994.)

Similar methods have been devised to attempt to measure Ω through the study of velocity correlations of galaxies around clusters of galaxies. They also point to large values of Ω (typically $0.3\pm.15$).

Geometry Measurements

A potentially powerful set of methods is to attempt to directly measure the geometry of the universe by measuring quantities sensitive to this geometry. For instance, the variation of the apparent luminosity of a "standard candle" (i.e., constant absolute luminosity) as a function of redshift would tell us about potential deviation from the $1/r^2$ law which is only valid in a flat space, or the change with redshift of the volume element traced by the density of galaxies would be a direct indication of spatial curvature. These cosmological tests directly address the question of the value of Ω (and Λ), but they require that one be able to differentiate between the variation due to cosmology and the evolution of the objects.

This problem has so far plagued the first type of measurement. For instance, in the classical test of Sandage and coworkers[20] using the brightest galaxy in a cluster as a standard candle, the luminosity evolution dominated cosmological effect. In recent years, a new idea has been developed: Type Ia supernovae appear to be reasonable standard candles "locally," they can be efficiently searched for[21] at intermediate redshifts (0.5), and they are probably the best hope to perform this luminosity test. The main problem appears to be the need for a very large amount of telescope time for photometric and spectroscopic follow-up.

The volume variation test was first attempted on radio galaxies, counting their number as a function of luminosity, but they evolve too rapidly with redshift. Recent attempts[22] to use the K band to measure the number density of optical galaxies as a function of luminosity gave ambiguous results, as the overall picture is complicated by mergers and "local" dwarf star-forming galaxies.

The real way to do the volume test would to directly measure the counts as a function of redshift, and not rely on luminosity which is not simply related to redshift. This approach was pioneered by Loh and Spillar[23], but their heroic attempt has been criticized on many grounds[24] and is not reliable. With the operation of the Keck 10-m telescope, it will soon be possible to do this measurement in a much more convincing way with a suitable multi-slit spectrograph which is now under construction at Lick under the auspices of the NSF Center for Particle Astrophysics and CARA. In particular, not only could the color and redshift of the object be measured, but also its mass through the measurement of the internal velocity dispersion. This would be a powerful method to detect merging, gravitational lensing and luminosity evolution, and should provide a much more reliable determination of the geometry.

Conclusion: Current Uncertainties of Ω

Figure 9 summarizes the effective values of Ω obtained by these various methods as a function of the scale on which they are performed. Most methods are in effect a measurement of the inhomogeneities on those scales and are lower limits. Three conclusions emerge from this graph.

Whatever the exact value of Ω is, it is tantalizingly close to 1. This is important as in General Relativity, the spatial curvature of a non-flat universe evolves extremely rapidly, and it is difficult to understand why our universe is so close to being flat without being exactly flat. This flatness problem is one of the reasons why inflation scenarios[25] are favored by many cosmologists (usually originating from particle physics), as it naturally predicts $\Omega = 1$. Inflation also explains why the cosmic microwave background can have basically the same temperature at points which naively appear causally disconnected. However, it should be noted that there may be a crisis on the horizon, as the most recent determinations of the Hubble parameters ($H \approx 75$ km/s/Mpc) may be incompatible $\Omega = 1$ and our current estimates of the age of the universe. If this discrepancy persists, we may have to accept the presence of a vacuum energy in the form of a cosmological constant Λ. Defining

$$\Omega_\Lambda = \frac{\Lambda}{3H^2},$$

spatial flatness requires then that

$$\Omega + \Omega_\Lambda = 1.$$

The second conclusion is that the effective value of Ω is rising very slowly as a function of the scale, contrary to what is predicted by conventional models of galaxy formation. Locally the effective value of Ω is surprisingly small for a flat universe. This profound problem is related to the quantitative failure of the standard cold dark matter model and its prediction of too much structure at small scale. This will discussed below in our analysis of the indirect evidences for nonbaryonic dark matter.

A third conclusion is that the current measurements imply both the existence of baryonic dark matter and of nonbaryonic dark matter. The density of visible objects (stars and high temperature clouds) is of the order of a few 10^{-3} of the critical density. This is much smaller than the density of baryons necessary in the standard scenario of (homogeneous) nucleosynthesis to account for the observational determinations of the primordial abundance of light elements (helium 4, deuterium, helium 3 and lithium 7)[26]. The limits on the baryonic density $\Omega_b = \rho_b/\rho_c$ provided by the most recent analysis are surprisingly narrow:

$$0.01 \leq \Omega_b h^2 \leq 0.016,$$

where h is the Hubble expansion parameter in the usual units of 100 km/s/Mpc. These limits are displayed in Figure 9 as a band, folding in the uncertainty on h. As h definitely seems to be observationally smaller than 1, Ω_b has to be larger than 1%, and considerably higher than the visible Ω. We can therefore conclude that we need at least *some* baryonic dark matter. Conversely, if we take seriously the values of Ω obtained on large scale with velocity flows for instance, they appear to be definitely larger than that the range allowed by the standard primordial nucleosynthesis. Attempts to increase this range by relaxing the homogeneity assumption made in these calculations have been so far unsuccessful. In particular, a possible quark hadron phase transition cannot modify significantly the aforementioned result[27]. Confirmation of a large value for Ω would definitely require the existence of *nonbaryonic* dark matter.

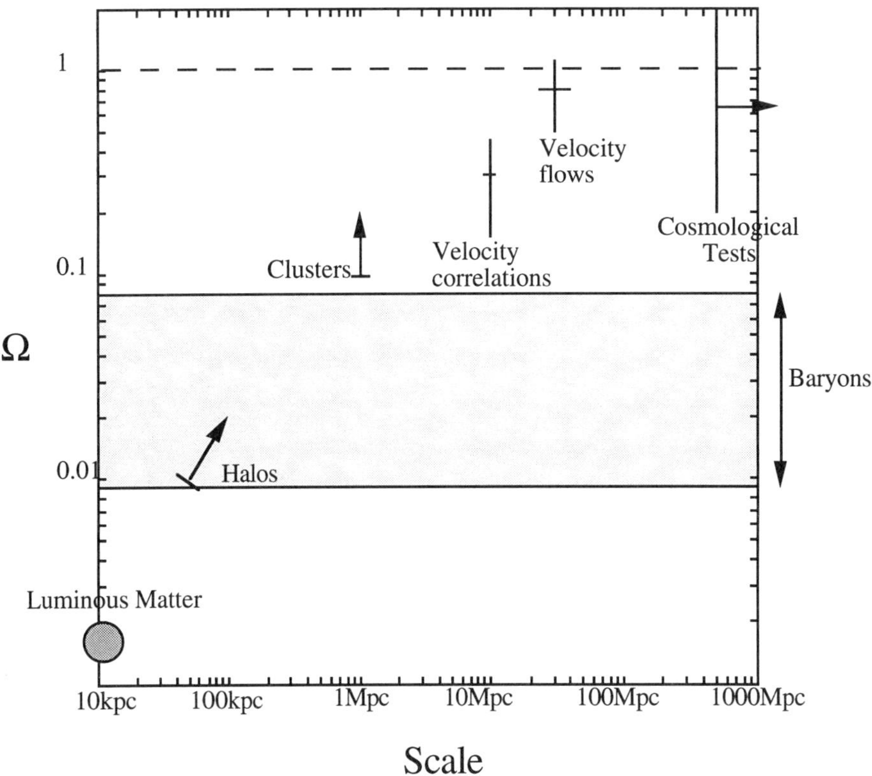

Figure 9. Measured effective values of Ω versus scale.

THE NATURE OF DARK MATTER

Even more than its existence and its abundance, the nature of dark matter is a central scientific question, especially if it turns out that it is not made of ordinary baryons. Figure 10 sketches the range of possibilities. One of the main current goals of experimental cosmology is to narrow down the choices. In this fundamental quest, we can obtain information both from classical astrophysical observations and from new attempts to directly observe dark matter, often using particle and nuclear physics methods. We analyze in this section what we can say today and what we may be able to find out in the near future from indirect astrophysical methods. We will describe direct searches in the following sections.

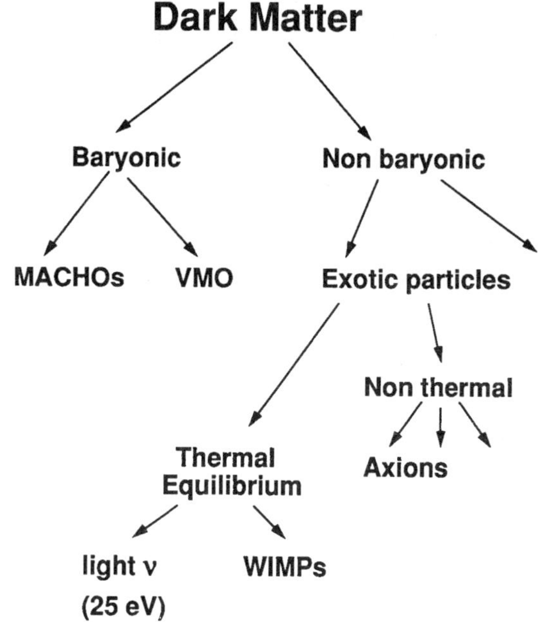

Figure 10. Nature of dark matter.

Astrophysical Constraints on Baryonic Dark Matter

What can we learn about dark baryons from astrophysics? As mentioned above, the main information comes from the measurements of primordial abundances of light elements which clearly indicate that there are dark baryons. But it is very difficult to prevent baryons from radiating or absorbing light.

From the absence of a Gunn-Peterson absorption trough in the spectrum of quasars, we know that baryonic dark matter cannot be in the form of cold gas[28], and too much hot gas will either impose Compton distortions on the cosmic microwave background, if it is diffuse, or emit too many x-rays, if it is clumped. The COBE results[29] put severe limits on the first possibility, while the second problem, related to the puzzle of the diffuse hard x-ray background[30], is still very much debated. Dust will absorb radiation and re-emit it in the infrared, and severe limits are put by IRAS on its contribution.

Instead of being in the form of gas or dust, dark matter could be made of condensed objects somehow formed in the very early universe. Here again, the possibilities are severely limited. There are not enough faint stars to account for our dark halo[31], and at present the only ones compatible with observations[32] are Very Massive Objects (VMOs) of at least 100 M_{sun} or Massive Compact Halo Objects (MACHOs, also called Brown Dwarfs). Because of their high mass, the VMOs very quickly underwent supernova explosions and formed black holes sufficiently massive to absorb all the material around them, therefore preventing contamination of the interstellar medium by high Z elements produced in the explosions. Constraints on these objects can in principle be obtained through the far infrared relic radiation they should have produced, and the DIRBE experiment aboard the COBE satellite is actively searching for such a signature. Too many of them will also destroy the disk of spiral galaxies that they would repeatedly cross. The second class of dark baryonic objects is formed by small jupiter-like objects that are not massive enough to burn hydrogen and are therefore not shining. We will discuss below the current status of their search.

Astrophysical Constraints on Nonbaryonic Dark Matter

As remarked above, if the observed large values for Ω are correct and Ω is larger than 0.2, we may be obliged within the current picture of primordial nucleosynthesis to postulate the existence of nonbaryonic dark matter.

A different line of argument based on the distribution of mass and the large scale structure that we observe in the universe may lead to similar conclusions. The formation of structure is certainly due, at least in part, to gravitational collapse of density fluctuations, and since dark matter is gravitationally dominant, its nature imprints characteristic features on the observed universe[33]. It has been known for some time that the mere fact that the 2.7K cosmic microwave background is so smooth and the large scale structure is so clumpy is difficult to understand without the presence of nonbaryonic dark matter. If the background photons tell us about the baryonic density fluctuations at time of recombination, without dark matter, there is simply not enough time to grow the large fluctuations observed today. With nonbaryonic dark matter, this is much easier. Before protons and electrons recombine into hydrogen, they are prevented from growing density fluctuations by their coupling to photons, which diffuse easily through the medium. Dark matter density fluctuations, on the contrary, can grow since they are uncoupled to photons. After recombination, the baryons fall into the potential wells prepared by dark matter, and there is enough time then to form the structure observed today. There are ways to circumvent this conclusion (e.g., the so-called "isocurvature" scenarios[34]), but they are less natural.

This argument has recently become much more precise with the detection by the COBE satellite of anisotropies[35] in the temperature of the cosmic microwave background. Figure 11 compares the COBE observations with the power spectrum of density fluctuations of galaxies as observed[36] by IRAS. In order to plot on the same graph the COBE results which refer to a redshift of a thousand, it is necessary to extrapolate the growth of density fluctuations since that time. The extrapolation shown here assumes adiabatic fluctuations, nonbaryonic dark matter, and a universe close to being flat. Note that it fits rather well with the galaxy power spectrum, giving strong evidence that the present large scale structure arises from the collapse of density fluctuations which have also induced the tiny anisotropies of the microwave background. It is important to remark that had we assumed only baryonic dark matter or an open universe (e.g., $\Omega \approx 0.1$), we would have dramatically under predicted the power spectrum at large scale.

Moreover, the smooth curve that we would be tempted to draw through the observations has the general shape of what is expected in the so-called cold dark matter model (CDM). Technically, cosmologists speak of "hot" or "cold" dark matter depending on whether it was relativistic or not at the beginning of galaxy formation. Cold dark matter includes, in addition to condensed baryonic objects (provided they are formed extraordinarily early), weakly interacting massive particles, axions (which are in most models created cold, in spite of their light mass), primordial black holes, etc. The prototype of hot dark matter is a light neutrino of, say, 25 eV, which would close the universe. In addition to cold dark matter, the cold dark matter model has many ingredients: it includes the assumption of a particular spectrum of initial density fluctuations (Harrison-Zel'dovich spectrum), which are also assumed to be Gaussian and uncorrelated on different scales. These assumptions are motivated by inflation. With a few parameters (overall amplitude, and biasing parameter), the resulting model[37] gives an amazingly good first approximation of the observed structure. In particular, the power spectrum has the right general shape.

This would not have been the case with a hot dark matter model, where basically no structure at small scale is able to form[33] as primordial density fluctuations are erased by neutrinos streaming below a scale of 40 h-1 Mpc. The larger structures would have formed first with the galaxies originating from instabilities inside large pancakes. This is incompatible with our observations which indicate that the clusters are still forming while most galaxies are relatively old objects. In addition, the velocity of cosmological neutrinos of 25 eV mass and 2K temperature would be larger than the escape velocities of most galaxies, and they could not congregate into galactic halos. Moreover, for smaller galaxies (dwarf spheroids), the halos appear too compact to be compatible with the observed velocity dispersion and the maximum phase space density expected for initially non-degenerate Fermi-Dirac particles[38].These arguments can be circumvented by assuming at least one additional element: a two component dark matter (e.g., hot + cold, or hot + baryonic) or topological singularities seeding galaxy formation[39].

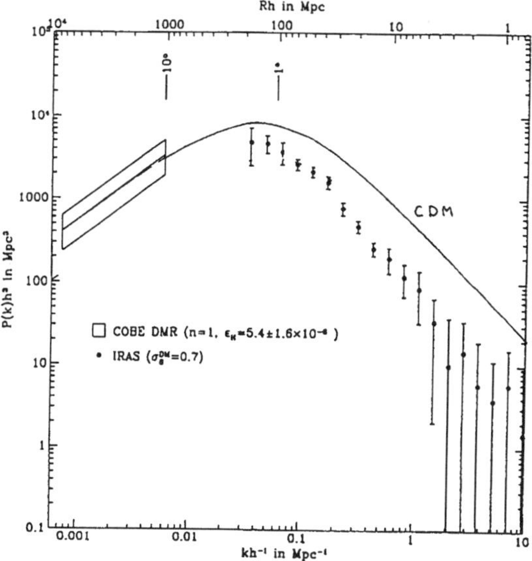

Figure 11. Measured power spectrum measured for IRAS galaxies and extrapolation of the COBE result assuming nonbaryonic dark matter and a flat universe. The curve indicates the prediction of the Cold Dark Matter model normalized to the COBE extrapolation. (After Fisher *et al.*, 1992.)

However, it is clear from Figure 11 that even though they qualitatively agree with the data, the predictions of the cold dark matter model are incorrect in the details. When normalized to COBE, the model predicts too much structure at small scale. The origin of this discrepancy is not understood and currently the object of numerous speculations. We may not understand the complex phenomena of star formation well enough and the feedback mechanisms which may slow down the formation of structure. The slope of the initial spectrum may be wrong or there may be a non-zero cosmological constant. Alternatively, a small admixture of 7 eV neutrinos would help to decrease the amount of structure at small scale as neutrinos stream out of larger density regions. Whatever the solution is, it is clear that the cold dark matter model, which has been sometimes presented as the "Standard Model" of cosmology, has at least to be enlarged, and the present difficulties may the sign of some exciting new physics.

It should be realized that in no way do the difficulties of the cold dark matter model imply that dark matter is not made of cold nonbaryonic particles. But we should also note that presumably because of the large number of available parameters, it is possible to fit the observations with isocurvature fluctuations without nonbaryonic dark matter.

In any case, it is clear that we need more data to close the gap between the COBE observations and the largest scales presently measured, and to check the extrapolation from the time of emission of the microwave background. The planned Sloan Digital Sky Survey and the DEEP program at the Keck promise to provide power spectrum at larger scale and probe regions of high redshift at a much earlier time in the structure formation. In parallel, measurements of the cosmic microwave at smaller scale will bridge the gap in the other direction and provide crucial tests of the detailed physics at play.

SEARCH FOR MASSIVE COMPACT HALO OBJECTS

We now turn our attention to direct searches, and will focus first on the search for baryonic dark matter in the form of massive compact halo objects. How can we look for Massive Compact Halo Object (MACHOs) if they do not emit any light?

The basic scheme was suggested by B. Paczynski[40]. Suppose that we observe a star, say in the Large Magellanic Cloud, a small galaxy in the halo of the Milky Way. If one MACHO assumed to be in the halo were to come close to the line of sight, it would gravitationally lens the light of the star and its intensity would increase. This object, however, cannot be static, lest it fall into the potential well. Therefore it will soon move out of the line of sight, and one would expect a temporary increase of the light intensity which, from the equivalence principle, should be totally achromatic.

These are exactly the characteristics of the events now observed[41] by three groups:

• The American-Australian MACHO collaboration (Lawrence Livermore National Laboratory, UC Berkeley, UC Santa Barbara, UC San Diego and Mount Stromlo Observatory, regrouped within the Center for Particle Astrophysics) which uses the largest Charge Coupled Device (CCD) array presently in operation at a telescope

• The French EROS collaboration working mostly with photographic plates.

• The Warsaw-Princeton-Carnegie OGLE collaboration, observing the bulge of the galaxy with a modest CCD.

In addition, two groups (A. Crotts, and P. Baillon *et al.*) are attempting observation of M31 (the Andromeda galaxy) which has the advantage of being fairly inclined (leading to a large difference in number of expected events between the near and far sides), but the obvious drawback of being too far for resolving stars individually.

Figure 12 shows a spectacular event with an amplification of 7 which has been seen toward the Large Magellanic Cloud. It is achromatic to high precision. From the last two years of observation, we could arrive at the following three conclusions.

The microlensing phenomenon has clearly been established. A total of some 70 events have been observed by the three groups (mostly towards the bulge of our galaxy). The distribution of amplification is compatible with expectation and microlensing is observed for all types of stars, dismissing the possibility that we are observing a new class of variable objects. Moreover, for one event which was recognized early enough, high precision spectra were taken several times during its intensity rise and fall; they were invariant, in contrast with what should happen for any flaring activity. A binary event was also observed by the OGLE and MACHO groups, and the complex light curve is very well understood.

The number of events seen towards the galactic Bulge are much more numerous than expected: the MACHO collaboration has observed some 45 of them while OGLE has 12 such events. This is more than a factor of three greater than the most optimistic estimation done before the observations. It could be the result of the presence of a bar, which was previously suspected, or of a disk denser than expected. Note that the mapping of the distribution of events in galactic coordinates will eventually allow to distinguish between these two models. In any case, these observations show that our previous modeling of the galaxy and its halo was too simple. This may also have an impact on the density assumed in searches for nonbaryonic dark matter.

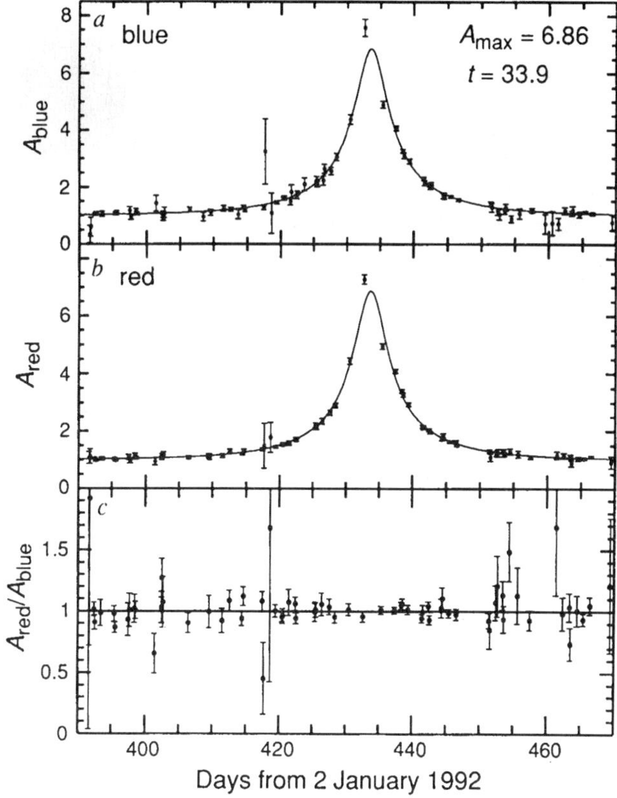

Figure 12. A microlensing candidate of the MACHO collaboration towards the Large Magellanic Cloud.

On the contrary, the number of events seen towards the Large Magellanic Cloud is somewhat low: two events were detected by the EROS group when they expected eight and the three events seen by the MACHO group represent[42] some 20% of the prediction of the simple halo model that everybody has used so far. This may indicate that MACHOs are indeed only a small fraction of the halo or due to a thick disk, and that our halo is mostly nonbaryonic. It may also be that our understanding of the halo is still too primitive. Before this fundamental question is answered, much more work has to be done on a set of self consistent models[43] of all components of the galaxy taking into account all the available observations.

In conclusion, we may indeed have observed the *baryonic* dark matter component. If and when this is firmly established, we may have solved the primordial nucleosynthesis puzzle of a baryonic density at the level of 3 to 5 percent of the critical density, but we may not have solved all of the other aforementioned cosmological problems (the mass in clusters, the velocity flows, the flatness and horizon problem, the comparison of COBE results with the IRAS spectrum, etc.). It is still likely that we need *nonbaryonic* dark matter. And we may claim that this nonbaryonic dark matter, if it exists, has also to be present in the halo of our galaxy; it is very difficult to prevent it from significantly accreting onto a preexisting baryonic halo, at least if it is non-relativistic! Therefore, within the present theoretical framework, not only is it not unlikely that nonbaryonic dark matter exists but also that it constitutes a significant fraction of the density in the halo of our galaxy.

NONBARYONIC DARK MATTER CANDIDATES

The above discussion clearly shows that it would at best be premature to stop the direct searches for nonbaryonic dark matter. It will take years to make all the tests which are necessary to fully establish the nature of the MACHO events, and there are still substantial arguments pointing to the presence of nonbaryonic dark matter in the halos of galaxies. It can even be argued that now that we begin to have one piece of the puzzle, it is important to aggressively attempt to identify the neighboring pieces.

What could this nonbaryonic dark matter be? If we discard exotica such as a shadow universe or primordial black holes (which, by the way, could appear as MACHOs), the most attractive hypothesis is that dark matter is made of particles that were created in the hot early universe and managed to stay around. In order to compute the relic abundances, it is necessary to distinguish between particles which have been produced in thermal equilibrium with the rest of the universe and those which were somehow produced out of equilibrium.

Axions

Axions[44] are an example of the second case, where we depend totally on the specific model considered to predict their abundances. These particles have been postulated in order to prevent the violation of CP in strong interactions in the otherwise extremely successful theory of quantum chromodynamics. Of course there is no guarantee that such particles exist, but the present laboratory and astrophysical limits on their parameters are such that if they do exist, they have to be cosmologically significant[45]. The first two searches[46] for cosmological axions performed a few years ago were missing a factor of 1000 in sensitivity. This is no longer the case, for an experiment is being prepared at Livermore which will reach the cosmologically interesting region at least for one generic type of axions (hadronic models[47]). Figure 13 compares the expected sensitivity with that of the two previous experiments. Although this experiment represents an exciting opportunity, it should be noted that the decade of frequency (and therefore of mass) that can be explored with the present method is only one out of the three which are presently allowed.

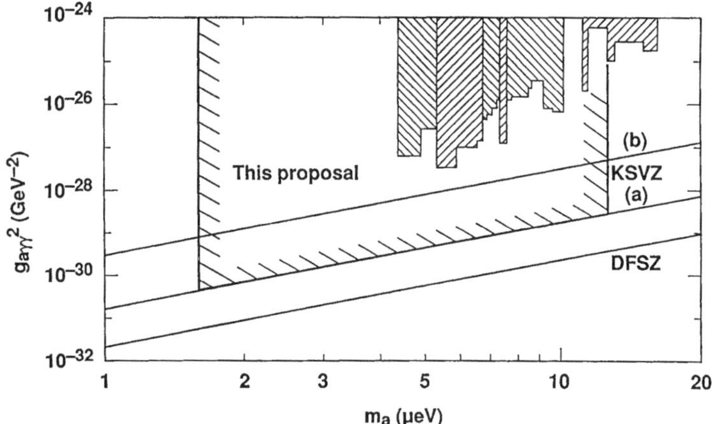

Figure 13. Expected sensitivity of the Livermore experiment. The lines labeled KSVZ and DFSZ refer to two generic species of axions. The shaded regions in the upper right are the previous experimental limits.

Neutrinos

In the opposite case, where particles have been in thermal equilibrium, the current density of dark matter particles depends on whether they were relativistic or not at the time of freeze out when they decoupled from the rest of the universe. Light massive neutrinos fall in the first category: If their mass is much smaller that 2 MeV/c^2, their relic density is related only to the decoupling temperature and is basically equal to that of the photons in the universe. The relic particle density is therefore directly related to its mass, and a neutrino species of 25 eV would give an Ω of the order of unity.

Unfortunately no good ideas exist to detect cosmological neutrinos, and one has to rely on the mass measurements of neutrinos in the laboratory, through the study of beta spectra, neutrinoless double beta decay, and oscillation experiments. Of particular importance in that respect are the two experiments assembled at CERN, NOMAD and CHORUS, which will explore the oscillation between ν_μ and ν_τ in the few eV. A recent claim[48] for neutrino oscillations by the LSND experiment at Los Alamos would also point to neutrino masses in this range. It is also possible to probe the mass of neutrinos in the astrophysical environment; we may be lucky enough to observe the neutrino flash of a supernova occurring in our galaxy which may allow a precise measurement[49], the apparent deficit of electrons in interaction of atmospheric neutrinos[50] may be an indication for massive neutrinos, as may the current MSW explanation[51] of the deficit of solar neutrinos.

Weakly Massive Interactive Particles

The second generic class of particles which were in thermal equilibrium in the early universe corresponds to the case where the decoupling occurred when they were non-relativistic. In that case it can be shown that their present density is inversely proportional to their annihilation cross section[52].

The argument is simple. Let us consider a species of particles that we will call δ, since our considerations cover equally well heavy neutrinos ν_H, supersymmetric neutralinos $\tilde{\gamma}$, $\tilde{h}$, $\tilde{\nu}$, $\tilde{z}$, Technicolor particles, etc. We assume that it has once been in thermal equilibrium with quarks (q) and leptons (l), presumably through the reactions

$$\delta\bar{\delta} \leftrightarrow \frac{q\bar{q}}{l\bar{l}}.$$

In the very early universe, at temperatures larger than the mass of the δs, the reactions above go both ways. As the universe expands and cools down below temperatures of about m$_\delta$/20, the equilibrium is displaced to the right. If the annihilation rate is much faster than the rate of expansion of the universe, and if there is *no initial asymmetry* between the δs and the $\bar{\delta}$s, they all disappear and cannot constitute the present dark matter. If, on the other hand, the annihilation rate is too small, the expansion quickly dilutes the δs, which soon cannot find an antiparticle to annihilate with, and their abundance now will be too large. In order to give the current ratio Ω_δ of the δ average density to the critical density, the annihilation cross section is, for δ masses in the few GeV/c^2 region:

$$\sigma v \approx \frac{10^{-26}\,\mathrm{cm}^3 s^{-1}}{\Omega_\delta\ h^2}$$

where h is the Hubble constant in units of 100 km/s/Mpc.

Such a result is interesting because of two facts:

• For $\Omega_\delta \approx 1$ this annihilation rate has roughly the value expected from weak interaction, while nowhere in the argument had we to assume a particular interaction scale. This may be a numerical coincidence, or a precious hint that physics at the W and Z scale is important for the problem of dark matter.

• The order of magnitude of interaction rate is given and allows the planning of experiments.

Moreover, it is a lower limit, for we could imagine *an initial asymmetry*, similar to the one usually assumed for baryons and antibaryons. In this case, the cross section could become large enough for all the pairs $\delta\bar{\delta}$ to disappear (as have happened for most u and d quarks), and the small excess of one component will make up the dark matter. Therefore, in the general case, we have

$$\sigma v \geq \frac{10^{-26}\,\mathrm{cm}^3 s^{-1}}{\Omega_\delta\; h^2}.$$

Inversely, physics at the W and Z^0 scale leads naturally to particles whose relic density is close to the critical density. In order to stabilize the mass of the vector intermediate bosons, one is led to assume the existence of a new family of particles in the 100 GeV mass range. Whether they are Technicolor or supersymmetric particles, the relic density of their lightest members tends to be in the region of interest for dark matter.

A Multi-component Dark Matter?

Before discussing in more detail the searches for these particles in the next section, it is worth pointing out that any model where the matter in the universe is constituted by more than one species of particle requires deep connections within microphysics. We are well accustomed to the idea that electrons and baryons should be in similar number, but this is not automatic and requires some conservation law (e.g., B-L) at play in the baryogenesis. More deeply, if it turns out that, say, Weakly Interacting Massive Particles (WIMPs) exist and bring the average density of the universe to the critical value, there should be some correlation between the baryogenesis phenomenon which fixes the amount of baryons today and supersymmetry, if this is what gives the WIMPs their interaction strength. The problem is compounded if a mixed dark matter model[53] with both massive neutrinos and WIMPs is more than simply an elegant way to introduce a new parameter in the modeling of structure formation but indeed describes reality. In that case, we have to explain why three species have similar densities in the universe! This requires additional relations between baryogenesis, physics at the weak scale, and the phenomena responsible for the neutrino masses. Does this mean that such possibilities are unlikely? Not within the general framework of grand unified theories; we are not speaking of an arbitrary fine tuning of parameters but of connections between physical phenomena occurring at different energy scales but certainly deeply related to each other. This is another example of the ways cosmology may teach us something very fundamental about the structure of forces.

SEARCHES FOR WIMPS

There are basically two methods for searching for WIMPs[54]: their elastic scattering rate on ordinary matter in the laboratory may be large enough to be detectable. Moreover, they can be trapped in the sun and the earth, leading to enhanced annihilations which may be detectable as a high energy neutrino flux[55]. We will focus here on the first method and briefly outline the challenges faced by the experimentalist.

The Challenges of the Detection of WIMPs

The known order of magnitude the annihilation cross section leads by crossing to a gross estimate of the elastic cross section of these particles on various nuclei, but a number of technical complications arise which makes rate calculations very delicate. Figure 14 attempts to summarize the situation for a germanium target[56]. The expected rate in a minimum supersymmetry model can be significantly less than 0.1 events per kilogram per day, much smaller than the limits (upper hatched region) that can be achieved[57] with state of the art techniques for low radioactivity background. The second challenge comes from the fact that the energy deposition is quite small, a few keV for the mass range of interest. For detectors based only on the collection ionization, this difficulty is compounded by the fact that the interaction produces a nuclear recoil which is much less efficient in ionizing than an electron of the same energy. It should be noted, however, that the accelerator experiments (LEP and the Tevatron) have tended to push up the interesting mass scale, at least in the case of supersymmetry, leading to larger energy deposition, but also to lower rates (as the number density of WIMPs decreases inversely proportionally to their mass). In our opinion, this leads the experimenter to a greater emphasis on redundancy and background rejection than on threshold. The third challenge comes from the fact that true signatures linking detected events to particles in the halo of the galaxy are extremely difficult to obtain. The best one would be the measurement of the direction of the scattered nucleus[58]. The halo has not collapsed as much as the disk and is expected to have a very small overall angular velocity. Because of the rotation of the sun inside the halo, dark matter particles will come preferentially from one direction. The expected asymmetry is sizable; however, measuring it is quite difficult. Low pressure gas counters is a potential technique pursued by the Masek group at UC San Diego, but it is expensive to get large target masses (see below). Another interesting possibility is that the ballistic phonons produced keep a memory of the initial direction of the momentum[59]. Short of that signature, it is in principle possible to look for a change in the event rate and the spectrum of energy deposition with the time of the year. The reason is simple[60]: While the sun goes around the galaxy and therefore through the halo at 250 km/s, the earth is adding or subtracting half of its velocity to the sun's velocity in the summer or the winter. The mean energy deposition varies by about ±4.5% and the rate varies by about ±2.5%. In order to observe such an effect at a 3 σ level, about 3700 events

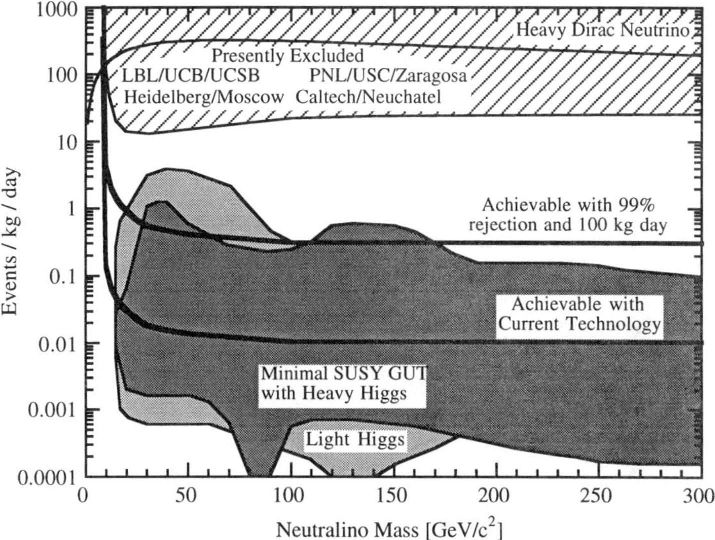

Figure 14. Expected sensitivity as a function of the WIMP mass. The hatched region at the top is excluded by direct detection experiments, and includes a wide range of masses for a heavy Dirac neutrino. The shaded regions in the middle are the rates predicted by the most general supersymmetric grand unified models with a light (50 GeV/c^2) and heavy (80–90 GeV/c^2) Higgs mass, including the constraints from LEP and CDF. The upper heavy curve, the event rate sensitivity the Center for Particle Astrophysics experiment should achieve by the end of 1995, will begin to constrain the models. The lower heavy curve is achievable with their present technology at a deep site and 10 kg of detectors.

are needed, and therefore very large mass detectors (of the order of 50 to 100 kg) will be required. Moreover, the detector sensitivity and background would have to be exceptionally stable.

The Need for Active Rejection of the Background

Before we examine the approaches pursued by the various groups involved, it may be worth commenting on the likely need for an active rejection of the background.

From the experience of the four groups using germanium detectors so far, it is clear that the main background is made of electron recoils, while the signal would be nuclear recoils. In principle, it is therefore possible to discriminate against the background. As efforts to further decrease the radioactive background become more difficult, this may be the only way to significantly improve the sensitivity of our searches.

This point is worth some expansion, as it is usually not fully appreciated. Active rejection is necessary not only because we would like to decrease the *magnitude* of the background under the signal, but also because it is essential to know the *shape* of the background (unless its magnitude is negligible). Otherwise it is always possible to "bury" a possible dark matter signal inside the background, for instance by assuming that the background contribution goes to zero at zero energy. The only constraint is that it cannot be negative. Therefore, an experiment without any identification of the background will have a sensitivity which, after an initial decrease with the product of the mass and the exposure time, will plateau as the sum of the potential signal and background becomes sufficiently well measured. This is one explanation (together with a relatively high threshold) of why the Heidelberg-Moscow group cannot give a limit significantly better than previous experiments. The only way to improve the sensitivity with exposure is to use the annual modulation[60], but this in effect is equivalent to an experiment with an efficiency of roughly the amplitude of the modulation, that is, 5%. We commented above on the careful control of systematics which is necessary.

In contrast, even a mediocre background rejection r can be enough for its identification and the measurement of its shape. This allows to subtract it and the sensitivity will improve as $[(1\text{-}r)/MT]^{1/2}$. For very good rejection, such as obtained with the simultaneous measurement[61] of ionization and phonons (we now get experimental numbers in excess of 98% around 20 keV), the gain can be spectacular, as exemplified by the upper full curves in Figure 14 corresponding to 100 kg per day exposure, an electron background of 2 events/keV/day, and a rejection factor of 99%. But the gain can be substantial, even for poorer rejections such as likely to be obtained with large scintillators. As shown by the Rome-Saclay team[62], the fall time is shorter for nuclear recoil, even close to their threshold of 3 keV (equivalent electron energy) and, although not many photoelectrons are available, it should be possible to estimate the background and subtract it. This is likely to be done in practice, through a simultaneous fit to the pulse-height, rise-time two-dimensional space in order to estimate the signal and the background. Another interesting development[63] is the possibility of using the strong wavelength dependence on the nature of the recoil which seems to occur for NaF cooled down to liquid nitrogen temperature. The differences may be sufficiently large to allow a few photoelectrons to provide excellent rejection.

It should be added that if the background rejection is large enough for no event to be observed in the region of interest, the sensitivity of the search increases linearly with MT, not with the square root.

Current Strategies in Searches for WIMPs

Given this experimental challenge, low expected rates and low energy depositions, the experimental teams interested in the direct detection are adopting five different strategies that we summarize in Figure 15:

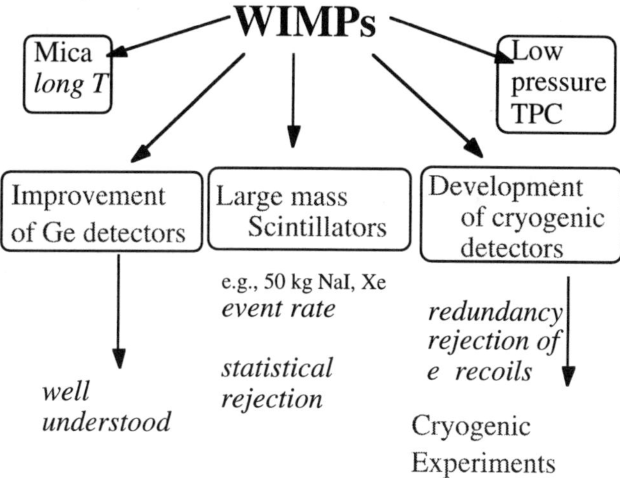

Figure 15. Current approaches to the search for Weakly Interacting Massive Particles.

Germanium Detectors. Four groups involved in the search for neutrinoless double beta decay in germanium have modified their apparatus to look for dark matter particles[10]: USC-PNL, now joined by Zaragoza, LBL-UCSB-UCB, Caltech-Neuchatel-PSI, and Heidelberg-Moscow. The results obtained so far are already interesting. As shown in Figure 1, the combination of these three experiments excludes the possibility that a heavy neutrino with a mass above 10 GeV/c^2 could form the major component of the dark matter in the halo of our galaxy. Combined with the recent results of SLC and LEP, which exclude a fourth generation of Dirac neutrinos below a mass of 40 GeV/c^2, we can safely conclude that heavy Dirac neutrinos cannot be the major component of our halo unless they are very heavy. Using the same technology with silicon detectors, the LBL/UCSB/UCB group have also been able to essentially exclude cosmions[64,] which have been proposed to explain the deficit of solar neutrinos.

These technologies have the advantage of being well known, and through an improvement of the radioactive environment and a rapid processing of the detector at the surface, it is probably possible to gain a factor of a few. With a considerable amount of care, background levels of 0.1 to 1 event/kg/keV/day at 20 keV can indeed be reached, but this is still a long way away from the background levels necessary to have a sensitivity of 0.1 event/kg/day, since the integration has to be made over an energy region of some 20 keV!

Large Mass Scintillators. A second approach is to strive for large target masses, trading off redundancy for simplicity and large event rates which, for instance, may allow the use of the annual modulation signature.

Detectors of NaI of masses between 30 and 100 kg are being assembled by at several groups: the Beijing-Rome-Saclay collaboration[65], the United Kingdom Dark Matter group (P. K. Smith *et al.*), the Osaka team of H. Ejiri[66], and the Zaragoza group. Such detectors are indeed very simple, and if manufactured properly have background levels only 2 to 5 times worse than conventional germanium detectors. Moreover, as explained above, they allow a statistical distinction between the nuclear recoils and the electron recoils, as the second type tend to have longer decay times. Given the number of photo-electrons, however, this cannot be done on an event-by-event basis. The Beijing-Rome-Saclay group estimates that such methods may allow an increase of sensitivity of roughly a factor 10 with respect to current limits.

A similar type of method using liquid Xenon is proposed by some of the same groups (Rome, UK) and the CERN-College de France team[67]. The simultaneous measurement of ionization and scintillation light, or alternatively the pulse shape of the scintillation light alone, may allow a distinction between nuclear recoils and electron recoils. However, the energy to produce an electron ion pair is about seven times higher than in semiconductors, and these detectors will have correspondingly higher thresholds. The loss of coherence for large nuclei is also a potential disadvantage.

Mica. Price and Snowden-Ifft are developing an elegant method using old mica to put limits on WIMP cross section on various elements. The basic idea is to replace a short integration of a year or so with a kilogram worth of material by a very long integration of one billion years with a very small quantity of material (10^{-6} grams). Heavy nuclei tracks damage the mica and with suitable chemical etching pits can be generated along a cleaved surface and measured with an atomic force microscope. Nuclear and alpha recoils can be distinguished by the absence of a coincident track on the other cleaved surface for the first type of recoil. So far Price *et al.* have convinced themselves that the material has not been annealed over the last billion years and they have not observed any background. Surprisingly, this very simple method gives sensitivity limits of the same order of magnitude for spin-dependent interactions as current germanium detectors which only contain a small amount of ^{73}Ge. Note, however, that in most models the spin-dependent cross section is totally negligible with respect to (second order) spin-independent terms; therefore in these cases the current results are not bringing any new information. The group is currently investigating how far they can improve the method before being limited by the inherent fission n background.

Low Pressure Time Projection Chambers. Another interesting approach is being pursued by the UC San Diego group of G. Masek. Following on an initial idea of J. Rich and M. Spiro, they are using a low pressure time projection chamber in a magnetic field. They use an optical readout which gives them a large amount of information at reasonable cost, and are able to distinguish between electron recoils (which are tight spirals) and nuclear recoils (which are more or less straight). Moreover, the nuclear recoil direction is a powerful tool to link possible interaction to the galaxy frame allowing the measurement suggested by Spergel[12].

Buckland and Masek are currently developing the technology, which is very interesting. As a first generation device, however, it seems to be very limited in mass (≈ 1 g for the current cubic meter volume) and would require a large extrapolation before being sensitive enough.

Cryogenic Detectors. A more ambitious approach, but probably necessary if we wish gains in sensitivity of a factor 100 or 1000, is to strive for maximum redundancy, and therefore to use systems with excitation energies much smaller than atomic electrons or electrons and holes in a semiconductor. This is the general idea behind the development of so-called "cryogenic detectors"[68]: Cooper pairs in a superconductor have binding energies of the order of 10^{-3} eV, and phonons in a crystal at 100 mK have energies of 10^{-5} eV. In order not to spontaneously excite these quanta, such detectors have to be operated at very low temperature, and the difficulties of low radioactivity are compounded by the difficulties of a sophisticated technology and of low temperature physics.

Table 1 summarizes the most recent results obtained with large mass cryogenic detectors. The r.m.s. baseline dispersion is the relevant parameter for the value of the threshold, while the resolution, for instance on the 60 keV line, gives an idea of the rejection power against the background.

It can be seen that there are a number of groups developing detectors of a few tens of grams, and their performance is now reaching a level sufficient to begin to search for WIMPs. And indeed many cryogenic detector experiments are being built.

Table 1. Massive calorimeter performance

Group	Mass (gram)	Temperature	Baseline σ	FWHM (60 keV)
Goddard-Wisconsin	10^{-5}g (Si)	100 mK	4 eV	7.5 eV (6 keV)
IAP (Coron)	25g(Sapphire)	60 mK	250 eV	3.6keV
Munich (Umlauf)	100g	30 mK	8 eV/√Hz	?
Berkeley	60g(Ge)	20 mK	Ionizat. 460 eV Thermal 200 eV	1.7 keV 700 eV
Munich (TU&MP)	30g(Sapphire)	15mK	40eV	100 eV (6keV)

The Milano group has operated an experiment in Gran Sasso for some time already, with double beta decay as a focus.

The Center for Particle Astrophysics (UC Berkeley, LBL, UC Santa Barbara, Stanford, Baksan) is putting together a cryogenic experiment using germanium detectors where both ionization and phonons are measured:

• As exemplified in Figure 16, this method provides a discrimination against electron recoils of better than 99% and should allow us to reach the detection limits shown in Figure 14. We envision crystals of typically 170 grams.

• In order to be sensitive to some Majorana dark matter particles, we need a nuclear spin, and will use 95% enriched ^{73}Ge target elements. The Russian Institute of Nuclear Research (Moscow and Baksan) has provided the basic material (850 g), which has been zone refined and grown as crystal at LBL. In order to have a control, we will also use enriched ^{76}Ge and natural germanium. We envision a total target mass of roughly 500 g for each of the three germanium types. We are also considering a natural silicon target.

• We have built a low background, ultra-low temperature environment: we have chosen to complement a commercial (Oxford 400 µW) side-access dilution refrigerator with an appendix (the "Icebox") that brings the cold into a radioactivity-controlled environment (Figure 17), providing us with roughly a cubic foot of space at 10 mK. The icebox has now reached 8 mK.

• Because of the likely difficulty of implementing this totally new technology, we have opted for an underground site in Stanford, which is close to our base but unfortunately relatively shallow (20 meter water equivalent). This small facility is now ready, and we have fully characterized the background, in particular the flux of neutrons produced by the cosmic ray muons. It is tolerable at least for the beginning of the experiment. We have moved the refrigerator, tested it in place, and are starting to remount the icebox.

• Because of this shallowness, we require a special shield which includes both a moderator and active vetoing of the muon traversing the shield. This shield is being constructed.

• Finally, we need the more conventional elements, low noise electronics, monitoring, and data acquisition system, similar to a high energy experiment. Real counting is expected in 1995.

Let us also quote other groups actively working on setting up experiments: The French collaboration is installing in Frejus a low background cryostat and implementing tests with the 25 g Coron detector. Munich has been approved for an experiment in Gran Sasso using a kilogram of Al_2O_3, focusing mostly on the low energy region where the absence of rejection is somewhat less critical (because of higher rates). The University of Tokyo will install in Kamiokande crystals of LiF for which the cross sections are expected to be more favorable than germanium for spin dependent interactions. Finally, the University of Bern is planning an experiment with 200 g of superconducting granules. This is a very active field! It remains that the full deployment of these technologies will certainly take some time, because of the complexity of the solid state physics and materials technology which has to be mastered, and the inconvenience of ultra-low temperature refrigerators in terms of turn around time and radioactivity environment.

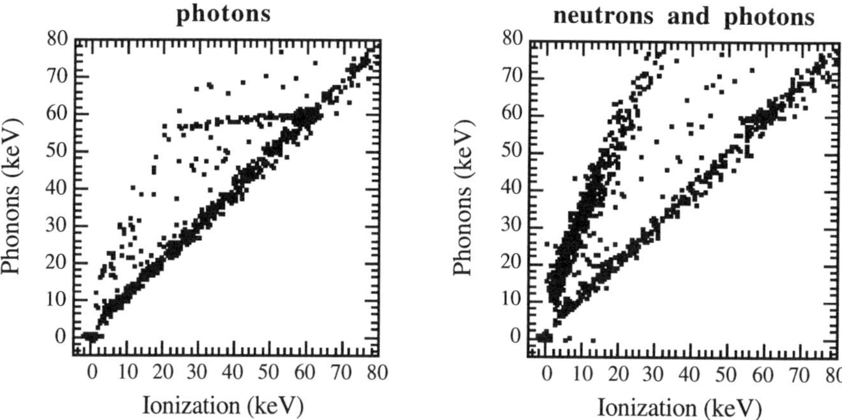

Figure 16. Phonon and ionization equivalent energies measured for (a) 59.5 keV photons and Compton scatters of background photons. (b) The same measurement as (a) with the addition of neutrons and photons from a ^{252}Cf source.

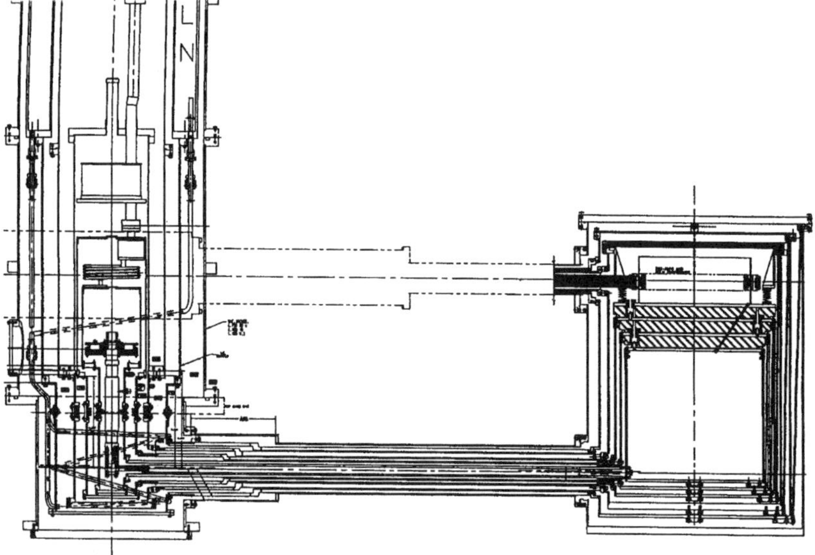

Figure 17. The "Ice Box" of the Center for Particle Astrophysics experiment. A volume of one cubic foot in a very low background environment is brought down to 10 mK by a commercial side access refrigerator. The Ice Box will be surrounded by a shield.

CONCLUSION

Dark matter is an example of the fascinating connections between the infinitely large and infinitely small. Beautiful pages have been written on the subject by Blaise Pascal[69] in the 17th century. They were, however, speculations mostly based on intuition. Modern cosmology and particle astrophysics are beginning to probe these connections experimentally, and the coming years will be particularly exciting, with qualitatively new data being gathered by detailed studies of large scale structure and the direct searches for dark matter, including the MACHOs and the WIMPs. These data are likely to increase even more the tension growing between the observations and our models, and this is likely to lead us to substantial changes in our paradigms. It may even lead, if dark matter can indeed be shown to be nonbaryonic, to the ultimate Copernican revolution[70]: not only are we not at the center of the universe, but we may not even be made of the same stuff as most of the universe!

ACKNOWLEDGMENTS

This review draws very much from many conversations with my colleagues, especially those of the Center for Particle Astrophysics.

This work was supported by the Center for Particle Astrophysics, a National Science Foundation Science and Technology Center operated by the University of California under Cooperative Agreement no. AST-912005.

REFERENCES

1. F. Zwicky, *Helv. Phys. Acta* 6:110 (1933).
2. S.M. Faber and J.S. Gallagher, *Ann. Rev. Astron. Ap.* 17:135 (1979); V. Trimble, "Existence and Nature of Dark Matter in the Universe," *Ann. Rev. Astron. Ap.* 25:245 (1987).
3. V.C. Rubin *et al.*, *Ap. J.* 238:471 (1980).
4. *See, e.g.*, A. Bosma, *Ap. J.* 86:1825 (1981).
5. *See, e.g.*, J.R. Mould, J.B. Oke, and J.M. Nemec, *Ap. J.* 92:153 (1987).
6. Fabian *et al.*, *MNRAS* 221:1049 (1986).
7. W. Forman and C. Jones, *Ann. Rev. Astron. Ap.* 20:547 (1982).
8. *See, e.g.*, L.L. Cowie, M. Henriksen, and R. Mushotzky, "Are the Virial Masses of Clusters Smaller Than We Think?," *Ap. J.* 317:593 (1987).
9. J.A. Tyson, F. Valdes, and R.A. Wenk, "Detection of Systematic Gravitational Lens Galaxy Image Alignments: Mapping Dark Matter in Galaxy Clusters," *Ap. J. Lett.* 349:L1 (1990).
10. M. Milgrom and J. Bekenstein, in *Dark Matter in the Universe*, J. Kormendy and G.R. Knapp, eds., Reidel, Dordrecht, 319 (1986); M. Milgrom, *Ap. J.* 270:365 (1983); R.H. Sanders, *Astron. Ap. Lett.* 136:L21 (1984); R.H. Sanders, *Astron. Ap.* 154:135 (1985).
11. A. Dar, preprint (1991).
12. D. Spergel, Princeton University preprint (1991).
13. *See, e.g.*, S. Weinberg, "Gravitation and Cosmology," (1972); E.W. Kolb and M.S. Turner, *The Early Universe*, Addison-Wesley, Redwood City, California, (1990).
14. J. Huchra and M. Geller, *Science* 246:891 (1989).
15. A. Dressler, S.M. Faber, D. Burstein *et al.*, "Spectroscopy and Photometry of Elliptical Galaxies: A Large Streaming Motion in the Local Universe," *Ap. J. Lett.* 313:L37 (1987); A. Dressler, D. Lynden-Bell, D. Burstein *et al.*, *Ap. J.* 313:42 (1987).
16. M.A. Strauss and M. Davis, "A Redshift Survey of IRAS Galaxies," in *Proceedings of IAU Symposium No. 130, Large Scale Structure of the Universe*, Balaton, Hungary, June, 1987; M. Davis, M.A. Strauss, and A. Yahil, "A Redshift Survey of Iras Galaxies: III Reconstruction of the Velocity and Density Fields," *UCB/SUNY/Cal Tech*, (July, 1990); W. Saunders *et al.*, "The Density Field of the Local Universe," *Nature* 349:32 (1991).

17. P.J.E. Peebles, "The Large Scale Structure of the Universe," Princeton University Press, section 14 (1980).

18. E. Bertschinger and A. Dekel, *Ap. J. Lett.* 336:l5 (1990); A. Dekel, E. Berstchinger, and S.M. Faber, *Ap. J.* 364 (1990); E. Bertschinger, A. Dekel, S.M. Faber *et al.*, *Ap. J.* 364 (1990).

19. A. Dekel, *Ann. Rev. Astr. Astrop.* 32:371 (1994).

20. A. Sandage, *Physics Today* 34 (1970).

21. H.U. Norgaard-Nielsen *et al.*, *Nature* 339:523 (1989); S. Perlmutter *et al.*, *Ap. J. Lett.* in press (1995).

22. L.L. Cowie, "Galaxy Formation and Evolution," *Physica Scripta* (1990).

23. E. Loh and Spillar, *Ap. J.* 303:154 (1986); *Ap. J. Lett.* 307:L1 (1988); E. Loh, *Ap. J.* 329:24 (1988).

24. *See*, e.g., Caditz and Petrosian, *Ap. J. Lett.* 337:L65 (1989); Bahcall and Tremaine, *Ap. J. Lett.* 326:L1 (1988); Omote and Yoshida, *Ap. J.* 361:27 (1990).

25. A. Guth, *Phys. Rev.* D23:347 (1981); A.D. Linde, "Chaotic Inflation," *Phys. Lett.* 129B:177 (1983); A. Albrecht and P.J. Steinhardt, "Cosmology for Grand Unified Theories with Radiatively-Induced Symmetry Breaking," *Phys. Rev. Lett.* 48:1220 (1982).

26. J. Yang *et al.*, "Primordial Nucleosynthesis: A Critical Comparison of Theory and Observation," *Ap. J.* 281:493 (1984); *see the recent reviews by* K.A Olive, D.N. Schramm, G. Steigman, and T. Walker, *Phys. Lett.* B426 (1990); D. Denegri, B. Sadoulet, and M. Spiro, "The Number of Neutrinos Species," *Rev. of Modern Physics* 62:1 (1990). *For a recent review, see* K.A. Olive, "The Quark Hadron Transition in Cosmology and Astrophysics," *Science* 251:1194 (1991).

27. H. Surki-Suonio, R.A Matzner, K.A Olive, and D.N. Schramm, *Ap. J.* 353:406 (1990).

28. J.E. Gunn and B. A. Peterson, *Ap. J.* 142:1633 (1965).

29. J.C. Mather *et al.*, *Ap. J. Lett.* 354:L37 (1990).

30. *See*, e.g., De Zotti,"The x-ray background spectrum," in *Proceedings of the 1991 Moriond Workshop*, Editions Frontières (1992).

31. J. Bahcall *et al.*, *Ap. J.* November (1994).

32. B. Carr and J.R. Primack, *Nature* 345:478 (1990).

33. *See, for instance,* J.R. Primack, "Dark Matter, Galaxies, and Large Scale Structure in the Universe," lectures presented at the International School of Physics "Enrico Fermi," Varenna, Italy, June 26– July 6, 1984, *SLAC-PUB-3387* (1984).

34. P.J.E. Peebles, *Nature* 327:210 (1987).

35. G. Smoot, C. Bennett, A. Kogut, E. Wright *et al.*, "Structure in the COBE DMR First Year Maps," *Ap. J. Lett.* 396:L1 (1992).

36. C. Fisher, M. Davis, M.A. Strauss, A. Yahil *et al.*, "The Power Spectrum of IRAS Galaxies," *Ap. J.*, (1992).

37. S.D.M. White, C.S. Frenk, M. Davis, and G. Efstathiou, *Ap. J.* 313:505 (1987); C.S. Frenk, S.D.M. White, G. Efstathiou, and M Davis, *Ap. J.* 351:10 (1990).

38. S.D. Tremaine and J.E. Gunn, *Phys. Rev. Lett.* 42:407 (1979); D.N Spergel., D.H. Weinberg, and J.R. Gott III, "Can Neutrinos be the Galactic Missing Mass?," Princeton Univ. Observatory preprint, (1988).

39. *See, e.g.,* A. Vilenkin, "Cosmic Strings and Domain Walls," *Phys. Rep.* 121:263 (1985); N. Turok, *Phys. Rev. Lett.* 63:2625 (1989); N. Turok and D.N. Spergel, *Phys. Rev. Lett.* 64:2736 (1990).

40. B. Paczynski, *Ap. J.* 301:503 (1992); K. Griest, C. Alcock, T. Axelrod *et al.*, "Gravitational Microlensing as a Method of Detecting Disk Dark Matter and Disk Stars," *Ap. J.* 366:412 (1991).

41. C. Alcock *et al.*, "Possible Gravitational Microlensing of a Star in the Large Magellanic Cloud," *Nature* 365:621 (1993); E. Aubourg *et al.*,"Evidence for Gravitational Microlensing by Dark Objects in the Galactic Halo," *Nature* 365:623 (1993); A. Udalski*et al.*, "The Optical Gravitational Lensing Experiment: Discovery of the First Candidate Microlensing Event in the Direction of the Galactic Bulge," *Acta Astronomica* 43:289 (1993).

42. C. Alcock *et al.*, "Experimental Limits on the Dark Matter Halo of the Galaxy from Gravitational Microlensing," submitted to *Phys. Rev. Lett.* (1995).

43. *For a first attempt see* E.I. Gates, G. Gyuk, and M.S. Turner, "Microlensing and Halo Cold Dark Matter," *Fermilab-Pub-941381A.*

44. R.D. Peccei and H. Quinn, *Phys. Rev. Lett.* 38:1440 (1977).

45. M.S. Turner, "Windows on the Axion," *Phys. Reports* 197 (1990).

46. S. DePanfilis *et al.*, "Limits on the Abundance and Coupling of Cosmic Axions at 4.5<ma<5.0meV," *Phys. Rev. Lett.* 59:839 (1987); S. DePanfilis *et al.*, *Phys. Rev.* D40:3153 (1989); C.A. Hagmann, "A Search for Cosmic Axions," University of Florida/thesis (1990).

47. KSVZ (Hadronic): J.E. Kim, *Phys. Rev. Lett.* 43:103 (1979); M.A. Shifman, A.I. Vainshtein, and V.I. Zakharov, *Nucl. Phys.* B166:493, (1980); DFFSZ: M. Dine, W. Fischler, and M. Srednicki, *Phys.*

Lett. 104B:199 (1981); A.P. Zhitniskii, *Sov. J. Nucl. Phys.* 31:260 (1980).

48. H. White, talk at the NSAC town meeting, Berkeley, February 4, 1995.

49. L. Krauss, P. Romanelli, and D. Schramm, "The Signal from a Galactic Supernova: Measuring the Tau Neutrino Mass," *Fermilab-Pub-91/293-A* (1991).

50. K.S. Hirata *et al.*, *Phys. Lett.* 280B:146 (1992); T. Kajita, in proceedings of the *Int. Conf. on Frontiers of Neutrino Astrophysics*, Y. Suzuki and K. Nakamura, eds., Takayama/Kamioka, Japan, 1992, Universal Academy Press, Tokyo, 293 (1993); R. Becker-Szendt *al.*, in proceedings of the *Int. Conf. on Frontiers of Neutrino Astrophysics*, Y. Suzuki and K. Nakamura, eds., Takayama/Kamioka, Japan, 1992, Universal Academy Press, Tokyo, 303 (1993).

51. S.P. Mikheyev and M.S. Smirnov, *Nuovo Cim.* 9C:17 (1986); L. Wolfenstein, *Phys. Rev.* D20:2634 (1979).

52. B.W. Lee and S. Weinberg, "Cosmological Lower Bound on Heavy-Neutrino Masses," *Phys. Rev. Lett.* 39:165 (1977). *For details about loopholes see, e.g.,* K. Griest and B. Sadoulet, "Model Independence of Constraints on Dark Matter Particles," in *Proceedings of the Second Particles Astrophysics School on Dark Matter,* Erice, Italy (1990).

53. *See, for instance,* reference 33.

54. B. Sadoulet, "Prospects for Detecting Dark Matter Particles by Elastic Scattering," in proceedings of the *13th Texas Symposium on Relativistic Astrophysics,* M.L. Ulmer, ed., Chicago, Dec. 14–19, 1986, World Scientific, Singapore, 260 (1987); K. Griest and B. Sadoulet, "Model Independence of Constraints on Dark Matter Particles," in proceedings of the *Second Particle Astrophysics School on Dark Matter,* Erice, Italy (1989); J.R. Primack, D. Seckel, and B. Sadoulet, "Detection of Cosmic Dark Matter," *Ann. Rev. Nucl. Part. Sci.* 38:751 (1988); P.F. Smith and J.D. Lewin, "Dark Matter Detection," *Physics Reports* 187:203 (1990); B. Sadoulet, "SUSY from the Sky: The Search for Weakly Interacting Massive Particles," in proceedings of the *Workshop on Supersymmetry*, CERN (1992).

55. L. Krauss, M. Srednicki, and F. Wilczek, "Solar System Constraints on Dark Matter Candidates," *Phys. Rev.* D33:2079 (1986).

56. K. Griest, G. Jungman, and M Kamionkowski, (1994) private communication.

57. S.P. Ahlen *et al.*, *Phys. Lett. B* 195:603 (1987); D.O. Caldwell *et al.*, "Laboratory Limits on Galactic Cold Dark Matter," *Phys. Rev. Lett.* 61:510 (1988); D. Reusser *et al.*, "Limits on Cold Dark Matter from the Gotthard Germanium Experiment," *Phys. Lett.* B235:143 (1991); Moscow-Heidelberg preprint (1993).

58. D.N. Spergel, "The Motion of the Earth and the Detection of WIMPs," *Phys. Rev. D* 37:353 (1988).

59. H.J. Maris and S. Tamura, "Anharmonic Decay and the Propagation of Phonons in an Isotopically Pure Crystal at Low Temperatures: Application to Dark Matter Detection," *Phys. Rev.* B47:727 (1993); T. More and H.J. Maris, "Directionality from Anisotropic Phonon Production in Solid State Dark Matter Detection," Fifth International Workshop on Low Temperature Detectors, Berkeley, 1993, proceedings published in *J. of Low Temperature Phys.* 93:387 (1993).

60. A.K. Drukier, K. Freese, and D.N. Spergel, "Detecting Cold Dark Matter Candidates," *Phys. Rev.* D33:3495 (1986); F. Freese, J. Frieman, and A Gould, "Signal Modulation in Cold Dark Matter Detection," SLAC preprint *SLAC-PUB-4427*, (1987).

61. T. Shutt, B. Ellman *et al.*, "Measurement of Ionization and Phonon Production by Nuclear Recoils in a 60 g Crystal of Germanium at 25 mK," *Phys. Rev. Lett.* 29:3425 (1992); T. Shutt, N. Wang, B. Ellman, Y. Giraud-Heraud *et al.*, "Simultaneous High Resolution of Phonons and Ionization Created by Particle Interactions in a 60 g Germanium Crystal at 25 mK," *Phys. Rev. Lett.* 29:3531 (1992).

62. R. Bernabei *et al.*, *Phys. Lett. B* 293:460 (1992); R. Bernabei *et al.*, *Phys. Lett. B* 295:330 (1992).

63. N. Spooner and P.F. Smith, *Phys. Lett. B* 314:430 (1993).

64. D.O. Caldwell *et al.*, "Searching for the Cosmion by Scattering in Si Detectors," *Phys. Rev. Lett.* 65:1305 (1990).

65. R. Bernabei *et al.*, *Phys. Lett. B* 293:460 (1992); R. Bernabei *et al.*, *Phys. Lett. B* 295:330 (1992).

66. H. Ejiri *et al.*, Osaka University preprint (1992).

67. J. Seguinot, G. Passardi, J. Tischhauser, and T. Ypsilantis, "Liquid Xenon Ionization and Scintillation. Studies for a Totally Active Vector Electromagnetic Calorimeter," CERN preprint *CERN-LAA 92-004* (1992); also D. Cline (1993) private communication.

68. The proceedings of the low temperature detector conferences are a useful source for the reader wanting to follow the recent evolution of the field: *Proceedings of the Workshop on Low Temperature Detectors for Neutrinos and Dark Matter,* K. Pretzl, N. Schmitz, and L. Stodolsky, eds., Springer-Verlag, Berlin, Heidelberg, 150 (1987); *Proceedings of the Third International Workshop on Low Temperature Detectors for Neutrinos and Dark Matter,* L. Brogiato, D.V. Camin, and E. Fiorini,

eds., Gran Sasso, L'Aquila, Italy, Sept. 20–23, 1989, Editions Frontières, Gif-sur-Yvette, France (1990); *Proceedings of the Fourth International Conference of Low Temperature Dark Matter and Neutrino Detectors,* N.E. Booth and G.L. Salmon, eds., Oxford, 1991, Frontières, 91192 Gif-sur-Yvette, France, 147 (1992); *Proceedings of the Fifth International Workshop on Low Temperature Detectors, LTD-5,* Berkeley, CA, July 29–August 3, 1993, *Journal of Low Temperature Physics* 93:393 (1993).

69. B. Pascal, "Les Pensées," #347, 348, 352 in Oeuvres Complètes, Bibliothèque de la Pléiade, NRF, Paris 1954.

70. Courtesy of J. Primack.

SELECTED EXPERIMENTAL RESULTS FROM NEUTRINO PHYSICS

M. SPIRO

C.E. Saclay, DAPNIA/SPP
91191 Gif-sur-Yvette, France

DOWN THE RABBIT HOLES

This is to remind the reader that a large fraction of the experiments aiming to measure neutrino properties require extremely low backgrounds. They are therefore performed in underground laboratories to reduce the cosmic muon flux (figure 1).

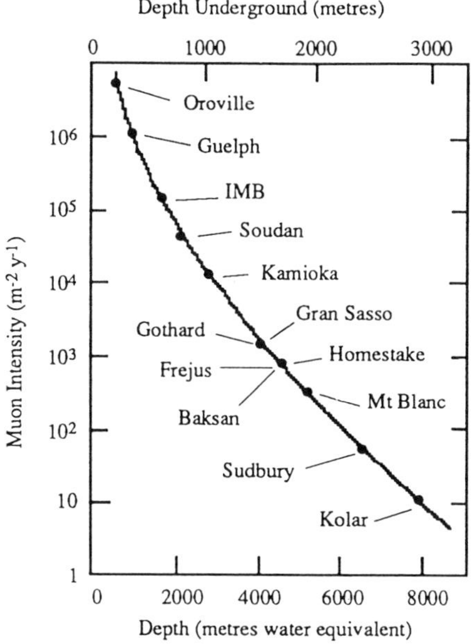

Figure 1. The vertical depth of world-wide underground laboratories (in metres equivalent water)

THROUGH THE LOOKING GLASS (DOUBLE BETA DECAY)

This has to do with the basic questions :
- Are neutrinos and antineutrinos the same particle (CP mirror image) ?
- Are neutrinos Dirac or Majorana particles ?

Frontiers in Particle Physics: Cargèse 1994
Edited by M. Lévy *et al.*, Plenum Press, New York, 1995

- Are neutrinos 4 component or 2 component objects ?
If neutrino are Majorana particles, as it is preferred in some Grand Unified Theories (not in the Standard Model), there exists only a left handed and a right handed neutrino. The neutrino is then its own CP mirror image. Since the weak interaction is mediated by a $(1-\gamma_5)$ current, and since $(1-\gamma_5)$ reduces to helicity only for massless particles, one expects that if the neutrino mass is non zero, there will be always a small admixture of antineutrinos (right handed neutrino) in a state produced by a $(1-\gamma_5)$ weak interaction current. The admixture of the wrong helicity is proportional to $\frac{m_\nu}{E}$ (amplitude).
The best prospect to look whether neutrinos are or are not Majorana particles is to search for ßß0v decay (figure 2b).

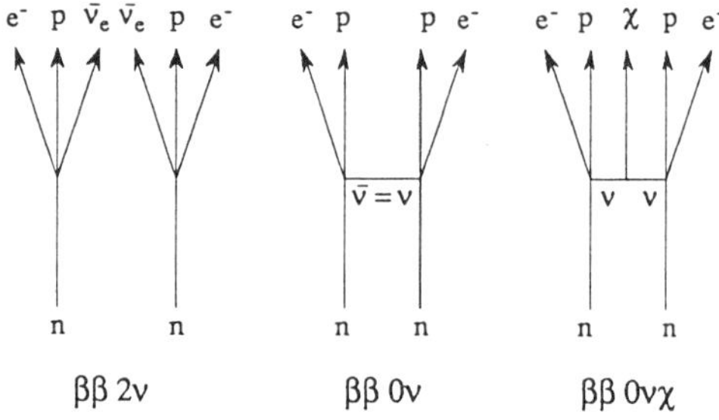

Figure 2. The processes of ßß2v (a), ßß0v (b) and ßß0vχ (c) decay.

In this process, the neutrino emitted at one vertex is absorbed as an antineutrino at the other one provided it is massive and it is a Majorana particle.
Taking into account the possibility of neutrino mixing of different flavours, the half life of the process is inversely proportional to the effective Majorana mass :

$$(T_{1/2})^{-1} \propto <m_\nu>^2 \qquad (1)$$

where, U_{ei} being the mass mixing matrix element for ν_e and the mass eigenstate ν_i :

$$<m_\nu> = \sum m_i U_{ei}^2 \qquad (2)$$

The ßß0v process is the reaction :

$$(A, Z) \rightarrow (A, Z+2) + e^- + e^- \qquad (3)$$

This reaction is only possible if :

$$Q_{\beta\beta} = (A,Z)_{mass} - (A,Z+2)_{mass} > 0 \qquad (4)$$

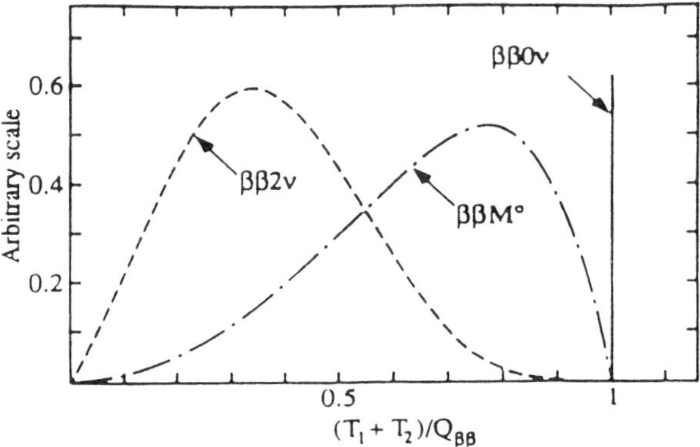

Figure 3. The 3 ßß spectra (T_1 and T_2 are the electron kinetic energies, $Q_{\beta\beta}$ the energy released in the process).

Evidence was also reported at Dallas'92 for ßß2v decay of ^{82}Se, ^{100}Mo and ^{150}Nd based on about 100 events almost background free for each channel [1]. Few events appear in excess of expectations near the endpoint of the sum energy spectra which could have been an evidence for the existence of a Majoron. The technique which is used is based on a TPC. The imaging of the two electrons allow a much more suppressed background than with a Ge crystal semiconductor detector. However the energy resolution is much poorer.

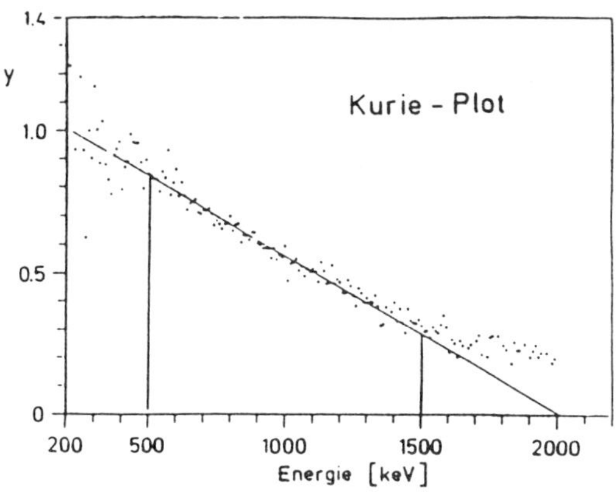

Figure 4a. The Heidelberg-Moscow experiment (^{76}Ge crystal) Kurie plot (≈ 4000 events)

The signature of such a process (figure 3) is a peak at $Q_{\beta\beta}$ for the sum energy of the two electrons.
While searching for this peak, one may encounter :
1. background coming from natural radioactivity inside or outside the detector. This is why the experiments are located underground.
2. background due to the Standard Model allowed transition ßß2v (figure 2a) with a continuum for the sum energy of the two electrons which is shown on figure 3.
3. "background" due to the possibility of Majoron emission (GUT violating lepton number object, $\Delta L = 2$, figure 2c), which should also induce a continuum for the sum energy of the two electrons (figure 3).

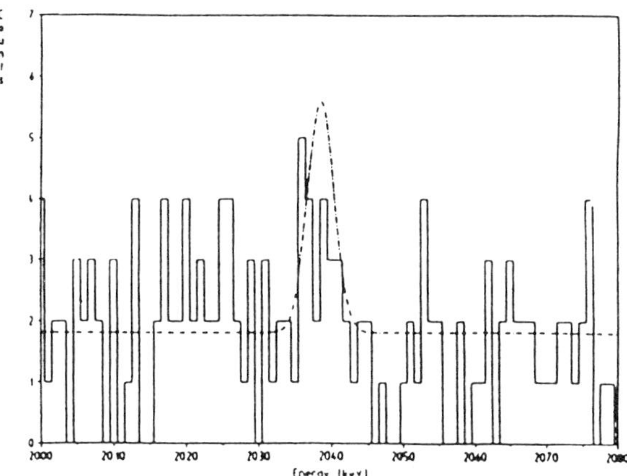

Figure 4b. 2σ excess of events in the endpoint region (2013 ± 31 keV) of the Heidelberg-Moscow experiment (^{76}Ge crystal).

At Dallas'92 [1] evidence for ßß2ν decay of ^{76}Ge (Ge crystal detector) was reported on the basis of ≈ 400 events. At this conference the Heidelberg-Moscow collaboration reported a signal based on ≈ 4000 events (after background subtraction with a signal to noise ratio of about one. The Kurie plot obtained is shown on figure 4a) which allow a detailed comparison of the energy spectrum with the expectations. A small excess (2σ) is seen near the endpoint of the Kurie plot which is not yet understood. For ßß0ν transition a lower limit on the lifetime is derived, $T_{1/2} > 1.5\ 10^{24}$ years at 90% C.L. together with an upper limit on <$m_ν$> of 1.4 eV. However one must add that in the precise region where we expect the signal, there is a 2σ excess of events (figure 4b).

At this conference new results were presented by the NEMO collaboration (in Fréjus) on ^{100}Mo (figure 5) obtained also by an imaging technique (Geiger, drift tubes and scintillator walls). The spectrum (≈ 455 events) is in good agreement with expectations and does not show any evidence for Majoron emission. The limit on <$m_ν$> from the absence of ßß0ν signal in this experiment is ≈ 7 eV.

In conclusion :

- ßß2ν signals are seen for 4 nuclei. They should help in better understanding the estimates of nuclear matrix elements.

- There is so far no convincing evidence of deviations of the spectra from expectations which could be attributed to Majoron emission.

- The absence of ßß0ν signal (although there is a 2σ excess in the Heidelberg-Moscow ^{76}Ge experiment) yields an upper limit of a few eV for <$m_ν$> if the neutrino is a Majorana particle.

- There are hopes to reach a 0.1 eV sensitivity in the next five years depending on the ultimate background limitation.

- This region of <$m_ν$> from 10^{-1} eV to few eV is of particular interest if one believes that, for instance, the $ν_τ$ is contributing significantly to the mass-energy density of the universe (≈ 10 eV), if it is a Majorana particle and if it has a mixing matrix element $U_{eτ}^2$ in the range of 10^{-2} to 10^{-1}, which is not yet excluded by any $ν_e \leftrightarrow ν_τ$ oscillation experiment.

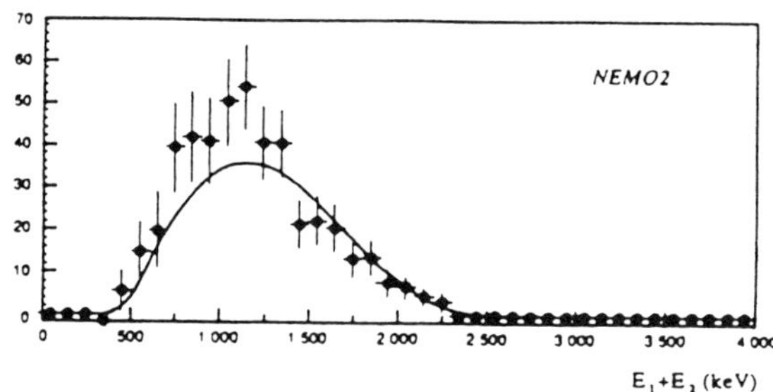

Figure 5. ^{100}Mo ßß2ν spectrum (≈ 455 events, E_1 and E_2 are the kinetic energies of the emitted electrons). The error bars take into account the external background.

A MAD TEA PARTY (ATMOSPHERIC NEUTRINOS)

The study of neutrinos produced in the high atmosphere offers a possibility to search for neutrino oscillations occurring during their travel through the earth and atmosphere in a range varying between about 10 and 13.000 km. As they are mainly produced in the decays of charged pions and muons, they consist of twice more $ν_μ$ or $\barν_μ$ than $ν_e$ or $\barν_e$.

Therefore, the fluxes, the energy and angular distributions of these particles are sensitive to $ν_μ \leftrightarrow ν_e$ or $ν_μ \leftrightarrow ν_τ$ oscillations if the mass squared difference between two neutrino flavours is larger than 10^{-4} eV^2 and, due to systematic uncertainties, if the corresponding mixing angles are such that $\sin^2 2θ$ is larger than about 0.4.

The interactions of atmospheric neutrinos are observed in large underground detectors where both $ν_μ$ and $ν_e$ interactions can be identified. Moreover high energy charged current $ν_μ$ interactions in the earth surrounding the detectors, producing upward or nearly horizontal going muons may be separated from the high rate of downward going atmospheric muons. During the recent years more than one thousand neutrino interactions occurring inside the underground detectors have been measured, and a similar amount of v_μ earth interactions were observed. Several reviews have been recently presented [2] [3] [4] on the results obtained by these experiments.

Atmospheric neutrino flux calculations

The atmospheric v_μ and v_e fluxes, their energy and angular distributions must be calculated as accurately as possible in order to find experimentally deviations which could be attributed to neutrino oscillations.
Starting from primary cosmic ray fluxes, their interactions on nitrogen and oxygen nuclei, and the propagation of the decay pions and muons in the atmosphere, many groups ([5] to [12]) have calculated the fluxes of v_μ and v_e reaching the earth, taking into account the geomagnetic effects and the time dependent solar activity. The results may be summarised as follows :

a) Mainly due to uncertainties on cosmic primary flux and composition, the total neutrino flux is known to about 20% at low energy ($\lesssim 10\,\mathrm{GeV}$) and to 30% at high energy ($\gtrsim 10$ GeV).

b) The composition of the atmospheric neutrinos agrees in all the models and the uncertainty on the flux ratio v_μ/v_e is probably smaller than 5% ([5] to [12])

c) The shapes of the energy and angular distributions of the atmospheric neutrinos are rather well predicted.

d) The interaction rate of the v_μ in the earth depends also on their cross section. This introduces an additional uncertainty of about 10% [13]. However, the shapes of the energy and angular distributions of the muons produced in the earth are reliably calculated [13].

Experimental studies

Detection of neutrino interactions inside underground detectors. These detectors originally designed to study the nucleon decays are in principle able to measure and to identify the v_μ and v_e charged current interactions and in some cases, with a smaller efficiency, the neutral current interactions. The interaction rate is of the order of 100 events/Kt.year. The large Cerenkov detectors experiments (1 to 5 Kt of fiducial volume for the Kamiokande ([14] to [19]) and I.M.B. [15]) have analysed 4.9 and 7.7 Kt.years of data respectively. The energy thresholds vary between 0.1 to 0.2 GeV for the v_e and 0.2 to 0.3 GeV for the v_μ interactions. In order to reach a good identification efficiency of the electrons and muons, the single ring events, fully contained in the detector are selected, limiting the energy to about 1.4 GeV. This allows the detection of the decay electron of the muons with a pulse delayed by the muon lifetime. The tracking calorimeter experiments have analysed 0.3, 1.6 and 0.5 Kt.year of data in the NUSEX [16], Fréjus [17] and Soudan 2 [18] detectors respectively. These experiments are in principle able to separate all v_μ and v_e interactions and, according to their atmospheric muon background, to make use of the events produced in the fiducial volume, but not necessarily fully contained in the detector. The threshold energy varies between 0.2 and 0.3 GeV for v_μ and v_e interactions.

Detection of the v_μ interactions in the earth surrounding the detector.
a) Upward going muons (zenith angle $\theta_z > 90°$, energy larger than $\approx 10\,\mathrm{GeV}$) : In order to separate the muons produced by v_μ interactions in the earth from the huge flux of downward going atmospheric muons an excellent separation in directionality is required. This is obtained by the water Cerenkov detectors Kamiokande [19] and IMB [20], and by time of flight in scintillator telescopes (Baksan experiment and MACRO). Until now, only results from the Baksan experiment have been published [21]. The IMB experiment [20] has also recorded the upward going muons stopping in the detector; they correspond to v_μ energy of the order of 50 GeV [13].
b) The nearly horizontal neutrinos may be separated without directionality information from the downward going muons in an angular region, which depends on the depth of the detector and the shape of the ground over the laboratory. This corresponds in the Fréjus experiment

[22] to $75° < \theta_z < 105°$ in which the muon rate amounts to about 60% of the upward going muons.

Results on atmospheric neutrino flavour composition and oscillations

v_μ and v_e **fluxes.** In each experiment, the data are compared to the expectation obtained by a Monte Carlo simulation, taking into account the response of the detector for each type of interaction, the calculated neutrino fluxes, their energy and angular distributions.
The published results give the following ratios ("v_μ" and "v_e" represent the rate of the interactions) :

$$R_T = \frac{(v_\mu + v_e)_{\text{simul.}}}{(v_\mu + v_e)_{\text{data}}} \qquad (5)$$

$$R_S = \frac{\dfrac{\text{single ring (prong)}}{\text{multi rings (prongs)}}\,\text{data}}{\dfrac{\text{single ring (prong)}}{\text{multi rings (prongs)}}\,\text{simul.}} \qquad (6)$$

$$R = \frac{(v_\mu / v_e)_{\text{data}}}{(v_\mu / v_e)_{\text{simul.}}} \qquad (7)$$

$$R_D = \frac{\dfrac{\mu\ \text{decay}}{\text{no}\ \mu\ \text{decay}}\,\text{data}}{\dfrac{\mu\ \text{decay}}{\text{no}\ \mu\ \text{decay}}\,\text{simul.}} \qquad \text{(in Cerenkov exp.)} \quad (8)$$

The values of these ratios are represented in figure 6. The calculated fluxes are all normalised to the Bartol flux predictions [7]. The errors shown are purely statistical. The following comments can be made on these results.
a) As indicated in the figure 6 caption, the event selections applied by the experiments to obtain these results are very different. In particular only 30% of the neutrino interactions are common in the Fréjus and the water Cerenkov experiments.

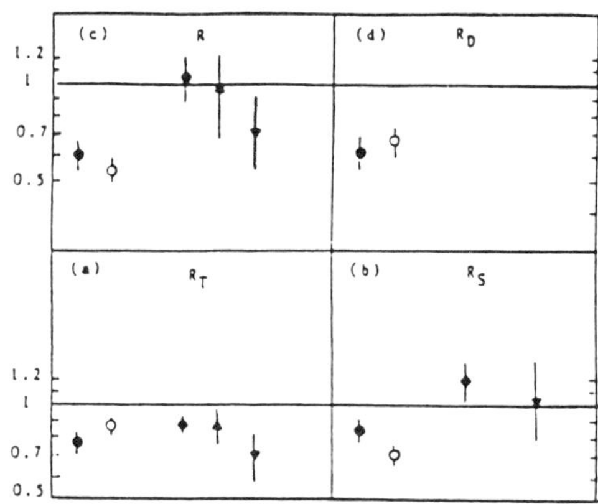

Figure 6. Data over simulation rate ratios for v interactions produced inside the detector.

● Kamiokande, 310 one ring, 147 mutiring fully contained events [19]
○ IMB, 610 one ring, 325 multiring fully contained events [15]
◆ Fréjus, 188 events (except 1 uncontained prong), 70 one prong events [17]
▲ NUSEX, 50 fully contained events [16]
▼ Soudan 2, 25 one prong, 12 multiprong, fully contained events [18]

b) The ratios R_T (figure 6a) are compatible and lower than unity in all experiments, suggesting that the Bartol flux [7] used in the simulations may be slightly overestimated but compatible with the estimated systematic error.

c) The flux independent ratios R_S (figure 6b) are lower than unity in the water Cerenkov experiments (especially in the IMB results where it differs by more than 5 standard deviations from unity). This shows that the fraction of the single ring events is not well described by the simulation in the IMB experiment. In case of a deficit of ν_μ, for which the energy threshold is higher than for ν_e, it is expected that the ratio R_S should be larger than unity.

d) The ratios R (figure 6c) are definitely lower than unity in the water Cerenkov but compatible with unity in the tracking calorimeter experiments. This has been interpreted by the Kamiokande group as an indication for neutrino oscillations $\nu_\mu \leftrightarrow \nu_e$ or $\nu_\mu \leftrightarrow \nu_\tau$. This deficit of ν_μ interactions is also visible in the ratio R_D (figure 6d) which is also lower than unity.

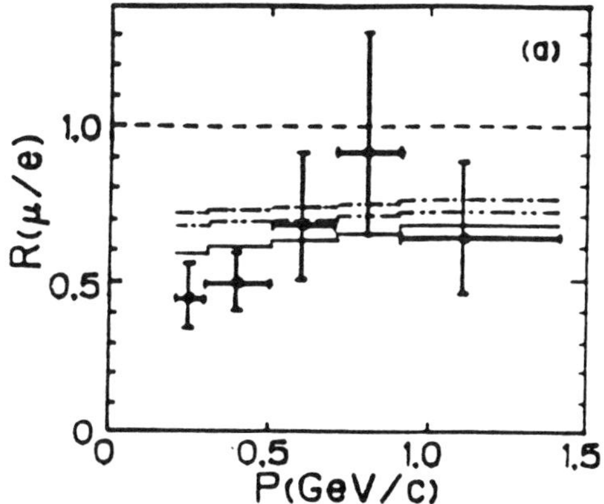

Figure 7a. Momentum dependence of the ratio R in the Kamioka experiment. The dashed and dashed-dotted lines correspond to no oscillation and some oscillation hypotheses

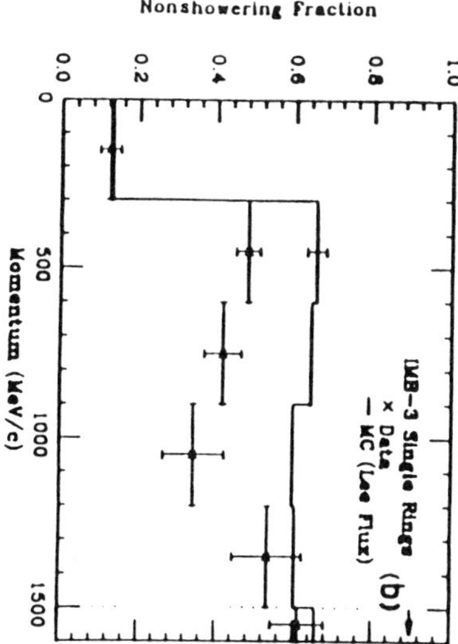

Figure 7b. Non showering fraction of events versus momentum in IMB experiment. The full line corresponds to no oscillation.

Beside the statistical errors on R plotted in figure 6c, some systematic errors have been evaluated to take into account the possible misidentification of muons and electrons which could be different in the data and in the simulation. These errors, R_{sys}, are 0.05, 0.12, 0.15 and 0.10 in the Kamioka, IMB, Fréjus and Soudan 2 experiments respectively (In the Fréjus experiment [23] they have been checked by making three independent simulations and analyses of the data). An important question is to know how reliable are these errors. An accelerator test of the electrons and muons identification efficiency will be performed in a water Cerenkov detector in a near future [4] to clarify this point.

Energy and zenith angle distribution. The v_μ deficit found in the water Cerenkov experiments is clearly visible in the shape of the energy distribution of R (figure 7a) in the Kamioka experiment and to a lesser degree in the IMB experiment (figure 7b). However the shape of the zenithal angle θ_z distributions do not show any evidence for an angular dependence of this deficit ([14] to [20]. The corresponding Fréjus distributions in energy and zenith angle are compatible, within the statistical accuracy with the Monte Carlo simulations [17].

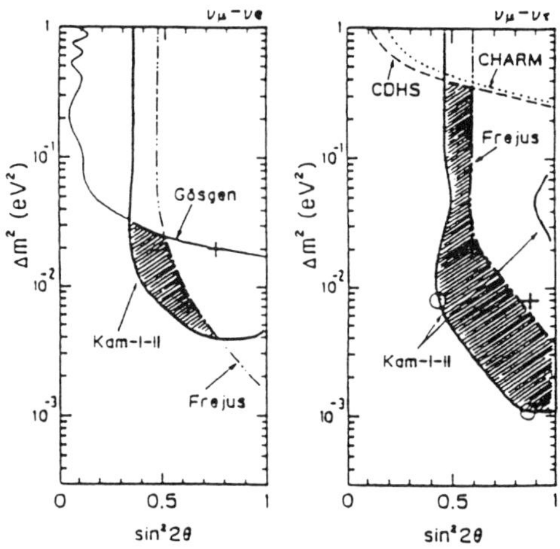

Figure 8. Allowed neutrino oscillation parameters from Kamioka experiment (right of the full line) and from Fréjus experiment (left of the dashed line) for $v_\mu \leftrightarrow v_e$ an $v_\mu \leftrightarrow v_\tau$.

Neutrino oscillation analyses. Assuming that the differences between the data and the simulations are due to neutrino oscillations, some exclusion plots in the diagrams Δm^2-$\sin^2 2\theta$ have been calculated by the Kamioka [14] and the Fréjus [23] experiments. The errors used are the quadratic sum of statistical and systematic errors and the 90% C.L. exclusion regions have been determined for $v_\mu \leftrightarrow v_e$ and $v_\mu \leftrightarrow v_\tau$ oscillations (figure 8a and 8b). The Kamioka analysis requires an oscillation in the region $\Delta m^2 \gtrsim 4.10^{-3} \text{eV}^2$, $\sin^2 2\theta \gtrsim 0.4$ for $v_\mu \leftrightarrow v_e$ or $\Delta m^2 \gtrsim 10^{-3} \text{eV}^2$, $\sin^2 2\theta \gtrsim 0.4$ for $v_\mu \leftrightarrow v_\tau$, while the Fréjus analysis excludes the region $\Delta m^2 \lesssim 3.10^{-3} \text{eV}^2$, $\sin^2 2\theta \lesssim 0.5$ for $v_\mu \leftrightarrow v_e$ and $\Delta m^2 \lesssim 6.10\text{-}3 \text{eV}^2$, $\sin^2 2\theta \lesssim 0.6$ for $v_\mu \leftrightarrow v_\tau$. The presence of oscillations which could explain the v_μ deficit in the Kamioka experiment does not change appreciably the shapes of the energy (figure 7a) and angular distributions [4].

Results on upward and horizontal going muons and neutrino oscillations.

A detailed review of the results obtained by the Kamioka [19], [4], IMB [20] and Baksan [21] experiments has been recently made by the Bartol-Penn. group [13]. A result on the rate of horizontal muons obtained in the Fréjus experiment [22] will also be included in the report.

Upward and horizontal going muons fluxes. In each experiment, the observed muon rate is compared to the predicted one and their ratio r,

$$r = \frac{\text{observed rate}}{\text{predicted rate}} \qquad (9)$$

has been measured using the energy and angle dependent ν_μ flux calculated by Volkova [5]. With other models of ν_μ flux and recent neutrino cross sections, the predicted rate may vary by about 30% [13]. The results on r with the statistical errors are presented in figure 9a. All values are compatible with unity, but somewhat lower than 1.3 expected by the larger Bartol flux. The shape of the angular distributions presented by Kamioka and Baksan are found to be compatible with the distributions calculated by the various models [13].

The IMB experiment has measured [20] the ratio of the muon rates stopping in the detector and crossing it :

$$f = \frac{\text{stopping muon rate}}{\text{through muon rate}} = 0.16 \pm 0.02 \qquad (10)$$

This ratio is almost independent of the ν_μ flux and is calculated to be 0.158 ± 0.050. Therefore the ratio :

$$r_s = \frac{f_{\text{data}}}{f_{\text{simul.}}} \qquad (11)$$

presented in figure 8b is found in agreement with unity.

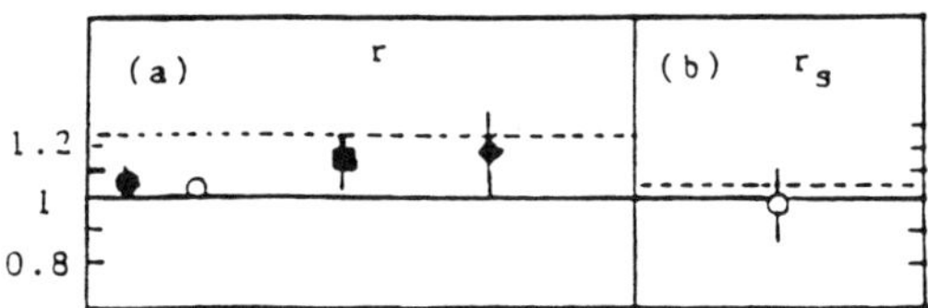

Figure 9. Data over simulation rate ratios over the ν_μ interactions produced in earth
● Kamiokande, 252 up going muons
○ IMB, 617 up going muons, 85 stopping up going muons
■ Baksan, 421 upgoing muons
◆ Fréjus, 55 horizontal muons
The dashed lines correspond to the maximum prediction.

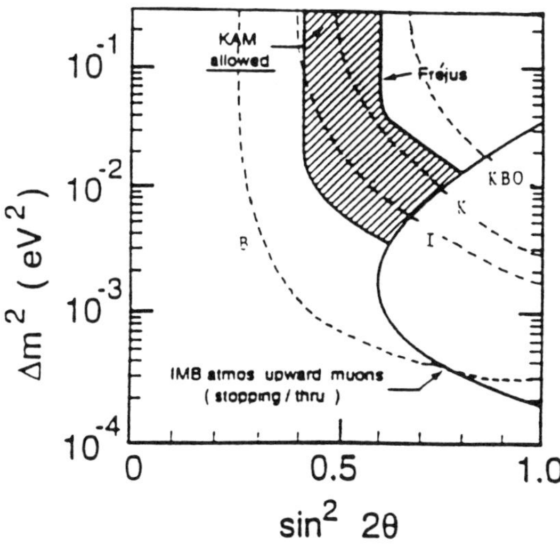

Figure 10. Allowed $\nu_\mu \leftrightarrow \nu_\tau$ oscillation parameter regions for the atmospheric interactions in detector (Kamioka and Fréjus experiments) and for interactions in earth. The dashed lines represent the B, K, and I limits obtained by Baksan, Kamioka, and IMB respectively and calculated with the Volkova flux. The line labelled KBO is calculated with the Kamioka data, the Bartol flux and Owen cross-section. The allowed regions lie to the left of the dashed curves. The full line is obtained by the IMB experiment, with the fraction of stopping up going muons.

Neutrino oscillation analyses. No evidence for ν_μ deficit is found in the results of upward going muon fluxes. In order to see whether or not these results were in contradiction with the ν deficit found in the composition of the neutrinos interacting in the detector, some exclusion zones in the oscillation parameter plot were calculated by the Kamioka, IMB and Baksan groups assuming various ν_μ flux models. Some of these zones exclude completely the previous allowed region (figure 10). However, the region remains untouched by the Kamiokande limit calculated with the Bartol flux and the Owen's cross section [13]. The excluded zone for $2.10^{-4}\,\mathrm{eV}^2 \lesssim \Delta m^2 \lesssim 2.10^{-2}\,\mathrm{eV}^2$ and $\sin^2 2\theta \gtrsim 0.7$ is due to the absence of discrepancy between the experimental and the calculated value of the ratio f.

Conclusion. The evidence for oscillation of the atmospheric neutrinos is still far to be compelling. No effect is found in the results of the ν_μ interaction in the earth. The deficit of ν_μ in the Cerenkov detector interactions might suggest an oscillation $\nu_\mu \leftrightarrow \nu_e$ or $\nu_\mu \leftrightarrow \nu_\tau$ in the region $\Delta m^2 \gtrsim 10^{-2}$ to $0.4\,\mathrm{eV}^2$, $\sin^2 2\theta \approx 0.5$. However, this conclusion must be confirmed by a check of the systematic uncertainties of the Monte Carlo simulations used to demonstrate this deficit, and by a better understanding of possible experimental systematic errors.

THE KING AND THE QUEEN OF THE SUN (SOLAR NEUTRINOS)

The most firm and solid prediction we have on the solar neutrino flux is based on energy conservation and steady state of the sun. We know that these two well admitted assumptions imply that the total power radiated by the solar surface (the luminosity $L.$) should be equal to the thermonuclear power generated by the fusion of hydrogen into helium. For four protons to combine into a ^{4}He nucleus, two electrons must be involved in the initial state for electric charge conservation, and then two ν_e must be emitted in the final state. The overall reaction is then:

$$4\,p + 2e^- \rightarrow {}^4He + 2\nu_e + 27\,\mathrm{MeV} \qquad (12)$$

where 27 MeV is the difference of the masses between the particles involved in the initial state and those involved in the final state (the energy of the neutrinos and the kinetic energy of the nuclei can be neglected in this approximate relation). It is then easy to derive the total flux of neutrinos expected to reach the earth :

$$N_\nu = \frac{2L.}{27\,\mathrm{MeV}\;4\pi\,d^2} = 6.5\;10^{10}\,\mathrm{cm}^{-2}\,\mathrm{s}^{-1} \qquad (13)$$

where d is the distance from the Earth to the Sun. Gallium target detectors are so far the most appropriate to measure the total number of neutrinos. This is because :
 - of the very low threshold (233 keV) of the capture reaction $\nu_e + {}^{71}Ga \rightarrow {}^{71}Ge + e^-$ which makes Gallium target detector sensitive to the bulk of the solar neutrino energy spectrum
 - of the high natural abundance of the stable ^{71}Ga isotope (40%)
 - of the relatively easy identification of even a few radioactive ^{71}Ge atoms in a large quantity of Gallium (30 tons).
However the firm prediction on the total number of solar neutrinos is not enough to compute the capture rate of solar neutrinos on a given target nucleus. To compute the energy spectrum one needs to go through solar modelling and through the exact chain of reactions which combine hydrogen into helium. There are mostly three cycles of reactions : ppI, ppII and ppIII.
 - In ppI the two neutrinos are coming from the $pp \rightarrow {}^2H + e^+ + \nu_e$ reaction ($2\,\nu_{pp}$)
 - In ppII one neutrino is a ν_{pp}, the other comes from the decay (through electron capture) of ^{7}Be (ν_{7Be})
 - In ppIII one neutrino is a ν_{pp}, the other comes from the ß decay of ^{8}B (ν_{8B}).
The ν_{pp} spectrum extends from 0 to 450 keV. Only the Gallium experiments are sensitive to those neutrinos. The ν_{7Be} are monoenergetic with a line at 860 keV. Both the Gallium and Chlorine ($\nu_e + {}^{37}Cl \rightarrow {}^{37}Ar + e^-$, threshold 820 keV) are sensitive to those neutrinos.
Finally the ν_{8B} neutrino spectrum extends from 0 to 15 MeV. All the presently running experiments (Gallium, Chlorine and Kamiokande) are sensitive to them. The Kamiokande experiment is based on the detection of the recoil electron in the elastic scattering of a ν_e with an experimental threshold of about 7 MeV on the energy of the recoil electron. From the

Solar Standard Model [24] the v_{pp}, v_{7Be} and v_{8B} intensities are computed to be 90%, 8% and 10^{-4} of the total flux. Although the relative intensity of the v_{8B} neutrino is very small, they contribute significantly to the capture rate, even in the Gallium experiment, due to their high energy. Notice however that the v_{7Be} and v_{8B} fluxes are highly sensitive to the ingredients of the SSM. If, for instance, one changes the input parameters, with, as a result, a change in the central temperature T_c prediction, it has been shown that the v_{8B} flux will vary as T_c^{18}, the v_{7Be} as T_c^8 and the v_{pp} flux only as $T_c^{-1.2}$.

The predictions of the SSMs are shown in Table 1, for the Gallium experiments, in terms of SNU (Solar Neutrino Units). One SNU corresponds to a capture rate of 10^{-36} per second per target nucleus (in this case for Ga).

We see that the Bahcall et al. SSM which is generally considered as giving high SNU values, predicts fluxes only slightly higher than the Turck-Chièze et al. SSM which is generally considered as giving low SNU values. So one might say that the predictions of the SSM for Gallium experiments are rather firm. Notice also, that although the v_{pp} are expected to represent 90% of the total flux of solar neutrinos, their contribution to the capture rate amounts only to 71 SNU out of 127. This is due to their low energy.

Table 1. Standard Solar Models predictions from Turck-Chièze et al, and Bahcall et al., for the gallium experiment.

	Capture rate (SNU)	
Source	Turck-Chièze et al	Bahcall et al
pp	70.6	71.3
pep	2.795	3.07
^{7}Be	30.6	32.9
^{8}B	9.31	12.31
^{13}N	3.87	2.68
^{15}O	6.50	4.28
^{17}F	-	0.04
Total	**124**	**127**

Are the predictions right ?

v_{8B} **flux.** The Kamiokande experiment uses a water Cerenkov detector. The basic process is neutrino scattering on electrons which then give detectable Cerenkov light. They measure two quantities, the energy of the recoil electron and its direction. A clear peak can be seen in the direction of the Sun and the excess in that direction is then taken as coming from solar neutrinos (figure 11). However the flux of v_{8B} they measure since January 1987 [25] is only 0.54 ± 0.08 of the Bahcall et al. SSM, so $0.5 \ 10^{-5}$ of the total flux.

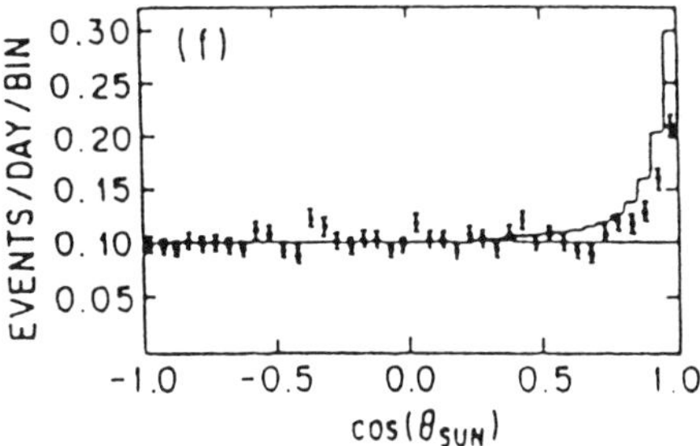

Figure 11. Counts of the Kamiokande detector plotted against the cosines of the angle of the electron to the sun's direction (cosθ_{SUN}, 1557 days of data, electron energies greater than 9.3 MeV for 449 days, 7.5 MeV for 794 days and 7.0 MeV for 314 days).

v_{7Be} **flux.** Since 1967 Davis and co-workers have performed a pioneering experiment by extracting ^{37}Ar from a tank of 615 tons of tetrachloroethylene (C_2Cl_4). The ^{37}Ar decays by electron capture. The resulting hole in the K shell can give X rays and Auger electrons with a total energy of 2.8 keV. The counter of 0.5 cm³ volume is designed to measure this energy. The half life of the decay is 35 days. A typical run consists of a 50 day exposure of the big tank followed by an extraction of the Argon atoms which are then introduced in the small counter. The counting lasts for 260 days. For the period 1970-1984 the data were analysed and give 339 counts of ^{37}Ar. This gives a non-corrected ^{37}Ar counting rate of 5 per run. The data are analysed by a maximum likelihood method assuming a flat background (as a function of time) plus a ^{37}Ar decaying component.

The result [26] is 3.6±0.4 times lower (figure 12) than expected in the Bahcall et al. SSM. This implies, taking into account the fact that the experiment is sensitive to both the v_{7Be} and v_{8B} components and taking into account the Kamiokande result (reduction of a factor 2 on the v_{8B} component) that the v_{7Be} flux is lower by a factor >4 than the prediction of Bahcall et al. SSM.

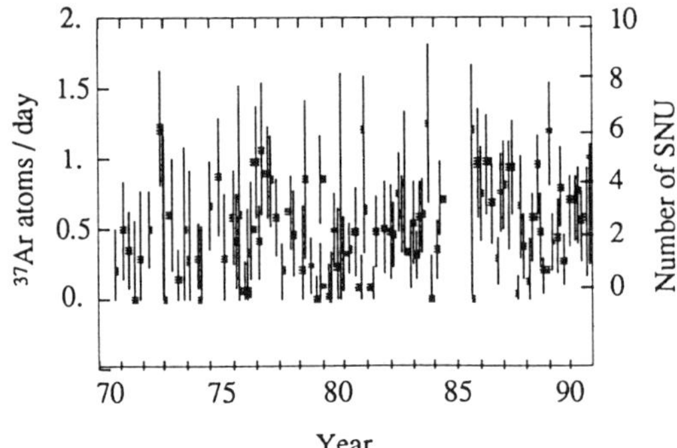

Figure 12. ^{37}Ar production rate in the Homestake chlorine solar neutrino detector (0.4 atom/d corresponds to 2 SNU).

These deficits are the basis of the solar neutrino problem. The reductions are very hard to reconcile with any modification of the SSM since we expect that any reduction on the v_{7Be} component should be accompanied by a stronger reduction for the v_{8B} component [30]. Neutrino masses and mixing could reconcile these reductions with the SSM through v_e, v_μ, v_τ oscillations. However, before invoking new physics in the neutrino sector, the results of the gallium experiments were eagerly awaited. The expectations are much less sensitive to SSM and we can derive absolute lower limits for the capture rate based only on energy conservation and steady state of the Sun.

Consistent predictions for gallium experiments

Since we know experimentally that the flux of v_{8B} is reduced by a factor two and that the flux of v_{7Be} is reduced by a factor greater than 4 compared to SSM predictions, we can deduce that the number of v_{pp} should be increased to 1.08±0.02 of the SSM to conserve the total number of neutrinos insuring energy conservation. One obtains expectations which range from 80 to 105 SNU. These are not SSM predictions but rather predictions which are consistent with the basic understanding of the Sun (energy conservation and steady state) and with the two experimental results coming from the Chlorine and Kamiokande experiments.

Results of the gallium experiments

Two experiments are now underway, SAGE in Russia which published the first results in January 1991 and GALLEX in Italy, which published their first results in June 1992. The recipes are the same : introduce 1 mg of inactive stable germanium in the 30 tons of Gallium, expose the Gallium to solar neutrinos in a low background environment, extract by a chemical method the solar neutrinos produced ^{71}Ge atoms together with the inactive

Germanium, transform into a counting gas (GeH$_4$), fill a proportional counter and count the decays of ^{71}Ge (11 d. half life). The main difference is that the SAGE experiment uses metallic liquid Gallium target while the GALLEX experiment uses an acidic aqueous Gallium Chloride solution. This induces important differences in the chemistry.

SAGE. The Soviet-American Gallium Experiment is located in the Baksan Valley in the Caucasus mountains (Russia) under about 4700 meter water equivalent. The expected rate for 30 tons target and 132 SNU is 1.2 ^{71}Ge atom created per day. Taking into account all the efficiencies, one expect only 3 counts per run (a run is 4 week exposure) due to ^{71}Ge K electron capture (^{71}Ge + e^-_K → ^{71}Ga + ν + X-rays + Auger electrons). Most of the runs in 1990 have preferred values of 0 SNU. Altogether they published in 1991 [27] a preferred value of 20 SNU with upper limits of 55 SNU (68% C.L.) and 79 SNU (90% C.L.). More recently they announced the results they obtained in the last runs when they increased the total mass of Gallium from 30 tons to 60 tons [1]. This is shown on figure 13. A signal seems now to emerge. In 1992 at Dallas, the quoted result was 58 ± 20(stat.)± 14(sys.) SNU [1]. It is now 70 ± 19 (stat.)± 10 (sys.) SNU [28] .

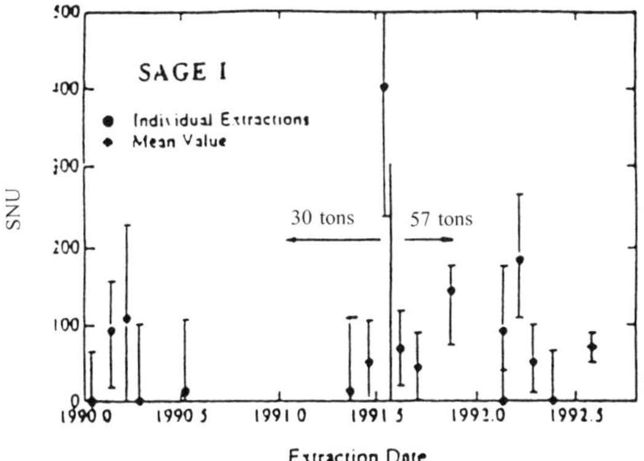

Figure 13. Results for all runs of the SAGE experiment.. The last point on the right shows the combined result.

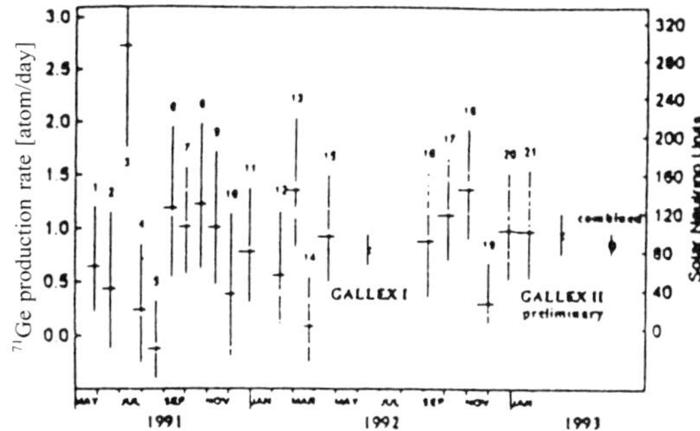

Figure 14. Final results the first period GALLEX I (before may 1992) and preliminary results for the second period GALLEX II (after august 1992). The combined values for GALLEX I, GALLEX II and GALLEX I + GALLEX II are also shown.

GALLEX. This experiment is located in the Gran Sasso Underground Laboratory in Italy. The 30 tons of Gallium are in the form of a solution of GaCl$_3$ acidified in HCl. The Ge atoms form the volatile compound GeCl$_4$. At the end of 3 week exposures, these molecules are swept out by bubbling a large flow of inert gas (N$_2$) through the solution. The experiment is sensitive to both K-shell and L-shell electron captures in the decay of ^{71}Ge atoms. Seven counts are then expected after each run, in the K and L regions. The data used in the analysis consist of 21 runs taken from may 1991 to may 1993. They are now published [29], [31]. There is compelling evidence for a signal : the peaks in energy at 1.2 keV and 10 keV for L and K electron capture are seen, the 11.3 half life of ^{71}Ge is well identified over a flat background. Figure 14 shows the results for all runs which have to be compared with the combined result of 83 ± 20 SNU, released in June 1992 and now updated at the level of 79 ± 13 (stat.) ± 5 (sys.) SNU [31]. Furthermore, GALLEX should be calibrated with an artificial neutrino source (2 MCi) in 1994.

Interpretations

Table 2. Summary table of solar neutrino experiment results (chlorine, Kamiokande and GALLEX) with the comparison to Turck-Chièze et al. and Bahcall et al. SSMs.

Experiment		Exp. Results	Turck-Chièze et al.	Bahcall et al.
Chlorine	(SNU)	2.33 ± 0.25	6.4 ± 1.4	7.2 ± 0.9
	(%)		36 ± 4	33 ± 3
Kamiokande	(%)		64 ± 8	54 ± 8
GALLEX	(SNU)	79 ± 15	123 ± 7	127^{-5}_{+7}
	(%)		64 ± 12	62 ± 12

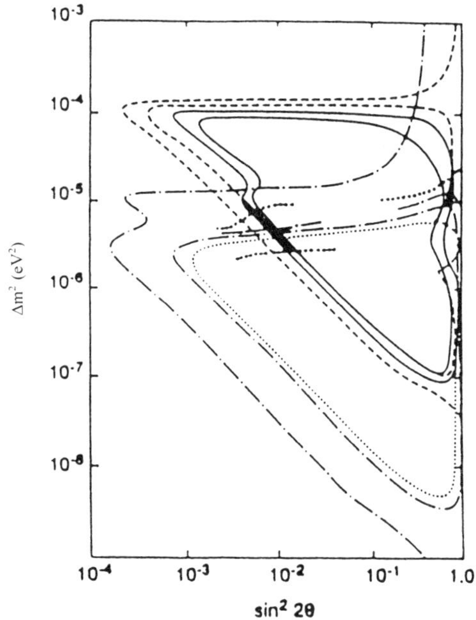

Figure 15. MSW diagram showing the preferred solution to reconcile experimental neutrino flux measurements and Standard Solar Models.

The SSM is unable to account for the deficit of solar neutrinos as observed by the Chlorine and Kamiokande experiments. However, on the basis of these experiments it is impossible to decide whether these discrepancies come from new physics in the neutrino sector or wrong ingredients in the Solar Standard Models. The Gallium experiments are in a much better position to do so. First, the predictions of the SSM are more stable to changes in the ingredients (120 to 140 SNU) and second it is impossible to have predictions below 80 SNU from basic simple principles. Consistent predictions for Gallium experiments which agree with these basic principles and with the deficits of solar neutrinos observed by the Chlorine and Kamiokande experiments are in the 80-105 SNU range, in agreement with the values measured by GALLEX.

By comparing the deficits of solar neutrinos as observed by the 3 experiments (table 2), the indications which favour neutrino oscillations are the facts that the chlorine experiment has a significantly larger suppression factor than the other experiments and that the Gallium experiments give results near the minimum needed to account for the sun luminosity, thus leaving not much room for ^{7}Be neutrinos. This would imply a more severe suppression for ^{7}Be neutrinos than for ^{8}B and pp neutrinos which cannot easily be accommodated by a modification of the SSM.

A decrease of the central temperature will produce a suppression factor for ^{8}B neutrinos which is larger than for ^{7}Be or pp neutrinos [30].

On the contrary, oscillations (MSW effect) could reconcile the SSM with all 3 experiments. Figure 15 shows the allowed range for neutrino masses and mixing angles. The preferred solution is for $\Delta m^2 \approx 7.10^{-6}\,\text{eV}^2$ and $\sin^2 2\theta \approx 6.10^{-3}$.

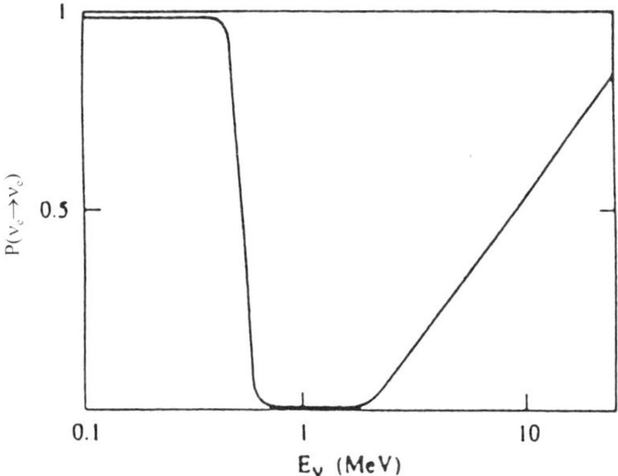

Figure 16. Suppression factor of neutrino flux as a function of the neutrino energy for the MSW preferred solution.

The suppression factor as a function of the neutrino energy is shown on figure 16 for this solution. It implies a distortion of the ^{8}B neutrino energy spectrum.

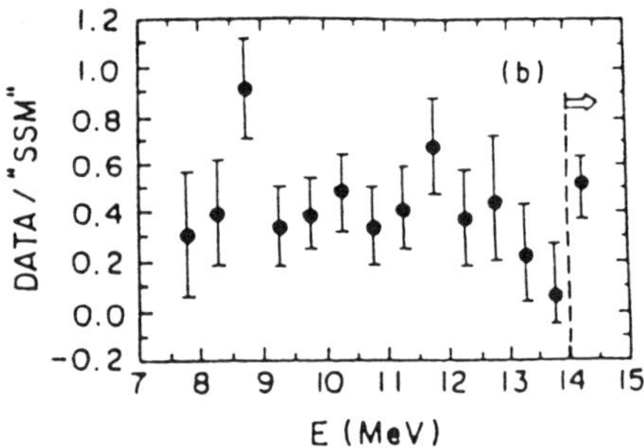

Figure 17. Recoil electron energy spectrum as observed in Kamiokande, normalised to the SSM prediction from Bahcall and Ulrich [33].

Figure 17 shows the recoil electron energy spectrum as observed in Kamiokande [32], normalised to the SSM prediction from Bahcall and Ulrich [33]. The recoil electron energy spectrum is only a smeared reflection of the neutrino energy spectrum.

- A flat suppression (no ν oscillation) gives a χ^2 of $16.3/13$.

- A suppression factor as predicted by the preferred MSW solution gives a χ^2 of $18.6/13$.

It is clear that one has to wait for the Superkamiokande and SNO experiments to establish or reject the small mixing angle MSW solution.

No firm conclusion on neutrino masses can yet be drawn from the present status of solar neutrino experiments and solar modelling. This may not be the case, hopefully, in few years from now when we may expect to have more input to solar models (nuclear cross sections, helioseismology...), better understanding of running experiments (calibrations) and more experiments (SNO, SuperKamiokande...).

ALICE'S EVIDENCE (COSMOLOGY)

The measurement of primordial fluctuations in the cosmic microwave background radiation energy by COBE combined with other measurements allows us to estimate the power spectrum of fluctuations in the universe from Gigaparsec scale down to Megaparsec scale (figure 18).

There seems to be now a consensus that a simple cold dark matter (CDM) cannot match the power spectrum on both large and small scales. Either there is too little power on large scales or too much power on small scales [34]. In vogue recently has been the idea that a mixed dark matter scenario with 70% WIMPs and 30% neutrinos (in energy density) might fit all scales. As a result we would have :

$$\Omega_{\mathrm{WIMPs}} = 0.65$$

$$\Omega_{\nu} = 0.30$$

$$\Omega_{\mathrm{baryons}} = 0.05$$

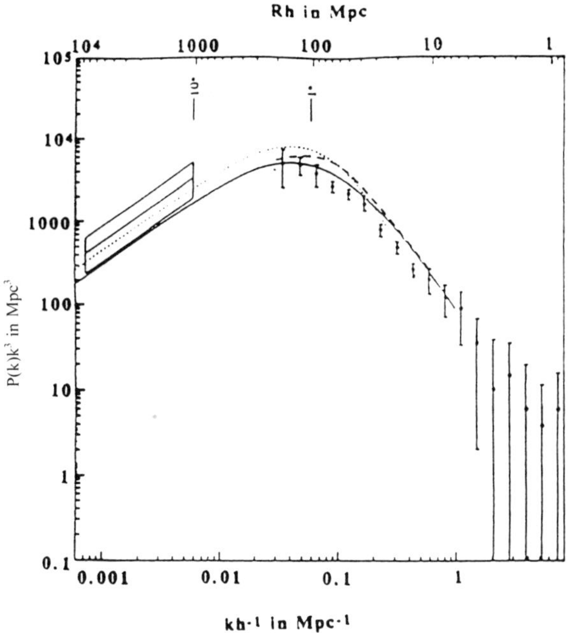

Figure 18. the power spectrum of fluctuations in the universe from Gigaparsec scale down to Megaparsec scale.

Although this fit is in agreement with all the data, one must recognise that it is now a 3 parameter fit which sounds a little bit artificial and suspicious. There are no known deep reasons to have Ω_{WIMPs} , Ω_v and $\Omega_{baryons}$ of the same magnitude.

Table 3. Present neutrino mass limits.

	Method	Reaction	Limit (95% C.L.)
ν_e	Tritium decay endpoint	$T \rightarrow {}^3He + e + \nu_e$	7.2 eV
ν_μ	π momentum + mass	$\pi_{stop} \rightarrow \mu + \nu_\mu$	270 keV
ν_τ	τ decay endpoint	$\tau \rightarrow 5\pi^{\pm} + \nu_\tau$	31 MeV

Nevertheless if we takes $\Omega_v = 0.30$ seriously, the prime candidate is a 7 eV neutrino. Present limits from direct mass measurements are shown in table 3.

Although ν_e, ν_μ and ν_τ are all compatible with a 7 eV mass, there is a theoretical prejudice that the ν_τ would be the heaviest and then the preferred candidate.

We are waiting now eagerly for the results of the NOMAD and CHORUS experiments which should be sensitive to such ν_τ masses, provided that the mixing angle $\sin^2 2\theta$ between the ν_μ and the ν_τ is larger than 10^{-4} or between the ν_e and the ν_τ is larger than 10^{-2} (figure 19).

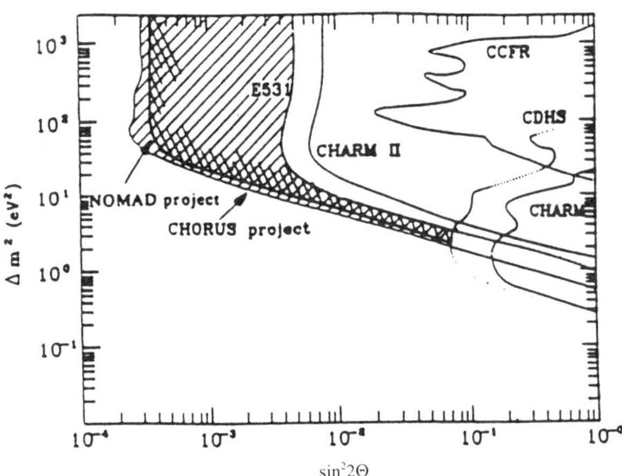

Figure 19. ($\sin^2 2\theta$, Δm^2) diagram with sensitivities of CHORUS and NOMAD (90% C.L.) for $\nu_\mu \leftrightarrow \nu_\tau$ oscillations.

HUMPTY DUMPTY (17 KEV NEUTRINO)

There has been in the past much evidence for a 17 keV neutrino. Unfortunately, it had a great fall and it will be hard to put it together again, since the reasons for the initial claim have been now understood [1], [35].

CONCLUSION

Neutrino physics is a rich and promising field. We have not yet reached compelling evidence for neutrino physics beyond the standard model. Neutrino masses can all be zero. Nevertheless there are hints for possible new phenomena which need further investigation.

ACKNOWLEDGEMENTS :

We are grateful to R.Barloutaud for providing us with the section on atmospheric neutrinos which can also be found in [36]. We are also grateful to J.Rich for illuminating discussions and corrections to the manuscript.

REFERENCES

[1]   R. G. H. Robertson, *Proceedings of the XXVI Int. Conf. on High Energy Physics*, Dallas 1992, p. 140
[2]   J. Schneps, *Neutrino '92 proceedings*, Granada 1992
[3]   E. W. Beier et al., *Phys. Lett.* B283 (1992)
[4]   Y. Totsuka, *Neutrino '92 proceedings*, Granada 1992
[5]   L. V. Volkova, *Yad. Fiz.* 31 (1980) 1510
[6]   T. K. Gaisser et al., *Phys. Rev.* D38 (1988) 85
[7]   G. Barr et al., *Phys. Rev.* D39 (1989) 3532
[8]   E. V. Bugaev et al., *Phys. Lett.* B232 (1989) 391
[9]   A. V. Butkevich et al., *Yad. Fiz.* 50 (1989) 142
[10] M. Honda et al., *Phys. Lett.* B248 (1990) 883
[11] H. Lee et al., *Nuovo Cimento* 105B (1990) 193
[12] M. Kawasaki et al., *Phys. Rev.* D43 (1991) 2900
[13] W. Frati et al., *Internal Report BA 92-71*, UPR 0218 E (1992)
[14] K. S. Hirata et al., *Phys. Lett.* B205 (1988) 416,
      K. S. Hirata et al., *Phys. Lett.* B280 (1992) 146
[15] R. Becker-Szendy et al., *Phys. Rev.* D46 (1992) 3720
[16] H. Aglietta et al., *Europhysics Lett.* 8 (1989) 611
[17] Ch. Berger et al., *Phys. Lett.* B227 (1989) 489

[18] M. Goodman (Soudan 2 coll.), *Internal Report FDK-540* (1992)
[19] Y. Oyama (Kamiokande coll.), *Moriond Jan. 92 proc.*, p.59
[20] R. Becker-Szendy et al., *Phys. Rev. Lett.* 69 (1992) 1010
[21] M. M. Boliev et al., *Venice Worshop Proceedings* (1991) 235
[22] H. Meyer (Fréjus coll.), *Moriond Jan. 92 proceedings*, p.169
      Y. Wei, *Thesis*, University of Wuppertal (1993)
[23] Ch. Berger et al., *Phys. Lett.* B245 (1990) 305
[24] S. Turck-Chièze et al, *Astrophys. J.* 335 (1988) 415
      J. N. Bahcall and W. H. Press, *Astrophys. J.* 370 (1991) 730
[25] A. Suzuki, *KEK preprint 93-96*, August 1993
[26] S. Turck-Chièze et al., *Phys. Rep.* 230 (1993) 59
[27] A. I. Abazov et al., *Phys. Rev. Lett.* 67 (1991) 3332
[28] V. Gavrin, *Communication at TAUP93*, Gran Sasso, sept. 1993
[29] P. Anselman et al., *Phys. Lett.* B285 (1992) 376
[30] P. Anselman et al., *Phys. Lett.* B285 (1992) 390
[31] P. Anselman et al., *Phys. Lett.* B327 (1994) 377
[32] Y. Totsuka, *Proceedings of Texas/PASCOS '92*, p. 344
[33] J. N. Bahcall and R.K.Ulrich, *Rev. of Mod. Phys.*,60 (1988) 297
[34] L. Krauss, *XXVIIIth rencontres de Moriond*, Villars-sur-Ollon, January 30- February 6, 1993
[35] A. Hime, *Phys. Lett.* B299 (1993) 165
[36] R. Barloutaud, *XXVIIIth rencontres de Moriond*, Villars-sur-Ollon, January 30- February 6, 1993

CONFORMAL FIELD THEORY

Vladimir Dotsenko

LPTHE
Université Pierre et Marie Curie
Université Denis Diderot
Bte 126, 4 Place Jussieu
75252 Paris CEDEX 05, FRANCE

INTRODUCTION

Conformal Field Theory (CFT) became a general technique in quantum field theory and its applications. One could say, in a sense, that for the critical phenomena in 2D statistical systems, and also for the string theory, the CFT plays the role similar to that which quantum mechanics plays for atomic physics. Other areas of theoretical physics where CFT is being used are 2D quantum field theory models, 2D quantum gravity, topological theories, condense matter physics (Kondo problem, quantum Hall effect being particular examples). In addition, there are numerous connections to pure mathematics: infinite dimensional Lie algebras and theory of their representations, quantum groups, etc.

These lectures are intended to provide an introduction and, at the same time, present all the basic structures of the CFT. This is by using the basic and simplest case, that of the Minimal Conformal Theory (MCT), which could be defined as a massless 2D quantum field theory with no extra (isotopic) symmetries except for the conformal ones (this is in a broad sense, including naturally the trivially conformal transformations like translation, rotation, dilatation). After the exposure of the general techniques in case of MCT it will be shown, in the last lecture, how the things generalize in case of conformal theories with extra symmetries. This will be done again by using the simplest nontrivial example, that of $SU(2)$ Wess-Zumino model which is CFT with extra, isotopic symmetries, generated by the corresponding current algebra.

For convenience, the presentation of CFT will use the framework of a critical phenomena theory of statistical physics, for which, to the present day, the application of CFT is most profound.

The contents of the lectures will be the following:

1.Minimal Conformal Theory

2.Free field representation for Minimal Conformal

3.Conformal Field Theory based on current algebras - $SU(2)$ Wess-Zumino theory and its free field representation.

Frontiers in Particle Physics: Cargèse 1994
Edited by M. Lévy *et al.*, Plenum Press, New York, 1995

1. MINIMAL CONFORMAL THEORY

At the critical point of a given statistical system the corresponding field theory is massless and is described by a set of local fields, or operators, like energy operator, local order parameter operators, with their scaling dimensions:

$$\{\Phi_i(x),\ \triangle_i\} \tag{1}$$

Massless theory, in general, is invariant w.r.t. global scaling transformation:

$$x \rightarrow \lambda x \tag{2}$$

$$\Phi_i(x) \rightarrow \tilde{\Phi}_i(x) = (\lambda)^{\Delta_i}\Phi(\lambda x) \tag{3}$$

Under this transformation the correlation functions of the theory

$$\langle \Phi_1(x_1)\Phi_2(x_2)...\Phi_N(x_N) \rangle \tag{4}$$

stay invariant. This fixes in particular the form of the two-point functions:

$$\langle \Phi_i(x)\Phi_i(x') \rangle = \frac{const}{|x-x'|}2\Delta_i \tag{5}$$

Generalization of scaling symmetry to the conformal one, for the critical phenomena theory, was suggested by A.M.Polyakov [1]. Conformal transformations of space are defined by the requirement that, locally, the infinitesimal lengths just scale:

$$x^\mu \rightarrow \tilde{x^\mu}(x) \tag{6}$$

Such that

$$(dx^\mu)^2 \rightarrow (d\tilde{x^\mu})^2 = (\lambda(x))^2(dx^\mu)^2 \tag{7}$$

Correspondingly, for the fields generalization of (3) will be:

$$\Phi(x) \rightarrow \tilde{\Phi}(x) = (\lambda(x))^\Delta \Phi(\tilde{x}(x)) \tag{8}$$

so that, locally, it is a scale transformation, but rescaling is different from point to point, λ is x dependent.

Apart from trivially conformal transformations which are translation, rotation, put together:

$$\tilde{x}^\mu = \omega^{\mu\nu}x^\nu + b^\mu \tag{9}$$

and global scaling

$$\tilde{x}^\mu = \lambda x^\mu, \lambda = const \tag{10}$$

one has inversion:

$$x^\mu \rightarrow \tilde{x}^\mu = \frac{x^\mu}{(x)^2} \tag{11}$$

which is conformal. One could check that (7) holds. One defines a special conformal transformation of space as a combination (successive, i.e. performed one after another) of inversion, translation by a vector α^μ, and inversion again. This gives, as one easily checks:

$$\tilde{x}^\mu = \frac{\frac{x^\mu}{x^2} + \alpha^\mu}{\left(\frac{x^\mu}{x^2} + \alpha^\mu\right)^2} \tag{12}$$

or equivalently

$$\frac{\tilde{x}^\mu}{(\tilde{x})^2} = \frac{x^\mu}{x^2} + \alpha^\mu \tag{13}$$

By expending in α^μ, assuming it is small, and keeping terms linear in α^μ one gets infinitesimal form of the transformation. For variations of points of the space

$$\delta x^\mu = \tilde{x}^\mu - x^\mu \tag{14}$$

One gets from (12)

$$\delta x^\mu \approx x^2 \alpha^\mu - 2(\alpha x) x^\mu \tag{15}$$

One checks that

$$(d\tilde{x})^2 \approx (1 - 4(\alpha x))(dx)^2 \tag{16}$$

which means that

$$\lambda(x) \approx 1 - 2(\alpha x) \tag{17}$$

By assuming this symmetry for the correlation functions one could show [1] that, in particular, two and three point functions have to have the following form:

$$\langle \Phi_i(x)\Phi_j(x') \rangle = const \frac{\delta_{ij}}{|x - x'|^{2\Delta_i}} \tag{18}$$

which is a kind of orthogonality, in which it is assumed that $i \neq j$ means $\Delta_i \neq \Delta_j$, and

$$\langle \Phi_1(x_1)\Phi_2(x_2)\Phi_3(x_3) \rangle = \frac{const}{|x_{12}|^{\Delta_1+\Delta_2-\Delta_3}|x_{13}|^{\Delta_1+\Delta_3-\Delta_2}|x_{23}|^{\Delta_2+\Delta_3-\Delta_1}} \tag{19}$$

where $|x_{12}| = |x_1 - x_2|$, etc.

For conformal transformations the two-dimensional space is special because the group of conformal transformations, presented above, could be extended to an infinite-dimensional one. In fact, in 2D one could introduce the complex coordinates

$$z = x_1 + ix_2, \bar{z} = x_1 - ix_2 \tag{20}$$

and then transformation of space points

$$z \to \tilde{z} = f(z) \tag{21}$$

$$\bar{z} \to \tilde{\bar{z}} = \overline{f(z)} \tag{22}$$

where $f(z)$ is any analytic function, is conformal. In fact:

$$(dx^\mu)^2 = dz d\bar{z} = |dz|^2 \tag{23}$$

$$|d\tilde{z}|^2 = |\frac{df(z)}{dz}|^2 |dz|^2 \tag{24}$$

$$\lambda(x) \equiv \lambda(z, \bar{z}) = |\frac{df(z)}{dz}| \tag{25}$$

For operators one has

$$\Phi_i(z, \bar{z}) \to \tilde{\Phi}_i(z, \bar{z}) = |\frac{df(z)}{dz}|^{\Delta_i} \Phi_i(f(z), \overline{f(z)}) \tag{26}$$

Conformal field theory 2D which is based on the assumption of symmetry w.r.t. these transformations was formulated by A.A.Belavin, A.M.Polyakov, A.B.Zamolodchikov

[2]. From now on we restrict ourselves to two dimensions and begin to outline this theory. But let us first notice that the special conformal transformation in 2D takes simple form of

$$\tilde{z} = \frac{z}{1 - \alpha z} \tag{27}$$

($\alpha = \alpha_1 + i\alpha_2$). Infinitesimally, from (27) one gets

$$\delta z = \alpha z^2 \tag{28}$$

Combined with translation, rotation and dilatation global scaling (dilatation) one gets

$$\tilde{z} = \frac{az + b}{cz + d} \tag{29}$$

This represents transformation of the finite dimensional subgroup in 2D.

In the general case, the infinitesimal conformal transformation could be presented as:

$$f(z) = z + \alpha(z) \tag{30}$$

$$\tilde{z} = z + \alpha(z), \ \delta z = \alpha(z) \tag{31}$$

Here $\alpha(z)$ is an analytic function which could be expanded into a series, regular in the origin $z = 0$:

$$\alpha(z) = \sum_{n=-1}^{\infty} \alpha_n z^{n+1} \tag{32}$$

The coefficients $\{\alpha_n\}$ could be regarded as an infinite set of parameters of the transformation. For the operators one gets

$$\Phi_{\Delta,\bar{\Delta}}(z,\bar{z}) \to \tilde{\Phi}_{\Delta,\bar{\Delta}}(z,\bar{z}) = (f'(z))^{\Delta}(\overline{f'(z)})^{\bar{\Delta}}\Phi_{\Delta,\bar{\Delta}}(f(z),\overline{f(z)}) \tag{33}$$

One assumes here that, as the scaling factor factorizes in 2D an z and $\bar{z}$ parts, that the scaling (conformal) dimensions Δ and $\bar{\Delta}$ of a given operator could, in principle, be different.

As, for (30),

$$f' = 1 + \alpha'(z) \tag{34}$$

One gets for

$$\delta\Phi = \tilde{\Phi} - \Phi \tag{35}$$

$$\delta\Phi_{\Delta,\bar{\Delta}} = (\alpha(z)\partial_z + \alpha'(z)\Delta + \overline{\alpha(z)}\partial_{\bar{z}} + \overline{\alpha'(z)}\bar{\Delta})\Phi_{\Delta,\bar{\Delta}}(z,\bar{z}) \tag{36}$$

To keep $\alpha(z)$ really small, and transformation infinitesimal, one could use the following trick: to perform the transformations (31),(36) with a given analytic $\alpha(t)$, with coefficients $\{\alpha_n\}$ in (32) being small, – this is just in a finite region D around the origin, Fig.1. Outside D $\alpha(z)$ is set equal to zero. So the transformation is actually singular at the boundary C.Consequently, in the following, there will be boundary terms.

Now we shall look for the consequences of this symmetry, w.r.t. transformation defined above, for the correlation functions of the theory. Let us assume that they are given by a functional integral (F I)

$$\langle \Phi_1(z_1\bar{z}_1)\Phi_2(z_2\bar{z}_2)...\Phi_N(z_N,\bar{z_N})$$
$$= \frac{\int D\varphi \exp -A[\varphi]\Phi_1\Phi_2...\Phi_N}{\int D\varphi \exp\{-A[\varphi]\}} \tag{37}$$

where φ is some basic field and $A[\varphi]$ is its action; the conformal operators (fields) $\Phi_1, \Phi_2, ..., \Phi_N$ arc assumed to be some composites of it. Next we do variation of $\varphi(z, \bar{z})$ under FI, the one which corresponds to conformal transformation, infinitesimal one, confined to the region D, Fig.1, as discussed above. This should not change the value of the FI, as we only redefine the integration variable. But there will be two pieces, one coming from variation of the action $A[\varphi]$, for the reason that boundary C of D moves under conformal transformation of space points (31), and the second term is produced by variations of the fields $\{\Phi_i\}$. Together they should give zero. In short, one gets Wards Identity (WI):

$$\frac{1}{2\pi} \oint ds^\mu \alpha^\nu \langle T_{\mu\nu} \Phi_1 \Phi_2 ... \rangle = \sum_{k=1}^{N} \langle \Phi_1 \Phi_2 ... \delta\Phi_k ... \rangle \tag{38}$$

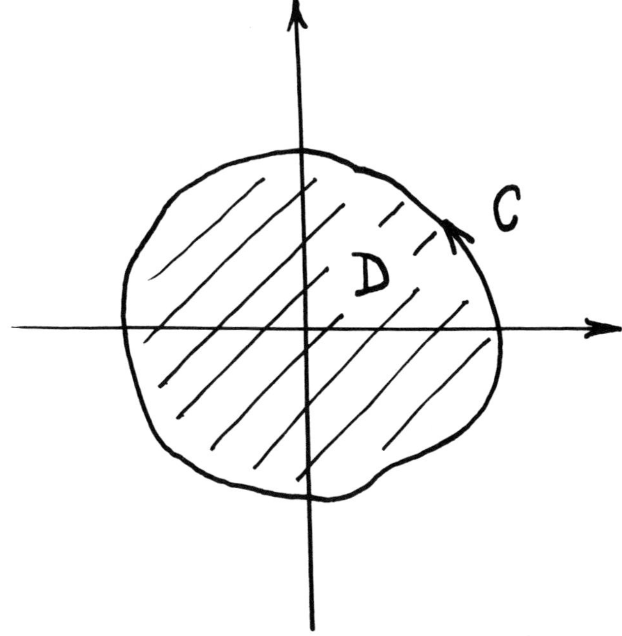

Figure 1: Finite region D where $\delta z = \alpha(z)$, $\delta\bar{z} = \overline{\alpha(z)}$ are defined.

Here $T_{\mu\nu}$ is the stress-energy tensor which is due to the variation of the action, eventually due to variation of the boundary C. The coefficient $\frac{1}{2\pi}$ is a particular choice of normalization of $T_{\mu\nu}$. Variations of fields $\{\delta\Phi_k\}$ are defined above (36). In complex coordinates the WI becomes:

$$\frac{1}{2\pi i} \oint_c d\xi [\alpha(\xi)\langle T_{zz}(\xi, \bar{\xi})\Phi_1 \Phi_2 ...\rangle + \overline{\alpha(\xi)}\langle T_{z\bar{z}}(\xi, \bar{\xi})\Phi_1 \Phi_2 ...\rangle]$$

$$- \frac{1}{2\pi i} \oint d\bar{\xi} [\overline{\alpha(\xi)}\langle T_{\bar{z}\bar{z}}(\xi, \bar{\xi})\Phi_1 \Phi_2 ...\rangle + \alpha(\xi)\langle T_{\bar{z}z}(\xi, \bar{\xi})\Phi_1 \Phi_2 ...\rangle] =$$

$$\sum_{k=1}^{N} [\alpha(z_k)\partial_k + \alpha'(z_k)\Delta_k + \overline{\alpha(z_k)}\bar{\partial}_k + \alpha'(\bar{z}_k)\bar{\Delta}_k]\langle \Phi_1 \Phi_2 ...\rangle \tag{39}$$

Here

$$T_{zz} = T_{11} - T_{22} - 2iT_{12} \tag{40}$$

$$T_{\bar{z}\bar{z}} = T_{11} - T_{22} + 2iT_{12} \tag{41}$$

$$T_{z\bar{z}} = T_{\bar{z}z} = T_{11} + T_{22} \tag{42}$$

If one takes a particular choice of, first,

$$\alpha(z) = a \tag{43}$$

and, secondly,

$$\alpha(z) = b(z - z_0) \tag{44}$$

with a, b, z_0 being constant (z independent) parameters, one gets from (39)

$$\partial_{\bar{z}} \langle T_{zz}(z, \bar{z})\Phi_1\Phi_2...\rangle = 0 \tag{45}$$

and

$$\langle T_{\bar{z}z}(z, \bar{z})\Phi_1\Phi_2...\rangle = 0 \tag{46}$$

and similar expressions for the conjugate components. As this holds for correlation functions of T with any set of operators one has, as a consequence of (45),(46), that in general, in operator sense:

$$T_{zz} = T(z), T_{\bar{z}\bar{z}} = \overline{T}(\bar{z}) \tag{47}$$

i.e., in particular, T_{zz} is holomorphic (only z dependent) and

$$T_{\bar{z}z} = T_{z\bar{z}} = 0 \tag{48}$$

Using this partial information of WI, obtained for a particular choice of $\alpha(z)$, the general WI simplifies. One could see that it actually breaks on independent z and $\bar{z}$ dependent parts. Then we could keep just the holomorphic part which is

$$\frac{1}{2\pi i} \oint_c d\xi\alpha(\xi)\langle T(\xi)\Phi_1\Phi_2...\rangle = \sum_{k=1}^{N} (\alpha(z_k)\partial_k + \alpha'(z_k)\Delta_k)\langle\Phi_1\Phi_2...\rangle \tag{49}$$

It should be remarked that important thing happened: the $z, \bar{z}$ mixed WI got decoupled on z and $\bar{z}$ parts. One achieves something like reduction of $2D \to 1D$. Eq.(49) could also be written as:

$$\frac{1}{2\pi i} \oint_c d\xi\alpha(\xi)\langle T(\xi)\Phi_1\Phi_2...\rangle = \frac{1}{2\pi i} \oint_c d\xi\alpha(\xi) \sum_k (\frac{\Delta_k}{(\xi - z_k)^2} + \frac{1}{\xi - z_k}\partial_k)\langle\Phi_1\Phi_2...\rangle \tag{50}$$

In fact, by calculating the residues for the contour integral in r.h.z. of (50) one recovers the r.h.s. of (49). Finally, as $\alpha(z)$ is an arbitrary function the integration could be lifted, to give

$$\langle T(z)\Phi_1\Phi_2...\rangle = \sum_{k=1}^{N} (\frac{\Delta_k}{(z - z_k)^2} + \frac{1}{z - z_k}\partial_k)\langle\Phi_1\Phi_2...\rangle \tag{51}$$

This is the conformal WI in its local form, i.e. without integrations involved.

We shall proceed next with the analysis of the spectrum of operators in conformal field theory. The operators Φ, in like T, they depend both on z and $\bar{z}$. But as the conformal transformations actually factorize on z and $\bar{z}$ parts and the conformal WI decouple on z and $\bar{z}$ parts we can suppress for the moment the $\bar{z}$ dependence of the fields Φ and study only the z dependence, z conformal structures involved. Formally, for the time being

$$\Phi_k(z_k, \bar{z}_k) \to \Phi_k(z_k) \tag{52}$$

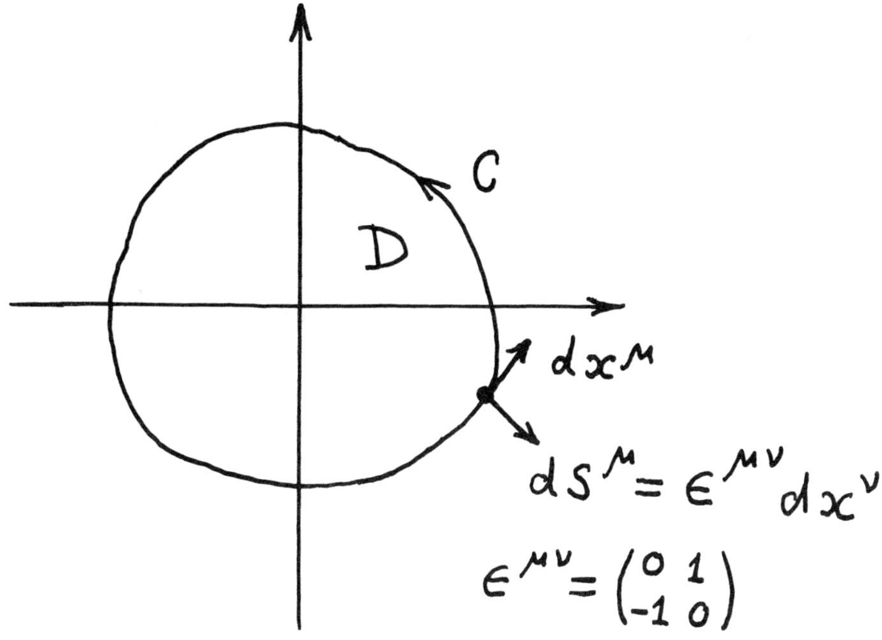

Figure 2: Definition of the contour integral.

From conformal WI (51) it follows that the operator product expansion (OPE) of T and any $\Phi, e.g.\Phi_1$, starts as:

$$T(z)\Phi_1(z_1) = \frac{\Delta_1}{(z-z_1)^2}\Phi_1(z_1) + \frac{1}{z-z_1}\partial_1\Phi_1(z_1) + \Phi_1^{(-2)}(z_1) + (z-z_1)\Phi_1^{(-3)}(z_1) + ... \quad (53)$$

The first two terms are evident from WI. The operators $\Phi_1^{(-2)}, \Phi_2^{(-3)}$ are new ones. So, from the product of T and Φ one gets new operators, an infinite family of them:

$$T\Phi \to \{\Phi^{(-2)}, \Phi^{(-3)}, ...\} \quad (54)$$

The corresponding expansion of the r.h.s. of the WI (51), in $(z-z_1)$, defines correlation functions of these extra operators with the rest. In this way the new operators get defined so far by their correlation functions. The definition of new operators could be organized better. Let us develop $T(z)$ in formal Laurent series around $z = z_1$:

$$T(z) = \sum_{n=-\infty}^{+\infty} \frac{L_n(z_1)}{(z-z_1)^{n+2}} \quad (55)$$

The coefficients $L_n(z_1)$ are themselves operators. They are Laurent series components of $T(z)$. Now the product $T\Phi$ takes the form

$$T(z)\Phi_1(z_1) = \sum_{n=-\infty}^{+\infty} \frac{1}{(z-z_1)^{n+2}} L_n(z_1)\Phi(z_1) \quad (56)$$

Comparing with (53) one gets:

$$L_n(z_1)\Phi_1(z_1) = 0, n \geq 1 \quad (57)$$

$$L_0(z_1)\Phi_1(z_1) = \Delta_1\Phi_1(z_1) \tag{58}$$

$$L_{-1}(z_1)\Phi_1(z_1) = \partial_1\Phi_1(z_1) \tag{59}$$

$$L_{-n}(z_1)\Phi_1(z_1) = \Phi^{(-n)}(z_1), n \geq 2 \tag{60}$$

The last two lines correspond to new operators produced from the product T and Φ:

$$T(z)\Phi_\Delta(z_1) \to \{\Phi_\Delta^{(-n)}(z_1) = L_{-n}\Phi_\Delta(z_1), n \geq 1\} \tag{61}$$

(We have replaced here the index-number of Φ by its conformal dimension Δ; we shall be switching between these types of notations also in the following, or leaving Φ without index at all).

The procedure could be continued. Out of the product

$$T(z)T(z')\Phi_\Delta(z_1) \tag{62}$$

one gets again a new infinite set of operators:

$$\{\Phi_\Delta^{(-n_1,-n_2)}(z_1) = L_{-n_1}L_{-n_2}\Phi_\Delta(z_1)\} \tag{63}$$

In general, one gets an infinite family:

$$\{\Phi_\Delta^{(-n_1,-n_2,\ldots,-n_k)} = L_{-n_1}L_{-n_2}\ldots L_{-n_k}\Phi_\Delta\} \tag{64}$$

At this point one clearly needs to know the commutation relation between $L'_n s$, in order to organize the space of operators (64). The definition of $L'_n s$ as coefficients in the series (55) could clearly be inverted, to define $L'_n s$ as:

$$L_n(z_1)\Phi(z_1) = \frac{1}{2\pi i} \oint_{C_{z_1}} d\xi(\xi - z_1)^{n+1}T(\xi)\Phi(z_1) \tag{65}$$

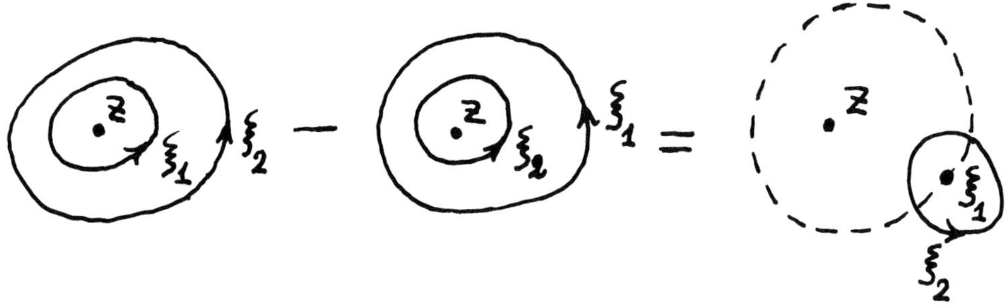

Figure 3: Difference of the contour integrals corresponding to the commutator $[L_n, L_m]$.

The contour C_{z_1} encircles the point z_1. We keep $\Phi(z_1)$ in this definition to show explicitly the way the operator L_n is applied to Φ. The commutation of $L'_n s$ could be calculated from the difference of the two expressions:

$$L_nL_m\Phi(z) = \frac{1}{(2\pi i)^2} \oint_{c_2} d\xi_2 \oint_{c_1} d\xi_1(\xi_2 - z)^{n+1}(\xi_1 - z)^{m+1}T(\xi_2)T(\xi_1)\Phi(z) \tag{66}$$

$$L_m L_n \Phi(z) = \frac{1}{(2\pi i)^2} \oint_{c_1} d\xi_1 \oint_{c_2} d\xi_2 (\xi_1 - z)^{m+1}(\xi_2 - z)^{n+1} T(\xi_1)T(\xi_2)\Phi(z) \tag{67}$$

One obtains, see also Fig.3,

$$[L_n, L_m]\Phi(z) = \oint_{c_z} d\xi_1 (\xi_1 - z)^{m+1} \oint_{c_{\xi_1}} d\xi_2 (\xi_2 - z)^{n+1} T(\xi_2)T(\xi_1)\Phi(z) \tag{68}$$

To take the integral inside, over ξ_2, one just needs to know the singular terms of the OPE of

$$T(\xi_2)T(\xi_1) \tag{69}$$

(comp. Fig.3) similar to OPE of $T\Phi$ considered above, these could be deduced from WI involving $T(z)T(z')$. In a way similar to that described above, to get WI (51), one could get the following WI:

$$\langle T(z)T(z')\Phi_1\Phi_2...\rangle = \frac{c/2}{(z-z')^2}\langle \Phi_1\Phi_2...\rangle + \left(\frac{2}{(z-z')^2} + \frac{1}{z-z'}\partial_{z'}\right)\langle T(z')\Phi_1\Phi_2...\rangle$$
$$+ \sum_k \left(\frac{\Delta_k}{(z-z_k)^2} + \frac{1}{z-z_k}\partial_k\right)\langle T(z')\Phi_1\Phi_2...\rangle \tag{70}$$

The only new ingredients is to start with $\langle T(z')\Phi_1\Phi_2...\rangle$, instead of $\langle \Phi_1\Phi_2...\rangle$, and to use the following form of infinitesimal variation of $T(z')$:

$$\delta T(z') = (2\alpha'(z') + \alpha(z')\partial_{z'})T(z') + \frac{c}{12}\alpha'''(z') \tag{71}$$

The first term here corresponds to the fact that the conformal dimension of T is 2, which is evident from the definition of T as the corresponding variation of the action $A[\varphi]$ see eq. (38) or (49); compare also with $\delta\Phi_\Delta$ in eq.(36). The last term in (71), which produces the first term in WI(70), is actually due to nonvanishing two-point function of T:

$$\langle T(z)T(z')\rangle = \frac{c/2}{(z-z')^4} \tag{72}$$

In fact, one gets (72) from (70) in a special case when all $\Phi's$ are removed. One could argue this way: in a field theory, or in a critical phenomena, it is natural to have the two-point function $\langle T(z)T(z')\rangle$ nonvanishing, which just means that $c \neq 0$ in (72). By the way of WI (70) one needs then the last term in δT, eq.(71). Notice also that by checking the dimensions of the terms in (71) one finds that the last term is the only possible modification of δT, linear in $\alpha(z)$, as compared to variations of the conformal fields Φ_Δ, eq.(36).

After these remarks, we deduce from eq.(70) that

$$T(\xi_2)T(\xi_1) = \frac{c/2}{(\xi_2 - \xi_1)^4} + \frac{2}{(\xi_2 - \xi_1)^2}T(\xi_1) + \frac{1}{\xi_2 - \xi_1}\partial_{\xi_1}T(\xi_1) + Regular Terms \tag{73}$$

Knowledge of singular terms, as $\xi_2 \to \xi_1$, is sufficient to take the integral $\oint_{c_{\xi_1}} d\xi_2$ in (67). This way, by substituting (73) in eq.(67), one gets:

$$[L_n, L_m]\Phi(z) = \frac{1}{2\pi i}\oint_{c_z} d\xi_1 (\xi_1 - z)^{m+1}\frac{c}{12}n(n^2-1)(\xi_1 - z)^{n-2}\Phi(z)$$
$$+ (n-m)\frac{1}{2\pi i}\oint_{c_z} d\xi_1 (\xi_1 - z)^{n+m+1}T(\xi_1)\Phi(z)$$
$$= \frac{c}{12}n(n^2-1)\delta_{n_1-m}\Phi(z) + (n-m)L_{n+m}\Phi(z) \tag{74}$$

Eventually, one gets:

$$[L_n, L_m] = (n - m)L_{n+m} + \frac{c}{12}n(n^2 - 1)\delta_{n_1 - m} \tag{75}$$

Conclusion: the Laurent series components of $T(z)$, the operator L_n, they commute as in (75), which is known as the Virasoro algebra. (In physics it originally appeared in dual amplitudes and string theory at the start of 70's).

We could finish now the general classification of the space of operators in conformal field theory. For a given conformal field Φ_Δ, which is called a primary one, by annilihing $L'_{-n}s$ one gets an infinite set of operators, which are called descendents [2],

$$\left\{\Phi^{(-n_1, -n_2, \ldots, -n_k)} = L_{-n_1} L_{-n_2} \ldots L_{-n_k} \Phi_\Delta\right\}$$
$$n_1 \leq n_2 \leq \ldots \leq n_k \tag{76}$$

evidently, the descendent operators different ordering of the indices $\{n_i\}$ could be related to those in eq.(76) by using the commutation relations (75).

One can check that, by using the Virasoro algebra and finally eq.(58), that all the operators (75) are eigenvectors w.r.t. L_0:

$$L_0 \Phi_\Delta^{(-n_1, -n_2, \ldots -n_k)} = (\Delta + N)\Phi_\Delta^{(-n_1, -n_2, \ldots, -n_k)} \tag{77}$$

$$N = \sum_{i=1}^{k} n_i \tag{78}$$

N is called the level number. In representation theory of the Virasoro algebra the set of operators (76) is called the Verma module. According to eq.(77) the operators in this module could be classified into levels.

Finally, to finish the general classification of the operators the conformal theory, it contains a certain number of primary fields $\{\Phi_i, \Delta_i\}$, which could be finite or infinite depending on a particular theory, and then the descendent operators which finds themselves in the Verma modules of primaries – each primary field has its own descendents. This general classification is pictured in Fig.4.

Remark existence in any conformal theory of an identity operator I. This is a trivial case of a primary operator, the one with $\Delta = 0$:

$$\Phi_{\Delta=0} = I = const, \; z \; independent \tag{79}$$

Notice that the operator $T(z)$, being not a primary one according to its transformation properties (71), it finds itself in the module of the identity operator. In fact, one gets, for the first descendents in the module of I:

$$I^{(-1)} = L_{-1}I = \partial I = 0 \tag{80}$$

$$I^{(-1,-1)} = L_{-1}^2 I = \partial^2 I = 0 \tag{81}$$

but

$$I^{(-2)} = L_{-2}I = \frac{1}{2\pi i} \oint_{c_z} d\xi (\xi - z)^{-1} T(\xi) = T(z) \tag{82}$$

Notice also that the operator L_{-1} acts always as a derivative, both on primaries and descendents (called sometimes secondaries).

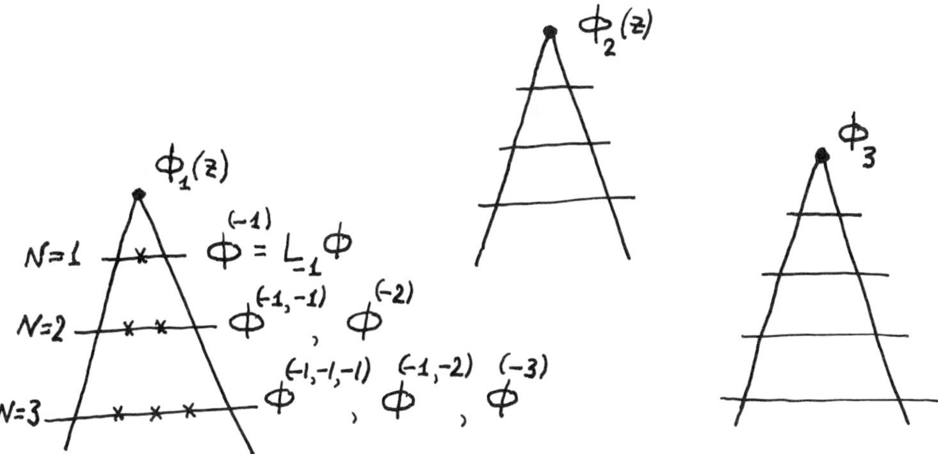

Figure 4: Verma modules of the primary operators Φ_1, Φ_2, Φ_3.

The classification of operators considered above implies that all the descendent operators, eq.(76), in a particular module, shown say in Fig.4, are linearly independent. In interesting cases, related to physics, this not the case in general. So to say, there are degeneracies in the modules, which, it turns out encodes extra important information on correlation functions of the theory. This we are going to discuss now.

To start, we are looking for nontrivial representations of the Virasoro algebra involving degeneracies. First, on level one of the module of Φ_Δ there is just one operator, $L_{-1}\Phi_\Delta$. Degeneracy on this level would mean

$$L_{-1}\Phi_\Delta = 0 \rightarrow \Phi_\Delta = I, \Delta = 0 \tag{83}$$

This picks up the identity operator as special, not more. Next, level two. One could in principle make a linear combination out of two operators there

$$\chi_2 = L_{-2}\Phi_\Delta + aL_{-1}^2\Phi \tag{84}$$

and look of it could vanish under certain conditions.

First, it should be remarked that linear combinations, like that in eq.(84), could be formed out of descendent operators of the same level N, otherwise vanishing of such combinations would be incompatible with the scaling symmetry. Second, one has to ensure that the linear combination (84), the operator χ_2, is a primary one, i.e. transforms conformaly as in eq. (36). If that is not the case, making it vanish would now be incompatible with the conformal symmetry. Equation

$$\chi_2 = 0 \tag{85}$$

would not be preserved under conformal transformation. This is because conformal transformation of descendent operators involve mixing of different operators. This could easily be checked for the simplest descendent $L_{-1}\Phi_\Delta = \partial\Phi_\Delta$. More generale cases could be found in [2].

This is somewhat similar to the requirement that equations, or lagrangians, should be generally covariant in the field theory which is supposed to have that symmetry. In the case of conformal theory the general covariance is replaced by conformal covariance.

The condition of being a primary is equivalent to

$$L_n\chi_2 = 0, n > 0 \tag{86}$$

– comp.eq.(57). In fact, it is sufficient to ensure that χ_2 is annihilated by L_{+1} and L_{+2}, the rest will follow by Virasoro algebra. So we have to equations:

$$L_{+1}(L_{-2}\Phi_\Delta + aL_{-1}^2\Phi_\Delta) = 0 \tag{87}$$

$$L_{+2}(L_{-2}\Phi_\Delta + aL_{-1}^2\Phi_\Delta) = 0 \tag{88}$$

By commuting, and finally using $L_0\Phi_\Delta = \Delta\Phi$, one gets from (87):

$$a = -\frac{3}{2(2\Delta + 1)} \tag{89}$$

and then from (88):

$$c = \frac{2\Delta(5 - 8\Delta)}{2\Delta + 1} \tag{90}$$

This, Δ gets determined by the central charge of the theory C. Then one has the equation:

$$L_{-2}\Phi_\Delta - \frac{3}{2(2\Delta + 1)}L_{-1}^2\Phi_\Delta = 0 \tag{91}$$

for the operator Φ_Δ with that Δ.

In other words, it is said that we have a singular operator χ_2 on the second level in the module of this Φ_Δ, singular in a sense that it turns out to be primary by its conformal transformation properties. And then we impose the condition that it vanishes, by removing in this way the degeneracy (presence of singular operators) of the module. Making vanish the singular operators in the modules is additionally justified, from the physical point of view, by the fact that they the states created by such operators have vanishing norms. Roughly, this is because the conjugation of Virasoro algebra operators, which one introduces to define scalar products, is $(L_{-n})^+ = L_n$, and because singular operators are annihilated by $L_n, n > 0$. For more details, see [2].

Suppose now we have such an operator Φ_Δ in our theory, which obeys the equation (91). We remind that its Δ is defined by eq.(90), could not be arbitrary. We shall show now that the consequence is the linear differential equation for the correlation functions involving Φ_Δ,

$$\langle\Phi_\Delta(z)\Phi_1(z_1)\Phi_2(z_2)...\rangle \tag{92}$$

By(91) we have:

$$\langle(L_{-2}(z)\Phi_\Delta(z))\Phi_1\Phi_2...\rangle - \frac{3}{2(2\Delta + 1)}\langle(L_{-1}^2(z)\Phi_\Delta(z))\Phi_1\Phi_2...\rangle = 0 \tag{93}$$

First of all:

$$\langle(L_{-1}^2(z)\Phi_\Delta(z))\Phi_1\Phi_2...\rangle = \partial_z^2\langle\Phi_\Delta(z)\Phi_1\Phi_2...\rangle \tag{94}$$

The first term in eq. (93) could also be transformed into a differential operator applied to the correlation function (92). This is achieved by transforming the contour integral defining $(L_{-2}\Phi_\Delta)$:

$$\langle (L_{-2}(z)\Phi_\Delta(z))\Phi_1\Phi_2...\rangle = \frac{1}{2\pi i}\oint_{C_z} d\xi (\xi - z)^{-1}\langle T(\xi)\Phi_\Delta(z)\Phi_1\Phi_2...\rangle \tag{95}$$

the way it is shown in Fig.5. The integral over the contour at infinity $C\infty$ vanishes, as the asymptotic behavior, the leading term, of $\langle T(\xi)...\rangle$ is given by $\langle T(\xi)T(0)\rangle \sim 1/\xi^4$. To take the integral over one of small contours, $C_k, k = 1, 2, ...$, one could use the OPE for $T\Phi_k$:

$$T(\xi)\Phi_k(z_k) = \frac{\Delta_k}{(z - z_k)^2}\Phi_k(z_k) + \frac{1}{(z - z_k)}\partial_k\Phi_k(z_k) + R.T'.s \tag{96}$$

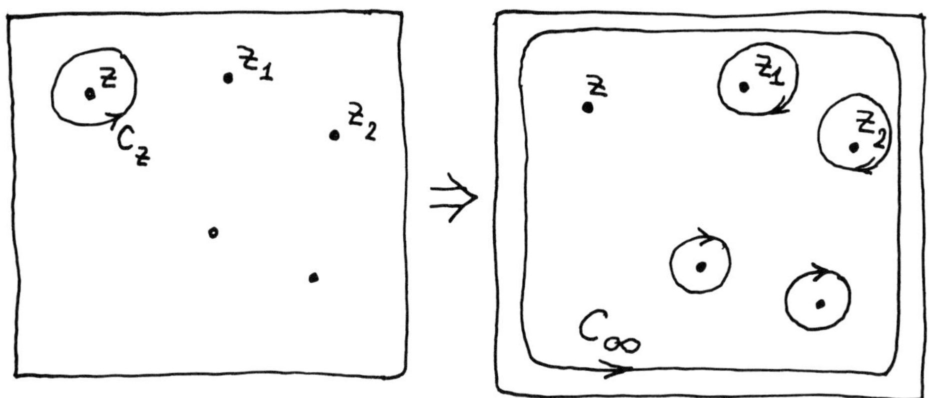

Figure 5: Taking the contour integral off the point z and shifting it to the rest of the points.

One obtains in this way:

$$\langle (L_{-2}(z)\Phi_\Delta(z))\Phi_1\Phi_2...\rangle = \sum_k \frac{1}{2\pi i}\oint_{C_k} d\xi (\xi - z)^{-1}\left(\frac{\Delta_k}{(\xi - z_k)^2} + \frac{1}{\xi - z_k}\partial_k\right)\langle \Phi_\Delta\Phi_1\Phi_2...\rangle$$

$$= \sum_k \left(\frac{\Delta_k}{(z - z_k)^2} + \frac{1}{z - z_k}\partial_k\right)\langle \Phi_\Delta(z)\Phi_1\Phi_2...\rangle \tag{97}$$

Using this result one gets from eq.(93) the linear second order differential equation for the correlation function $\langle \Phi_\Delta\Phi_1...\rangle$:

$$\frac{3}{2(2\Delta + 1)}\partial_z^2\langle \Phi_\Delta(z)\Phi_1\Phi_2...\rangle = \sum_{k=1,2,...}\left(\frac{\Delta_k}{(z - z_k)^2} + \frac{1}{z - z_k}\partial_k\right)\langle \Phi_\Delta(z)\Phi_1\Phi_2...\rangle \tag{98}$$

One could repeat, as exercise, the same reasoning for the descendents on the third level, in the module of an another operator Φ_Δ:

$$\chi_3 = L_{-3}\Phi_\Delta + aL_{-1}L_{-2}\Phi_\Delta + bL_{-1}^3\Phi_\Delta \tag{99}$$

$$L_{+1}\chi_3 = 0 \tag{100}$$

$$L_{+2}\chi_3 = 0 \tag{101}$$

One checks that (100) define the coefficients

$$a = -\frac{2}{\Delta + 2}, \quad b = \frac{1}{(\Delta + 1)(\Delta + 2)} \tag{102}$$

and (101) relates then Δ to C as:

$$\Delta^2 - \frac{7 - c}{3}\Delta + \frac{2 + c}{3} = 0 \tag{103}$$

In general now. One is looking for a singular operator in the module of Φ_Δ on the N th level:

$$\chi_N = \sum b_{\{-n_i\}} \Phi_\Delta^{\{-n_i\}} \tag{104}$$

Here notation is used $\Phi_\Delta^{\{-n_i\}}$ for $\Phi_\Delta^{(-n_1, -n_2, \ldots, -n_k)}$ and similarly for the coefficients $b_{\{-n_i\}}$. For this one defines the matrix of scalar products for the descendents on the N th level:

$$
\begin{aligned}
M^N_{\{n_i\}\{m_i\}} &= \langle (\Phi_\Delta L_{+n_k}...L_{+n_2}L_{+n_1})(L_{-m_1}L_{-m_2}...L_{-m_e}\Phi_\Delta) \rangle \\
&\equiv \langle \Delta | L_{+n_k}...L_{+n_2}L_{+n_1}L_{-m_1}L_{-m_2}...L_{-m_e} | \Delta \rangle \\
&\qquad\qquad\qquad\qquad \sum n_i = \sum m_i = N
\end{aligned} \tag{105}
$$

The determinant of this matrix $DetM^{(N)}(\Delta, c)$ (106) which is a function of Δ and c would signal degeneracies if it vanishes for some values of Δ, with c given. This is because the singular operators have also vanishing norms, as has been mentioned before. The general form of the determinant $DetM^{(N)}(\Delta, c)$ and its zeros has been conjectured by V.Kac [3] and proved by Feigin and Fuks [4]. The zeros could be given as:

$$\Delta_{n',n} = \frac{(\alpha_- n' + \alpha_+ n)^2 - (\alpha_- + \alpha_+)^2}{4} \tag{106}$$

where a specific parametrization of the central charge is used:

$$
\begin{aligned}
c &= 1 - 24\alpha_0^2 \\
\alpha_\pm &= \alpha_0 \pm \sqrt{\alpha_0^2 + 1}
\end{aligned} \tag{107}
$$

The indices n, n' of $\Delta_{n',n}$ in (106) are related to the level number:

$$N = n \times n' \tag{108}$$

This means that for N given $Det\,M$ possess as many zeros as there are ways to factorize N on a product of two integer factors. And the corresponding values of Δ 's for which the determinant vanishes are given by eq.(106). Examples:

$$N = 2, \ \Delta_{1,2}, \ \Delta_{2,1} \tag{109}$$

$\Delta_{1,2}, \Delta_{2,1}$ are given by eq.(106), and they correspond, as can be checked, to two solutions of the eq.(90) obtained above.

$$N = 3, \ \Delta_{1,3}, \ \Delta_{3,1} \tag{110}$$

$\Delta_{1,3}, \Delta_{3,1}$ are also the two solutions of the eq.(103).

$$N = 4, \ \Delta_{1,4}, \ \Delta_{4,1}, \ \Delta_{2,2} \tag{111}$$

etc.

Minimal conformal theory is made out of operators

$$\{\Phi_{n',n}, \; \Delta_{n',n}\} \tag{112}$$

with dimensions given by the Kac formula (106) and their descendents. The operators $\Phi_{n',n}$ have degenerate modules on the level $N = n' \times n$, and as a consequence their correlation functions satisfy N th order linear differential equation, which could in principle be derived after finding the explicit form of the corresponding singular operator in the module. In this sense the theory is solvable.

In fact the correlation functions could be calculated move directly, by using the free field representation for the minimal conformal theory, also called the Feigin-Fuks, or the Coulomb gas representation [5, 6]. This will be described in the next section. So to say, it gives directly the solutions of the differential equations discussed above.

Important point is that the operator product algebra of the operator (112) and its descendents closes. This means that the linear decomposition of the product of two operators taken from the family (112):

$$\Phi_1(z, \bar{z})\Phi_2(z', \bar{z}') = \sum_p \frac{D^p_{12}}{|z - z'|^{2(\Delta_1+\Delta_2-\Delta_p)}}[\Phi_p(z', \bar{z}') + descendents] \tag{113}$$

contains, in the r.h.s., just the operators from this same family (112). In (113) we have put back the $\bar{z}$ dependence of the operators. The coefficients D^p_{12} of the operator algebra (113) has been calculated in [7]. The technique of the calculation, which uses the free field representation, will be described in the next section.

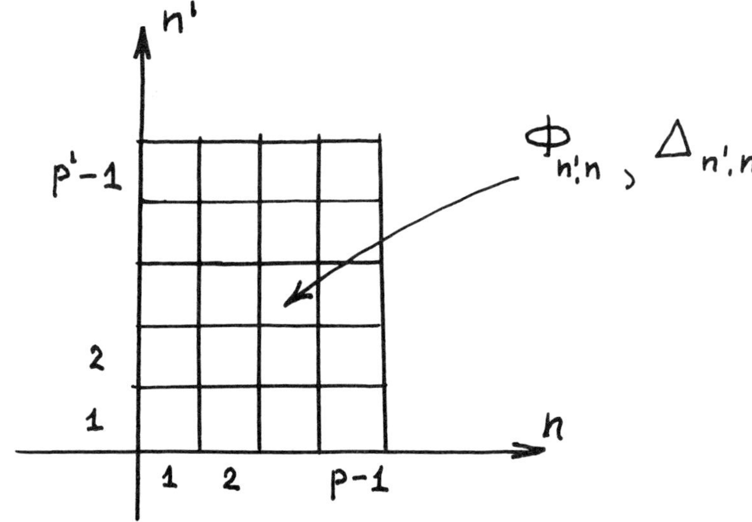

Figure 6: Table of primary operators for the conformal theory with $c = 1 - 6(p - p')^2/pp'$.

Next important point is that for special discrete values of the central charge of the theory

$$c = c_{p',p} = 1 - \frac{6(p' - p)^2}{p',p} \tag{114}$$

the operator algebra closes by a finite set of primary operators (always accompanied of course by an infinite set of their descendents):

$$\{\Phi_{(n'.n)}\} \; with \; 1 \leq n' \leq p' - 1, \; 1 \leq n \leq p - 1 \tag{115}$$

Having two indices, they could be represented by tables, see Fig.6.,[2].

For still special values of the central charge, those with $p' = p + 1$:

$$c_{p+1,p} = 1 - \frac{6}{p(p+1)} \tag{116}$$

the corresponding theories has been shown to be unitary [8]. In applications to critical statistical systems one encounters both unitary and non unitary minimal conformal theories.

We shall finish this section with a brief list of examples of two-dimensional statistical systems, at their critical points, and the corresponding conformal theories.

The $p = 3$ unitary minimal model, with $c = \frac{1}{2}$ by eq.(116), has been identified with the Ising model [2]. The corresponding tables of operators and their conformal dimensions are shown in Fig.7. It should be noted that all the finite tables possess a symmetry w.r.t. certain reflection. This follows from the Kac formula for $\Delta_{n'.n}$ in case of central charges (114). There is a doubling of operators, and actually the number different of primary operators is twice less that in eq.(115). (There is a way to show the decoupling of operators which are outside the the tables precisely by using this double appearance of each operator inside).

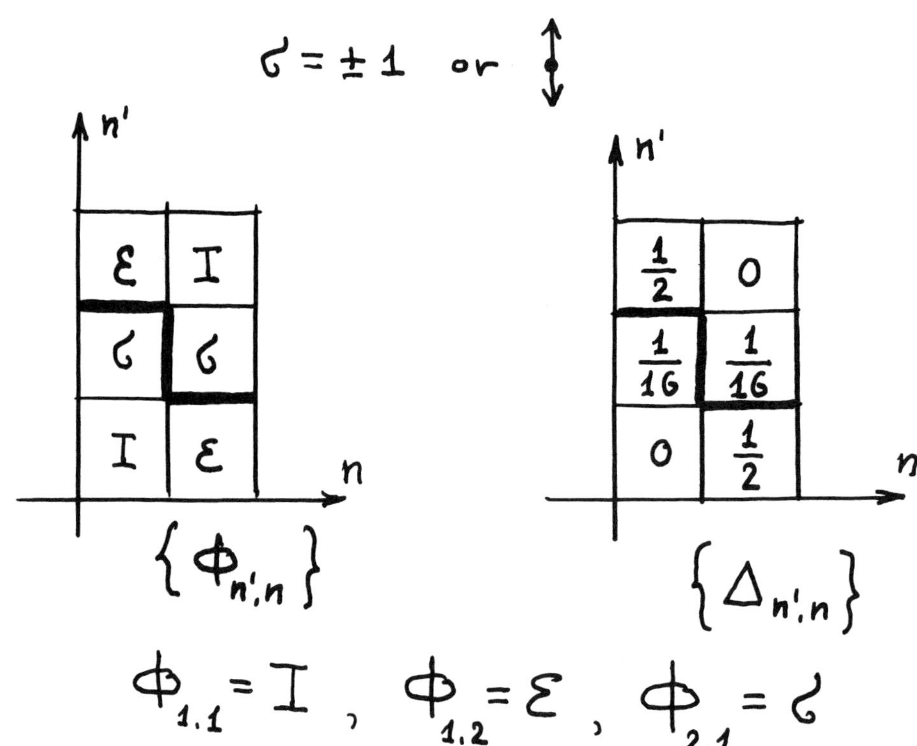

Figure 7: Ising model, $p = 3$; the spin variables, as they are defined on the lattice before the continuum limit is taken, take two values $\sigma = \pm 1$.

So, except for the identity operator I the Ising model contains just two nontrivial primary operators, that of energy ε and spin σ. Their conformal dimensions could be recalculated into the corresponding critical exponents of the correlation length and the magnetization, by using the standard scaling relations. This is true in general, all the measured or calculated critical exponents for global observables, like interval energy, magnetization, etc., follows from the set of conformal dimensions $\{\Delta_{n'.n}\}$ of primary operators of the corresponding conformal theory.

Next example is the three component Potts model, or z_3 model, which is defined on lattice similarly to the Ising model but with spin variables taking three different values instead of two. The classical energy is given by the nearest neighbor interactions of spin variables:

$$A[\bar{\sigma}] = J \sum_{x,\alpha} \bar{\sigma}_x \bar{\sigma}_{x+\alpha} \tag{117}$$

Here runs over sites of, say, square lattice, and α takes two unit vector values corresponding to two basic directions on the lattice. This model possesses, like the Ising model, the second order phase transition point. At the critical point it was shown to be described by the $p = 5$ unitary minimal conformal theory, with $c = \frac{5}{6}$ by eq.(116) [9]. Its tables of primary operators and their conformal dimensions are shown in Fig.8. Operators which are not shown, those in the second row, the decouple from the rest by the operator algebra. In principle they could be added, and they are contained in the theory, but they are not produced in operator products of the operators which are shown. This basic set of energy-like (ε, X, Y), and magnetization - like (z, σ) contains five primary operators. This is different from the Ising model, apart from physically obvious local energy and local order parameter operator there are three extra ones.

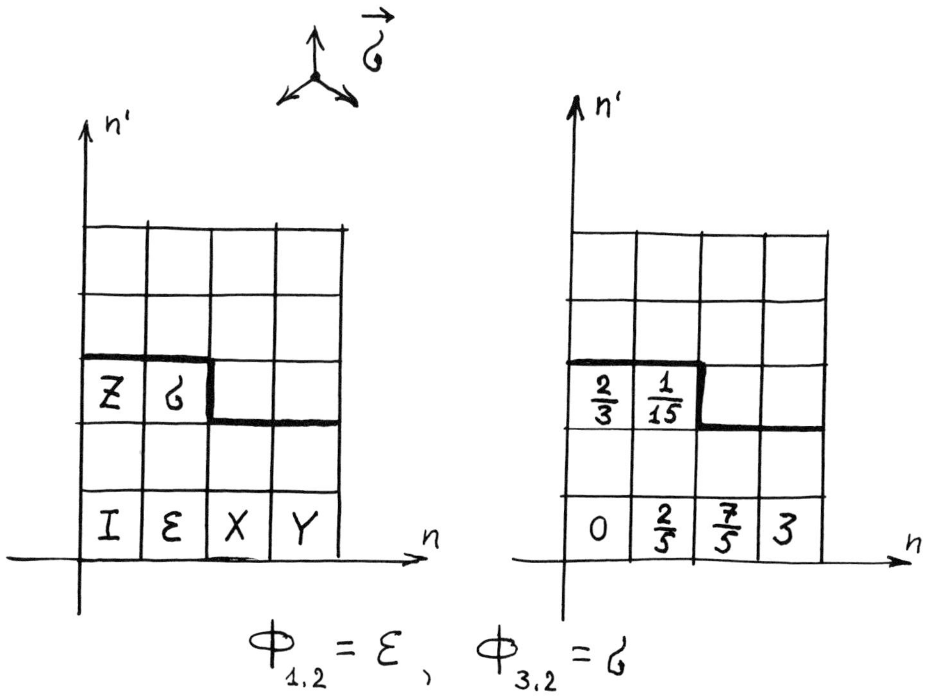

$$\Phi_{1,2} = \varepsilon, \quad \Phi_{3,2} = 6$$

Figure 8: Z_3, or the three-component Potts model.

The complete analysis of the z_3 model, of all its operators and symmetries is achieved in the parafermionic conformal theory [10]. This is outside the scope of these lectures.

We shall list next some other critical statistical systems (the list is by far incomplete) for which the corresponding conformal theories have been identified:

q component Potts models. q could take also fractional and continuous values.

q component tricritical Potts models. In both cases q is limited by $q \leq 4$. For $q > 4$ the model has first order phase transition point [11], so no continuum field theory applies.

$O(N)$ model, $N \leq 2$. It also could be defined for continuous values of N. For $N > 2$ it does not have ordering phase transition of finite temperatures, in two dimensions.

The limit of $q \to 0$ for the Potts model conformal theory describes percolation problem $N \to 0$ limit of $O(N)$ model describes $2D$ polymers.

Self - avoiding paths are also treated exactly by the corresponding conformal theory.

Spin model of restricted heights, so called Solid-on-Solid restricted (RSOS) model, being in fact an infinite series of statistical models, maps on the infinite series of minimal unitary models (116). The reader could further consult itself for the statistical physics applications of conformal field theory in refs. [12, 13].

2. FREE - FIELD REPRESENTATION OF MINIMAL CONFORMAL THEORY

Having given in the previous section the general structure of the minimal conformal theory we shall describe now its specific representation by operators made of free field, the representation which is most efficient to actually solve the theory, to calculate correlation functions and the operator algebra of primary operators. For analogy with problems in quantum mechanics, one knows that it is important, to solve a problem, to find proper variables. For conformal field theory representation by free fields. As this representation generalises also to the case of other conformal theories, i.e. these having extra symmetries, it is in fact an important ingredient of the theory in general. This justifies describing it in detail for the presently considered case of minimal theory.

One makes use of a single free scalar field $\varphi(z, z)$, massless, having action:

$$A[\varphi] = \frac{1}{4\pi} \int d^2x \partial_z \varphi \partial_{\bar{z}} \varphi = \frac{1}{16\pi} \int d^2x (\partial \varphi)^2 \tag{118}$$

Here $(\partial \varphi)^2 = (\partial_1 \varphi)^2 + (\partial_2 \varphi)^2 = 4\partial_z \varphi \partial_{\bar{z}} \varphi$, as $\partial_z = \frac{1}{2}(\partial_1 - i\partial_2), \partial_{\bar{z}} = \frac{1}{2}(\partial_1 + i\partial_2)$. The two-point function for $\varphi(z, \bar{z})$ could be defined by the functional integral:

$$\langle \varphi(z, \bar{z})\varphi(z', \bar{z}') \rangle = \frac{\int D\varphi exp\{-A[\varphi]\}\varphi(z, \bar{z})\varphi(z', \bar{z}')}{\int D\varphi exp\{-A[\varphi]\}} \tag{119}$$

One finds

$$\langle \varphi(z, \bar{z})\varphi(z', \bar{z}') \rangle = 4log\frac{R}{|z - z'|} \tag{120}$$

Here R is the size of the statistical system, of the space for the field theory. In fact, if one calculates (2) directly, by passing to Fourier components, gets finally log divergent series at long distances. To regulate this infrared divergence one introduces a big, i.e. $R \gg |z - z'|$, but finite size of the space. Then one gets eq.(3). Infrared problem for quantization of free massless scalar fields in $2D$ is well known. We shall see next that it disappears in case of specific operators, made of φ 's, which one uses in conformal theory. So one could safely take finally the limit of $R \to \infty$.

The primary operators of the conformal theory are represented by exponentials of ϕ:

$$\Phi_{\Delta,\bar{\Delta}}(z,\bar{z}) \sim V_\alpha(z,\bar{z}) = e^{i\alpha\varphi(z,\bar{z})} \tag{121}$$

We remark that, in general, operators made as exponentials of free fields are called also vertex operators. So, one is going to represent $\{\Phi_{\Delta,\bar{\Delta}}\}$ by vertex operators $\{V_\alpha(z,\bar{z})\}$. Let us calculate first the two-point function for vertex operators. One gets:

$$\langle V_\alpha(z'\bar{z})V_{-\alpha}(z',\bar{z}')\rangle = \frac{\int D\varphi\, exp\{-A[\varphi]\}e^{i\alpha\varphi(z,\bar{z})} \times e^{-i\alpha\varphi(z',\bar{z}')}}{\int D\varphi\, exp\{-A[\varphi]\}}$$

$$= exp\{-\frac{1}{2}\langle(\alpha\varphi(z,\bar{z}) - \alpha\varphi(z',\bar{z}'))^2\rangle\}$$

$$= exp\{-\frac{\alpha^2}{2}(\langle\varphi^2(z,\bar{z})\rangle + \langle\varphi^2(z',\bar{z}')\rangle - 2\langle\varphi(z,\bar{z})\varphi(z',\bar{z}')\rangle)\}$$

$$= exp\{-4\alpha^2(\log\frac{R}{a} - \log\frac{R}{|z-z'|})\} = |\frac{a}{z-z'}|^{4\alpha^2} \propto \frac{1}{|z-z'|^{4\alpha^2}} \tag{122}$$

We used here that for the Gaussian integrals one has in general:

$$\langle e^{\lambda x}\rangle = \frac{\int dx\, e^{-\frac{x^2}{2}}e^{\lambda x}}{\int dx\, e^{-\frac{x^2}{2}}} = e^{\frac{\lambda^2}{2}\langle x^2\rangle} \tag{123}$$

This generalizes to functional integrals over fields with Gaussian, i.e. free field action. Also we put a short distance cut-off for

$$\langle\varphi^2(z,\bar{z})\rangle = \langle\varphi^2(z',\bar{z}')\rangle = 4\log\frac{R}{a} \tag{124}$$

where when one calculates one counters log divergence both at short and at large distances.

In a similar way one would get for the four-point function

$$\langle V_{\alpha_1}(z_1,\bar{z}_1)V_{\alpha_2}(z_2,\bar{z}_2)V_{\alpha_3}(z_3,\bar{z}_3)V_{\alpha_4}(z_4,\bar{z}_4)\rangle \propto \prod_{i<j}|z_i - z_j|^{4\alpha_i\alpha_j} \tag{125}$$

– if $\sum\alpha_i = 0$. Otherwise one gets zero, in the limit of $R \to \infty$.

The stress-energy tensor for field φ with the action (1) has the form:

$$T_{zz} = -\frac{1}{4}\partial\varphi, T_{\bar{z}\bar{z}} = -\frac{1}{4}\bar{\partial}\varphi\bar{\partial}\varphi \tag{126}$$

$$T_{z\bar{z}} = T_{\bar{z}z} = 0 \tag{127}$$

Let us factorize away the $\bar{z}$ part of the correlation functions. One gets, formally

$$\langle V_\alpha(z)V_{-\alpha}(z')\rangle \propto \frac{1}{(z-z')^{2\alpha^2}} \tag{128}$$

$$\langle V_{\alpha_1}(z_1)V_{\alpha_2}(z_2)V_{\alpha_3}(z_3)V_{\alpha_4}(z_4)\rangle \propto \prod_{i<j}(z_i - z_j)^{2\alpha_i\alpha_j} \tag{129}$$

$$\sum\alpha_i = 0 \tag{130}$$

If compared with

$$\langle\Phi_\Delta(z)\Phi_\Delta(z')\rangle \propto \frac{1}{(z-z')^{2\Delta}} \tag{131}$$

one finds that the conformal dimensions Δ of the vertex operators are given by:

$$\Delta(V_\alpha) = \Delta(V_{-\alpha}) = \alpha^2 \qquad (132)$$

Let us remark on the analogue of the normal ordering procedure in the functional integral and correlation functions formalism which we use. It amounts to ruling out the dependence on the short - distance cut-off a. For $\partial\varphi(z)\partial\varphi(z)$ it amounts to replace it with

$$: \partial\varphi(z)\partial\varphi(z) := \lim_{\varepsilon \to 0}\{\partial\varphi(z+\varepsilon)\partial\varphi(z-\varepsilon) - \langle\partial\varphi(z+\varepsilon)\partial\varphi(z-\varepsilon)\rangle\} \qquad (133)$$

For the vertex operator $V_\alpha(z)$ it amounts to factoring (instead of subtracting) out the a dependent piece. $e^{i\alpha\varphi(z)}$ is to be replaced with

$$: e^{i\alpha\varphi(z)} := \frac{e^{i\alpha\varphi(z)}}{\langle e^{i\alpha\varphi(z)}\rangle} = \frac{1}{(a)^{\alpha^2}}e^{i\alpha\varphi(z)} \qquad (134)$$

So, the corrected expressions for the vertex operator and for the stress - energy tensor are:

$$V_\alpha(z) =: e^{i\alpha\varphi(z)} : \qquad (135)$$

$$T(z) = -\frac{1}{4} : \partial\varphi(z)\partial\varphi(z) : \qquad (136)$$

Notice that if we recalculate the two-point function (5) for "normal ordered" operators defined above then the $(a)^{4\alpha^2}$ factor in the resulting expression will be cancelled. Same is true for the multipoint functions, like (8),(12).

Operator product expansions are calculated by coupling φ' s, say in T and V_α, effectively it amount to using the Wick theorem for operators made as composites of free fields. One checks:

$$T(z)V_\alpha(z') = \frac{\Delta_\alpha}{(z-z')^2}V_\alpha(z') + \frac{1}{z-z'}\partial_{z'}V_\alpha(z') + Reg.Terms \qquad (137)$$

It should be noted that in fact, the normalisation of $A[\phi]$, an consequently of $T(z)$, was chosen so that the coefficient at $\partial_{z'}V_\alpha(z')$ in r.h.s. of (20) was 1. This is the standard normalization of T in conformal theory. Then first term contains conformal dimension of the operator V_α. Eq.(20) is to be compared with eq. (53) of the general theory. In particular, Δ_α calculated from OPE TV_α agrees with that obtained from the two-point function (11), eq.(15).

According to the general theory the central charge C of the Virasoro algebra could be found from

$$\langle T(z)T(z')\rangle = \frac{c/2}{(z-z')^4} \qquad (138)$$

Straightforward calculation for $T(z) = -\frac{1}{4} : \partial\varphi\partial\varphi :$ gives:

$$\langle T(z)T(z')\rangle = \frac{1/2}{(z-z')^4} \qquad (139)$$

which means that we have a representation for the special case of a conformal theory with $c = 1$.

Now we change the rules of calculation for the correlation function. Effectively we change the way the φ field is quantized to achieve deformation of the representation to the case of $c < 1$.

New rule will be

$$\langle V_{\alpha_1}(z_1)V_{\alpha_2}(z_2)...V_{\alpha_N}(z_N)\rangle_{(\alpha_0)} = \lim_{R\to\infty}\{(R)^{8\alpha_0^2}\langle V_{\alpha_1}(z_1)V_{\alpha_2}(z_2)...V_{\alpha_N}(z_N)\times V_{-2\alpha_0}(R)\rangle_{(0)}\} \tag{140}$$

New, extra operator is inserted $V_{-2\alpha_0}(R)$ in calculation of any correlation function. Effectively it is placed at infinity. This is called quantisation of φ with the background charge - $2\alpha_0$ placed at infinity.($\alpha'_i s$ are often called charges in analogy with $2D$ Coulomb gas system, and the representation itself is often called the Coulomb gas representation). Alternatively, this amounts to quantizing φ with nontrivial asymptotics of infinity. And one needs a compensation factor in this case, the factor $(R)^{8\alpha_0^2}$ in eq.(23).

Nonvanishing correlators are still of the form

$$\prod_{i<j}(z_i - z_j)^{2\alpha_i\alpha_j} \tag{141}$$

but now they are subject to the condition

$$\sum_{i=1}^{N}\alpha_1 = 2\alpha_0 \tag{142}$$

One finds in particular that the nonvanishing two-point function is

$$\langle V_\alpha(z)V_{2\alpha_0-\alpha}(z')\rangle = \frac{1}{(z-z')^{2\alpha(\alpha-2\alpha_0)}} \tag{143}$$

By orthogonality

$$\langle \Phi_{\Delta_i}(z)\Phi_{\Delta_j}(z')\rangle = \frac{\delta_{\Delta_i,\Delta_j}}{(z-z')^{2\Delta_i}} \tag{144}$$

we have to conclude that

$$\Delta_\alpha = \Delta_{2\alpha_0-\alpha} = \alpha^2 - 2\alpha\alpha_0 \tag{145}$$

Next we have to ensure, for consistency, that

$$T(z)V_\alpha(z') = \frac{\Delta_\alpha}{(z-z')^2}V_\alpha(z') + \frac{1}{z-z'}\partial_{z'}V_\alpha(z') + ... \tag{146}$$

with Δ_α given by eq.(28). This is achieved, as one could check, by deforming $T(z)$ in the following way:

$$T(z) = \frac{1}{4} : \partial\varphi\partial\varphi : +i\alpha_0\partial^2\varphi \tag{147}$$

Now one calculates $\langle TT\rangle$ and gets (21) with

$$c = 1 - 24\alpha_0^2 \tag{148}$$

So we achieved realisation of the conformal theory with $c < 1$. This is due to the quantization of φ with the background charge at infinity.

In the theory with $c < 1, \alpha_0 \pm 0$, the technique of calculation of the correlation functions gets further modified. The general multipoint functions are calculated with the action

$$A[\varphi] = \frac{1}{4\pi}\int d^2x\partial\varphi\bar{\partial}\varphi \tag{149}$$

which is also deformed (added) with a term:

$$\mu_+\int d^2xV_+(x) + \mu_-\int d^2xV_-(x) \tag{150}$$

Here

$$V_\pm \equiv V_\pm(z,\bar z) =: e^{i\alpha_\pm \varphi(z,\bar z)} : \tag{151}$$

We remind the definition of $\alpha_\pm$

$$\alpha_\pm = \alpha_0 \pm \sqrt{\alpha_0^2 + 1} \tag{152}$$

which have already been used in the first section, in the Kac formula (106). Deformation of the action (32) with a term (33) is allowed, in a sense that it does not break the conformal symmetry. This is because

$$\Delta(V_\pm) = \bar\Delta(V_\pm) = 1 \tag{153}$$

(Remind that $\Delta(V_\alpha) = \alpha^2 - 2\alpha\alpha_0$).

The representation for the general multipoint functions then takes the following form:

$$\langle \Phi_1(x_1)\Phi_2(x_2)...\Phi_N(x_N)\rangle \propto \lim_{R\to\infty}\{R^{8\alpha_0^2}\langle V_{\alpha_1}(x_1)V_{\alpha_2}(x_2)...V_{\alpha_N}(x_N) \times$$

$$V_{-2\alpha_0}(R) \times exp\{\mu_+ \int d^2x V_+ + \mu_- \int d^2x V_-\}\rangle_{(0)}\} \tag{154}$$

Expanding in μ_+, μ_- one gets

$$\langle \Phi_1\Phi_2...\Phi_N\rangle \propto \lim_{R\to\infty}\{(R)^{8\alpha_0^2}\sum_{k=0}^{\infty}\sum_{l=0}^{\infty}(\mu_+)^k(\mu_-)^l \times$$

$$\langle V_{\alpha_1}V_{\alpha_2}...V_{\alpha_N} \times V_{-2\alpha_0}(R) \times (\int d^2x V_+)^k(\int d^2x V_-)^l\rangle_{(0)}\} \tag{155}$$

In the minimal conformal theory one considers operators with $\Delta's$ given by the Kac formula (106). By using $\Delta(V_\alpha) = \alpha^2 - 2\alpha\alpha_0$ one checks that the corresponding values of $\alpha's$ are:

$$\alpha_{n',n} = \frac{1-n'}{2}\alpha_- + \frac{1-n}{2}\alpha_+ \tag{156}$$

– to have $\Delta(V_{\alpha_{n',n}}) = \Delta_{n',n}$. In the μ_+, μ_- series above, eq.(38), one has to have

$$\sum_{i=1}^{N}\alpha_i + l\alpha_- + k\alpha_+ = 2\alpha_0 \equiv \alpha_+ + \alpha_- \tag{157}$$

to have a nonvanishing result. $\{\alpha_i\}$ belong to the set of values (39), they are made as combinations of α_- and α_+. Next we assume that for general values of α_0, and so of α_+, α_-, there is no compensation between α_+, α_-. Then their numbers have to satisfy the condition (40) separately. This fixes l and k.

We summarize. For $\{\alpha_i\}$ given, in the series (38) just one term remains, the one with l and k found from the condition (40).

In this way one arrives at the general formula for multipoint functions in minimal conformal theory:

$$\langle \Phi_1\Phi_2...\Phi_N\rangle \propto (\mu_-)^l(\mu_+)^k \times \lim_{R\to\infty}\{(R)^{8\alpha_0^2}\prod_{i=1}^{l}\int d^2u_i \prod_{j=1}^{k}\int d^2v_j \times$$

$$\langle V_{\alpha_1}(x_1)V_{\alpha_2}(x_2)...V_{\alpha_N}(x_N) \times V_-(u_1)...V_-(u_e) \times V_+(v_1)...V_+(v_k) \times V_{-2\alpha_0}(R)\rangle_{(0)}\} \tag{158}$$

Using

$$\langle \prod_i V_{\alpha_i}(x_i)\rangle_{(0)} = \prod_{i<j} |x_i - x_j|^{4\alpha_i\alpha_j} \tag{159}$$

and taking the limit $R \to \infty$ one obtains from eq.(41):

$$\langle \Phi_1(x_1)\Phi_2(x_2)\Phi_3(x_3)\Phi_4(x_4)\rangle \propto \prod_{i=1}^{l}\int d^2u_i \prod_{j=1}^{k}\int d^2v_j \times$$

$$\prod_{i=1}^{l} |x_1 - u_i|^{4\alpha - \alpha_1}|x_2 - u_i|^{4\alpha - \alpha_2}|x_3 - u_i|^{4\alpha - \alpha_3}|x_4 - u_i|^{4\alpha - \alpha_4} \times$$

$$\prod_{j=1}^{k} |x_1 - v_j|^{4\alpha + \alpha_1}|x_2 - v_j|^{4\alpha + \alpha_2}|x_3 - v_j|^{4\alpha + \alpha_3}|x_4 - v_j|^{4\alpha + \alpha_4} \times$$

$$\prod_{i<i'} |u_i - u_{i'}|^{4\alpha_-^2} \prod_{j<j'} |v_j - v_{j'}|^{4\alpha_+^2} \prod_{i,j} |u_i - u_j|^{-4} \tag{160}$$

In the above expression we have limited ourselves to the case of four-point functions. We used also

$$\alpha_+\alpha_- = -1 \tag{161}$$

Next simplification one achieves by using the invariance w.r.t.

$$z \to \tilde{z} = \frac{az+b}{cz+d} \tag{162}$$

(eq.(29) in the first section). With this transformation the position of three points could be fixed. The standard choice is:

$$x_1 = 0, x_2 = (z, \bar{z}), x_3 = 1, x_4 \to \infty \tag{163}$$

The reduced expression for the four-point functions becomes

$$\langle \Phi_1(0)\Phi_2(z, \bar{z})\Phi_3(1)\Phi_4(\infty)\rangle \propto \prod_{i=1}^{l}\int d^2u_i \prod_{j=1}^{k}\int d^2v_j \times$$

$$\prod_{i=1}^{l} |u_i|^{4\alpha - \alpha_2}|u_i - z|^{4\alpha - \alpha_2}|u_i - 1|^{4\alpha - \alpha_3} \times \prod_{j=1}^{k} |v_j|^{4\alpha + \alpha_1}|v_j - z|^{4\alpha + \alpha_2}|v_j - 1|^{4\alpha + \alpha_3} \times$$

$$\prod_{i<i'} |u_i - u_{i'}|^{4\alpha_-^2} \prod_{j<j'} |v_j - v_{j'}|^{4\alpha_+^2} \prod_{i,j} |u_i - v_j|^{-4} \tag{164}$$

Now we shall demonstrate the way in which these multiple integrals over the two-dimensional plane could be reduced to well defined analytic functions. We shall do this for the case of a single integral:

$$G(z, \bar{z}) = \int d^2u |u|^{2a}|u - z|^{2c}|u - 1|^{2b} \tag{165}$$

But first, to demonstrate the technique, we shall define even simpler integral:

$$G(0) \equiv I(a, b) = \int d^2u |u|^{2a}|u - 1|^{2b} \tag{166}$$

Using Euclidian coordinates this could be given as

$$I(a, b) = \int_{-\infty}^{+\infty} du_1 \int_{-\infty}^{+\infty} du_2 (u_1^2 + u_2^2)^a ((u_1 - 1)^2 + u_2^2)^b \tag{167}$$

We make now a turn of the contour of integration over u_2 (assuming that the integral converges at ∞). This amounts to:

$$u_2 \to ie^{-2i\varepsilon}u_2 \approx i(1 - 2i\varepsilon)u_2 \tag{168}$$

– see Fig.9. ε is assumed to be small and eventually will be set to 0. The integral (50) takes the form:

$$\begin{aligned}
I(a,b) &= i\int_{-\infty}^{+\infty} du_1 \int_{-\infty}^{+\infty} du_2 \times (u_1^2 - u_2^2 e^{-4i\varepsilon})^a((u_1 - 1)^2 - u_2^2 e^{-4i\varepsilon})^b \\
&= \frac{i}{2}\int_{-\infty}^{+\infty} du_+(u_+ - i\varepsilon(u_+ - u_-))^a(u_+ - 1 - i\varepsilon(u_+ - u_-))^b \\
&\quad \times \int_{-\infty}^{+\infty} du_-(u_- + i\varepsilon(u_+ - u_-))^a(u_- - 1 + i\varepsilon(u_+ - u_-))^b
\end{aligned} \tag{169}$$

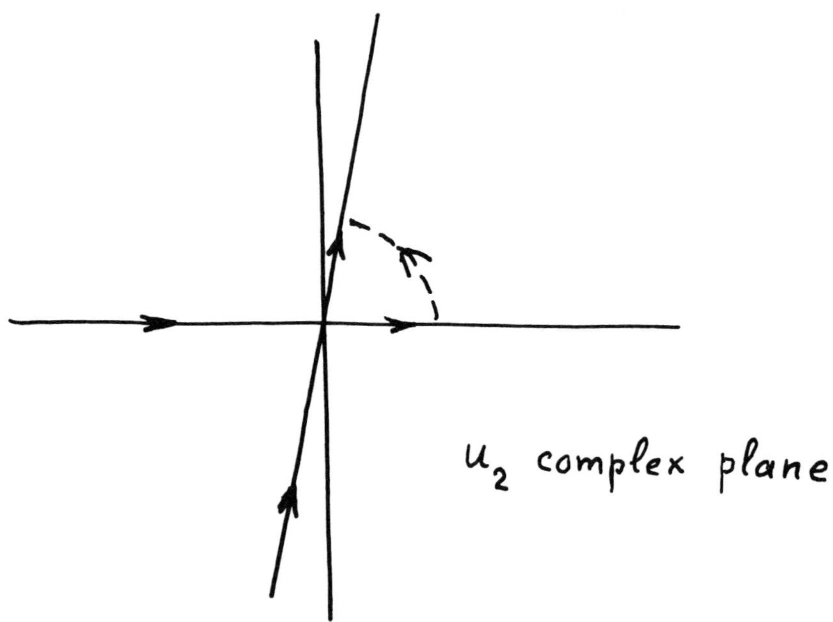

Figure 9: Turning the contour of integration in the complex plane of u_2 variable.

Here $u_\pm = u_1 \pm u_2$. Next we break the integration over u_+ into three intervals:

$$\begin{aligned}
I(a,b) &= \frac{i}{2}\int_{-\infty}^{0} du_+(u_+)^a(u_+ - 1)^b \times \int_{-\infty}^{+\infty} du_-(u_-)^a(u_- - 1)^b \\
&\quad + \frac{i}{2}\int_{0}^{1} du_+(u_+)^a(u_+ - 1)^b \times \int_{-\infty}^{+\infty} du_-(u_-)^a(u_- - 1)^b \\
&\quad + \frac{i}{2}\int_{1}^{+\infty} du_+(u_+)^a(u_+ - 1)^b \times \int_{-\infty}^{+\infty} du_-(u_-)^a(u_- - 1)^b
\end{aligned} \tag{170}$$

The contours of integration over u_- are shown in Fig.10. The ways the u_- contours go around points $u_- = 0$ and $u_- = 1$ are defined by the small ε parts in the integrals (52), by the signs of $(u_+ - u_-)$. According to Fig.10 the u_- contours in the first and the third integrals could be shifted to ∞.

386

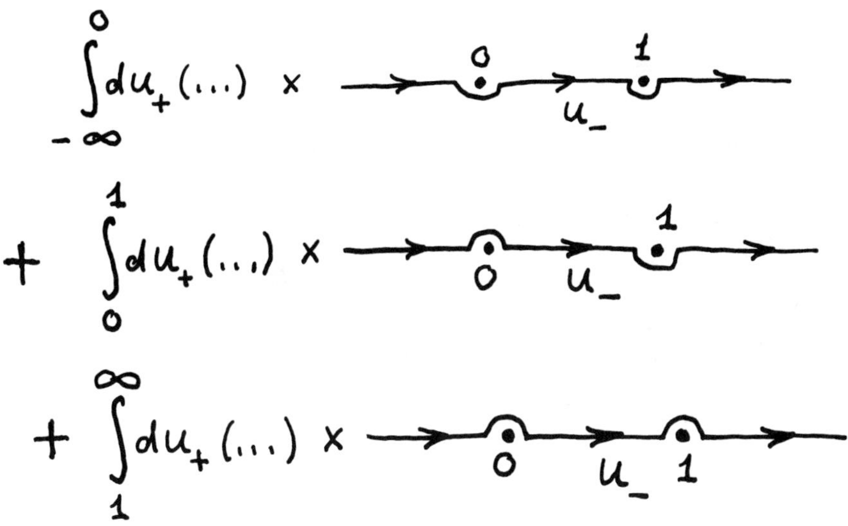

Figure 10: Decomposition of the $2D$ integral $G(0)$ into a sum of products of u_+, u_- contour integrals.

As we assume convergence at ∞, the corresponding integrals vanish. There remains the second integral. This could be given as:

$$I(a,b) = \frac{i}{2} \int_0^1 du_+ (u_+)^a (1 - u_+)^b \int_c du_- (u_-)^a (u_- - 1)^b \tag{171}$$

with the contour c given in Fig.11.

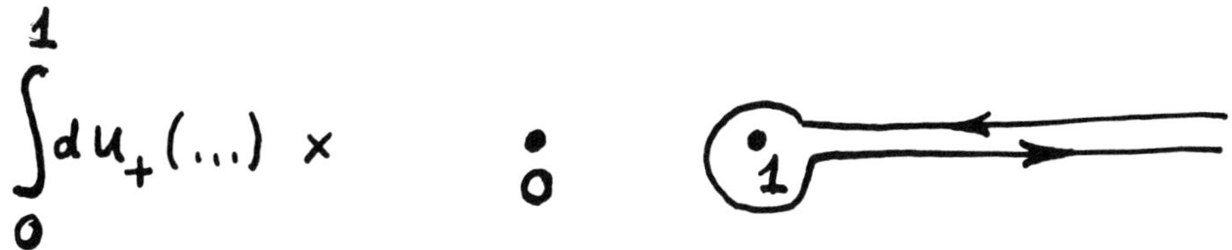

Figure 11: Deformation of the u_- contour of integration.

Finally one gets:

$$I(a,b) = \int_0^1 du_+ (u_+)^a (1 - u_+)^b \times (-\sin \pi b) \times \int_1^\infty du_- (u_-)^a (u_- - 1)^b$$

$$= -\sin \pi b \frac{\Gamma(1 + a)\Gamma(1 + b)}{\Gamma(2 + a + b)} \times \frac{\Gamma(-1 - a - b)\Gamma(1 + b)}{\Gamma(-a)} \tag{172}$$

Here $\Gamma(z)$ is Euler's Γ function. It should be noted also that in passing from eq.(53) to eq.(54) the overall phase (factor) has been adjusted so that the final result would be real, as real is the original integral (49). Using

$$\Gamma(z)\Gamma(1 - z) = \frac{\pi}{\sin \pi z} \tag{173}$$

and $\Gamma(1+z) = z\Gamma(z)$, the result (55) could be given in a symmetrized form:

$$I(a,b) = \int d^2 u |u|^{2a} |u-1|^{2b} = \pi \frac{\Gamma(1+a)}{\Gamma(-a)} \frac{\Gamma(1+b)}{\Gamma(-b)} \frac{\Gamma(-1-a-b)}{\Gamma(2+a+b)} \tag{174}$$

We come back now to the integral (48). In a similar way one gets in this case:

$$G(z,\bar{z}) = \int_0^z du_+ (u_+)^a (z - u_+)^c (1 - u_+)^b \times \int_{-\infty}^{+\infty} du_- (u_-)^a (u_- - z)^c (u_- - 1)^b$$

$$+ \int_z^1 du_+ (u_+)^a (u_+ - z)^c (1 - u_+)^b \times \int_{-\infty}^{+\infty} du_- (u_-)^a (u_- - z)^c (u_- - 1)^b \tag{175}$$

The contours of integration for the first and second integrals over u_- are given in the Fig.12. With some extra manipulation with the contours of integration one gets finally:

$$G(z,\bar{z}) = \frac{s(a)s(c)}{s(a+c)} |I_1(z)|^2 + \frac{s(b)s(a+b+c)}{s(a+c)} |I_2(z)|^2 \tag{176}$$

For more details see ref.[6]. Above the following notation is used: $s(a) = \sin \pi a, ...,$

$$I_1(z) = \int_0^z du (u)^a (z - u)^c (1 - u)^b \tag{177}$$

$$I_2(z) = \int_1^\infty du (u)^a (u - z)^c (u - 1)^b \tag{178}$$

The integrals $I_1(z)$ and $I_2(z)$ are proportional to the hypergeometric functions.

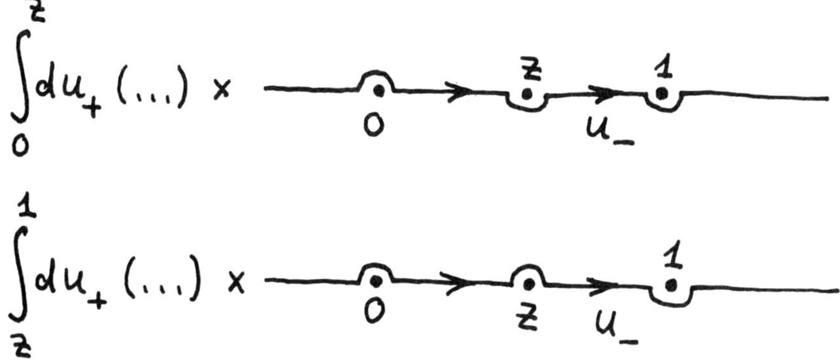

Figure 12: Sum of products of the u_+, u_- contour integrals for $G(z,\bar{z})$.

The technique and the result (58) generalizes in a straight-forward way. For the most general four-point function, given by a multiple integral, one gets:

$$\langle \Phi_1(0)\Phi_2(z,\bar{z})\Phi_3(1)\Phi_4(\infty)\rangle \equiv G(z,\bar{z}) = \sum_p X_p |I_p(z)|^2 \tag{179}$$

Here $I_p(z)$ are certain generalization of the hypergeometric functions. They are defined by multiple contour integrals [6]. They are also solutions of the differential equations that we have seen in the general formulation of minimal conformal theory.

It could be shown also that the coefficients $\{X_p\}$ factorize on the coefficients of the operator algebra. If we define

$$\Phi_1(O)\Phi_2(z,\bar{z}) = \sum_p \frac{D_{12}^p}{|z|^{2(\Delta_1 + \Delta_2 - \Delta_p)}} (\Phi_p(O) + ...) \tag{180}$$

388

and similar for the product of $\Phi_3\Phi_4$, then

$$X_p = D^p_{12}D^p_{34} \tag{181}$$

For details see ref.[7].

3. CONFORMAL FIELD THEORIES WITH EXTRA SYMMETRIES

We shall describe now how the basic structure of the conformal field theory generalizes in case when the theory possesses extra symmetries. This we shall do for the case of conformal theories based on current algebras.

The principal operators which generate the symmetries are now currents, instead of the stress-energy tensor:

$$T(z) \rightarrow J^\alpha(z) \tag{182}$$

with notations: $J^\alpha(z)$ for $J^\alpha_z(z)$, and similar for the $\bar{z}$ component. Instead of OPE

$$T(z)T(z') = \frac{c/2}{(z-z')^4} + \frac{2}{(z-z')^2}T(z') + \frac{1}{z-z'}\partial_{z'}T(z') + Reg.T.'s \tag{183}$$

which leads to the Virasoro algebra, one has OPE for currents:

$$J^\alpha(z)J^\beta(z') = \frac{k/2}{(z-z')^2}q^{\alpha\beta} + \frac{f^{\alpha\beta}_\gamma}{z-z'}J^\gamma(z') + Reg.T.'s \tag{184}$$

Here $f^{\alpha\beta}_\gamma$ are the structure constant of the corresponding classical Lie algebra; $q^{\alpha\beta}$ is its metric tensor, $q^{\alpha\beta} = tr(\lambda^\alpha\lambda^\beta)$, $\{\lambda^\alpha\}$ are the generators; k is called the level number (k corresponds to c). Introducing the decomposition into components:

$$J^\alpha(z) = \sum_{n=-\infty}^{+\infty} \frac{J^\alpha_n}{(z)^{n+1}} \tag{185}$$

$$J^\alpha_n = \frac{1}{2\pi i} \oint dz(z)^n J^\alpha(z) \tag{186}$$

one gets for them the so called Kac-Moody algebra

$$[J^\alpha_n, J^\beta_m] = f^{\alpha\beta}_\gamma J^\gamma_{n+m} + \frac{k}{2}q^{\alpha\beta}n\delta_{n,-m} \tag{187}$$

The operators which realize representations, $\Phi(z)$, are defined by their OPE with currents:

$$J^\alpha(z)\Phi(z') = \frac{\lambda^\alpha}{z-z'}\Phi(z') + Reg.T.'s \tag{188}$$

$\{\lambda^\alpha\}$ are generators of the classical Lie algebra, they are matrices corresponding to the representation of $\Phi(z')$:

$$\lambda^\alpha\Phi(z) \equiv (\lambda^\alpha)^a_b\Phi^b(z) \tag{189}$$

This is in place of OPE

$$T(z)\Phi_\Delta(z') = \frac{\Delta}{(z-z')^2}\Phi_\Delta(z') + \frac{1}{z-z'}\partial_{z'}\Phi(z') + Reg.T.'s \tag{190}$$

in minimal conformal theory. In a sense

$$\Delta \to \lambda^\alpha \tag{191}$$

Ward Identity, which is derived in way similar to minimal theory, takes the form:

$$\langle J^\alpha(z)\Phi_1(z_1)\Phi_2(z_2)...\Phi_N(z_N)\rangle = \sum_{i=1}^{N} \frac{\lambda_{(i)}^\alpha}{z - z_i}\langle \Phi_1(z_1)\Phi_2(z_2)...\Phi_N(z_N)\rangle \tag{192}$$

Here $\lambda_{(i)}^\alpha$ is in the representation of $\Phi_i(z_i)$.

In case of conformal theories based on currents the conformal (Virasoro) algebra is obtained for a composite operator:

$$T^{(Sug)}(z) = \frac{1}{k+2} : J^\alpha(z)J^\beta(z) : q_{\alpha\beta} \tag{193}$$

It is called the Sugawara stress-energy tensor. It should be noted that starting with eq.(12) we shall limit ourselves to the simplest non-trivial case, that of SU(2) current algebra. Most of the general expressions generalize in a straightforward way to the case of theories based on other classical Lie groups, see [14].

By using the OPE for currents one could check that $T^{(Sug)}$ produces the OPE for $T^{(Sug)}(z)T^{(Sug)}(z')$ with

$$c = \frac{3k}{k+2} \tag{194}$$

It is simpler in fact to check the Virasoro algebra for the components

$$L_n^{(Sug)} = \frac{q_{\alpha\beta}}{k+2}\sum_m : J_{-m}^\alpha J_{n+m}^\beta : \tag{195}$$

by using the Kac-Moody algebra for $\{J_n^\alpha\}$. Normal ordering for products of $\{J_m^\alpha\}$ means placing $\{J_m^\alpha\}$ with a negative index m to the left.

One gets differential equation for correlation functions as follows: Take

$$L_{-1}^{(Sug)} = \frac{q_{\alpha\beta}}{k+2}\sum_m : J_{-m}^\alpha J_{-1+m}^\beta : \tag{196}$$

Put this into a correlation function

$$\langle (L_{-1}^{(Sug)}\Phi_1(z_1))\Phi_2(z_2)\Phi_3(z_3)...\rangle = \frac{q_{\alpha\beta}}{k+2}\sum_m \langle (: J_{-m}^\alpha J_{-1+m}^\beta : \Phi_1(z_1))\Phi_2(z_2)\Phi_3(z_3)...\rangle \tag{197}$$

We have

$$q_{\alpha\beta}\sum_m : J_{-m}^\beta J_{-1+m}^\beta := q_{\alpha\beta}\sum_{m=+2}^{\infty} J_{-m}^\alpha J_{-1+m}^\beta + 2q_{\alpha\beta}J_{-1}^\alpha J_0^\beta + q_{\alpha\beta}\sum_{m=-1}^{-\infty} J_{-1+m}^\beta J_{-m}^\alpha \tag{198}$$

As $J_n^\alpha\Phi_1(z_1) = 0, n \geq 1$, we get:

$$\langle (L_{-1}^{(Sug)}\Phi_1(z_1))\Phi_2(z_2)\Phi_3(z_3)...\rangle = \frac{2q_{\alpha\beta}}{k+2}\langle (J_{-1}^\alpha J_0^\beta\Phi_1(z_1))\Phi_2(z_2)\Phi_3(z_3)...\rangle \tag{199}$$

Next, it is assumed that $T^{(Sug)} = T$, the stress-energy tensor which generates conformal transformations of the space. In this case

$$L_{-1}^{(Sug)}\Phi_1(z_1) = L_{-1}\Phi_1(z) = \partial_1\Phi_1(z_1) \tag{200}$$

Also we have, according to eqs.(5),(7),

$$J_0^\beta \phi_1(z_1) = \lambda_{(1)}^\beta \Phi_1(z_1) \tag{201}$$

Now, by taking the contour integral defining J_{-1}^α off $\Phi_1(z_1)$ and applying it to the rest of the operators, like this has been done in minimal conformal theory, one obtains the Knizhnik - Zamolodchikov equation for correlation functions:

$$\partial_{z_1}\langle\Phi_1(z_1)\Phi_2(z_2)\Phi_3(z_3)...\Phi_N(z_N)\rangle = \frac{2q_{\alpha\beta}}{k+2}\sum_{i=2}^{N}\frac{\lambda_{(i)}^\alpha \lambda_{(1)}^\beta}{z-z_i}\langle\Phi_1(z_1)\Phi_2(z_2)\Phi_3(z_3)...\Phi_N(z_N)\rangle \tag{202}$$

FREE-FIELD REPRESENTATION

We limit ourselves to SU(2). The fields

$$\omega(z),\ \omega^t(z),\ \varphi(z,\bar{z}) \tag{203}$$

are being employed, with actions

$$A[\omega,\omega^+] \propto i\int d^2x\,\omega^+\bar\partial\omega \tag{204}$$

$$A[\varphi] \propto \int d^2x\,\partial\varphi\bar\partial\varphi \tag{205}$$

The two-point functions are given by:

$$\langle\omega(z)\omega^+(z')\rangle = \frac{i}{z-z'} \tag{206}$$

$$\langle\varphi(z,\bar{z})\varphi(z',\bar{z}')\rangle = \log\frac{1}{|z-z'|^2} \tag{207}$$

Then one could check that the following composite operators realize the SU(2) currents:

$$J^+(z) = \omega^+(z) \tag{208}$$

$$J^0(z) = -i(\omega\omega^+ + \frac{1}{2\alpha_0}\partial\varphi) \tag{209}$$

$$J^-(z) = \omega\omega\omega^+ + ik\partial\omega + \frac{1}{\alpha_0}\partial\varphi\omega \tag{210}$$

with

$$k = -2 + \frac{1}{2\alpha_0^2} \tag{211}$$

It is called the Wakimoto representation [15]. With the normalization of two-point functions chosen above one finds the following current algebra:

$$J^\alpha(z)J^\beta(z') = \frac{k/2}{(z-z')^2}q^{\alpha\beta} + \frac{f_\gamma^{\alpha\beta}}{z-z'}J^\gamma(z') + Reg.T.'s \tag{212}$$

Here

$$q^{00} = 1,\ q^{+-} = q^{-+} = 2 \tag{213}$$

$$f_+^{+0} = f_-^{0-} = -1,\ f_0^{+-} = 2 \tag{214}$$

The rest of components are zero.

The operators which realize representations with respect to these currents are given by [16]:

$$\Phi_m^j(z) = (\omega)^{j-m} \times e^{i\alpha_j \varphi} \tag{215}$$

$$\alpha_j = -2\alpha_0 j \tag{216}$$

One checks:

$$J^+(z)\Phi_m^j(z') = \frac{-i}{z-z'}(j-m)\Phi_{m+1}^j(z') + Reg.T.'s \tag{217}$$

$$J^0(z)\Phi_m^j(z') = \frac{1}{z-z'}m\Phi_m^j(z') + Reg.T.'s \tag{218}$$

$$J^-(z)\Phi_m^j(z') = \frac{i}{z-z'}(j+m)\Phi_{m-1}^j(z') + Reg.T.'s \tag{219}$$

This corresponds to spin j representation of SU(2) algebra.

Like for minimal conformal theory, this free-fields representation could be used to calculate correlation functions. For details and further references see [17].

CONCLUDING REMARKS

The conformal field theory based on current algebras provide solutions for $2D$ Wess-Zimino models. The basic fields $\Phi(z, \bar{z})$ take values in Lie groups:

$$\Phi(z, \bar{z}) \sim g_{ab}(z, \bar{z}) \tag{220}$$

$$g_{ab} \in G, \; Lie\,group \tag{221}$$

In general, the algebra of the symmetry generators, like Virasoro or Kac-Moody, is called the chiral algebra of a respective conformal theory.

Other known chiral algebras are SUSY ones, W algebra, algebra of parafermions. The generate corresponding conformal theories.

One more large class of conformal theories is provided by various coset constructions. These are certain reductions of current algebra theories. But in general they are poorly studied. Even the corresponding symmetries are unknown. I.e. unknown are, in general, the basic chiral algebras of these theories. This results in absence of classification of the fields (operators).

Finally we mention some references which might be useful for further study of conformal field theory. For statistical physics applications they are [12, 13]. For connection to integrable models see [18]. Some references on application to $2D$ quantum gravity could be found in [19].

Acknowledgements

The efforts of the organizers of the School are gratefully acknowledged.

References

[*]   Also at the Landau Institute for Theoretical Physics, Moscow

[†]   Laboratoire associé No. 280 au CNRS.

[1] A. M. Polyakov, *JETP Lett.* **12**,(1970) 381.

[2] A. A. Belavin, A. M. Polyakov and A. B. Zamolodchikov, *J.Stat.Phys.* **34**, (1984) 333; *Nucl.Phys.* **B241**,(1984) 333.

[3] V. Kac, *Lecture Notes in Physics* **94**,(1979) 441.

[4] B. L. Feigin and D. B. Fuks, *Functional Anal.Appl.* **16**,(1982) 114.

[5] B. L. Feigin and D. B. Fuks, private communication (1983), unpublished

[6] Vl. S. Dotsenko and V. A. Fateev, *Nucl.Phys.* **B240**,(1984) 312; **B251**(1985) 691.

[7] Vl. S. Dotsenko and V. A. Fateev, *Phys.Lett.* **B154**,(1985) 291.

[8] D. Friedan, Z. Qiu and S. Shenker, *Phys.Rev.Lett.* **52**,(1984) 1575.

[9] Vl. S. Dotsenko, *J.Stat.Phys.* **34**,(1984) 781; *Nucl.Phys.* **235**,(1984) 54.

[10] V. A. Fateev and A. B. Zamolodchikov, Sov.Phys. JETP **62**,(1985) 215.

[11] R. J. Baxter, *J.Phys.* **C6**,(1973) L445.

[12] J. L. Cardy, *Phase Transitions and Critical Phenomena* **11**, C. Domb, J. L. Lebowitz eds., Academic Press, N.Y.(1987).

[13] B. Duplantier, *Physica* **D38**,(1989) 71.

[14] V. G. Knizhnik and A. B. Zamolodchikov, *Nucl.Phys.* **B247**, (1984) 83.

[15] M. Wakimoto, *Commun.Math.Phys.* **104**, (1986) 604.

[16] A. B. Zamolodchikov, *talk given at Montreal*(1984), unpublished.

[17] Vl. S. Dotsenko, *Nucl.Phys.* **B358**,(1991) 547.

[18] P. Di Francesco, *Int.J.Mod.Phys.* **A7**,(1992) 407.

[19] Vl. S. Dotsenko, *Mod.Phys.Lett.* **A6**,(1991) 3601.

ON THE BASICS OF TOPOLOGICAL QUANTUM FIELD THEORY

Laurent Baulieu

LPTHE
Université Pierre et Marie Curie
Université Denis Diderot
Laboratoire associé No. 280 au CNRS
BP 126, 4 Place Jussieu
75252 Paris CEDEX 05, FRANCE

1   INTRODUCTION

During the last years, Topological Quantum Field Theories have emerged as possible realizations of general coordinates invariant symmetries [1][2].

One of the special features of these theories is their ability to produce space-time metric independent correlations functions, although they are defined from a local action.

In Topological Quantum Field Theories, an important symmetry operator which is at disposal is the BRST operator Q, such that the Hamiltonian is $H = \frac{1}{2}[Q, \overline{Q}]$. Q and $\overline{Q}$ can be often understood as "twisted" deformations of $N = 2$ supersymmetry generators.

An attractive scheme is to introduce Topological Quantum Field Theories by the path integral quantization of topological terms. The techniques relies on the BRST formalism. More precisely, one can often start from a topological term, expressed as the integral over a manifold of a Lagrangian locally equal to a pure divergency which is a function of a set of given fields. Such a "classical" action is for instance a characteristic number, or any given invariant depending only of the topology of field configurations and/ or the space over which the fields are defined. No classical dynamics is generated. However, the existence of a gauge symmetry of the Lagrangian, namely the group of arbitrary infinitesimal deformations of fields, permits the quantization of the theory through the general formalism of BRST invariant gauge fixing. Our present knowledge makes this construction quite generic, provided one gets the intuition of (i) which manifold should be studied, and (ii) which fields should be introduced for this purpose. Actually, it is interesting to speculate that the symmetries of nature could be fundamentally of the topological type, and that the observed gauge symmetries would be obtained by gauge-fixing the huge topological symmetry in a BRST invariant way, leaving therefore an $N = 2$ supersymmetric theory of particles.

Frontiers in Particle Physics: Cargèse 1994
Edited by M. Lévy *et al.*, Plenum Press, New York, 1995

Not surprisingly, the problem of computing observables in Topological Quantum Field Theories is often technically complicated. The basic idea is the introduction of fields with positive and negative degrees of freedom (classical and ghost fields) which permit the exploration of topological properties through the computation of Green functions whose coefficient turn out to be topological invariants. Once the theory has been defined, dimensional reductions may appear as the only possible technical way to perform realistic computations. The complexity of these computations may hide the beautiful simplicity of the idea! As an example, to compute the knot polynomials associated to the Chern-simons theory, one reduces the $3-D$ theory into $2-D$ conformal theories [3].

One usually defines the physical Hilbert space of Topological Quantum Field Theories as the cohomology of Q (states which are annihilated by Q without being the Q transformation of other states). This definition of the physical Hilbert space is perfectly suited for ordinary gauge theories. For Topological Field Theories there are doubts on the general validity of this definition. Due to properties of the vacuum, other relevant observables than those defined by the BRST cohomology could exist. In particular, $Q-$exact observables with non vanishing mean values can exist. This is for instance the case in topological models of the type of those introduced by Witten in [1]. In other topological ones, based on first order actions like the Chern-Simons action [3], formal arguments show that the situation is similar. In all these models, one sees furthermore that a local version of of the topological BRST symmetry seems to single out the form of the supersymmetric potential [5][6].

The simplest examples of Topological Quantum Field Theories are zero-dimensional and turn out to be are $N = 2$ supersymmetric quantum mechanics models. Interestingly enough, two models exist which illustrate both extreme cases: (i) the Hilbert space is made of pure topological observables and (ii) the Hilbert space is made of particle degrees of freedom. In these notes we find it interesting to detail them. Indeed they provide elementary examples showing the basic rules of the BRST invariant topological gauge fixing procedure. In particular, they address the questions of the selection of gauge functions and of the computability of observables (which can be completely worked out in the case (i)). The example (ii) is intriguing since it might be generalized to other particle or string models with N=2 supersymmetry.

These notes are the result of joint works with R. Attal and E. Rabinovicci.

2 SUPERSYMMETRIC QUANTUM MECHANICS ON A PUNCTURED PLANE

We wish to work with a simple topological classical Lagrangian that is a candidate to generate a topological quantum mechanics. We consider as a target space a plane from which we exclude the origin, so that one has a non trivial, although very simple topological structure defined by the winding number around the origin of the trajectories of a particle. We denote the time by the real variable t and the Euclidian time by τ, with $t = i\tau$ and τ real. The cartesian coordinates on the plane are q_i, with $i = 1, 2$. We select trajectories with periodic conditions, namely such that between the initial and final times $t = 0$ and $t = T$ the particle ends up at its starting point so an integer value of the winding number can be assigned to its trajectory.

From our understanding of the nature of a topological field theory [4], we start from a topological classical action $\mathcal{I}_{cl}[\vec{q}]$. $\mathcal{I}_{cl}[\vec{q}]$ must not depend on the time metric. This condition is satisfied if it is the integral of a locally closed form. The natural candidate is

$$
\begin{aligned}
\mathcal{I}_{cl}[\vec{q}] = \int f d\tau &= \int_0^T d\tau \, f \dot{\tau}(\tau) \\
&= \int_0^T d\tau \, f \frac{\epsilon^{ij} \dot{q}_i q_j}{\vec{q}^2}
\end{aligned}
\tag{1}
$$

where f is a real number. This action measures the winding number of the particle times $f/2\pi$. It shares analogy with the second Chern class $\int d^4x \, \mathrm{tr}\, F \wedge F$ where F is the curvature of a Yang-Mills field. Here and in what follows the symbol $\dot{X}$ denotes $\frac{dX}{d\tau}$.

To obtain the Topological Quantum Theory associated to our space, we need to give sense to the Euclidian path integral

$$
\int \mathcal{D}[\vec{q}] \exp -\mathcal{I}_{cl}[\vec{q}]
\tag{2}
$$

as well as to compute topological quantities from Green functions

$$
\text{Topological information} = \int \mathcal{D}[\vec{q}] O \exp -\mathcal{I}_{cl}[\vec{q}]
\tag{3}
$$

where O is a well chosen composite operator.

The difficulty for realizing this objective is that our action is different from that of conventional quantum mechanics where classical degrees of freedom exist at the classical level and quantum fluctuations occur around the solutions of equations of motion. Here the Lagrangian is locally a pure derivative, the Hamiltonian vanishes and one has no equation of motion. On the other hand, one observes that the action $\mathcal{I}_{cl}[\vec{q}]$ is invariant under the gauge symmetry

$$
\vec{q(t)} \rightarrow \vec{q(t)} + \vec{\epsilon(t)}
\tag{4}
$$

where $\epsilon(t)$ is any given local shift of the particle position $q(t)$ which does not change the winding number of the trajectory. Using the BRST technique it is then possible to define the path integrals (2) and (3) by a conventional gauge fixing of the action $\mathcal{I}_{cl}[\vec{q}]$.

The BRST transformation laws associated to the symmetry (4) are of the simple form

$$s\vec{q} = \vec{\Psi} \quad s\vec{\Psi} = 0 \quad s\vec{\overline{\Psi}} = \vec{\tau} \quad s\vec{\tau} = 0 \tag{5}$$

The anticommuting fields $\vec{\Psi}(t)$ and $\vec{\overline{\Psi}}(t)$ are the topological ghosts and antighosts associated to the particle position $\vec{q}(t)$. $\vec{\tau}(t)$ is a Lagrange multiplier. s acts on field functions as a differential operator graded by the ghost number.

To get a gauge fixed action with a quadratic dependence on the velocity $\dot{\vec{q}}$, one chooses a gauge function of the type $\dot{q}_i + \frac{\delta V}{\delta q_i}$, where the prepotential V is an arbitrary given function of $\vec{q}$. This yields the following gauge fixed BRST invariant action $\mathcal{I}_{gf}$ which is supersymmetric

$$
\begin{aligned}
\mathcal{I}_{gf}[\vec{q}, \vec{\Psi}, \vec{\overline{\Psi}}, \vec{\tau}] &= \int_0^T d\tau \left(f\dot{\tau} - s\overline{\Psi}_i(\frac{1}{2}\tau_i - i\dot{q}_i + \frac{\delta V}{\delta q_i}) \right) \\
&= \int_0^T d\tau \left(f\dot{\tau} - \frac{1}{2}\tau_i^2 + i\tau_i \left(\dot{q}_i + \frac{\delta V}{\delta q_i} \right) - i\overline{\Psi}_i \left(\dot{\Psi}_i + \frac{\delta^2 V}{\delta q_i q_j}\Psi_j \right) \right)
\end{aligned} \tag{6}
$$

The BRST symmetry $s\mathcal{I}_{gf}[\vec{q}, \vec{\Psi}, \vec{\overline{\Psi}}, \vec{\tau}] = 0$ holds true independently of the choice of the function $V(\vec{q})$ and the partition function and the mean values of BRST invariant observables

$$Z = \int \mathcal{D}[\vec{q}]\mathcal{D}[\vec{\Psi}]\mathcal{D}[\vec{\overline{\Psi}}]\mathcal{D}[\vec{\tau}] \exp -\mathcal{I}_{gf} \tag{7}$$

$$< O > = \int \mathcal{D}[\vec{q}]\mathcal{D}[\vec{\Psi}]\mathcal{D}[\vec{\overline{\Psi}}]\mathcal{D}[\vec{\tau}] O \exp -\mathcal{I}_{gf} \tag{8}$$

are now well defined Euclidian path integrals. To understand $\dot{q}_i + \frac{\delta V}{\delta q_i}$ as a gauge function for the quantum variable $\vec{q}$, one may interpret the result of the integration over the ghosts as a determinant. The BRST invariance of the field polynomial O allows one to prove, at least formally, the topological properties of $< O >$. On the other hand our knowledge of supersymmetric quantum mechanics tells us that this mean value may depend on the class of the function V. What happens is that in the case of topological field theories, the Euclidian path integral explores the moduli-space of the equation $\dot{q}_i + \frac{\delta V}{\delta q_i} = 0$, as a result of the gauge fixing.

The question of finding a symmetry principle which would select the prepotential $V(\vec{q})$ leading to interesting topological information was investigated in [5]. The idea is to ask for the invariance of the action under a symmetry which is more restrictive than the topological BRST symmetry, namely a local version of it, for which the parameter becomes an affine function of the time, with arbitrary infinitesimal coefficients. One requires

$$\delta_l \mathcal{I}_{gf}[\vec{q}, \vec{\Psi}, \vec{\overline{\Psi}}, \vec{\tau}] = 0 \tag{9}$$

where the "local" BRST transformations δ_l are

$$\delta l \vec{q} = \eta(t)\vec{\Psi} \quad \delta_l \vec{\Psi} = 0 \quad \delta_l \vec{\overline{\Psi}} = \eta(t)\vec{\tau} - \dot{\eta}(t)\vec{q} \quad \delta_l \vec{\tau} = \dot{\eta}(t)\vec{\Psi} \tag{10}$$

and $\eta(t) = a + bt$ where a and b are constant anticommuting parameters. The idea of local BRST symmetry was considered in [9] for the sake of interpreting higher order cocycles which occurs when solving the anomaly consistency conditions, and has been shown to play a role in topological field theories in [6].

Imposing this local symmetry implies that V satisfies the constraint [5]

$$\frac{\delta V}{\delta q_i} + q_j \frac{\delta^2 V}{\delta q_i \, \delta q_j} = 0 \tag{11}$$

This constraint is solved for $V(\vec{q}) = f\tau$ where τ is the angle such that $q_1 + iq_2 = |\vec{q}| \exp i\tau$ and f is a number [5][1]. By putting this value of $\dot{q}_i + \frac{\delta V}{\delta q_i}$ in (6) and eliminating the Lagrange multiplier τ by its equation of motion we obtain

$$\mathcal{I}_{gf}[\vec{q}, \vec{\Psi}, \vec{\overline{\Psi}}] = \int_0^T \, d\tau \left(\frac{1}{2}\dot{q}_i^2 + \frac{f^2}{2\vec{q}^2} - \overline{\Psi}_i \left(\dot{\Psi}_i \pm f \frac{\delta^2\tau}{\delta q_i \delta q_j} \Psi_j \right) \right) \tag{12}$$

Notice that

$$\begin{aligned}
\frac{\delta^2\tau}{\delta q_i \delta q_j} &= \frac{1}{\vec{q}^2}\begin{pmatrix} -\sin 2\tau & \cos 2\tau \\ \cos 2\tau & \sin 2\tau \end{pmatrix}_{ij} \\
&= \frac{1}{\vec{q}^2}\begin{pmatrix} \cos\tau & -\sin\tau \\ \sin\tau & \cos\tau \end{pmatrix}\begin{pmatrix} 0 & -1 \\ -1 & 0 \end{pmatrix}\begin{pmatrix} \cos\tau & \sin\tau \\ -\sin\tau & \cos\tau \end{pmatrix}_{ij}
\end{aligned} \tag{13}$$

The superconformal potential $1/\vec{q}^2$ has been already studied in [7][10]. We shall shortly compute the observables which seems interesting to us from the topological point of view in the canonical quantization formalism. We will show that a very specific supersymmetry breaking mechanism occurs and implies the existence of non vanishing Q exact observables which are metric independent as well as of a fractional Witten index.

We believe that the signal that the theory truly carries some topological information is the existence of an interesting instanton structure. Let us remember that, from our gauge fixing in the Euclidian time region, we have obtained an action whose bosonic part is the square of the gauge function. It follows that the solutions to the Euclidien equations of motion can be written as

$$\dot{q}_i + \frac{\epsilon^{ij} q_j}{\vec{q}^2} = 0 \tag{14}$$

$$\dot{\Psi}_i \pm f\frac{\delta^2\tau}{\delta q_i \delta q_j}\Psi_j = 0 \tag{15}$$

If we introduce $z = q_1 + iq_2$ and $\Psi_z = \Psi_1 + i\Psi_2$, with $sz = \Psi_z$, we can write these equations as

$$\dot{z} + \frac{i}{z^*} = 0 \tag{16}$$

$$\dot{\Psi}_z - \frac{i}{(z^*)^2}\Psi_z^* = 0 \tag{17}$$

---

[1]For the case of one variable x we would obtain $V = \log x$, with quite similar properties of the supersymmetric system, but the geometrical interpretation would be less clear and no meaningful observable exists

Assuming periodic boundary conditions, the solutions for $\vec{q}$ are circles described at constant velocities and indexed by an integer n

$$z^{(n)} = \sqrt{\frac{T}{2n\pi}} \exp -i\frac{2nt}{T} \qquad n \in Z \tag{18}$$

while for the ghost

$$\Psi_z^{(n)} = \eta \exp i\frac{2nt}{T} \tag{19}$$

where η is a constant fermion. The Euclidian energy and angular momentum of the action evaluated for these field configurations vanish for all values of n.

Due to the existence of these degenerate zero modes of the action we expect that BRST invariant observables should exist and that their mean values should be non zero as well as energy and time reparametrization independent. The corresponding numbers should to be expressible as a series over an integer related to the one which label the instanton solutions. This is the conjecture that we shall now verify.

To compute the meam values of observables, we will use the canonical formalism. We do a Wick rotation to recover the real Minkowski time t by setting $\tau = it$, and change the quantum mechanical variables into operators. The Hamiltonian associated to the action $\mathcal{I}_{gf}$ is

$$H = \frac{1}{2}\varphi^2 + \frac{f^2}{2\vec{q}^2} - f\overline{\Psi}_i \frac{\delta^2 \tau}{\delta q_i \delta q_j}\Psi_j \tag{20}$$

where the quantization rules are (remember that $q_i = (x, y)$ stands for the cartesian coordinates on the plane)

$$[p_i, q_j] = -i\delta_{ij} \qquad [\overline{\Psi}_i, \Psi_j]_+ = \delta_{ij}$$

$$[\Psi_i, \Psi_j]_+ = [\overline{\Psi}_i, \overline{\Psi}_j]_+ = [\Psi_i, p_j] = [\overline{\Psi}_i, p_j] = [\Psi_i, q_j] = [\overline{\Psi}_i, q_j] = 0 \tag{21}$$

By construction H can be written as

$$H = \frac{1}{2}\left\{Q, \overline{Q}\right\} \tag{22}$$

with

$$Q = \Psi_i(p_i + if\frac{\delta\tau}{\delta q_i}) \qquad \overline{Q} = \Psi_i(p_i - if\frac{\delta\tau}{\delta q_i}) \tag{23}$$

Following [10], we use the following matricial representation for the ghost and antighost operator

$$\Psi_1 = \begin{pmatrix} 0 & 1 & 0 & 0 \\ 0 & 0 & 0 & 0 \\ 0 & 0 & 0 & 1 \\ 0 & 0 & 0 & 0 \end{pmatrix} \qquad \Psi_2 = \begin{pmatrix} 0 & 0 & -1 & 0 \\ 0 & 0 & 0 & 1 \\ 0 & 0 & 0 & 0 \\ 0 & 0 & 0 & 0 \end{pmatrix} \tag{24}$$

One has $\overline{\Psi} = \Psi^\dagger$ and $p_i = -i\delta/\delta q_i$. In this representation

$$H = \begin{pmatrix} H_0 & 0 & 0 & 0 \\ 0 & H_{11} & H_{12} & 0 \\ 0 & H_{21} & H_{22} & 0 \\ 0 & 0 & 0 & H_2 \end{pmatrix} \tag{25}$$

where

$$H_0 = H_2 = -\frac{1}{2r}\frac{\delta}{\delta r}r\frac{\delta}{\delta r} - \frac{1}{2r^2}\frac{\delta^2}{\delta\tau^2} + \frac{f^2}{2r^2} \tag{26}$$

and

$$\begin{pmatrix} H_{11} & H_{12} \\ H_{21} & H_{22} \end{pmatrix} = H_0\delta_{ij} + f\frac{\delta^2\tau}{\delta q_i \delta q_j}$$

$$= R_{-\tau}\left(-\frac{1}{2r}\frac{\delta}{\delta r}r\frac{\delta}{\delta r} + \frac{f^2 + 1 - \frac{\delta^2}{\delta\tau^2}}{2r^2} - \frac{1}{r^2}\begin{pmatrix} 0 & f + i\frac{\delta}{\delta\tau} \\ f - i\frac{\delta}{\delta\tau}\tau & 0 \end{pmatrix}\right)R_\tau \tag{27}$$

where r and τ are the polar coordinates on the plane and

$$R_\tau = \begin{pmatrix} \cos\tau & \sin\tau \\ -\sin\tau & \cos\tau \end{pmatrix} \tag{28}$$

The spectrum of H is straightforward to derive in this representation. One uses the usual strategy based on the fact that if an eigenstate of H has energy E, its Q and $\overline{Q}$ transforms are either zero or an eigenstate of H with the same energy. States are labelled by their non negative energy E, angular momentum n and fermion number α, that is ghost number. We denote them as $|E, n, \alpha >$. For each value E and n, one has four states labelled by $\alpha = 1, 2, 3, 4$. The states with $\alpha = 1$ and $\alpha = 4$ are respectively annihilated by Q and $\overline{Q}$. This is due to the fact that states $|\phi >$ which are BRST invariant, $Q|\phi >= 0$, are such that

$$|\phi >= \begin{pmatrix} |E, n > \\ 0 \\ 0 \\ 0 \end{pmatrix} \qquad \overline{Q}|\phi >= \begin{pmatrix} 0 \\ (p_1 - if\frac{\delta\tau}{\delta q_1})|E, n > \\ -(p_2 - if\frac{\delta\tau}{q_2})|E, n > \\ 0 \end{pmatrix}$$

$$H|\phi >= \begin{pmatrix} H_0|E, n > \\ 0 \\ 0 \\ 0 \end{pmatrix} \tag{29}$$

One has similar relations for states $|\overline{\phi} >$ satisfying $\overline{Q}|\overline{\phi} >= 0$.

Let us define $g_{E,n} =< r, \tau|E, n >$. This function is the solution of the equation

$$< r, \tau|H_0|E, n >= \left(-\frac{1}{2r}\frac{\delta}{\delta r}r\frac{\delta}{\delta r} - \frac{1}{2r^2}\frac{\delta^2}{\delta\tau^2} + \frac{f^2}{2r^2}\right)g_{E,n} = Eg_{E,n} \tag{30}$$

$g_{E,n}$ is also the solution of the ghost number 2 equation $< r, \tau|H_2|E, n >= E|E, n >$. Its knowledge is sufficient to get the full spectrum for $E \neq 0$. One has indeed

$$|E, n, 1 >= \begin{pmatrix} |E, n > \\ 0 \\ 0 \\ 0 \end{pmatrix} \qquad |E, n, 2 >= \frac{1}{\sqrt{E}}\overline{Q}|E, n, 1 >$$

$$|E, n, 4 >= \begin{pmatrix} 0 \\ 0 \\ 0 \\ |E, n > \end{pmatrix} \qquad |E, n, 3 >= \frac{1}{\sqrt{E}}\overline{Q}|E, n, 4 > \tag{31}$$

The diagonalization of the part with ghost number one of the Hamiltonian (27) amounts to solve the equations

$$\left(-\frac{1}{2r}\frac{\delta}{\delta r}r\frac{\delta}{\delta r} + \frac{f^2+1-\frac{\delta^2}{\delta\tau^2}\pm 2\sqrt{f^2-\frac{\delta^2}{\delta\tau^2}}}{2r^2}\right)g_{E,n,\pm} = Eg_{E,n,\pm} \tag{32}$$

which are of the same type as (30).

To solve (30) and (32) we set

$$g_{E,n} = \frac{1}{\sqrt{2\pi}}\exp in\tau\, f_{E,n}(r) \qquad n \in Z$$

$$g_{E,n,\pm} = \frac{1}{\sqrt{2\pi}}\exp in\tau\, f_{E,n,\pm}(r) \qquad n \in Z \tag{33}$$

For $E \neq 0$, $f_{E,n}(r)$ and $f_{E,n,\pm}$ are expressible as a Bessel function $J_\nu(\sqrt{2}Er)$ of order ν, with

$$f_{E,n}(r) = \frac{1}{\sqrt{2}}J_{\sqrt{n^2+f^2}}(\sqrt{2}Er) \tag{34}$$

and

$$f_{E,n,\pm}(r) = \frac{1}{\sqrt{2}}\exp in\tau J_{\sqrt{f^2+1+n^2\pm 2\sqrt{f^2+n^2}}}(\sqrt{2}Er) \tag{35}$$

These states are normalizable as plane waves in one dimension. This is a consequence of the continuity of the spectrum in the radial direction. They build an appropriate basis of stationary solutions since, with the normalization factor which is explicit in (34), one has $\sum_n \int_{E>0} dE|E,n><E,n| = 1$. On the other hand, for $E = 0$, the Schrödinger equations (30) and (32) have no admissible normalizable solution. Thus we have a continuum spectrum, bounded from below, with a spin degeneracy equal to 4 and an infinite degeneracy in the angular momentum quantum number n. The peculiarity of this spectrum is that there is no ground state, since we have states with energy as little as we want, but we cannot have $E = 0$. This is a consequence of the conformal property of the potential $\frac{1}{|\vec{q}|^2}$.

Since we cannot reach the energy zero which would be the only Q and $\overline{Q}$ invariant state, we conclude that supersymmetry is broken.

It is useful for what follows to redefine the ghost and antighost operators into

$$\begin{pmatrix}\Psi_r \\ \Psi_\tau\end{pmatrix} = \begin{pmatrix}\cos\tau & \sin\tau \\ -\sin\tau & \cos\tau\end{pmatrix}\begin{pmatrix}\Psi_1 \\ \Psi_2\end{pmatrix} \quad \begin{pmatrix}\overline{\Psi}_r \\ \overline{\Psi}_\tau\end{pmatrix} = \begin{pmatrix}\cos\tau & \sin\tau \\ -\sin\tau & \cos\tau\end{pmatrix}\begin{pmatrix}\overline{\Psi}_1 \\ \overline{\Psi}_2\end{pmatrix} \tag{36}$$

These rotated ghost operators satisfy similar anticommutation rotations as the Ψ_i and $\overline{\Psi}_i$. On the other hand, notice that

$$[\frac{\delta}{\delta\tau}, \begin{pmatrix}\overline{\Psi}_r \\ \overline{\Psi}_\tau\end{pmatrix}]_+ = \begin{pmatrix}\overline{\Psi}_\tau \\ -\overline{\Psi}_r\end{pmatrix} \qquad [\frac{\delta}{\delta\tau}, \begin{pmatrix}\Psi_r \\ \Psi_\tau\end{pmatrix}] = \begin{pmatrix}\Psi_\tau \\ -\Psi_r\end{pmatrix} \tag{37}$$

$$[\frac{\delta}{\delta r}, \begin{pmatrix}\overline{\Psi}_r \\ \overline{\Psi}_\tau\end{pmatrix}] = [\frac{\delta}{\delta r}, \begin{pmatrix}\Psi_r \\ \Psi_\tau\end{pmatrix}]_+ = 0 \tag{38}$$

One has the following expression of Q and $\overline{Q}$ which will be used shortly

$$Q = -i\Psi_r \frac{\delta}{\delta r} - i\frac{1}{r}\Psi_\tau(\frac{\delta}{\delta \tau} - f) \qquad \overline{Q} = -i\overline{\Psi}_r \frac{\delta}{\delta r} - i\frac{1}{r}\overline{\Psi}_\tau(\frac{\delta}{\delta \tau} + f) \tag{39}$$

These expressions in curved coordinates could be obtained from the general formalism of [11]].

We now turn to computation of BRST Invariant Observables. We have just seen that supersymmetry is broken in a very special way. This opens the possibility of having non vanishing BRST-exact Green functions which are topological in the sense that they are scale independent, that is independent of time, or energy, rescalings.

From dimensional arguments the candidates for such commutators are

$$O_\tau = [Q, r\overline{\Psi}_\tau]_+ = [\overline{Q}, r\Psi_\tau]_+^\dagger \qquad O_r = [Q, r\overline{\Psi}_r]_+ = [\overline{Q}, r\Psi_r]_+^\dagger \tag{40}$$

The mean values of these operators between normalized states are

$$\frac{< E, n|[Q, r\overline{\Psi}_\tau]_+|E, n >}{< E, n|E, n >} = n + if \tag{41}$$

and

$$\frac{< E, n|[Q, r\overline{\Psi}_r]_+|E, n >}{< E, n|E, n >} = \lim_{L\to\infty} \frac{L^2 J^2_{\sqrt{n^2+f^2}}(L)}{\int_0^L dr J_{\sqrt{n^2+f^2}}(r)} \tag{42}$$

The last quantity is bounded but ill-defined, so we reject it. We get therefore that for any normalized state $|\phi_n >= \int dE\rho(E)|E, n >$ with a given angular momentum n, the expectation value of $[Q, r\overline{\Psi}_\tau]_+$ is

$$< \phi_n|[Q, r\overline{\Psi}_\tau]_+|\phi_n >= n + if \tag{43}$$

independently of the weighting function ρ.

If we now sum over all values of n, what remains is the topological number

$$< O_\tau >= \sum_n < \phi_n|[Q, r\overline{\Psi}_\tau]_+|\phi_n >= \sum_n n + if \sum_n 1 \tag{44}$$

¿From a topological point of view, our result mean that there are two observables, organized in a complex form, in the cohomology of the punctured plane. The summation over the index n, that is the angular momentum, could have expected from the formal argument that in the path integral one gets a single finite contribution from each instanton solution to the mean value of a topological observable, so that

$$\text{Topological information} = \int \mathcal{D}[\vec{q}] O_f \exp -\mathcal{I}_{cl}[\vec{q}] \sim \sum_n f(n) \tag{45}$$

Our computation shows the existence of a BRST invariant observable with non zero mean value which is Q-closed. The supersymmetry breaking mechanism made possible by our potential choice (on the basis of local BRST symmetry) is responsible of this situation. With other potentials than the one that we have chosen , either supersymmetry would be unbroken, or a mass gap would occur. In the previous case all Q-exact observable would vanish; in the latter case they could be nonzero but they would be scale dependent.

As another topological observable of the theory, we may consider the Witten index [12] [13]. The idea is that although there is no normalizable vacuum in the theory, we can consider the trace

$$\Delta = \mathrm{Tr}(-)^F \exp -\beta H \tag{46}$$

where the trace means a sum over angular momentum as well as over all energy including energy zero, and $(-)^F$ is the ghost or fermion number operator. The result should be finite because, although the state with energy zero is not normalizable, it contributes only over a domain of integration with zero measure. Indeed, since supersymmetric compensations occur for $E \neq 0$ and provided one uses a BRST symmetry preserving regularization, the full contribution to Δ should come from the domain of integration concentrated at $E \sim 0$, while the topological nature of the theory should warranty that Δ is non zero and independent on β.

By using the suitably normalized eigen-functions of the Hamiltonian, eqs.(34) and (35), one can write the index Δ as follows

$$\Delta = \sum_n \int_0^\infty dE \quad \exp \quad -\beta E \int r dr \frac{1}{2} (2J^2_{\sqrt{n^2+f^2}}(\sqrt{2E}r)$$
$$- \quad J^2_{\sqrt{f^2+1+n^2+\sqrt{f^2+n^2}}}(\sqrt{2E}r) - J^2_{\sqrt{f^2+1+n^2-\sqrt{f^2+n^2}}}(\sqrt{2E}r))$$

$$\tag{47}$$

To compute this double integral one needs a regularization. Following for instance [13], we can use a dimensional regularization . Thus we change dr into $r^\epsilon dr$. Then, the analytic continuation of the result when $\epsilon \to 0$ is

$$\Delta = \sum_n \frac{1}{2} \left(2\sqrt{f^2+n^2} - \sqrt{f^2+1+n^2+\sqrt{f^2+n^2}} - \sqrt{f^2+1+n^2-\sqrt{f^2+n^2}} \right)$$

$$\tag{48}$$

As announced this result is independent on β. As a series, it diverges logarithmically as $\sum 1/n$ which is presumably the consequence of the conformal invariance of the potential. We see that the contribution of each topological sector is n dependant.

Let us now summarize what we understood from this model. We have shown an example for which the requirement of local BRST symmetry for topological quantum mechanics results in selecting a superconformal quantum mechanichal system. As a result, the spectrum of the theory has no ground state and a supersymmetry breaking mechanism occurs, without the the presence of a dimensionful parameter. Our goal was to understand the mechanism which provide topological observables. We observed that the special properties of the potential allows the computation of energy independent quantities although they are of mean values of BRST exact observables between non zero energy states. These quantities deserve to be called topological and they get a contribution from the whole spectrum of the theory. We have also singled out the Witten index, in a computation which includes a contribution from the non normalizable state of zero energy. The generalization of these observations to quantum field theory is an interesting open question.

3 The SUPERSYMMETRIC LAGRANGIAN FOR SPIN-ONE PARTICLES

Supersymmetric quantum mechanics can be used to describe the dynamics of spinning point particles. The use of anticommuting variables to describe spinning particles was introduced in [14]. Then, it was found that local supersymmetry of rank $2S$ on the worldline is necessary to describe consistently a particle of spin S. The resulting constrained system [17] [18] requires a careful gauge-fixing of the einbein and the gravitini. One obtains eventually a tractable Lagrangian formulation [19], [20]. (There are many references on the subject, of which we quote very few) as well as to compute a certain number of topological invariants of the target space [12].

Using these facts, we will now point out an example showing that topological quantum theories may exhibit a phase with a Hilbert space made of particle degrees of freedom. We will interpret local supersymmetry on the worldline as a residue of a more fundamental topological symmetry, defined in a target-space with two extra dimensions. One of the coordinates is eventually identified as the einbein on the worldline. Other fields must be introduced to enforce the topological BRST invariance. They can be eliminated by their equations of motion and decouple from the physical sector. To obtain in a natural way a nowhere vanishing einbein, we use a disconnected higher dimensional target-space where the hyperplane $\{e = 0\}$ is a priori extracted. Thus, one introduces some topology before any gauge-fixing. Two disconnected topological sectors exist, $\{e > 0\}$ and $\{e < 0\}$, which correspond to the prescription $\pm i\epsilon$ for the propagators. It is fundamental that the gauge functions be compatible with the topology of space: they must induce a potential which rejects the trajectories from the hyperplane $\{e = 0\}$.

We will first review the supersymmetric description of a relativistic spinning particle in a Riemannian space-time. Then we will consider the case of $N = 2$ supersymmetry and show a link between the supersymmetric description of scalar or spin-one particles and topological quantum mechanics in a higher dimensional target-space. Finally, we will verify that the constraints of the theory identify its physical content and illustrate the result by computing the deviation of the trajectories from geodesics due to the interactions between geometry and spin.

Consider a spin-S particle in a D-dimensional space-time. Classically, it follows a worldline whose coordinates $X^\mu(\tau)$ are parametrized by a real number τ. If the particle is massive, a natural choice of this parameter is the proper-time. The idea originating from [14] is to describe the spin of the particle by assigning to each value of τ a vector with anticommuting coordinates $\Psi_i^\mu(\tau)$ where the vector index μ runs between 1 and D and i between 1 and $2S$. Indeed, in the case of a flat space-time and spin one-half, the Lagrangian density introduced in [14] is

$$\mathcal{L} = \frac{1}{2}(\dot{X}^2(\tau) - \Psi^\mu(\tau)\dot{\Psi}_\mu(\tau)) \tag{49}$$

where the dot $\dot{\ }$ means ∂_τ, τ being a parametrization of the worldline. Upon canonical quantization $\Psi^\mu(\tau)$ is replaced by a τ-independent operator $\hat{\Psi}^\mu$ which satisfies anticommutation relations

$$\{\hat{\Psi}^\mu, \hat{\Psi}_\nu\}_+ = 2\delta_\nu^\mu \tag{50}$$

The Hamiltonian is

$$H = \frac{1}{2}p^2 = \frac{1}{2}Q^2 \tag{51}$$

with $Q = p_\mu \hat{\Psi}^\mu$. Due to (50) the $\hat{\Psi}$'s can be represented by Dirac matrices and Q is the free Dirac operator. Q commutes with H and it makes sense to consider the restriction of the Hilbert space to the set of states $|\varphi>$ satisfying

$$Q|\varphi>= 0 \tag{52}$$

By definition of Q, this equation means that the $|\varphi>$ are the states of a massless spin one-half particle. The extension to the case of a massive particle implies the introduction of an additional Grassmann variable Ψ^{D+1} and the generalization of $\mathcal{L}$ to

$$\mathcal{L} = \frac{1}{2}(\dot{X}^2(\tau) - \Psi^\mu(\tau)\dot{\Psi}_\mu(\tau) - \Psi^{D+1}(\tau)\dot{\Psi}^{D+1}(\tau) + m^2) \tag{53}$$

(Formally, $\dot{X}^{D+1} \to m$), so that

$$H = \frac{1}{2}(p^2 - m^2) = \frac{1}{2}Q^2 \tag{54}$$

with

$$Q = p_\mu \hat{\Psi}^\mu + m\hat{\Psi}^{D+1} \tag{55}$$

and one has in addition to (50)

$$\{\hat{\Psi}^{D+1}, \hat{\Psi}^{D+1}\}_+ = -2 \qquad \{\hat{\Psi}^\mu, \hat{\Psi}^{D+1}\}_+ = 0 \tag{56}$$

The condition (52) is now the free Dirac equation for a spin one-half particle of mass m, multiplied by $\hat{\Psi}^{D+1}$. The generalization to the case of an arbitrary spin is obtained by duplicating $2S$ times the components of Ψ, $\Psi^\mu \to \Psi_i^\mu, 1 \leq i \leq 2S$, as can be seen by constructing the representations of $SO(D)$ by suitable tensor products of spin one-half representations [21] [22].

To understand the constraint (52), it is in fact necessary to promote the global supersymmetry of the action, corresponding to the commutation of H and Q, into a local supersymmetry. Indeed, when time flows, the state of the particle must evolve from a solution of the Dirac equation to another solution of this equation, without any possibility to collapse in an unphysical state (out of $Ker(Q)$). A natural way to reach such a unitarity requirement is to impose the supersymmetry independently for all values of τ, that is, to gauge the supersymmetry on the worldline. In this way, the condition (52) appears as the definition of physical states in a gauge theory with generator Q which ensures unitarity, like the transversality condition of gauge bosons in ordinary Yang-Mills theory. For consistency, the diffeomorphism invariance on the worldline must be also imposed since the commutator of two supersymmetry transformations contains a diffeomorphism. One thus introduces gauge fields for these symmetries, the einbein $e(\tau)$ and the (anticommuting) gravitino $\alpha(\tau)$. By minimal coupling on the worldline, (53) is thus generalized to the following Lagrangian which is locally supersymmetric and reparametrization invariant, up to a pure derivative with respect to τ :

$$\mathcal{L} = \frac{1}{2}\left(e^{-1}\dot{X}^2 - \Psi(\dot{\Psi} + \alpha e^{-1}\dot{X}) - \Psi^{D+1}(\dot{\Psi}^{D+1} + m\alpha) + em^2\right) \tag{57}$$

(we will now omit the vector and spin indices). Formally, $\dot{X}^{D+1} \to me$. The transformation laws of e and α are those of one-dimensional supergravity of rank $2S$.

The gauge-fixing $e(\tau) = 1$ and $\alpha(\tau) = 0$ identifies (2.5) and (2.9), up to Faddeev-Popov ghost terms. These ghost terms have a supersymmetric form $b\dot{c} + \beta\dot{\gamma}$. They decouple effectively, since their effect is to multiply all the amplitudes by a ratio of determinants, independent of the metric in space-time. This gauge-fixing is however inconsistent because it is too strong, since the Lagrangian is gauge invariant only up to boundary terms. Therefore, given a general gauge transformation, one must put restrictions on its parameters to get the invariance of the action, and there are not enough degrees of freedom in the symmetry to enforce the gauge $e(\tau) = 1$ and $\alpha(\tau) = 0$. One can at most set $e(\tau) = e_0$ and $\alpha(\tau) = \alpha_0$, letting the constants $e_0 > 0$ and α_0 free, that is, doing an ordinary integration over e_0 and α_0 in the path integral after the gauge-fixing [19]. This yields the following partition function for the theory

$$Z = \int_0^\infty de_0 \int d\alpha_0 \int [dX(\tau)][d\Psi(\tau)] \exp - \int_0^1 d\tau \mathcal{L}_0 \tag{58}$$

with

$$\mathcal{L}_0 = \frac{1}{2}\left(e_0^{-1}\dot{X}^2 + e_0 m^2 - \Psi(\dot{\Psi} + \alpha_0 e_0^{-1}\dot{X}) - \Psi^{D+1}(\dot{\Psi}^{D+1} + m\alpha_0)\right) \tag{59}$$

Using the Lagrangian (53) instead of (59) implies that one misses crucial spin-orbit interactions described by the Grassmann integration over the constant α_0 which induces the fermionic constraint $\int d\tau(\Psi\dot{X} + me_0\Psi^{D+1}) = 0$. The use of (53) leads indeed to a spin-zero particle propagator while (2.11) leads to the expected spin one-half propagator. One gets the $\pm i\epsilon$ propagators depending on the choice of the integration domain $\{e_0 > 0\}$ or $\{e_0 < 0\}$. Notice that the e-dependence of the Lagrangian (53) gives a negligible weight in the path integral (58) to the trajectories with points near the hyperplane $\{e_0 = 0\}$. The integration over e_0 and α_0 has a simple interpretation in Hamiltonian formalism. The Hamiltonian associated to (59) is

$$\begin{aligned} H &= \frac{e_0}{2}(p^2 - m^2) + \frac{\alpha_0}{2}(p_\mu\hat{\Psi}^\mu + m\hat{\Psi}^{D+1}) \\ &= \frac{e_0}{2}(p^2 - m^2) + \frac{\alpha_0}{2}Q \end{aligned} \tag{60}$$

The constants e_0 and α_0 are thus Lagrange multipliers which force the particle to satisfy the Klein-Gordon equation and the Dirac equation (or its higher spin generalizations $Q_i|\varphi> = 0$). Observe that in Lagrangian formalism, the Klein-Gordon equation is not a consequence of the Dirac equation, due to the anticommutativity of Grassmann variables, and the two constraints $Q|\varphi> = 0$ and $H|\varphi> = 0$ must be used separately. Therefore, we have a theory where the Hamiltonian is a sum of constraints, which leads to known technical difficulties [17][18]. In Lagrangian formalism, supergravity on the worldline and its correct gauge-fixing take care of all details [19].

The above description is valid for a flat space-time. It can be generalized to the case where the particle moves in a curved space-time and/or couples to an external electromagnetic field, by minimal coupling in the target-space. The compatibility between the worldline diffeomorphism invariance and local supersymmetry with reparametrization invariance in the target-space for a general metric $g_{\mu\nu}$ is however possible only for $N \leq 2$ [21]. This phenomenon is possibly related to the limited number of consistent supergravities [23].

We will now consider the case $N = 2$ and show n the link of the theory with a topological model.

The $N = 2$ supersymmetric Lagrangian with a general background metric $g_{\mu\nu}$ is

$$
\begin{aligned}
\mathcal{L}_{SUSY} &= \frac{1}{2e}g_{\mu\nu}\dot{X}^\mu\dot{X}^\nu - \overline{\Psi}^\mu(g_{\mu\nu}\dot{\Psi}^\nu + e\Gamma_{\mu\nu\rho}\dot{X}^\nu\Psi^\rho) + e^{-1}g_{\mu\nu}\dot{X}^\mu(\overline{\Psi}^\nu\alpha + \overline{\alpha}\Psi^\nu) \\
&+ \frac{em^2}{2} - \overline{\Psi}^{D+1}\dot{\Psi}^{D+1} + m(\overline{\Psi}^{D+1}\alpha + \overline{\alpha}\Psi^{D+1}) \\
&- e^{-1}\overline{\alpha}\alpha\overline{\Psi}\Psi + \frac{e}{2}R_{\mu\nu\rho\sigma}\overline{\Psi}^\mu\Psi^\nu\overline{\Psi}^\rho\Psi^\sigma
\end{aligned}
\tag{61}
$$

where Ψ and $\overline{\Psi}$ are independent Grassmann coordinates. (Compare with [21]). The Lagrangian (61) has two local supersymmetries, with generators Q and $\overline{Q}$. An $O(2)$ symmetry between Ψ and $\overline{\Psi}$ can be enforced by introducing a single gauge field $f(\tau)$ and adding a term $f\overline{\Psi}\Psi$. However, no new information is provided, since one increases the symmetry by one generator, which is compensated by the introduction of the additional degree of freedom carried by f. The latter can indeed be gauge-fixed to zero and one recovers (61). Moreover, in view of identifying Ψ and $\overline{\Psi}$ as ghosts and antighosts, one wishes to freeze the symmetry between these two fields. We thus ignore the possibility of gauging the $O(2)$ symmetry. We will check shortly that the Hilbert space associated to the Lagrangian (3.1) contains spin-one particles.

The Lagrangian (61) can be conveniently rewritten in first order formalism by introducing a Lagrange multiplier $b^\mu(\tau)$. One gets the equivalent form

$$
\begin{aligned}
\mathcal{L}_{SUSY} &\sim -\frac{e}{2}(g_{\mu\nu}b^\mu b^\nu - m^2) + g_{\mu\nu}b^\mu(\dot{X}^\nu + e\Gamma^\nu_{\rho\sigma}\overline{\Psi}^\rho\Psi^\sigma + \overline{\Psi}^\nu\alpha + \overline{\alpha}\Psi^\nu) \\
&\quad - \overline{\Psi}^\mu(g_{\mu\nu}\dot{\Psi}^\nu + e\partial_\rho g_{\mu\nu}\dot{X}^\nu\Psi^\rho) - \Gamma_{\nu\rho\sigma}\overline{\Psi}^\rho\Psi^\sigma(\overline{\Psi}^\nu\alpha + \overline{\alpha}\Psi^\nu) \\
&\quad - \overline{\Psi}^{D+1}\dot{\Psi}^{D+1} + m(\overline{\Psi}^{D+1}\alpha + \overline{\alpha}\Psi^{D+1}) - \frac{e}{2}\partial_\nu\Gamma_{\mu\rho\sigma}\overline{\Psi}^\mu\Psi^\nu\overline{\Psi}^\rho\Psi^\sigma
\end{aligned}
\tag{62}
$$

(The symbol $\sim$ means that the two Lagrangians differ by a term which can be eliminated using an algebraic equation of motion, and, consequently, define the same quantum theory). For $e = 1$, $\alpha = \overline{\alpha} = 0$ and $\Psi^{D+1} = \overline{\Psi}^{D+1} = 0$, the Lagrangian (3.2) can be interpreted as the gauge-fixing of zero or of a term invariant under isotopies of the curve X [4]. In this interpretation the Ψ are topological ghosts and the $\overline{\Psi}$ are antighosts. The BRST graded differential operator s of the topological symmetry is defined by

$$
\begin{aligned}
sX^\mu &= \Psi^\mu \\
s\Psi^\mu &= 0 \\
s\overline{\Psi}^\mu &= b^\mu \\
sb^\mu &= 0
\end{aligned}
\tag{63}
$$

and the gauge-fixing Lagrangian is s-exact modulo a pure derivative

$$
\mathcal{L}_{GF} = s\left(\overline{\Psi}_\mu\left(-\frac{1}{2}b^\mu + \dot{X}^\mu + \frac{1}{2}\Gamma^\mu_{\rho\sigma}\overline{\Psi}^\rho\Psi^\sigma\right)\right)
\tag{64}
$$

(Since $s^2 = 0$, $\mathcal{L}_{GF}$ is s-invariant.) To identify (3.1) as a topological Lagrangian, we must introduce new ingredients. We will enlarge the target-space with two additional components, and add a ghost of ghost. We will eventually identify one of the extra coordinates with the einbein e and the other one will be forced to vary in a Gaussian way around an arbitrary scale, with an arbitrary width. The gravitini α and $\overline{\alpha}$ of the effective worldline supergravity will be interpreted as ghosts of the topological symmetry. The $O(2)$ invariance corresponds to the ghost number conservation.

We consider a $(D+2)$-dimensional space-time with coordinates $X^A = (X^\mu, X^{D+1} = e, X^{D+2})$. We exclude from the space the hyperplane $\{X^{D+1} = 0\}$ which yields two separated half-spaces, characterized by the value of $sign(e)$. We wish to define a partition function through a path integration over the curves $X^A(\tau)$, with a topological action which is invariant under the BRST symmetry associated to isotopies of this curve in each half-space. In other words we wish to construct an action by consistently gauge-fixing the topological Lagrangian $sign(e)$. In a way which is analogous to the case of topological Yang-Mills symmetry, where one gauge-fixes the second Chern class $\int Tr\, F^2$ [4], we combine the pure topological symmetry, with topological ghosts $\Psi^A_{top}(\tau)$, to the diffeomorphism symmetry on the curve, with Faddeev-Popov ghost $c(\tau)$. The apparent redundancy in the number of ghost variables $\Psi^A_{top}(\tau)$ and $c(\tau)$, which exceeds the number of bosonic classical variables, is counterbalanced by the introduction of a ghost of ghosts $\Phi(\tau)$ with ghost number two. The action of the BRST differential s is defined by

$$
\begin{aligned}
sX^\mu &= \Psi^\mu_{top} + c\dot{X}^\mu = \Psi^\mu \\
se &= \Psi^e_{top} + c\dot{e} = 2\eta = \alpha + \dot{\Psi}^{D+1} \\
sX^{D+2} &= \Psi^{D+2}_{top} + c\dot{X}^{D+2} = \Psi^{D+2} \\
s\Psi^\mu &= 0 \\
s\Psi^{D+2} &= 0 \\
s\Psi^{D+1} &= \Phi \\
s\alpha &= -\dot{\Phi} \\
s\Phi &= 0
\end{aligned}
\tag{65}
$$

In agreement with the art of BRST invariant gauge-fixing, we introduce $D+2$ antighosts with ghost number (-1) and the associated Lagrange multipliers for the gauge conditions on the X^A's. We also introduce an antighost $\overline{\Phi}$ with ghost number (-2) and its fermionic partner $\overline{\eta}$ with ghost number (-1) which we will use as a fermionic Lagrange multiplier for the gauge condition in the ghost sector. In this sector the action of s is

$$
\begin{aligned}
s\overline{\Psi}^A &= b^A \\
sb^A &= 0 \\
s\overline{\Phi} &= \overline{\eta} \\
s\overline{\eta} &= 0
\end{aligned}
\tag{66}
$$

The gauge-fixing Lagrangian must be written as an s-exact term

$$
\mathcal{L}^X + \mathcal{L}^{D+1} + \mathcal{L}^{D+2} + \mathcal{L}^\Phi = s\left(\overline{\Psi}^A(\ldots)_A + \overline{\Phi}(\ldots)\right)
\tag{67}
$$

For the gauge-fixing in the X-sector, we choose

$$
\begin{aligned}
\mathcal{L}^X = {}& s\left(-\frac{e}{2}g_{\mu\nu}\overline{\Psi}^\nu b^\mu + g_{\mu\nu}\overline{\Psi}^\mu(\dot{X}^\nu + \overline{\eta}\Psi^\nu + \frac{e}{2}\Gamma^\nu_{\rho\sigma}\overline{\Psi}^\rho\Psi^\sigma)\right) \\
= {}& -\frac{e}{2}g_{\mu\nu}b^\mu b^\nu + g_{\mu\nu}b^\mu(\dot{X}^\nu + e\Gamma^\nu_{\rho\sigma}\overline{\Psi}^\rho\Psi^\sigma + \overline{\Psi}^\nu\eta + \overline{\eta}\Psi^\nu) \\
& - \overline{\Psi}^\nu(g_{\mu\nu}\dot{\Psi}^\mu + e\partial_\rho g_{\mu\nu}\dot{X}^\nu\Psi^\rho) - \frac{e}{2}\partial_\nu\Gamma_{\mu\rho\sigma}\overline{\Psi}^\mu\Psi^\nu\overline{\Psi}^\rho\Psi^\sigma - \Gamma_{\mu\rho\sigma}\overline{\Psi}^\mu\eta\overline{\Psi}^\rho\Psi^\sigma
\end{aligned}
\tag{68}
$$

For the gauge-fixing in the e-sector, we choose

$$\mathcal{L}^{D+1} = -s\left(\overline{\Psi}^{D+1}e(m + \frac{b^{D+1}}{2})\right) = -e\frac{(b^{D+1})^2}{2} + b^{D+1}(-me + \overline{\Psi}^{D+1}\eta) + 2m\overline{\Psi}^{D+1}\eta \tag{69}$$

After elimination of the field b^{D+1}, we obtain

$$\mathcal{L}^{D+1} \sim \frac{em^2}{2} + m\overline{\Psi}^{D+1}\eta \tag{70}$$

For the gauge-fixing in the X^{D+2}-sector, we choose

$$\begin{aligned}
\mathcal{L}^{D+2} &= s\left(\overline{\Psi}^{D+2}\left(-\frac{a}{2}b^{D+2} + X^{D+2} - C - \frac{1}{a}\frac{\overline{\Psi}^{D+1}\dot{\Psi}^{D+1}}{X^{D+2} - C}\right)\right) \\
&= -\frac{a}{2}(b^{D+2})^2 + b^{D+2}\left(X^{D+2} - C - a\frac{\overline{\Psi}^{D+1}\dot{\Psi}^{D+1}}{X^{D+2} - C}\right) \\
&\quad - \overline{\Psi}^{D+2}\left(\Psi^{D+2} - as\left(\frac{\overline{\Psi}^{D+1}\dot{\Psi}^{D+1}}{X^{D+2} - C}\right)\right)
\end{aligned} \tag{71}$$

a and C are arbitrarily chosen real numbers. After elimination of the field b^{D+2}, we find

$$\mathcal{L}^{D+2} \sim -\overline{\Psi}^{D+1}\dot{\Psi}^{D+1} + \frac{1}{2a}(X^{D+2} - C)^2 - \overline{\Psi}^{D+2}\left(\Psi^{D+2} - as\left(\frac{\overline{\Psi}^{D+1}\dot{\Psi}^{D+1}}{X^{D+2} - C}\right)\right) \tag{72}$$

The variable X^{D+2} can be eliminated by its algebraic equation of motion as well as the corresponding ghosts Ψ^{D+2} and $\overline{\Psi}^{D+2}$, after some field redefinitions. X^{D+2} is concentrated in a Gaussian way around the arbitrary scale C, with an arbitrary width a. We are thus left with the propagating term for Ψ^{D+1} and $\overline{\Psi}^{D+1}$ which was missing in $\mathcal{L}^X$ and $\mathcal{L}^{D+1}$

$$\mathcal{L}^{D+2} \sim -\overline{\Psi}^{D+1}\dot{\Psi}^{D+1} \tag{73}$$

We finally choose the gauge-fixing in the ghost sector. To recover the full Lagrangian (62) and eventually identify the coordinate e as the einbein of the projection of the particle trajectory in the D-dimensional physical space-time, we need a term linear in $\overline{\eta}$ as well as another term to get rid of unwanted higher order fermionic terms. We define

$$\begin{aligned}
\mathcal{L}^{\Phi} &= s(\overline{\Phi}(m\Psi^{D+1} - \Gamma_{\nu\rho\sigma}\overline{\Psi}^{\rho}\Psi^{\sigma}\Psi^{\nu})) \\
&= \overline{\eta}(m\Psi^{D+1} - \Gamma_{\nu\rho\sigma}\overline{\Psi}^{\rho}\Psi^{\sigma}\Psi^{\nu}) + \overline{\Phi}(m\Phi - s(\Gamma_{\nu\rho\sigma}\overline{\Psi}^{\rho}\Psi^{\sigma}\Psi^{\nu}))
\end{aligned} \tag{74}$$

The dependence on the ghosts of ghosts Φ and $\overline{\Phi}$ is trivial: these fields decouple after a Gaussian integration. One has thus

$$\mathcal{L}^{\Phi} \sim m\overline{\eta}\Psi^{D+1} - \Gamma_{\nu\rho\sigma}\overline{\Psi}^{\rho}\Psi^{\sigma}\overline{\eta}\Psi^{\nu} \tag{75}$$

Adding all terms (68), (70), (73) and (75), we finally recognize that $\mathcal{L}^X + \mathcal{L}^{D+1} + \mathcal{L}^{D+2} + \mathcal{L}^{\Phi}$ is equivalent to the Lagrangian (3.2), modulo the elimination of auxiliary fields and the change of notation $(\eta, \overline{\eta}) \to (\alpha, \overline{\alpha})$. We have therefore shown the announced result: the $N = 2$ local supersymmetry of the Lagrangian describing spin-one particles is a residual symmetry coming from a topological model after a suitable gauge-fixing.

To verify the physical content of the model presented just above, we consider a flat space-time, and choose the gauge where the einbein and gravitini are constants over which we integrate. The Hamiltonian is

$$H = \frac{e_0}{2}(p^2 - m^2) + \overline{\alpha}_0 Q + \alpha_0 \overline{Q} \tag{76}$$

with

$$Q = p_\mu \Psi^\mu + m \Psi^{D+1}$$
$$\overline{Q} = p_\mu \overline{\Psi}^\mu + m \overline{\Psi}^{D+1} \tag{77}$$

The matrices Ψ and $\overline{\Psi}$ satisfy the Clifford algebra

$$\{\Psi^A, \overline{\Psi}^B\}_+ = \eta^{AB} \quad , \quad \{\Psi^A, \Psi^B\}_+ = \{\overline{\Psi}^A, \overline{\Psi}^B\}_+ = 0 \tag{78}$$

for $A, B = 1, ..., D+1$. Since the underlying gauge symmetry has Q and $\overline{Q}$ as generators, the physical states satisfy

$$Q|\phi> = 0 \qquad \overline{Q}|\phi> = 0 \tag{79}$$

in addition to

$$(p^2 - m^2)|\phi> = 0 \tag{80}$$

The Ψ and $\overline{\Psi}$ are generalizations of the Pauli matrices, and it is convenient to use a Schwinger type construction, in order to exploit directly their Clifford algebra structure. One introduces a spin vacuum $|0>$ annihilated by the Ψ's. Then, the $\overline{\Psi}$'s can be identified as their adjoints and act as creation operators. In the X representation, we can write a general state as

$$|\phi> = \left(\varphi_0 + \varphi_\mu \overline{\Psi}^\mu + \varphi_{\mu_1 \mu_2} \overline{\Psi}^{\mu_1} \overline{\Psi}^{\mu_2} + \ldots + \varphi_{\mu_1 \ldots \mu_D} \overline{\Psi}^{\mu_1} \ldots \overline{\Psi}^{\mu_D}\right) |0>$$
$$+ \overline{\Psi}^{D+1} \left(\overline{\varphi}_0 + \overline{\varphi}_\mu \overline{\Psi}^\mu + \overline{\varphi}_{\mu_1 \mu_2} \overline{\Psi}^{\mu_1} \overline{\Psi}^{\mu_2} + \ldots + \overline{\varphi}_{\mu_1 \ldots \mu_D} \ldots \overline{\Psi}^{\mu_D}\right) |0> \tag{81}$$

The wave functions $\varphi_{\mu_1 \ldots \mu_p}(X)$ and $\overline{\varphi}_{\mu_1 \ldots \mu_p}(X)$ are antisymmetric and it is useful to consider the differential forms

$$\varphi_p = \frac{1}{p!} dX^{\mu_1} \ldots dX^{\mu_p} \varphi_{\mu_1 \ldots \mu_p}(X)$$
$$\overline{\varphi}_p = \frac{1}{p!} dX^{\mu_1} \ldots dX^{\mu_p} \overline{\varphi}_{\mu_1 \ldots \mu_p}(X) \tag{82}$$

for $0 \leq p \leq D$. The constraints (79) can be conveniently written as

$$d\varphi_p + im\overline{\varphi}_{p+1} = 0 \tag{83}$$

$$d^*\overline{\varphi}_p + im\varphi_{p-1} = 0 \tag{84}$$

$$d\overline{\varphi}_p = 0 \tag{85}$$

$$d^* \varphi_p = 0 \tag{86}$$

Where $d = dx^\mu \partial_\mu$ and d^* is its Hodge dual. One has also

$$(d^* d + dd^*)\varphi = -m^2 \varphi \qquad (d^* d + dd^*)\overline{\varphi} = -m^2 \overline{\varphi} \tag{87}$$

These equations determine the independent degrees of freedom. When $m \neq 0$, they couple the two sectors of opposite chiralities. Moreover, when D is even, the first one contains $\frac{D}{2}$ forms, namely one scalar (φ_0), one vector (φ_1), ..., and one ($\frac{D}{2} - 1$)-form ($\varphi_{\frac{D}{2}-1}$). The other one has a dual structure ($\overline{\varphi}_{\frac{D}{2}+1},...,\overline{\varphi}_D$). For ($\varphi_1$), the constraints (4.4) can be rewritten:

$$\partial_\mu \overline{\varphi}^{\mu\nu} + im\varphi^\nu = 0$$
$$\partial_\mu \varphi^\mu = 0$$
$$\partial_{[\mu}\varphi_{\nu]} + im\overline{\varphi}_{\mu\nu} = 0 \tag{88}$$

Thus the vector wave function φ_1 satisfies Proca's equations, and describes a spin-one particle with mass m. It follows that the field equations of φ_1 and $\overline{\varphi}_1$ can be derived by minimizing Proca's Lagrangian

$$\mathcal{L}_{Proca} = \frac{m}{2}\overline{\varphi}_{\mu\nu}\varphi^{\mu\nu} - \frac{i}{2}\varphi^{\mu\nu}(\partial_\mu\varphi_\nu - \partial_\nu\varphi_\mu)$$
$$- \frac{i}{2}\overline{\varphi}^{\mu\nu}(\partial_\mu\overline{\varphi}_\nu - \partial_\nu\overline{\varphi}_\mu) + m\overline{\varphi}_\mu\varphi^\mu \tag{89}$$

When $m = 0$, the two sectors of opposite chiralities decouple. In each sector, the independent degrees of freedom are now one 0-form A_0 (with $\varphi_1 = dA_0$), one 1-form A_1 (with $\varphi_2 = dA_1$),..., one (D-2)-form A_{D-2} (with $\varphi_{D-1} = dA_{D-2}$). The φ_p's are closed and co-closed, i.e. the A_p's satisfy Maxwell's equations and are defined up to gauge transformations. Consequently, φ_2 can be identified with the field strength of a photon. If we consider the case $D = 4$ and $m \neq 0$, the spectrum reduces to two scalars and two massive spin-one particles, and contains 8=2(1+3) degrees of freedom. For $m = 0$, we have two massless scalars and two massless vectors, so that we still have 8=2(1+1+2) independent degrees of freedom.

As an application of this formalism, we study the classical behavior of spinning particles in a curved space-time. We are interested in the approximation where the trajectory of the particle is classical, while the spin effects are visible as it would be the case in a Stern-Gerlach experiment. This situation occurs if the order of magnitude of the interaction energy between the spin and the curvature, which is essentially proportional to the space-time curvature times $\hbar$ (analogously to the interaction between the the magnetic field and a magnetic moment due to the spin), is comparable to the kinematical energy of the particle. One must also measure the position of the particle on a domain much larger than its Compton wavelength. In this limit the position X^μ and momentum P_μ are ordinary numbers and the quantum Hamiltonian becomes simply a matrix built from the Ψ's and $\overline{\Psi}$'s acting in the spin-space with coefficients depending on the classical position X and momentum P. The τ-dependence of the classical dynamics of the particle can be expressed by applying Hamilton-Jaccobi's method with this matricial Hamiltonian. The only quantum effects are due to the spin interaction with the space-time curvature. (In a fully classical approximation, $\hbar = 0$, and the spin effects disappear, since all the fermionic operators are proportional to $\sqrt{\hbar}$.) One can always find a basis for the spin states, which depends on the space-time position and such that the Hamiltonian is diagonal. In this basis the spin value is conserved through evolution, i.e. the spin observables are paralelly transported along the trajectory. By diagonalization in spin space, H determines independent Hamilton-Jaccobi's equations for each spin degree of freedom of the particle. For the spin-one case, we expect three different trajectories corresponding to the values 1, 0 and -1 for the projection of the spin on a spatial axis in the rest frame of the particle.

We consider the case of a Schwarzschild gravitational field in four dimensional space-time ($ds^2 = (1 - \frac{r_0}{r})dt^2 - (1 - \frac{r_0}{r})^{-1}dr^2 - r^2(d\theta^2 + sin^2\theta\ d\phi^2)$ with $r_0 = 2GM/c^2$.) We will compute the correction, due to the spin, to Einstein's formula predicting the shift of the perihelion of a spinless point particle. For the other classical test of general relativity, i.e. the bending of light rays in a gravitational field, we will find that the wave vector of a polarized photon deviates from geodesic motions by a relative shift proportional to $\hbar$. These results are in agreement with the fact that a particle with an angular momentum interacts with the space-time curvature, as first pointed out by Papapetrou for a rotating body [25]. The advantage of a supersymmetric Hamiltonian is that it defines unambiguously the spin effects. Since we work to first non-trivial order in $\hbar$, we restore from now on the $\hbar$ dependence in the formulae. The matricial Hamilton-Jacobi's equation is obtained by replacing in the supersymmetric Hamiltonian the classical momentum p_μ by $\frac{\delta S}{\delta X^\mu}$ where $S[X^\mu, \tau]$ is the action of the classical trajectory of the particle in a given spin state, with arbitrarily chosen initial and final boundary conditions. Notice that keeping the lowest order in $\hbar$ means that we only retain the covariant derivative of the fermionic variables and not the curvature term. This yields

$$g^{\mu\nu}\frac{\delta S}{\delta X^\mu}\frac{\delta S}{\delta X^\nu} - m^2 + 2\hbar\frac{\delta S}{\delta X^\mu}\omega^\mu_{ab}\Sigma^{ab} + O(\hbar^2) = 0 \tag{90}$$

The space-time spin-connection ω is related to the space-time vierbein E and to Christoffel's symbol Γ

$$\omega_{\mu ab} = E_a^\alpha E_b^\beta \Gamma_{\alpha\mu\beta} \tag{91}$$

$$E_\alpha^a E_\beta^b \eta_{ab} = g_{\alpha\beta} \tag{92}$$

$$\Gamma_{\alpha\mu\beta} = \frac{1}{2}(\partial_\mu g_{\alpha\beta} + \partial_\beta g_{\alpha\mu} - \partial_\alpha g_{\beta\mu}) \tag{93}$$

The $\Sigma^{ab} = \frac{i}{2}\left(\overline{\Psi}^a\Psi^b - \overline{\Psi}^b\Psi^a\right)$ are the generators of the (reducible) 32-dimensional representation of the Lorentz group defined by the algebra (4.3) and acting on the states solving (4.6). If the matrix form of $\overline{\Psi}^5$ is chosen diagonal, the spin operators Σ^{ab} become block-diagonal with two independent sectors of opposite chiralities, corresponding to the eigen-values 0 and 1 of $\overline{\Psi}^5\Psi^5$, so the 32-dimensional representation splits into two independent 16-dimensional representation, each one containing five sectors of dimensions 1,4,6,4,1 corresponding respectively to 0-forms, 1-forms, 2-forms, 3-forms, and 4-forms. As explained above, the constraints imply that only two block-sectors made of one 0-form and one 1-form sectors are independent wave-functions. The one-form sector, and the corresponding 4×4 Hamiltonian matrix, determine the dynamics of spin-one particles. Moreover, in a Schwarzschild metric with characteristic radius r_0, the motion is planar, so one can separate the variables and write

$$S = -Et + L\varphi + S_r(r) \tag{94}$$

The spin-dependent part of Hamilton-Jacobi's equation is obtained by the substitution

$$p_\mu\omega^\mu_{ab}\Sigma^{ab} = \frac{r_0 E}{r^2}\Sigma^{01} - \frac{2L}{r^2}\left(1 - \frac{r_0}{r}\right)^{1/2}\Sigma^{13} \tag{95}$$

where

$$\Sigma^{01} = \frac{1}{2} \begin{pmatrix} 0 & 1 & 0 & 0 \\ -1 & 0 & 0 & 0 \\ 0 & 0 & 0 & 0 \\ 0 & 0 & 0 & 0 \end{pmatrix}, \quad \Sigma^{13} = \frac{1}{2} \begin{pmatrix} 0 & 0 & 0 & 0 \\ 0 & 0 & 0 & i \\ 0 & 0 & 0 & 0 \\ 0 & -i & 0 & 0 \end{pmatrix} \tag{96}$$

By inserting (94) and (95) into Hamilton-Jaccobi's equation (90), one obtains a matricial equation for $\frac{\partial S}{\partial r}$. The diagonalization can be done easily, and one gets three possibilities S_ϵ for the classical action, indexed by $\epsilon = 0, \pm 1$

$$\frac{E^2}{c^2} \left(1 - \frac{r_0}{r}\right)^{-1} - \left(\frac{L^2}{r^2} + m^2\right) - \left(1 - \frac{r_0}{r}\right)\left(\frac{\partial S_\epsilon}{\partial r}\right)$$
$$+ 2\epsilon \frac{\hbar}{r^2} \left(L^2\left(1 - \frac{r_0}{r}\right) - \left(\frac{r_0 E}{2c}\right)^2\right)^{1/2} = 0 \tag{97}$$

(We have restored the dependence in the speed of light c.) The energy E and the angular momentum L are constants of motion of the particle. The values $\epsilon = 0, \pm 1$ correspond to the three possible projections of the spin along a given spatial axis in the rest frame of the particle. The case $\epsilon = 0$ corresponds to the geodesic trajectory followed by the scalar particle. Far from the Schwarzschild horizon, we can use the standard techniques of integration of Hamilton-Jaccobi's equation to determine the three possibilities for the shift of the perihelion over a quasi-periodic trajectory. This amounts to replace L in the classical formulas [26] by an effective angular momentum L_ϵ defined by

$$L_\epsilon^2 = L^2 + 2\epsilon\hbar L \sqrt{1 - \left(\frac{r_0 E}{2Lc}\right)^2} \tag{98}$$

(Notice that near the horizon, unitarity breaks down). In the case of a massive particle, the shift of the perihelion is thus given by:

$$\delta\phi_\epsilon = \frac{3\pi}{2}\left(\frac{mcr_0}{L_\epsilon}\right)^2 \sim \delta\phi_0\left(1 - \epsilon\frac{\hbar}{L}\sqrt{1 - \left(\frac{r_0 E}{2Lc}\right)^2}\right) \tag{99}$$

For a non-relativistic Z^0 orbiting quasi-tangentially to the sun at a speed of $10^5 m/s$, which is approximately the circular velocity around the sun , we find $|\delta\phi_+ - \delta\phi_0|/\delta\phi_0 \sim \hbar/L \sim 10^{-21}$, which is much to small to be detected.

The solutions of Hamilton-Jaccobi's equation are continuous when $m \to 0$. However, in this limit the interpretation of its solution S is different. The particle is a photon following the laws of the geometrical optics, S is the eikonal of the light ray, and $\frac{\delta S}{\delta X^\mu}$ is its wave-vector. The solution $\epsilon = 0$ must then be rejected. In this massless case, one finds for the deflections of the two helicities $\epsilon = \pm 1$ the following formula

$$\delta\phi_\epsilon = \frac{2r_0\omega}{cL_\epsilon} \sim \delta\phi_0\left(1 - \epsilon\frac{\hbar}{2L}\sqrt{1 - \left(\frac{r_0 E}{2Lc}\right)^2}\right) \tag{100}$$

where $\omega = \frac{E}{\hbar}$. For an optical photon of wavelength $\lambda = 7 \times 10^{-7} m$ (red) skimming past the sun, we find $|\delta\phi_+ - \delta\phi_0|/\delta\phi_0 \sim \frac{\hbar}{2L} = \frac{\lambda}{2R_{sun}} \sim 10^{-15}$. (Note that this ratio does not depend on $\hbar$: the gravitational field interacts classically with the two polarizations of the electromagnetic field.) However, this doubling of Einstein's rings is to small to be detected.

References

[1] E. Witten, *Comm. of Math. Phys.* **117**, (1988), 353; *Comm. of Math. Phys.* **118**, (1988) 601; *Phys. Lett.* **B206** , 1988.

[2] For a review see O. Birmingham, M. Blau, M. Rakowski and G. Thomson, *Physics Reports* **209**, (1991), 129 and references therein.

[3] E. Witten, *Comm. of Math. Phys.* **121**, (1989), 351.

[4] L. Baulieu and I.M. Singer, *Nucl. Phys. Proc. Suppl.* **5B**, (1988), 12; *Comm. of Math. Phys.* **125**, (1989), 227; *Comm. of Math. Phys.* **135**, (1991), 253.

[5] L. Baulieu and C. Aragao de Carvalho *Phys. Lett.* **B275** (1991)323; *Phys. Lett.* **B275** (1991)335 .

[6] D. Birmingham, M. Rakowski and G. Thompson, *Nucl. Phys.* **B329** (1990) 83; D. Birmingham and M. Rakowski *Mod. Phys. Lett.* **A4** (1989) 1753; F. Delduc, F. Gieres and S.P. Sorella *Phys. Lett.* **B225** (1989) 367.

[7] S. Fubini and E. Rabinovici *Nucl. Phys.* **B245** (1984) 17; V. de Alfaro, S. Fubini and G. Furlan *Nuovo Cimento* **34 A**, (1976), 569.

[8] E. Witten, *Nucl. Phys.* **B323** (1989) 113.

[9] L. Baulieu, B. Grossman and R. Stora, *Phys. Lett.* **B180** (1986) 95.

[10] A. Forge and E. Rabinovici *Phys. Rev.* **D32**, (1985), 927.

[11] A. C. Davis, A. J. Macfarlane, P. C. Popat, and J. W. Van Holten *J. Phys. A Math. Gen.* **17**, (1984), 2945.

[12] E. Witten, *Nucl. Phys.* **B202**, (1982), 253. L. Alvarez Gaumé, *Comm. of Math. Phys.* **90**, (1983), 161; D. Friedan and P. Windey *Nucl. Phys.* **B235**, (1984), 395.

[13] N.A. Alvez, H. Aratyn and A.H. Zimmerman *Phys. Rev.* **D31**, (1985), 3298; R. Akhoury and A. Comtet *Nucl. Phys.* **B246**, (1984), 253.

[14] F.A. Berezin and M.S. Marinov, *JETP Lett.* **21** (1975) 320 and *Ann. Phys. NY* **104** (1977) 336; R. Casalbuoni, *Nuovo Cimento* **33A** (1976) 389 and *Phys. Lett.* **62B** (1976) 49; A. Barducci, R. Casalbuoni and L. Lusanna, *Nuovo Cimento* **35A** (1976) 377; L. Brink, S. Deser, B. Zumino, P. di Vecchia and P.S. Howe *Phys. Lett.* **64B** (1976) 43.

[15] L. Brink and J.H. Schwarz, *Nucl. Phys.* **B121** (1977) 285; L. Brink, P. di Vecchia and P.S. Howe, *Nucl. Phys.* **B118** (1977) 76 and *Phys. Lett.* **65B** (1976) 471; S. Deser and B. Zumino *Phys. Lett.* **65B** (1976) 369.

[16] A.M. Polyakov, *Phys. Lett.* **103B** (1981) 211.

[17] P.A.M. Dirac. Lectures on Quantum Mechanics, (Belfer Graduate School of Science, Yeshiva University; New York; 1964); M. Henneaux and C. Teitelboim, Quantization of Gauge Systems (Princeton University Press; 1992).

[18] W. Siegel, Introduction to String Field Theory. (World Scientific; 1988).

[19] A.M. Polyakov, Gauge Fields and Strings (Harwood Academic Publishers; 1987); Vl. S. Dotsenko, *Nucl. Phys.* **B285** (1987) 45.

[20] R.H. Rietdijk and J.W. van Holten, *Class. Quantum Grav.* **7** (1990) 247.

[21] P. Howe, S. Penati, M. Pernici and P. Townsend, *Phys. Lett.* **215B** (1988) 555.

[22] R. Marnelius and U. Martensson, *Nucl. Phys.* **B335** (1990) 395; U. Martensson, Preprint Goteborg-92-3 (Jan. 92).

[23] P. Van Nieuwenhuyzen, *Phys. Rep.* **68C** (1981) 189.

[24] R.H. Rietdijk and J.W. van Holten, *Class. Quantum Grav.* **10** (1993) 575.

[25] A. Papapetrou, *Proc. Roy. Soc. London* **A209** (1951) 248.

[26] L.D. Landau, The Classical Theory of Fields. (Pergamon Press; London; 1971).

SOME PROPERTIES OF (SUPER)P-BRANES

Paul Demkin[1]

Institute of Theoretical Physics
Uppsala University
Box 803, S-75108
Uppsala, Sweden

INTRODUCTION

Nowadays not only one-dimensional relativistic objects, strings, but also the objects of higher dimension, p-dimensional branes, are suggested as substantial physical and mathematical objects. Such a membrane model naturally appears (i) when generalizing the known shell-electron model that has been suggested by Dirac [1, 2, 3] ; (ii) as a cosmic domain wall in the post-inflationary universe [4, 5] ; (iii) as an effective model of supergravity [6] ; and (iv) as, like superstring, a model unifying fundamental interactions [8, 7].

Let us turn to the last point. Unlike the properties of strings, those of p-branes are much less investigated so far [9, 10]. A possible correlation between ordinary and rigid (super)p-branes and, in particular, the correlation between the rigid string and the ordinary membrane at $p=2$ has been considered in [11, 12]. The calculation of static potential for the p-brane compactified on space-times of various forms has been considered in [13, 14] The quantum properties of supermembrane are known only on the semiclassical level. The spectrum continuity of supermembrane has been treated as its quantum and even classical instability, as its degenerate turning into an infinite string without changing its energy [19, 20] or even as its total instability. Therefore, the term "instability" in this context means asymptotical behaviour of the p-brane solution at $t \rightarrow \infty$. An important role belongs to the algebra of area-preserving diffeomorphysms as a rest symmetry abgebra of p-brane in a light-cone gauge [21] and its possible deformations [22].

The further development of the (super)p-branes was inspired by finding intriguing interrelations between strings and 5-branes. They might imply that string theory is an

---

[1]On leave from Department of Physics, Vilnius University, Saulėtekio al.9, 2054, Vilnius, Lithuania
e-mail: paul@rhea.teorfys.uu.se or povilas.demkinas@FF.VU.Lt

equivalent of the theory of 5-branes in $D10$ [23, 24, 25, 26]. As a consequence of this approach, a) the strong-weak duality transformation in string theory is related with the target-space duality trasformation in 5-brane theory; b) the corresponding theories compactified on a six-dimensional torus are equivalent. Then S-duality transformation in string theory interchanges the states carrying non-zero momenta in the internal direction of the string with the 5-brane vinding modes on the torus.

This short review based on author's works presents only some of the aspects of supermembrane theory.

The first part "Super$p$-branes" is a brief introduction. The definitions of the action and symmetries of the (super)p-brane are given. The general form of the equation of motion of the super$p$-brane gives the idea of the complex character of this dynamic system and of the difficulties that might arise during quantization. A very natural desire that comes out at this point is to try to simpify this system. As is well known, the one-dimensional membrane (p=1), i.e. the string, is a rather simple object. In the "orthogonal" gauge of the equation of motion, the string becomes linear and may be easily quantized. This is not the case with multidimensional objects. The (super)p-brane is not symmetric enough, i.e. at $p > 1$ the condition "the numbers of independent parameters on the world-sheet equals the number of intrinsic symmetries" is not observed; as a consequence, the equation of motion remains non-linear at any choice of the gauge. Investigating a linearized model is the first and often very useful step when studying a non-linear system.

In the second part a linearized model of the (super)p-brane is presented. The bosonic and the fermionic cases are given a special consideration. The solutions of the equations of motion of the linearized model for the p-brane with arbitrary topology and massless eigenstates as well as with not integer critical dimensions D_{cr} after quantization are presented. Physical aspects of the constraint condition of the linearized model are discussed.

Another very important question is that of the class being not empty. We must be sure that at least one physically meaningful example of the theory is available to further develop this theory. The (super)p-brane must meet a number of physical requirements, "stability" being one of the most important among them. If the (super)p-brane in $D4$ allows to be stabilized by compactification on the compact space K^6, in $D10$ or $D11$ it must be stable by itself. This point acquired special importance with revealing the correlation between heterotic superstring and heterotic 5-brane, where both objects are studied in $D10$. One of the natural ways is to investigate the so-called Bogomolnyi bound for the (supr)p-brane [10, 27]. In the part "On the stability of bosonic p-brane" the bosonic p-brane is considered and some explicit solutions of the equations of motion are presented in flat and curved spacetimes, these solutions being stable in the sense of both asymptotical behaviour of the solution at $t \to \infty$ and the character of the phase diagram of the corresponding system.

Some authors [33, 34, 36] belive that for the purposes of the QCD the string model with rigidity is more acceptable. Its peculiarity is an additional term in the action that depends on the worldvolume curvature and to which corresponds a new additional parameter K of string rigidity. The part "Correlation between ordinary and rigid (super)p-branes" shows that when considering a special type of compactification the (super)p-brane with rigidity may be obtained from the (super)$(p+1)$-brane. This means that the bosonic string with rigidity may be obtained from the bosonic membrane.

The aspects presented reveal only some traits of a very attractive image of the (super)p-brane, the latter being a very promising physical and mathematical object.

SUPER P-BRANES

For the supermembrane $(p = 2)$, action is a direct multidimensional generalization of the string action [7]:

$$S = -\frac{T}{2}\int d^3\xi[\sqrt{h}h^{ij}\Pi_i^a\Pi_j^b\eta_{ab} - \sqrt{h} + 2\varepsilon^{ijk}\Pi_i^A\Pi_j^B\Pi_k^C B_{CBA}], \tag{1}$$

where T is the parameter of tension with the dimension $(mass)^{(p+1)}$ or $(length)^{-(p+1)}$, ξ^i $(i = 0, 1, ..., p)$ are the worldvolume coordinates, h_{ij} is the metric of the worldvolume, $h = -\det(h_{ij})$, η_{ab} is the Minkowski spacetime metric, and $\Pi_i^A = \partial_i Z^M E_M^A$, $A = a, \alpha$; $M = \mu, \dot{\alpha}$. Here, Z^M are the coordinates of the D-dimensional curved superspace, and E_M^A is the supervielbein. The 3-form $B = \frac{1}{6}E^A E^B E^C B_{CBA}$, $E^A = dZ^M E_M^A$ is the potential for the closed 4-form $H = dB$.

Action (1) is invariant respecting the global D-dimensional Poincaré transformations and also respecting local parametrizations of the worldvolume with the parameters $\eta^i(\xi)$:

$$\delta Z^M = \eta^i(\xi)\partial_i Z^M, \quad \delta h_{ij} = \eta^k\partial_k h_{ij} + 2\partial_{(i}\eta^k h_{j)k} . \tag{2}$$

It is also invariant under local fermionic "k-transformations":

$$\delta Z^M E_M^a = 0, \tag{3}$$

$$\delta Z^M E_M^\alpha = (1 + \Gamma)_\beta^\alpha k^\beta, \tag{4}$$

$$\delta(\sqrt{h}h^{ij}) = -2i(1 + \Gamma)_\beta^\alpha k^\beta(\Gamma_{ab})_{\alpha\gamma}\Pi_n^\gamma h^{n(i}\varepsilon^{j)kl}\Pi_k^a\Pi_l^b -$$

$$-\frac{2i}{3\sqrt{h}}k^\alpha(\Gamma_c)_{\alpha\beta}\Pi_k^\beta\Pi_l^c h^{kl}\varepsilon^{mn(i}\varepsilon^{j)pq}(\Pi_m^a\Pi_{pa}\Pi_n^b\Pi_{qb} + \Pi_m^a\Pi_{pa}h_{nq} + h_{mp}h_{nq}), \tag{5}$$

with an anticommuting spacetime spinor $k^\alpha(\xi)$ and the matrix Γ defined by

$$\Gamma = \frac{1}{6\sqrt{h}}\varepsilon^{ijk}\Pi_i^a\Pi b_j\Pi_k^c\Gamma_{abc} . \tag{6}$$

Unlike the two-dimensional string action, action (1) at $p \neq 1$ is not invariant respecting local conformal transformations with parameter $\Lambda(\xi)$:

$$\delta Z^M = 0; \tag{7}$$

$$\delta h^{\alpha\beta} = \Lambda(\xi)h^{\alpha\beta} . \tag{8}$$

Varying the initial action leads to essentially nonlinear field equations

$$\partial_i(\sqrt{h}h^{ij}\Pi_j^a) + \sqrt{h}h^{ij}\Pi_j^b\Pi_i^C\Omega_{Cb}^a + i\varepsilon^{ijk}\Pi_{ib}(\Pi_j^\alpha\Gamma_{\alpha\beta}^{ab}\Pi_k^\beta) +$$

$$+\varepsilon_{ijk}\Pi_i^b\Pi_j^c\Pi_k^d H_{bcd}^a = 0, \tag{9}$$

$$[(1 - \Gamma)h^{ij}\Pi_i^\mu\Gamma_\mu]_\beta^\alpha\Pi_j^\beta = 0, \tag{10}$$

where Ω_B^A is the 1-form connection in the D-dimensional curved superspace, and to the "embedding" equation

$$h_{ij} = \Pi_i^a\Pi_j^b\eta_{ab} , \tag{11}$$

which remains non-linear at any gauge. Their solution is known for certain simplest cases [15].

For open membranes, or for the existing open dimensions, at $\sigma_i = \sigma_i^a, \sigma_i = \sigma_i^b$ the border condition is observed on the coordinates $Z^M(\xi)$:

$$\int d^3\xi \partial_i(\delta Z^a \sqrt{h} h^{ij}\Pi_{ja} + 3\varepsilon^{ijk}\delta Z^A \Pi_j^B \Pi_k^C B_{CBA}) = 0, \tag{12}$$

where h_{ij} is given by equation (11).

Any new solution of equations of motion (9) and (10) describing the motion of a multidimensional relativistic object, on the one hand, is of interest in itself, and on the other hand, it serves as a starting point for semiclassical quantization, when the minor variations respecting the known classical solution are investigated.

We have considered a mathematically simpler case at $p=2$. There is a p-dimensional generalization of the supermembrane action with similar properties [9].

A LINEARIZED MODEL

In the case when a complicated, nonlinear dynamic system is investigated, it seems reasonable to start from its linearized model. This section aims to investigate a special type of action corresponding to a linerized model of the relativistic (super)p-brane. Such approach is possible in all cases when the (super)p-brane model appears.

Bosonic p-branes

Let us consider, as a less complicated, the case of the bosonic relativistic p-brane. This means that we are considering the action

$$S = -T \int d^{p+1}\xi |det(\partial_\alpha X^\mu \partial_\beta X^\nu g_{\mu\nu})|^{\frac{1}{2}}, \tag{13}$$

where $\xi = (\tau, \sigma_1, \ldots, \sigma_p)$, $\quad \xi_\alpha \in [\xi_\alpha^a, \xi_\alpha^b]$, $\quad X^\mu = X^\mu(\tau, \sigma_1, \ldots, \sigma_p)$, $\mu = 0, \ldots, D-1$, where D is the dimension of the Minkowski spacetime with metric $g_{\mu\nu}$; $\alpha = 0, \ldots, p$, where p is the space dimension of p-brane.

The equation of motion

$$\partial_\alpha(\sqrt{h} h^{\alpha\beta}\partial_\beta X^\mu) = 0, \tag{14}$$

resulting from (13), in the case the border conditions are taken into account, may be obtained from the classically equivalent action

$$S = -\frac{T}{2} \int d^{p+1}\xi \sqrt{h}[h^{\alpha\beta}\partial_\alpha X^\mu \partial_\beta X^\nu g_{\mu\nu} - (p-1)], \tag{15}$$

where an auxiliary metric $h_{\alpha\beta}$ on the worldvolume of the membrane is introduced. Actions (13) and (15) to be equivalent, the metric $h_{\alpha\beta}$ must obey the imbedding condition:

$$h_{\alpha\beta} = \partial_\alpha X^\mu \partial_\beta X^\nu \eta_{\mu\nu}, \tag{16}$$

like the embedding condition (11) in the supersymmetric case.

Besides, we must check if the constraint conditions $p + 1$ are observed:

$$P_\tau^\mu X_{\mu;i} = 0, \qquad P^2 + T^2 det h_{ij} = 0, \tag{17}$$

where $P_\tau^\mu = \delta\mathcal{L}/\delta\dot{X}^\mu$, $\quad 1 \le i, j \le p$.

There are at least two ways: to investigate small variations respecting the classical solutions and introduce the quadratic action as a new independent action of the linearized version of p-brane. Let us consider both of these possibilities.

We cannot quantize action (15) at $p > 1$, but we can introduce a certain simplification. Let Y^μ be a variation respecting the classical solution X_0^μ:

$$X^\mu = X_0^\mu + \varepsilon Y^\mu \, . \tag{18}$$

The requirement of the X^μ-solution of the equation of motion being the first order in ε leads to the equation $\partial_\alpha C^{\mu\alpha} = 0$:

$$\partial_\alpha \frac{\partial A}{\partial X_{\mu;\alpha}} + \frac{3}{2h^0} \sum_{\alpha=0}^{p} \partial_\alpha (A \frac{\partial h^0}{\partial \dot{X}_{0\mu}}) = 0, \tag{19}$$

where $A = \sum_{i,j=0}^{p} \partial_i X_0^\mu \partial_j Y_\mu h_{ij}^0$, $h_{ij}^0 = \partial_i X_0^\mu \partial_j X_{0\mu}$, $h^0 = det h_{ij}^0$.

The exact expression for the equation of motion (19) depends on the solution $X_0^\mu(\xi)$. As an example, the solution for the toroidal membrane on the spacetime with the topology $R^{D-2} \times S^1 \times S^1$ is

$$X^1 = l_1 R_1 \sigma, \quad X^2 = l_2 R_2 \rho, \quad X^I = 0, \quad I = 3, ..., D, \tag{20}$$

where $0 \leq \sigma \leq 2\pi$, $0 \leq \rho \leq 2\pi$, R_1 and R_2 are the radii of the two circles, and l_1 and l_2 are the integers characterizing the winding numbers of the membrane around the two circles.

If we consider the fluctuations Z^μ of the transverse coordinate around this classical solution

$$X^1 = \sigma + Z^1, \quad X^2 = \rho + Z^2, \quad X^I = Z^I, \, I = 3, ..., D, \tag{21}$$

then, keeping only the terms of the linear order in Z, we find from equation (7)

$$\ddot{Z}^m = \partial_\sigma^2 Z^m + \partial_\rho^2 Z^m, \quad \ddot{Z}^I = \partial_\sigma^2 Z^I + \partial_\rho^2 Z^I, \tag{22}$$
$$m = 1, 2 \, , \quad I = 3, ..., D \, .$$

Equations of motion (28) and (29) are a special case of the equations (20). But here it should be noted that, as follows from (22) and (26), there is a special gauge condition, in which the general equation (20) turns into the ordinary wave equation.

The way described above is the investigation of small variations considering the classical solution. We may as well try to investigate the original action (13).

Let us introduce new variables $\bar{X}^\mu$:

$$\partial^\alpha \bar{X}^\mu = \sqrt{|h|} h^{\alpha\beta} \partial_\beta X^\mu \, . \tag{23}$$

This means that

$$\bar{h} = det(\partial_\alpha \bar{X}^\mu \partial_\beta \bar{X}_\mu) = sign(h) |h|^{(p+1)^2+1}. \tag{24}$$

With these variables, the equation of motion (14) turns into the wave equation

$$\partial_\alpha \partial^\alpha \bar{X}^\mu = 0 \, , \tag{25}$$

and the conditions of the constrains (17) turn into

$$\bar{P}^2 + T^2 |\bar{h}|^{-\frac{p^2}{(p+1)^2+1}} det(\partial_i \bar{X}^\mu \partial_j \bar{X}_\mu) = 0, \tag{26}$$

where $\bar{P}^\mu \equiv \dot{\bar{X}}$ and $i, j = 1, ..., p$ are space indexes of the membrane.

For the sake of convenience, the space parameter of the membrane $\xi_i \in [\xi_i^a; \xi_i^b]$ is considered $\sigma_i \in [0; \pi]$ and for the open dimension

$$X^\mu(\tau, ..., \sigma_i = 0, ...) \neq X^\mu(\tau, ..., \sigma_i = \pi, ...), \tag{27}$$

unlike for the closed dimension, where the condition of periodicity is observed:

$$X^\mu(\tau, ..., \sigma_i, ...) = X^\mu(\tau, ..., \sigma_i + \pi, ...), \tag{28}$$

or

$$X^\mu(\tau, ..., \sigma_i, ...) = X^\mu(\tau, ..., \sigma_i + 2\pi, ...), \tag{29}$$

depending on the spheric or toroidal type of compactification.

The border conditions for the p-brane in bar variables $\bar{X}^\mu$ are the same like ordinary variables X^μ. If we can express the motion of the p-brane in X^μ variables obeying the equation of motion (32), then the solution of this equation will be the solution of the corresponding linearized model.

In the general case, for the membrane with an arbitrary topology, when there are p_0 open dimensions, p_1 closed dimensions with the period π, and p_2 closed dimensions with the period 2π $(p_0 + p_1 + p_2 = p)$, equation (25) may be solved in the following way:

$$
\begin{aligned}
X^\mu(\xi) = X^\mu &+ \frac{1}{\pi^p T} p^\mu \tau + \\
+ i\sqrt{\frac{2^{p-1}}{\pi^p T}} \sum_{\mathbf{n}} n^{-1}(\alpha_{\mathbf{n}}^\mu e^{-in\tau} &- \alpha_{\mathbf{n}}^{*\mu} e^{in\tau}) \prod_{i=1}^{p} \cos n_i \sigma_i + \\
+ i\sqrt{\frac{2^{p-1}}{\pi^p T}} \sum_{\mathbf{m}} m^{-1}[(\alpha_{\mathbf{m}}^\mu e^{-2im\tau} &- \alpha_{\mathbf{m}}^{*\mu} e^{2im\tau}) e^{-2i\bar{m}\bar{\sigma}} + \\
+ (\beta_{\mathbf{m}}^\mu e^{-2im\tau} &- \beta_{\mathbf{m}}^{*\mu} e^{2im\tau}) e^{2i\bar{m}\bar{\sigma}}] + \\
+ i\sqrt{\frac{2^{p-1}}{\pi^p T}} \sum_{\mathbf{k}} k^{-1}[(\alpha_{\mathbf{k}}^\mu e^{-ik\tau} &- \alpha_{\mathbf{k}}^{*\mu} e^{ik\tau}) e^{-i\bar{k}\bar{\sigma}} + \\
+ (\beta_{\mathbf{k}}^\mu e^{-ik\tau} &- \beta_{\mathbf{k}}^{*\mu} e^{ik\tau}) e^{i\bar{k}\bar{\sigma}}],
\end{aligned}
\tag{30}
$$

where X^μ are the initial coordinates of the mass centrum and p^μ is the impuls of the mass centrum of the membrane at

$$
\begin{aligned}
\mathbf{n} \in \mathbf{N}^{p_0} \backslash 0, \quad & n = \sqrt{n_1^2 + ... + n_{p_0}^2} \; ; \\
\mathbf{m} \in \mathbf{N}^{p_1} \backslash 0, \quad & m = \sqrt{m_1^2 + ... + m_{p_1}^2} \; ; \\
\mathbf{k} \in \mathbf{N}^{p_1} \backslash 0, \quad & k = \sqrt{k_1^2 + ... + k_{p_2}^2} \; ; \\
\bar{m}\bar{\sigma} \equiv m_{p_0+1}\sigma_{p_0+1} &+ ... + m_{p_0+p_1}\sigma_{p_0+p_1} \; , \\
\bar{k}\bar{\sigma} \equiv k_{p_0+p_1+1}\sigma_{p_0+p_1+1} &+ ... + k_p\sigma_p \; .
\end{aligned}
\tag{31}
$$

Quantization of the model

To investigate the quantum properties of the p-brane we would like to have at our disposal the appropriate classical properties of the original p-brane. The motion of the

p-brane in the $\bar{X}^\mu$ variables is the same as described by the original action (13), where all difficulties are hidden in the constraint conditions (33). Finding the solution of the wave equation obeying these constraint conditions is an intricate task in itself, and its solution is yet unknown. As a first step, let us consider the quadratic action under X^μ variables that may be interpreted as an action in the original variables X^μ:

$$S = -\frac{T}{2}\int d^{p+1}\xi h^{\alpha\beta}\partial_\alpha X^\mu \partial_\beta X^\nu g_{\mu\nu},\tag{32}$$

where $h^{\alpha\beta} = \eta^{\alpha\beta}$, $\alpha,\beta = 0,...,p$; $g_{\mu\nu} = \eta_{\mu\nu}$, $\mu,\nu = 0,...,D-1$.

Action (32) is invariant respecting the global D-dimensional Poincaré transformations, but not invariant under local conformal and reparametrization transformations.

The absence of reparametrization transformations means the absence of the constraints. This allows easy quantization of the quadratic action.

Consider $X^\mu(\xi)$ the open p-brane. Then the solution of the equation of motion which follows from (32) is like that of (25), and the density of the energy-momentum tensor

$$P_\tau^\mu = -\frac{\partial \mathcal{L}}{\partial \dot{X}_\mu} = -i\sqrt{\frac{2^{p-1}T}{\pi^p}}\sum_{\mathbf{n}}(\alpha_{\mathbf{n}}^\mu e^{-in\tau} - \alpha_{\mathbf{n}}^{*\mu}e^{in\tau})\prod_{i=1}^{p}cosn_i\sigma_i,\tag{33}$$

$$\alpha_0^\mu = \frac{1}{\sqrt{2^{p+1}\pi^p T}}p^\mu, \qquad \mathbf{n}\in \mathbf{N}^p.$$

In the light-cone coordinates with the assumption that tangent components $\alpha_{\mathbf{n}}^{\pm}$ are physically meaningless, like in the string case, we have from the commutation relations

$$[X^\mu(\tau,\sigma), P_\tau^\nu(\tau,\sigma')] = i\eta^{\mu\nu}\delta(\sigma-\sigma')\tag{34}$$

on the quantum level

$$[\alpha_{\mathbf{m}}^i, \alpha_{\mathbf{n}}^j] = n\eta^{ij}\delta_{\mathbf{m},\mathbf{n}} \quad,\tag{35}$$

where $\alpha_{\mathbf{n}}^{*\nu} \to \alpha_{\mathbf{n}}^{+\nu}$.

The quantum Hamiltonian $H = \int_0^\pi d^p\sigma(P_\tau^\mu \dot{X}_\mu - \mathcal{L})$ is

$$H = \frac{T}{2}\int_0^\pi (\dot{X}^2 + X_1^2 + ... + X_p^2)d^p\sigma =$$

$$= \alpha_0^2 + \sum_{\mathbf{n}}\alpha_{\mathbf{n}}^+\alpha_{\mathbf{n}} + \frac{D-p-1}{2}\sum_{\mathbf{n}}n, \quad \mathbf{n}\in \mathbf{N}^p\backslash 0.\tag{36}$$

As could be expected, the excitations of the linearized model are an ordinary sum of the infinite number of harmonic oscillations described by creating and annihilating operators.

The zero-point energy of the infinite number oscillators (the Casimir energy) diverges, and for correct definition it must be regularized.

Consider the regularization by the contracted Riemann zeta-function [16]:

$$\zeta_p'(s) = \sum_{\mathbf{n}}(n_1^2 + n_2^2 + ... + n_p^2)^{-s}, \qquad \mathbf{n}_i \in \mathbf{N}^p\backslash 0,\tag{37}$$

p	$\zeta'_p(-\tfrac{1}{2})$	D_{cr}
2	0.026	79.623
3	0.053	42.080
4	0.048	46.610
5	0.036	61.603
6	0.249	15.032
7	0.017	128.829
8	0.011	199.398

Then, substituting the quantities $\zeta'_p(-\tfrac{1}{2})$ in (36), we obtain undiverging meanings of the Casimir energy and, correspondingly, the desired properties of the Hamiltonian H.

We remember that in the quantum case we have no constraints for this model. But we may impose "by hand" an additional condition $H|\phi\rangle = 0$. In this case, we obtain that for the existence of a massless vector, the coefficients at the third term in (36) must be equal to one. This condition gives $D = D_{cr} = 1 + p + 2(\sum_{n>0} n)^{-1}$. Hence, the ground state of this model is a tachyon.

In the case of closed toroidal or spherical types of the p-branes we have the same properties: new fractional critical dimensions D_{cr} and the tachyonic ground state [16].

A supersymmetric linearized model

It is of interest to examine the supersymmetric case of the bosonic p-brane in the GS and NSR approaches. Let us consider the supersymmetric linearized model in the NSR approach. Let $p = 2$. Passing over to the $p \geq 2$ will be simple.

The direct generalization of the linearized model of the bosonic action is

$$S = -\frac{T}{2} \int d^3\xi (\partial_\alpha X^\mu \partial^\alpha X_\mu + i\bar{\psi}^\mu \gamma^\beta \partial_\beta \psi_\mu), \tag{38}$$

where ψ^μ is the Majorana spin-vector, $\{\gamma^\alpha, \gamma^\beta\} = -2\eta^{\alpha\beta}$. We shall use the basis for γ^α:

$$\gamma^0 = \begin{pmatrix} 0 & -i \\ i & 0 \end{pmatrix}, \quad \gamma^1 = \begin{pmatrix} 0 & i \\ i & 0 \end{pmatrix}, \quad \gamma^2 = \begin{pmatrix} i & 0 \\ 0 & -i \end{pmatrix}. \tag{39}$$

This action is invariant under the global transformations

$$\delta X^\mu = i\bar{\eta}\psi^\mu, \qquad \delta\psi^\mu = (\gamma^\alpha \partial_\alpha X^\mu)\eta \quad . \tag{40}$$

The equations of motion that follow from the action of the super p-brane, are

$$\partial_\alpha \partial^\alpha X^\mu = 0, \quad \partial_2 \psi_1^\mu + (\partial_1 - \partial_0)\psi_2^\mu = 0, \quad (\partial_1 + \partial_0)\psi_1^\mu - \partial_2 \psi_2^\mu = 0. \tag{41}$$

For the variables X^μ we may use the same solution as in the bosonic case (25). Let the solutions of the equations of motion for the fermionic part hold as follows:

$$\psi_1^\mu(\xi) = \sum_{\mathbf{n}} d_{\mathbf{n}}^{\mu(1)} e^{-i(n_0\tau + n_1\sigma_1 + n_2\sigma_2)}, \tag{42}$$

$$\psi_2^\mu(\xi) = \sum_{\mathbf{n}} d_{\mathbf{n}}^{\mu(2)} e^{-i(n_0\tau - n_1\sigma_1 + n_2\sigma_2)}, \tag{43}$$

where $\mathbf{n} \in \mathbf{Z}^3$

In this case the equation of motion imposes restrictions on the coefficients $d_{\mathbf{n}}^{\mu(1)}$ and $d_{\mathbf{n}}^{\mu(2)}$:

$$d_{\mathbf{n}}^{\mu(2)} = n_2/(n_0 - n_1)d_{\mathbf{n}}^{\mu(1)} = (n_0 + n_1)/n_2 d_{\mathbf{n}}^{\mu(1)} \tag{44}$$

The Hamiltonian of such system equals to

$$H = \frac{T}{2} \int d^3\xi [\dot{X}^2 + X_1^2 + X_2^2 + \frac{i}{2}(\bar{\psi}^\mu \gamma^0 \dot{\psi}_\mu - \dot{\bar{\psi}}^\mu \gamma^0 \psi_\mu)] \tag{45}$$

In the quantum case, the coefficients $d_{\mathbf{n}}^{\mu(i)}$ obey the anticommutation relations:

$$\{d_{\mathbf{m}}^{\mu(i)}, d_{\mathbf{n}}^{\nu(j)}\} = \eta^{\mu\nu}\delta_{\mathbf{m},-\mathbf{n}} \quad . \tag{46}$$

Therefore, the Hamiltonian is the sum of the bosonic and fermionic oscillations:

$$H = \alpha_0^2 + \sum_{\mathbf{n}}[\alpha_{\mathbf{n}}^+ \alpha_{\mathbf{n}}] + \sum_{i,\mathbf{n}}[n d_{\mathbf{n}}^{+(i)} d_{\mathbf{n}}^{(i)}], \tag{47}$$

for which there is no Casimir energy. This is what must be the case with the supersymmetric model.

The initial action does not contain any auxiliary metric on the worldvolume, hence the constraints in the system are absent, too. Like in the bosonic case, we may impose an additional condition $H = 0$ and consider it in the quantum case as well.

In this case, we find that in the supersymmetric model the condition $H = 0$ gives us massless ground states and no critical dimensions whatsoever.

Some remarks on the linearized model

In this part we have considered the simplest case of the bosonic and fermionic membranes, when they contain only linear terms in their equations of motion. The general situation is much more complicated.

An essential point of our consideration is imposing additional conditions like $H = 0$. But in the case of the linearized model we can consider these conditions as a certain remnant constraint condition like $L_n = 0$.

One would remark that D_{cr} in the bosonic case is not an integer and, consequently, has no physical meaning. Indeed, in all considered cases $D_{cr} \neq \mathbf{N}$. But even in the case when $D_{cr} \in \mathbf{N}$, D_{cr} has no physical meaning. The point is that we cannot pick out physical states among all possible states in the Hilbert space, as we have not enough constraints or the conditions like those and can not obtain the physical sector. On the other hand, the discrete values of the spacetime dimension D_{cr} imply the existence of the fractal properties of the extended objects. Some of the aspects of these properties are considered in [17].

In the supersymmetric case we have additional possibilities to impose a condition, at which the supercurrent $J^\alpha = K\gamma^\beta\gamma^\alpha\psi^\mu\partial_\beta X_\mu$ vanishes. In this case the condition $J^\alpha = 0$ is equivalent to six conditions $\partial_\alpha X^\mu \psi_\mu^i = 0$ or their Fourier transformation $F_{\mathbf{n}}^{\alpha i} = \int_{-\pi}^{\pi} d^2\sigma e^{i\bar{n}\bar{\sigma}}\partial_\alpha X^\mu \psi_\mu^i$. The supersymmetric action contains the constraints $F_{\mathbf{n}}^{\alpha i} = 0$. We may also express this quantity in the $\alpha_{\mathbf{n}}, d_{\mathbf{n}}^{(i)}$ variables and consider the quantum case, but this will be also not enough to distinguish the physical sector. Nevertheless, due to the quadratic action we can analytically calculate the partition function and transition amplitude for this model.

The linearized model allows us to separate linear and nonlinear effects in the general (super)p-brane. For instance, in [18] , due to the restriction of the constraint condition for the bosonic p-brane, D_{cr} has been obtained, whereas the purely linearized model has no critical dimensions. This means that in [18] a nontrivial conformity between the linearized model and the imposed constraint condition was obtained.

We may try to impose sufficient constraint conditions as an additional condition, but in this case a very important question arises: how to conform the solution of the equation of motion with the constraint conditions? We can make it sure that in the bosonic sector the simplest quadratic constraints $\dot{X}^2 + X_{;1}^2 + ... + X_{;p}^2 = 0$, $\dot{X}^\mu X_{;i\mu} = 0$, which are a natural generalization of the string constraints, cannot coexist with the solutions of the linear wave equation of motion for the bosonic p-brane. Thus, the conformity between the solution of the equation of motion in the linearized model and the additional constraint conditions is nontrivial and of interest in itself.

On the other hand, we may not only use global supersymmetry and vanishing of the supercurrent J^α, but also the condition of local supersymmetry may be imposed. Indeed, we may use the linearized model of the (super)p-brane with local supersymmetry and try to find the conformity between the solutions and constraints. However, (1) it is not clear how to do it even in a less complicated case without supersymmetry, and (2) this will be not enough to distinguish the physical sector, either.

Thus, we may consider the linearized model an auxiliary model of the (super)p-brane. An important aspect of this consideration is the possibility to separate the physical properties belonging to the linearized model from other properties characteristic of the essentially nonlinear behaviour of the relativistic (super)p-brane.

ON THE STABILITY OF P-BRANE

The existence of the stable p-brane solution is important to the general theory of extended objects. For further development of the theory (quantization, perturbation theory and so on) we have to be sure that there is at least one example of stable solution of the equation of motion.

We shall show that the class of stable p-brane solutions is not empty. For this purpose we shall consider some of the solutions of the p-brane equations of motion, new as well as the known ones [28, 29], in curved and flat spacetimes.

It is necessary to agree upon the main term 'stability' in advance. There are many kinds of stability that are known in mathematics (structural, Poisson, Lagrange, conditional, absolute *etc.*). Solutions of equations of motion are their stable points respecting their mappings. We shall restrict our consideration only to the asymptotical behaviour of the p-brane solution at $t \to \infty$ and to Lyapunov and asymptotical stabilities. The stable point x_0 of the mapping A is Lyapunov-stable (and, respectively, asymptotically stable), if $\forall \varepsilon > 0$, so that if $|x - x_0| < \delta$, then $|A^n x - A^n x_0| < \varepsilon$ for all $0 < n < \infty$ (correspondingly, $A^n x - A^n x_0 \to 0$, as $n \to \infty$).

Bosonic membrane in curved spacetime

We may separate the bosonic part of the membrane in $D = 11$ by extinguishing fermionic degrees of freedom of the supermembrane. The action for the supermembrane becomes the action for a purely bosonic membrane in a curved spacetime:

$$S = -\frac{T}{2} \int d^3\xi \sqrt{h}[h^{ij}\partial_i x^M \partial_j x^N g_{MN}(x) -$$

$$-\frac{1}{3}\varepsilon^{ijk}\partial_i x^M \partial_j x^N \partial_k x^P B_{MNP}(x) - 1], \tag{48}$$

where $h_{ij} = \partial_i x^M \partial_j x^N g_{MN}$, $\quad h = |\det h_{ij}|$, $\quad g_{MN}(x)$ is a metric of curved spacetime, $B_{MNP}(x)$ is an antisymmetric tensor field of rank 3 that couples with the membrane via a Wess-Zumino term [8, 7, 9].

In this case, the equations of motion for the membrane turn into

$$\partial_i(\sqrt{h}h^{ij}\partial_j x^N g_{MN}) - \frac{1}{2}\sqrt{h}h^{ij}\partial_i x^N \partial_j x^P \partial_N g_{MP} -$$
$$-\frac{1}{6}\varepsilon^{ijh}\partial_i x^N \partial_j x^P \partial_k x^Q F_{MNPQ} = 0, \tag{49}$$

where $F_{MNPQ} = 4\partial_{[M}B_{NPQ]}$ is the tension of potential B_{MNP}.

Let us consider the solution for the bosonic membrane in $D = 11$ with a special spherical spacetime symmetry metric $g_{MN}(x)$

$$ds^2 = -e^{2a}dt^2 + e^{2b}[dr^2 + r^2 d\Omega^2] + e^{2c}[dy + g\cos\theta d\varphi - qdt]^2 +$$
$$+(dx^5)^2 + ... + (dx^{10})^2 \tag{50}$$

in the form

$$\xi^0 = t,\ \xi^1 = r,\ \xi^2 = y;\quad X^3 = \theta,\ X^4 = \varphi,\ X^N \neq f(\xi),\ N = 5, ..., 10. \tag{51}$$

Using the explicit expression for metric (50) and choosing coordinates (51) we obtain the equations of motion for the parameters a, b, c, q with the extra condition $a + b + c = const$ [30].

The solutions to these equations are:

1. The stable monopole solution found by D.J.Gross and M.J.Perry [31]:

$$exp(2b) = exp(-2c) = 1 + \bar{g}/r,\quad a = 0,\quad q = 0,\quad g^2 = \bar{g}^2. \tag{52}$$

2. Its electrically charged analog

$$exp(-2a) = exp(2c) = 1 + \bar{A}/r,\quad b = 0,$$
$$q = -(A/r)(1 + \bar{A}/r)^{-1},\quad g = 0,\quad A^2 = \bar{A}^2. \tag{53}$$

3. Its dyon analog

$$exp(-2a) = exp(2b) = (1 + \bar{g}/r)^2,\quad c = 0,$$
$$q = -(g/r)(1 + \bar{g}/r)^{-1},\quad g^2 = 2\bar{g}^2. \tag{54}$$

When considering a purely bosonic membrane, in the supersymmetric action we can extinguish fermionic degrees of freedom, *i.e.* spacetime gravitino $\psi_M(X)$, and fermionic coordinates $\theta(\xi)$ by equalling them to zero. In this case we obtain a bosonic sector of the supermembrane in a curved spacetime. We have just considered such a way and obtained stable monopol-like solutions. The Nambu-Goto action in a curved spacetime is another way to consider a purely bosonic membrane. We may start from the Nambu-Goto action

$$S_D = -T \int d^{p+1}\xi \sqrt{|\det \partial_\alpha X^\mu \partial_\beta X^\nu g_{\mu\nu}|}, \tag{55}$$

where $g_{\mu\nu} = g_{\mu\nu}(X)$.

The equations of motion derived from this action have the following form:

$$\partial_\alpha \left(\sqrt{\gamma}\gamma^{\alpha\beta}\partial_\beta x^\nu g_{\mu\nu}\right) = \frac{1}{2}\sqrt{\gamma}\gamma^{\alpha\beta}\partial_\alpha X^\nu \partial_\beta X^\lambda \frac{\partial g_{\nu\lambda}}{\partial X^\mu}, \tag{56}$$

where $\gamma_{\alpha\beta}$ is completely defined by the induced metric

$$\gamma_{\alpha\beta} = \partial_\alpha X^\mu \partial_\beta X^\nu g_{\mu\nu}. \tag{57}$$

Let us consider the simplest spherically simmetric non-flat metric, *i.e.* the Schwarzschild solution of the metric

$$g_{\mu\nu} = diag(1 - q, -(1 - q)^{-1}, -r^2, -r^2 \sin\theta) , \tag{58}$$

where $q = 2GM/r$, G is Newton's constant, and M is the total gravitational mass of the membrane.

For the $p = 2$ membrane in the spherical coordinate system in $D = 4$, $X^\mu = (\tau, r(\tau), \theta, \phi)$, the equation of motion (56) becomes

$$(1 - q)r\ddot{r} - 2(1 - q)\dot{r}^2 + \frac{1}{4}q\dot{r}^3 + \frac{1}{4}q(1 - q)^2\dot{r} + 2(1 - q)^3 = 0. \tag{59}$$

At $q = 1$ it gets into the black-hole region. In the case when $q \leq 1$ (close to unity), we may decompose the solution $r(\tau)$ near the Schwarzschild radius $r_0 = 2GM(r = r_0 + \rho)$. Then the equations of motion (59) turn into

$$\left[(r_0 + \rho)\rho\ddot{\rho}/r_0 + \frac{1}{4}\dot{\rho}^3 - 2\rho\dot{\rho}^2/r_0\right] (1 + \rho/r_0)^{-1} + \tag{60}$$
$$\left(\frac{1}{4}\rho^2\dot{\rho}/r_0^2 + 2\rho^3/r_0^3\right) (1 + \rho/r_0)^{-3} = 0.$$

We may consider this equation accounting for the terms that make the greatest contribution into the evolution of $\rho = \rho(\tau)$:

$$\rho\ddot{\rho} + 2\rho^3/r_0^3 = 0. \tag{61}$$

The solution of this equation is

$$\rho(\tau) = \left[\rho^{-\frac{1}{2}}(0) + \frac{1}{\sqrt{3}r_0^{\frac{3}{2}}}(\tau - \tau_0)\right]^{-2} . \tag{62}$$

The velocity of such membrane motion is

$$\dot{\rho}(\tau) = -\frac{2}{\sqrt{3}}(\rho/r_0)^{\frac{3}{2}}, \tag{63}$$

i.e. we obtain an asymptotical fall of the membrane on the Schwarzschild sphere during the infinite time.

Bosonic p-brane in flat spacetime

The equations of motion for bosonic relativistic p-brane in a flat spacetime are

$$\partial_\alpha(\sqrt{h}h^{\alpha\beta}\partial_\beta X^\mu) = 0. \tag{64}$$

These equations and the constraint conditions

$$P_\tau^\mu X_{\mu;i} = 0, \qquad P^2 + T^2 deth_{ij} = 0, \tag{65}$$

where $P^\mu_\tau = \delta\mathcal{L}/\delta\dot{X}^\mu$, $1 \leq i,j \leq p$,   and the border conditions for open dimensions

$$\frac{\partial h(\sigma_i^{in})}{\partial X^\mu_{;i}} = \frac{\partial h(\sigma_i^f)}{\partial X^\mu_{;i}} = 0, \qquad \sigma_i^\in \left[\sigma_i^{in}, \sigma_i^f\right] \tag{66}$$

may be obtained from the action

$$S = -T \int d^{p+1}\zeta \sqrt{|det(\partial_\alpha X^\mu \partial_\beta X_\mu)|}. \tag{67}$$

Equations of motion (64) are equivalent to the ordinary Laplace operator $\triangle$ on a non-compact manifold Σ^{p+1}, *i.e.* on the worldvolume of the p-brane. We know that the Laplace operator in the Euclidean spacetime is a stable operator under small perturbations near the solution X_0^μ as a linear elliptic operator. However, the worldvolume metric $h_{\alpha\beta}$ is a pseudo-Euclidean metric, and the stability in this case is yet unclear.

In order to have the physical picture of surface Σ^{p+1}, we may choose $X^0 = \tau$ and obtain

$$X^\mu = (\tau, X^m(\tau, \sigma_1, ..., \sigma_p)), \qquad m = 1, ..., D - 1. \tag{68}$$

Without any restriction we may use the reparametrization invariance of the action and suggest $h_{0\beta} = \dot{X}^m \partial_b X_m = 0$. Then the metric tensor is

$$h_{\alpha\beta} = \begin{pmatrix} \dot{X}^2 - 1 & 0 \\ 0 & \partial_a X^m \partial_b X_m \end{pmatrix}. \tag{69}$$

The equations of motion become

$$\ddot{X}^m + \frac{1}{2}\partial_a \dot{X}^2 h^{ab} \partial_b X^m + (\dot{X}^2 - 1)\triangle X^m = 0, \tag{70}$$

$$\partial_\tau \left(\frac{\sqrt{h}}{\sqrt{1 - \dot{X}^2}}\right) = 0, \qquad \dot{X}^2 \leq 1, \tag{71}$$

where $\triangle = \frac{1}{\sqrt{h}}\partial_a(\sqrt{h}h^{ab}\partial_b)$ is the Laplace operator on the space part of the metric tensor h_{ab}. Now that we have defined the Laplace operator on an appropriate Euclidean subspace, we may treat these equations of motion as a general dynamic system.

Let us consider a special solution to the equations of motion for a closed surface that is a mixture of pulsation and rigid rotation:

$$X^m(\tau, \sigma_1, ..., \sigma_p) = x(\tau)(\cos \varphi(\tau)n^k, \sin \varphi(\tau)n^l, 0, ..., 0), \tag{72}$$

where $n^k, n^l = (n_1, ..., n_d)$ is a d-dimensional unit vector describing the embedding of a p-dimensional closed surface in S^{d-1} and $d \leq (D - 1)/2$. Let $L = x^2\dot{\varphi} = const$ corresponding to the conservation of the angular momentum. Then the equations of motion (70) (71) become

$$(\triangle + p)n^k = 0, \tag{73}$$
$$\ddot{x} - L^2/x^3 + px^{2p-1}/C^{2p} = 0. \tag{74}$$

The first equation is the Laplace operator in Euclidean space R^d. If the metric of the surface is known, this is an ordinary linear operator on a compact manifold, and it is stable. In the general case, this is the equation of minimal surface in Euclidean space. From the general theory of minimal surfaces we know that solutions to this equation do exist, and some of the solutions n_0^k are stable under small variations near n_0^k [32]. The second equation is an ordinary differential equation equivalent to

$$y = \dot{x}, \tag{75}$$
$$\dot{y} = f(x), \quad f(x) = L^2/x^3 - px^{2p-1}/C^{2p}, \quad f'(x_0) < 0. \tag{76}$$

The characteristical equation $\lambda^2 - f'(x_0) = 0$ has purely imaginary solutions. The energy conservation law in the form

$$\left(\frac{x}{C}\right)^{2p} = 1 - \dot{x}^2 - \frac{L^2}{x^2} \geq 0 \tag{77}$$

following from (76) is the holomorphous integral $F(X,Y) = C$ in the neighbourhood of the point (X_0, Y_0), *i.e.* the necessary and sufficient condition for the centre to exist. Its Poincaré index equals to unity.

Equation of motion (74) for the radial part $x(\tau)$ depends on two parameters, but in fact we may neglect one of them. Condition (77) is valid for both each time moment τ and initial conditions. From (77) it follows that

$$\mathbf{y}^2 = 1 - k\mathbf{x}^{2p} - \frac{1}{\mathbf{x}^2}, \tag{78}$$

where $\mathbf{x} = \frac{x}{L}$ is a new dimensionless variable, $\mathbf{y} = \dot{x}$, $k = \left(\frac{L}{x_0}\right)^{2p}\left(1 - \dot{x}_0^2 - \frac{L^2}{x_0^2}\right)$. The phase diagrams of this equation are represented in Fig.1.

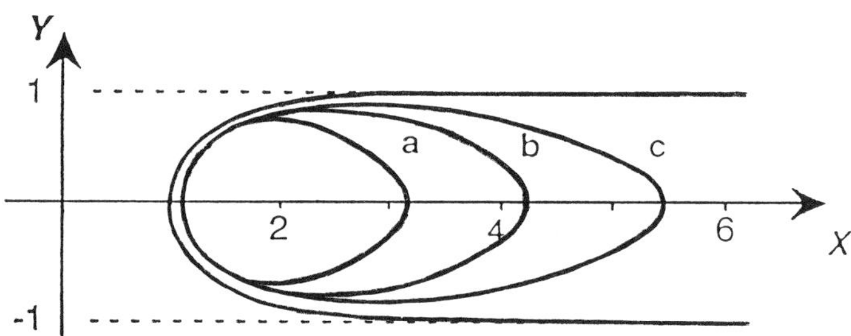

Fig.1. The phase trajectories $y^2 = 1 - kx^{2p} - \frac{1}{x^2}$ of equation of motion (76) at $p = 2$ and (a) $k = 10^{-2}$; (b) $k = 3 \cdot 10^{-3}$; (c) $k = 10^{-3}$. Stability region is $y^2 < 1 - \frac{1}{x^2}$.

Thus, we have stable solutions for bosonic p-brane in curved and flat spacetime. All of them obey the constraints (65). They prove the existence of stable solutions for further development of the theory of relativistic extended objects.

CORRELATION BETWEEN ORDINARY AND RIGID (SUPER)P-BRANES

In the last years surface models with extrinsic curvature received much attention. The first model was suggested by Polyakov [33] on the grounds of general symmetry properties. A term of this kind was also found in the effective bosonic theory obtained from the action of both the Neveu-Schwarz-Ramond and Green-Schwarz strings by the integration over fermionic fields in the ghostless covariant light-cone gauge [34]. The appearance of an additional term in the bosonic string action is necessary for restricting the outgrrow of fluctuation spikes [35].

The thermodynamic properties of such systems are also interesting. The high temperature partition function of the rigid string was studied in ref.[36], and the high temperature limit of free energy per length unit for the rigid string was found to agree dimensionally with that of the QCD string, unlike in the case of the Nambu-Goto string.

Not only strings and particles have dynamical properties that depend on the extrinsic curvature. It is also possible to construct a bosonic membrane with rigidity. Some of the classical properties of the rigid membrane were investigated in ref.[37]. The one-loop Casimir energy for rigid membrane was calculated and compared with that of the standard bosonic membrane in ref.[38].

In this section the correlation between ordinary super $(p+1)$-branes and rigid super p-branes in flat superspace is presented. We shall consider here the double spacetime and worldvolume compactification in a less complicated case when the supersymmetric membrane is presented without local k-supersymmetry. As a consequence, in a pure bosonic sector we obtain rigid particles and the rigid Polyakov string [12].

The action of the super $(p+1)$-brane in flat $D+1$ spacetime is

$$S = T' \int d^{p+2}\xi \sqrt{|\, det\Pi_{\bar{i}}^{\bar{\mu}}\Pi_{\bar{j}\bar{\mu}}\,|}\,, \tag{79}$$

where $\xi^{\bar{i}}(\bar{i},\bar{j} = 1, ..., p+2)$ - worldvolume coordinates, $\Pi_{\bar{i}}^{\bar{\mu}}\Pi_{\bar{j}\bar{\mu}} = \gamma_{\bar{i}\bar{j}}(\xi)$ - worldvolume metric, $\Pi_{\bar{i}}^{\bar{\mu}} = \partial_{\bar{i}}x^{\bar{\mu}} - i\bar{\theta}\Gamma^{\bar{\mu}}\partial_{\bar{i}}\theta$, $\bar{\mu} = 1, ..., D+1$.

We shall consider the double spacetime and worldvolume compactification and make a p-one split of the worldvolume coordinates

$$\xi^{\bar{i}} = (\xi^i, \rho),\ i = 1, ..., p+1;\ \rho \in [0; 2\pi r] \tag{80}$$

and a D-one split of the spacetime coordinates

$$\bar{x}^{\bar{\mu}} = (\bar{x}^{\mu}, x^{D+1}),\ \mu = 1, ..., D\,; \tag{81}$$

and make the partial gauge choice

$$x^{D+1} = \rho\,, \tag{82}$$

which identifies the D-th spacetime dimension with the p-th dimension of the worldvolume.

The essential point of this consideration is giving the condition

$$\bar{x}^{\mu} = \bar{x}^{\mu}(\xi^i, \rho),\quad \partial_{\rho}\theta = 0\,. \tag{83}$$

By expanding $\bar{x}^{\mu}$ in a series and restricting the consideration by the first term in decay we have

$$\bar{x}^{\mu} = x^{\mu} + k\rho y^{\mu}\,, \tag{84}$$

where $\partial_{\rho}x^{\mu} = \partial_{\rho}y^{\mu} = 0$.

Substituting this expression into the induced worldvolume metric $\gamma_{\bar{i}\bar{j}} = \Pi_{\bar{i}}^{\bar{\mu}}\Pi_{\bar{j}\bar{\mu}}$ yields

$$\gamma_{\bar{i}\bar{j}} = \begin{pmatrix} \gamma_{ij} + \omega_i\omega_j + 2k\rho H_{ij} & \omega_i + ke_i + k^2\rho g_i \\ \omega_j + ke_j + k^2\rho g_j & 1 + k^2 y^{\mu}y_{\mu} \end{pmatrix}, \tag{85}$$

where $\gamma_{ij} = \Pi_i^\mu \Pi_{j\mu}$, $\omega_i = -i\bar\theta\Gamma^{(D+1)}\partial_i\theta$, $e_i = \Pi_i^\mu y_\mu$, $g_i = y^\mu\partial_i y_\mu$, $H_{ij} = \Pi_i^\mu\partial_j y_\mu$, $i,j,=1,...,p$ and terms of an order higher than $k\rho$ were neglected.

From the latter expression, the determinant to the same order is

$$\bar\gamma = \gamma - 2k\gamma\gamma^{ij}e_i\omega_j + k^2 y^2\gamma(1+\gamma^{ij}\omega_i\omega_j) - \omega^i\omega^j e_i e_j - k^2\gamma\gamma^{ij}e_i e_j +$$
$$+2k\rho\gamma\gamma^{ij}H_{ij} - 2k^2\rho\gamma\gamma^{ij}g_i\omega_j - 4k^2\rho H H^{ij}e_i\omega_j \tag{86}$$

$\bar\gamma = |det\bar\Pi_i^{\bar\mu}\bar\Pi_{j\bar\mu}|$, $\gamma = |det\Pi_i^\mu\Pi_{j\mu}|$, $\gamma^{ik}\gamma_{kj} = \delta_j^i$, $H = det H_{ij}$.

Substituting this expression into the action for super$p$-brane we can eliminate non-physical fields ω_i in D-dimensional spacetime according to its equations of motion in a linear order under $k\rho$. Here we also make the last assumption about the properties of the field y^μ :

$$e_i = \Pi_i^\mu y_\mu = 0 , \tag{87}$$

which means the orthogonal condition for y^μ to the vectors Π_i^μ on the worldvolume.

Now we have

$$S = T' \int d^{p+2}\xi\sqrt{\gamma}(1 + \frac{1}{2}k^2 y^\mu y_\mu + k\rho\gamma^{ij}\Pi_i^\mu\partial_j y_\mu) . \tag{88}$$

The integration over $\rho \in [0; 2\pi r]$ and introducition of new fields and constants

$$T = 2\pi r T', \quad H^\mu = \sqrt{2K}ky^\mu, \quad K = \frac{1}{2}\pi^2 r^2 , \tag{89}$$

yields the action

$$S = T \int d^{p+1}\xi\sqrt{\gamma}(1 + \frac{1}{4K}H^\mu H_\mu + \gamma^{ij}\Pi_i^\mu\partial_j H_\mu) . \tag{90}$$

Substituting the equation of motion of the auxiliary field H^μ into the latter action we obtain a rigid super$p$-brane in D dimensional spacetime

$$S = T \int d^{p+1}\xi\sqrt{\gamma}\left[1 - K(\nabla^i\Pi_i^\mu)(\nabla^j\Pi_{j\mu})\right] . \tag{91}$$

The essential point of this approach is the initial assumptions. Indeed, condition (83) plays a crucial role here. The other condition (87) for variables y^μ is an orthogonal condition for bosonic variations of the x^μ, and it agrees with the requirement of transversality for relativistic super p-branes.

Parameter r is apparently connected with the radius of the compactification dimension. Note that the rigid constant K according to (89) depends on r^2.

As a consequense of this approach, in a pure bosonic sector at $p = 1$ and $p = 2$ we obtain the rigid particle and the rigid Polyakov string.

Correlation between the actions of $(p + 1)$ and p-branes leads to the correlation between their solutions. As an application of this correlation between the solutions of soft and rigid super p-branes we may consider the behaviour of the Regge trajectories $J(E^2)$ for a set of classical soft and rigid bosonic membranes [12].

Acknowledgements

Author wants to express his gratitude to Prof.A.Bytsenko, Prof.S.Odintsov, Prof.R.Zaikov for the useful comments and the Swedish Institute for Grant 304/01 GH/MLH, which gave him the possibility to enjoy the kind hospitality of Prof. A. Niemi, Doc. S. Yngve and all the staff of the Institute of Theoretical Physics, Uppsala University.

References

[1] P.A.M.Dirac, Proc.R.Soc.(London)A 268 (1962) 57.

[2] D.H.Hartley, M.Önder and R.W.Tucker, Class.Quantum Grav. 6 (1989) 1301.

[3] A.O.Barut and M.Pavšič, Phys.Lett.B 306 (1993) 49.

[4] A.Linde and D.H.Lyth, Phys.Lett.B 246 (1990) 353.

[5] Z.Lalak, S.Thomas and B.A.Ovrut, Phys.Lett.B 306 (1993) 10.

[6] J.Hughes, J.Liu, J. Polchinski, Phys.Lett.B 180 (1986) 370.

[7] E.Bergshoeff, E.Sezgin and P.K.Townsend, Phys.Lett.B 189 (1987) 75.

[8] E.Bergshoeff, E.Sezgin and P.K.Townsend, Phys.Lett.B 180 (1986) 370.

[9] M.Duff, Class. Quantum Grav. 6 (1989) 1577.

[10] M.Duff, J.X.Lu, Nucl.Phys.B416(1994)301.

[11] U.Lindström, Phys. Lett. B 218 (1989) 315.

[12] P.Demkin, Phys.Lett.B 305 (1993) 230.

[13] A.Bytsenko, S.Odintsov, Class. Quantum Grav. 9(1992) 391.

[14] A.Bytsenko, S.Zerbini, Nuovo Cimento 105A (1992) 1275.

[15] E.Bergshoeff, E.Sezgin and P.K.Townsend, Ann.of Phys. 185 (1988) 330.

[16] P.Demkin, *Some properties of the linearized model of the (super)p-brane*, UUITP 11/94, hep-th/9411148.

[17] P.Demkin, L.Žukauskaitė, Phys.Lett.A 146 (1990) 155.

[18] U.Marquard, M.Scholl, Phys.Lett.B 227 (1989) 227.

[19] B.de Wit, J.Hoppe and H.Nicolai, Nucl.Phys.B 305[FS23] (1988) 545.

[20] K.Fujikawa, J.Kubo, Nucl.Phys.B 356(1991)208.

[21] B.de Wit, U.Marquard and H.Nicolai, Commun. Math. Phys. 128 (1990) 39.

[22] P. Demkin, *Special functions as structure constants for new infinite-dimensional algebras*, UUITP 4/94; hep-th/9405102.

[23] M.Duff and J.Lu, Nucl.Phys.B354 (1991) 141.

[24] M.Duff and J.Lu, Phys. Rev. Lett 66 (1991) 1402.

[25] M.Duff and J.Lu, Class. Quantum Grav. 9 (1991) 1.

[26] M.Duff, R.Khuri and J.Lu, Nucl Phys.B377 (1992) 281.

[27] R.Gregory and R.Laflamme, *Evidence for the stability of extremal black p-branes* Preprint DAMTP/R-94/40, LA-UR-94-3323; hep-th/9410050.

[28] X.Li, Phys.Lett.B 205 (1988) 451.

[29] J.Hoppe and H.Nicolai, Phys.Lett.B.196 (1987) 451.

[30] P.Demkin, *On the stability of the p-brane*, to be published in Class. Quant. Gravity.

[31] D.J.Gross and M.J.Perry, Nucl.Phys.B 226 (1983) 29.

[32] Dao Chong Thi, A.T.Fomenko, *Minimal surfaces and the Plateau problem*,(Moscow, Nauka, 1987),(In Russian).

[33] A.Polyakov, Nucl.Phys.B 268 (1986) 406.

[34] P.Weigman, Nucl.Phys. B 323 (1989) 330.

[35] J.Ambjörn, B.Durhuus, J.Fröhlich and T.Jonsson, Nucl.Phys. B 290 [FS20] (1987) 480.

[36] J.Polchinski and Zhu Yang. *High temperature partition function of the rigid string.* Preprint UTTG-08-92. hep-th/9205043.

[37] D.Hartley, M.Önder and R.W.Tucker, Class.Quantum Grav. 6 (1989) 1301.

[38] S.D.Odintsov, Phys.Lett. B 247 (1990) 21.

ACCELERATOR EXPERIMENTS FOR THE NEXT CENTURY

Pierre Darriulat

CERN
1211 Geneva 23, Switzerland

A NEW LAND TO BE EXPLORED

The mass spectrum of the elementary particles which we know today (Fig. 1) sets the scale of our future explorations. It is dominated in the boson sector by the W–Z triplet, $m_W = 80.2 \pm 0.2$ GeV and $m_Z = 91.189 \pm 0.004$ GeV, and in the fermion sector by the top quark, $m_t = 174 \pm 16$ GeV. The two other mass scales which we know of, the grand unification and Planck scales, are many orders of magnitude higher and exclude direct exploration using accelerator experiments.

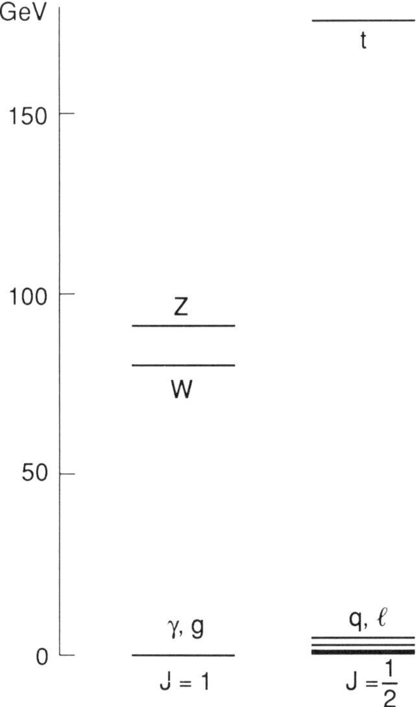

Figure 1. The mass spectrum of the known elementary bosons and fermions.

Frontiers in Particle Physics: Cargèse 1994
Edited by M. Lévy *et al.*, Plenum Press, New York, 1995

The mere fact that the weak bosons have a non-zero mass – in contrast with photons and gluons – calls for a mechanism preventing the divergence of the theory. The Higgs mechanism plays precisely this role. In the form requiring the minimal number of additional particles it describes SU(2) × U(1) symmetry breaking with the help of a single neutral scalar corresponding to a vacuum expectation value

$$v = \left(\sqrt{2}\, G_F\right)^{-1/2} = 2m_W / \left(e / \sin\theta_W\right) \approx 250 \text{ GeV}.$$

The mass of the Higgs scalar m_H is a free parameter of the theory, only its couplings are constrained. It exceeds ≈ 60 GeV, otherwise it would have been discovered at LEP, and unitarity implies that it be less than

$$\sqrt{16\pi / 3}\ \ v \approx 900 \text{ GeV}.$$

Beyond this unitarity limit its width (which grows as the mass to the third power) would exceed its mass and weak interactions would become strong in the corresponding q^2 range.

The Higgs sector may be much richer than implied by this simple scenario. Theories including supersymmetry predict the existence of several Higgs bosons. They generally require a second doublet of complex Higgs fields, with vacuum expectation values v_1 and v_2, coupling to down-type and up-type fermions respectively, and related by

$$1 \lesssim \tan\beta = v_2/v_1 \lesssim m_t/m_b.$$

In particular the simplest form of such theories, the minimal supersymmetric standard model (MSSM), predicts the existence of five physical Higgs bosons: two CP-even neutrals, mixing into h^0 and H^0 with mixing angle α; one CP-odd neutral A^0; and a pair of charged bosons, $H^\pm$. At tree level the following inequalities are obeyed:

$$m_{h^0} < m_Z < m_{H^0}$$
$$m_{h^0} < m_{A^0} < m_{H^0}$$
$$m_W < m_{H^\pm}.$$

However, large corrections are expected from higher-order terms and in particular m_{h^0} could be as large as ≈ 130 GeV.

Supersymmetric models imply the existence of many undiscovered particles that would be the partners of the particles which we know today. The two families are related by R-parity which transforms bosons into fermions and fermions into bosons. While their masses are essentially unconstrained by the theory, most models predict that several of them should populate the 100–1000 GeV mass range in which the Higgs mechanism is expected to operate.

The large top mass is intriguing. The fermion mass spectrum, excluding neutrinos, spans nearly six orders of magnitude and the coupling constant of the top quark to the standard Higgs boson is of order unity:

$$m_t v \approx 0.7.$$

This may be trying to tell us something, a possibility which has recently been triggering renewed interest in theories which assign a dynamical role to the Yukawa couplings. If such were the case, we might expect new phenomena to occur at higher masses. Here again our current knowledge of the particle world points to a mass range extending typically an order of magnitude above the W–Z mass for future exploration.

While some of these predictions are highly speculative, others seem more difficult to escape, in particular those directly associated with the mass-generation mechanism. The importance of making new tools available for the exploration of this new land has become a clear and urgent goal of particle physics. The present lecture aims at making some simple comments of general relevance to this topic. In Figure 2 I have collected a few data illustrating currently available limits on the masses of possible new particles.

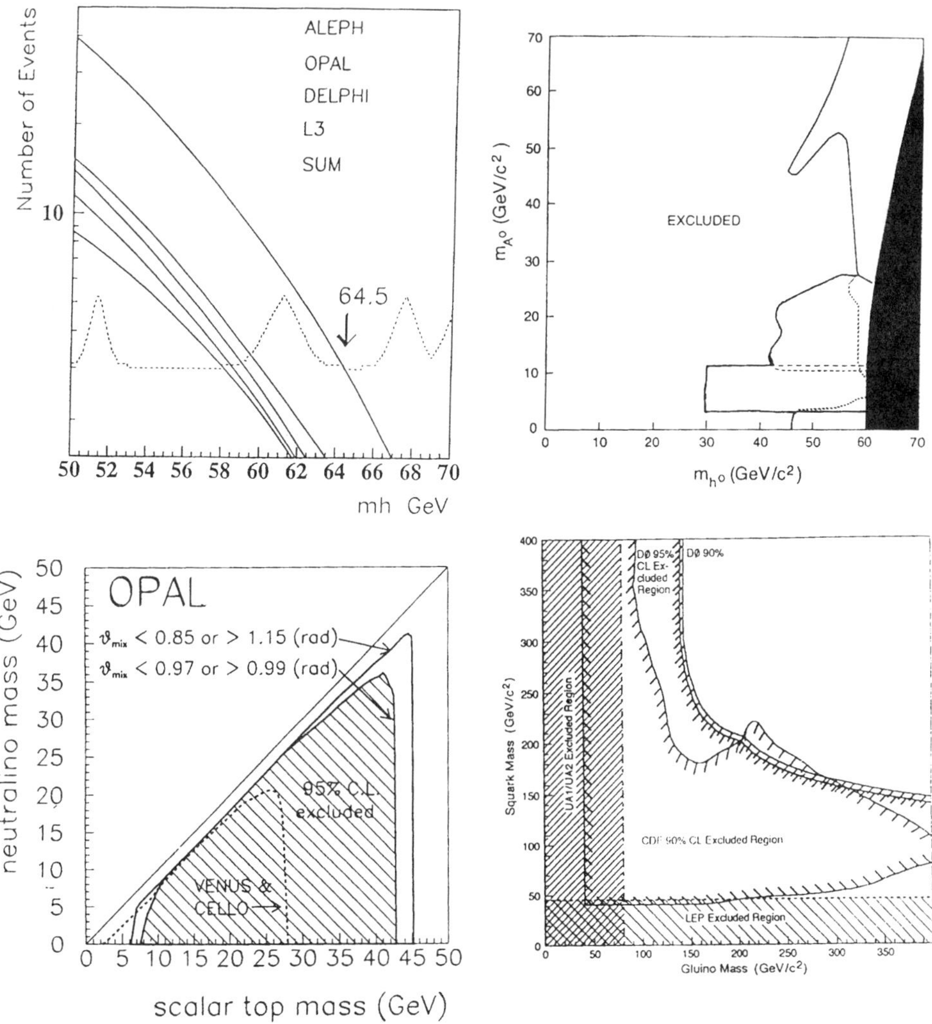

Figure 2. Some current limits on the masses of possible new particles. (a) Limits on the standard Higgs boson mass obtained by each of the four LEP experiments (≈ 60 GeV) and by all four combined (≈ 64.5 GeV). (b) Limits in the h^0–A^0 plane obtained by OPAL. The dark region on the right-hand side is excluded by the MSSM model. (c) Limits in the $\tilde{m}_{\chi_0} - \tilde{m}_t$ plane obtained by OPAL for different values of θ_{mix} which describes the mixing between the right-handed and left-handed stops. (d) Limits in the squark–gluino plane.

PROTON COLLIDERS AND ELECTRON COLLIDERS

In order to reach the large masses at which we aim, the new accelerators must be operated in the collider mode rather than in the fixed-target mode. The energy available for the production of new particles is $\approx \sqrt{2E_{\text{beam}}/m_{\text{target}}}$ times higher in the former case (E_{beam} is the beam energy and m_{target} the target mass, of order 1 GeV). The low values of the production cross-sections of interest – in the picobarn range – call for very high luminosities at the limit of present-day technology. In practice pp and e^+e^- colliders are the only tools able to reach such luminosities. Other schemes such as $p\bar{p}$ or $\mu^+\mu^-$ colliders have also been considered, but were unable to compete.

Protons are composite particles made of partons (quarks and gluons) of different longitudinal densities (Fig. 3). The production of a new particle implies the interaction of two such partons having fractional longitudinal momenta x_1 and x_2. Their invariant mass m and rapidity y obey the relations

$$m^2 = x_1 x_2 s$$
$$\text{th}\, y = \frac{x_1 - x_2}{x_1 + x_2}$$

with $\sqrt{s} = 2E_{\text{beam}}$ and where transverse momenta have been neglected. Similarly, the proton remnants have fractional longitudinal momenta $(1 - x_1)$ and $(1 - x_2)$ and their invariant mass M and rapidity Y obey the relations

$$M^2 = (1 - x_1)(1 - x_2)s$$
$$\text{th}\, Y = \frac{x_2 - x_1}{2 - x_1 - x_2}\ .$$

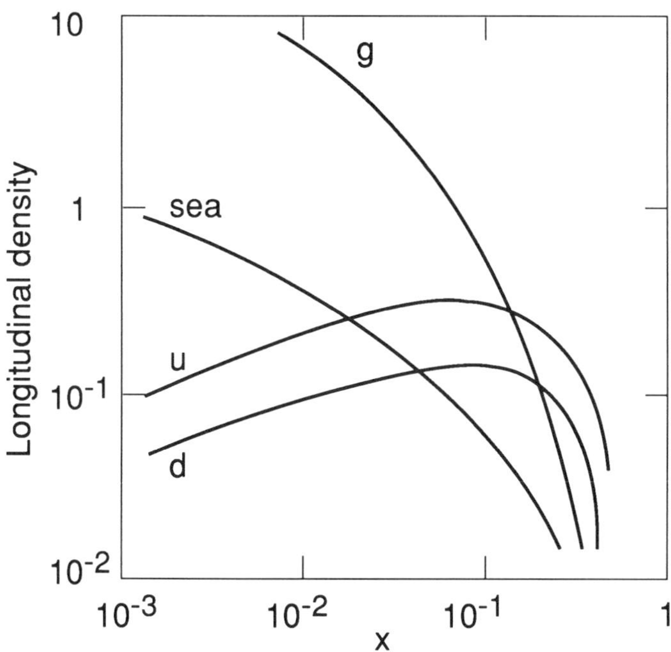

Figure 3. Longitudinal parton densities in the proton.

As a result, the operation of proton colliders as parton colliders has to face two essential difficulties:

- the energy effectively available for the production of new particles is significantly lower than the total energy $\sqrt{s}$ of the colliding protons;
- the analysis of the final state is obscured by the presence of low-transverse-momentum particles resulting from the hadronization of the proton remnants and foreign to the process under study.

This situation (Fig. 4) can be coped with by working in a regime where x_1 and x_2 are as close to unity as possible. When x_1 and x_2 approach unity, the energy of the colliding partons approaches $\sqrt{s}$ and their rapidity becomes more and more central. At the same time, the remnants carry less energy and are diluted over a larger rapidity range. However, the density of large-x partons is a steeply falling function of x and the luminosity effectively available for the parton–parton collisions decreases rapidly as $x_1 x_2$ increases. This means that x_1 and x_2 should not exceed the limit beyond which the production rate would become negligibly small.

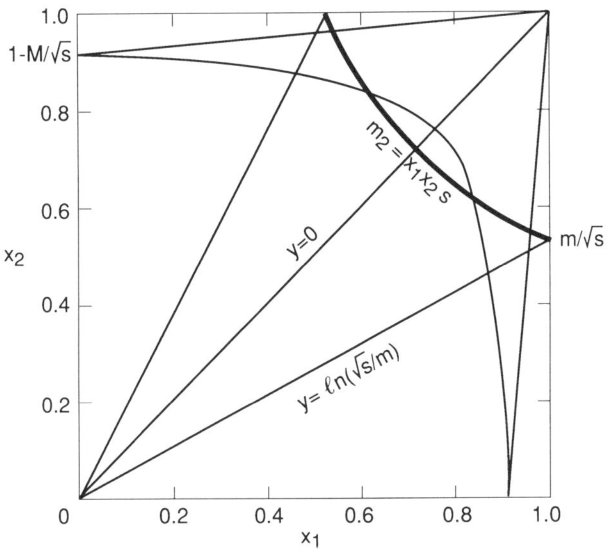

Figure 4. Kinematics of parton–parton collisions.

We see from the above considerations that the effective mass reach of a proton collider depends upon both its luminosity and its energy. In practice, an optimal balance needs to be sought between these parameters when designing a proton collider. Figure 5 illustrates this feature. It shows the dependence of the collider luminosity upon $\sqrt{s}$ for a fixed parton luminosity of 10^{31} cm^{-2} s^{-1} (Fig. 5a) and for the discovery of standard Higgs bosons of various masses (Fig. 5b). Similar discovery potentials are obtained at $\sqrt{s} = 17$ GeV and 40 GeV when the luminosity is 4–10 times higher in the former case than in the latter.

Such arguments have their limits. In particular the optimization of a collider design in the luminosity–energy plane must take into consideration the severe constraints imposed on the operation of the detectors in a high-luminosity environment.

In contrast with protons, electrons are elementary particles: e$^+$e$^-$ collisions make full use of the available energy and their final states are not obscured by uninteresting remnants. However, the smaller electron mass is the source of much larger synchrotron radiation losses. In order to keep them at a reasonable level, the curvature of the guide field must be as small as possible, imposing large dimensions – and therefore a high cost – on the collider design. Moreover, the particularly favourable mechanism of resonant annihilation, which has made the spectacular success of e^+e^- colliders on the J/Ψ, Υ and Z masses, is no longer present at higher energies. Already at LEP the study of W bosons calls for pair production, imposing a doubling of the beam energy. The production of Higgs bosons is expected to proceed predominantly from the fusion of two weak bosons bremsstrahled from the colliding electrons, a far less favourable configuration.

We often hear statements such as "Electron physics is clean, hadron physics is dirty" or "Hadron colliders make the discoveries, electron colliders study the details." While understanding the rationale behind such statements, we must avoid accepting them as dogmas. Each particular case must be considered separately. A few examples are illustrated in Figure 6. While production cross-sections are usually higher in pp collisions than in e^+e^- collisions (because of the colour factor and the richer parton content, including gluons), their discovery potential is generally smaller for the same value of $\sqrt{s}$ (because of the dilution in rapidity and the more important background induced by the strong interactions of the constituent partons).

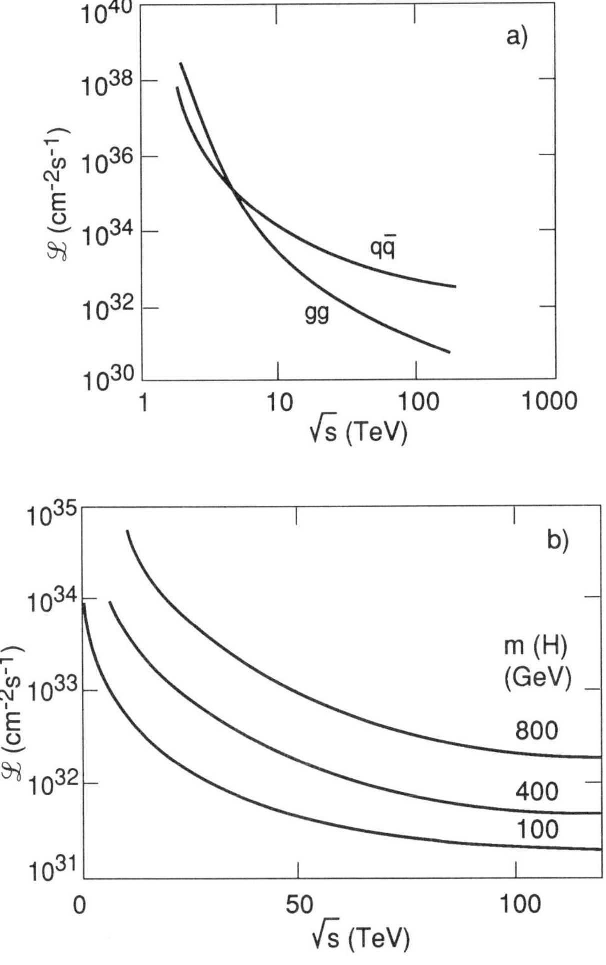

Figure 5. Dependence of the luminosity upon energy for (a) a parton–parton effective luminosity of 10^{31} cm^{-2} s^{-1} in the TeV region; (b) the detection of standard Higgs bosons of various masses.

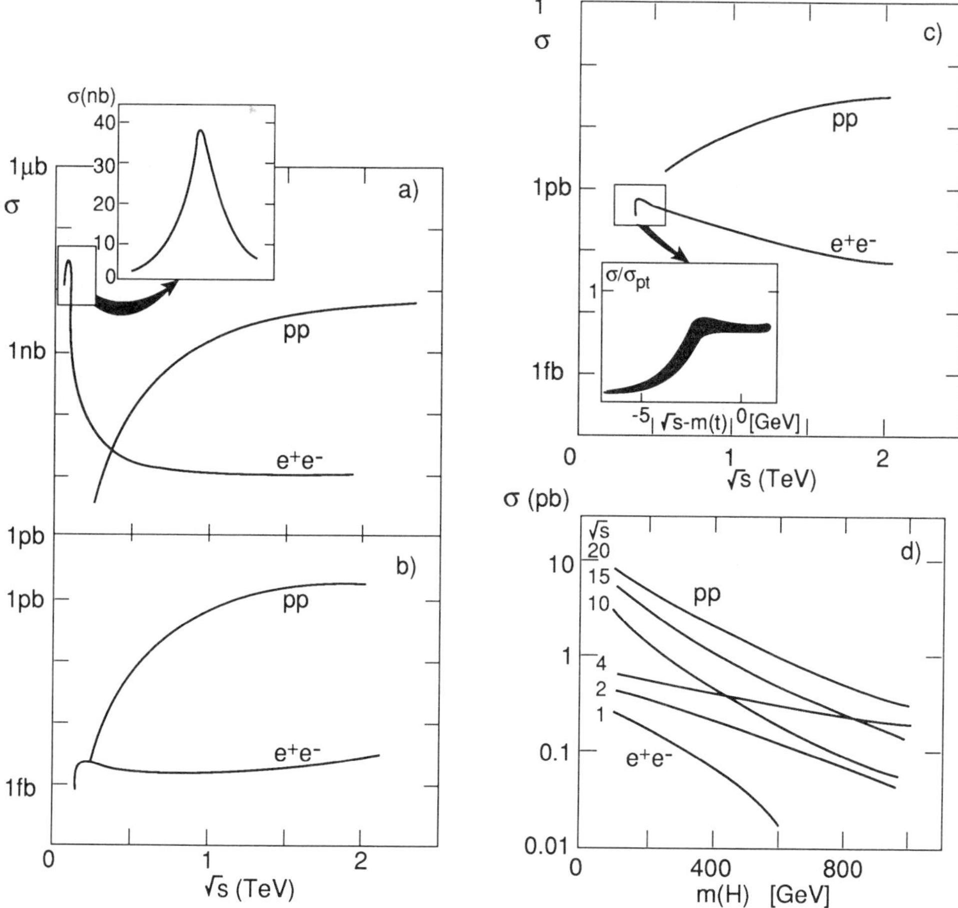

Figure 6. Production cross-sections in e^+e^- and pp collisions: (a) inclusive Z production. The LEP regime is illustrated in the insert; (b) inclusive W production; (c) inclusive $t\bar{t}$ production. The top quark decays promptly to Wb before having the time to fragment or to form toponium. The threshold behaviour is illustrated in the insert for e^+e^- collisions; (d) standard Higgs boson. The cross-section is shown as a function of its mass for various values of $\sqrt{s}$.

We may ask the question: Which are the minimal energies $\sqrt{s_{ee}}$ and $\sqrt{s_{pp}}$ that an e^+e^- collider and a pp collider must have in order to discover a given particle? This exercise is illustrated in Figure 7. On the average we find $\sqrt{s_{pp}} \approx 3\sqrt{s_{ee}}$. A notable exception is the τ lepton which a pp collider could only have discovered as a Z decay product. It is indeed the only particle, among those shown in Figure 7, which has been discovered exclusively in e^+e^- collisions (the J/Ψ was simultaneously discovered in e^+e^- and pp collisions).

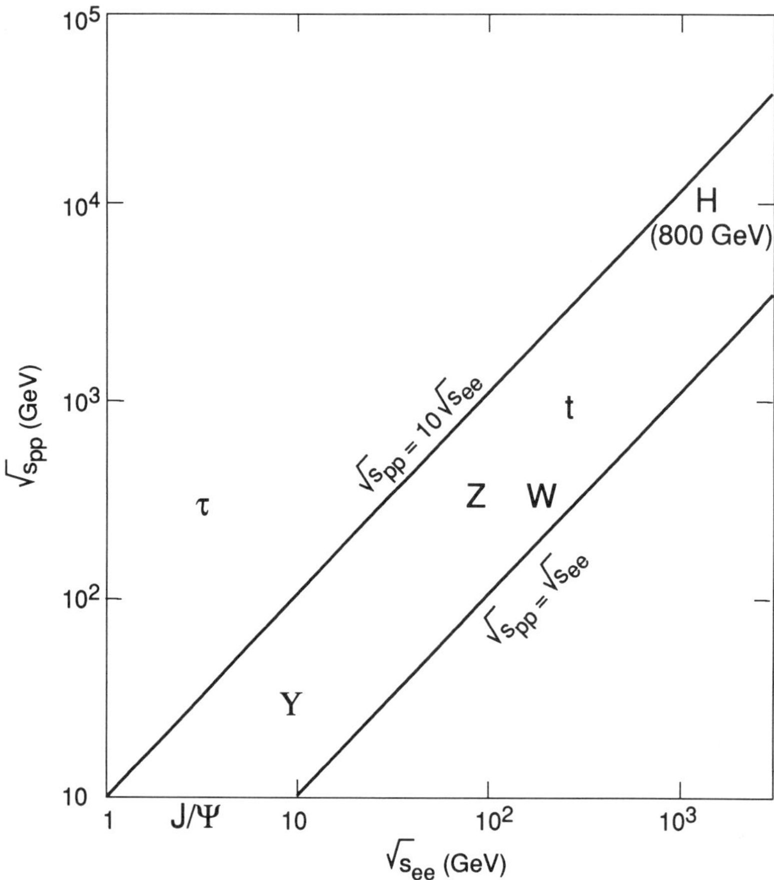

Figure 7. The values of $\sqrt{s_{pp}}$ and $\sqrt{s_{ee}}$ necessary to discover a given particle in pp and e^+e^- collisions.

SOME PRACTICAL LIMITATIONS IN ACCELERATOR PERFORMANCE

Synchrotron radiation losses prevent the operation of e^+e^- colliders above the LEP energy range. This is illustrated in Table 1 where some caricatural scenarios are displayed. The scale of the energy spectrum of the radiated photons is set by the critical energy $E_{crit} \propto \gamma^3/\rho$ where γ is the Lorentz factor of the beam particles and ρ the bending radius, and the energy loss per turn is proportional to γ^4/ρ. This imposes a linear collider design in the e^+e^- case, a very high price to pay: it means giving up two major assets of circular colliders, namely the effective increase of the beam current and the modest requirements in terms of RF power, both of which result from the multiple traversal of the same particle bunches in the accelerating cavities and the experiments' detectors (≈ 10 kHz in LEP). In the pp case, synchrotron radiation losses remain small and a circular design can be maintained.

Table 1. Some collider scenarios.

Collider	Particles	E_{beam}	Tunnel length (km)	Sync. rad. losses per turn	Feasibility
LEP 200	ee	100 GeV	27	2.5 GeV	+
'LEP 2000'	ee	1 TeV	27	25 TeV	–
'LEP 2000'	ee	1 TeV	27 000	25 GeV	–
LHC	pp	7 TeV	27	10 keV	+
Linear ee[1]	ee	1 TeV	40	0[2]	+
Linear pp[1]	pp	7 TeV	280	0[2]	–

[1] Assuming an accelerating gradient of 50 MeV/m.

[2] Neglecting quadrupole losses.

In the energy range of interest we are therefore talking about pp circular colliders or e^+e^- linear colliders, implying very different strategies for their operation. In the pp case the same bunches are reused many times and the collisions have to be gentle enough to make this reuse possible. In the e^+e^- case a given bunch pair is used once only and we can afford very brutal collisions in order to reach the highest possible luminosity.

It may be useful at this stage to recall a few elementary properties of colliding beams. Unless otherwise stated, we shall assume that each beam is made of identical cylindrical particle bunches, with length ℓ and transverse area A, equally spaced by a time interval Δt, perfectly aligned, and each containing N particles.

The interaction rate (Fig. 8a) for a given physical process of cross-section σ is

$$\mathcal{L}\sigma = N\frac{N\sigma}{A}\frac{1}{\Delta t} ,$$

where the luminosity $\mathcal{L} = (N/\Delta t)\,(N/A)$ is proportional to the beam current $(N/\Delta t)$ and to the transverse bunch density (N/A). Both parameters need to be simultaneously optimized in the collider design.

At the collision points, large transverse densities are achieved with the help of strongly focusing (low-β) quadrupoles which reduce the beam dimensions while accordingly increasing its angular divergence. The beam envelope at the collision point can be approximated by a parabola (Fig. 8b) and

$$A = \beta^* \varepsilon = \beta^* \varepsilon_n/\gamma .$$

Here β^* characterizes the strength of the focusing quadrupole and the beam emittance ε decreases with energy (γ) in such a way that $\varepsilon_n = \varepsilon\gamma$ is an invariant, independent of beam optics, and defined by the conditions at injection. The Liouville theorem prevents adiabatic reductions of ε_n (a notable exception is beam cooling) but care must be taken to avoid increases which could be induced by non-linear effects. An increase of ε_n means a deterioration of the luminosity, and emittance preservation at its original injection level is essential.

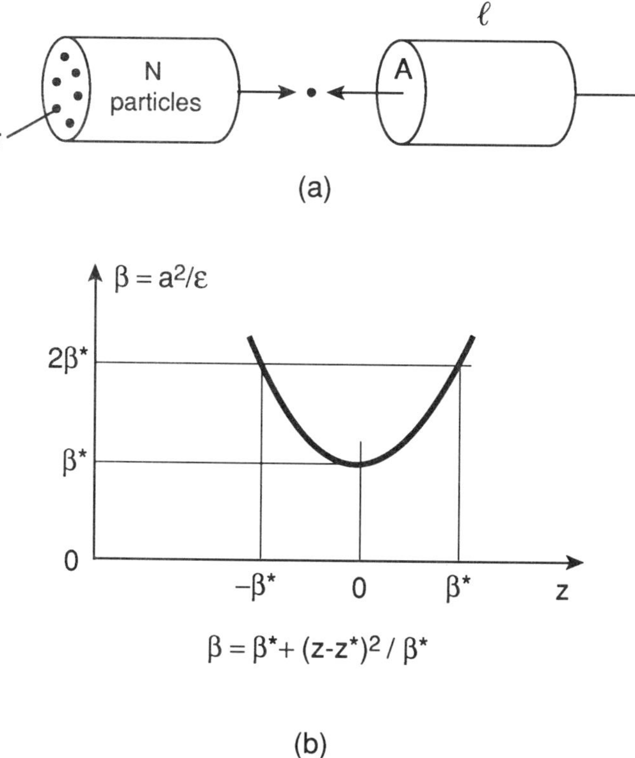

Figure 8. Bunch collisions: (a) the main parameters. A fraction $N\sigma/A$ of the bunch cross-section is available for collisions with the N particles of the other bunch; (b) the low-β geometry at the collision point.

The mutual action of colliding bunches on each other is – to first order – that of a lens of convergence $C \propto N/\gamma A$ as the Coulomb force deflects particles in proportion to their distance from the axis. The luminosity may be rewritten as $\mathcal{L} \propto CP$ where $P \propto \gamma N/\Delta t$ is the effective beam power. Bunch–bunch interactions have therefore the effect of altering the focusing properties of the optics in the lattice; they change the Q value, i.e. the number of betatron oscillations per turn, by a quantity

$$\Delta Q = C\beta^*/4\pi \propto N/4\pi\varepsilon_n \,,$$

an effect which must be taken into consideration in the lattice design and during machine operation. However, to higher orders, they induce non-linear perturbations which cannot be compensated for. The fluctuations $\delta(C)$ across the bunch result in fluctuations $\delta(\Delta Q) = \delta(C)\,\beta^*/4\pi$ which generally cause the bunches to blow up, the beam emittance to increase, and the luminosity to decrease accordingly. To a good approximation the size of the first-order effect ΔQ sets the scale of the higher-order terms $\delta(\Delta Q)$ which must be kept small enough to preserve the emittance. In circular colliders ΔQ, summed over the collision points around the ring, must not exceed a few per cent.

This beam–beam limit is a major limitation on the achievable luminosity. Other sources of emittance blow-up – such as beam–gas collisions or non-Coulomb interactions at the collision points – are of lesser importance and circular colliders are generally operated as close as possible to the beam–beam limit.

A linear e^+e^- collider can afford much higher ΔQ values as the emittance does not need to be preserved after the collision (bunches are not reused). However, other effects become significant when the bunches become denser, setting new upper limits to the luminosity achievable. Their scale is set by the disruption factor $D = \ell C$ that measures the bunch length in units of the focal length of the effective beam–beam lens. As long as $D \lesssim 1$, the focusing action of one beam on the other (pinch effect) is beneficial and amplifies the luminosity. But when D exceeds unity, disastrous disruptive effects appear that prevent a further reduction of the transverse beam size and set a limit on the luminosity. Such effects are difficult to calculate reliably but can be simulated approximately. Other effects set in near this high-luminosity limit, such as the deterioration of the beam-energy resolution, proportional to N^3/DA^2 and the emission of a strong synchrotron-radiation flux with critical energy proportional to $N^2 E_{\text{beam}}/DA^{3/2}$, large enough to create e^+e^- pairs.

A NEW pp COLLIDER: THE LHC

In order to explore the new mass range, CERN, the European Laboratory for Particle Physics, has proposed the construction of a proton collider reaching an energy $\sqrt{s} = 14$ TeV and a luminosity $\mathscr{L} = 10^{34}$ cm^{-2} s^{-1}. The new machine, called LHC for Large Hadron Collider, will make use of major elements of the existing LEP infrastructure, such as the injection chain, the tunnel and the cryogenic plant. This will make it possible to keep the construction budget well below $3 \cdot 10^9$ Swiss francs and the construction time below seven years. If the project is approved before the end of the current year (1994) the collider should be available for physics in 2003.

The beam energy is defined from the size of the existing LEP tunnel and from the state of the art in superconducting magnet technology. The guide field, of up to 9 T, is produced by 10-m-long magnets equipped with niobium–tin superconducting coils cooled down to 2 K in a superfluid helium cryostat. The field is limited upwards by the critical current density (Fig. 9) above which a transition to the normal conducting state would occur. Both beams are guided by the same set of magnets, the two coils being inserted in a single yoke (the so-called two-in-one technology, Fig. 10). Several prototypes have been successfully constructed and shown to reach a field of 9 T without difficulty.

The machine will be operated near the beam–beam limit with bunches containing $N = 10^{11}$ protons each and distant by $\Delta t = 25$ ns. This corresponds to a stored energy of 700 MJ, implying a sophisticated fast-ejection scheme to prevent accidents in the event of beam instabilities. The beam current reaches 0.5 A corresponding to an effective power of nearly 7 TW. The high luminosity, 10^{34} cm^{-2} s^{-1}, is obtained with the help of low-β quadrupoles which bring the beam diameter down to 16 μm at the collision points. With such parameters a collision lifetime of the order of 10 h should be obtained (the ramping time should not exceed 20 min). The synchrotron radiation losses, amounting to ≈ 0.44 W/m, require the presence of a radiation shield cooled down to ≈ 10 K in order to protect the vacuum chamber which must be kept at the superfluid helium temperature (2 K) at which heat cannot easily be removed (Fig. 11).

In addition to pp collisions, LHC will also provide for heavy-ion collisions to study the deconfined phase of matter, and, at a later stage, ep collisions with the LEP beam. Two major experiments, ATLAS and CMS are being designed to operate in the pp mode and a smaller detector, ALICE, would be dedicated to the study of heavy-ion collisions. A summary of the main relevant parameters is given in Table 2.

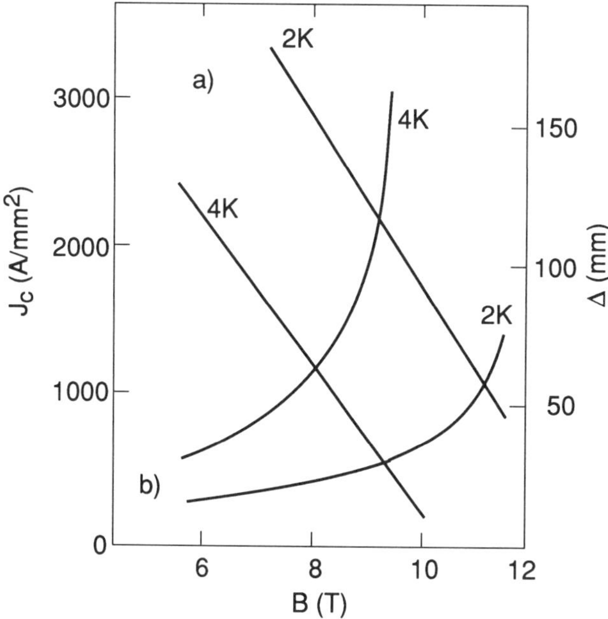

Figure 9. Dependence of the critical current J_c (a) and of the magnet coil thickness Δ (b) upon the magnetic field B for niobium–tin cables with a copper-to-superconducting ratio of 1.7. The curves are for normal (4 K) and superfluid (2 K) liquid-helium temperatures.

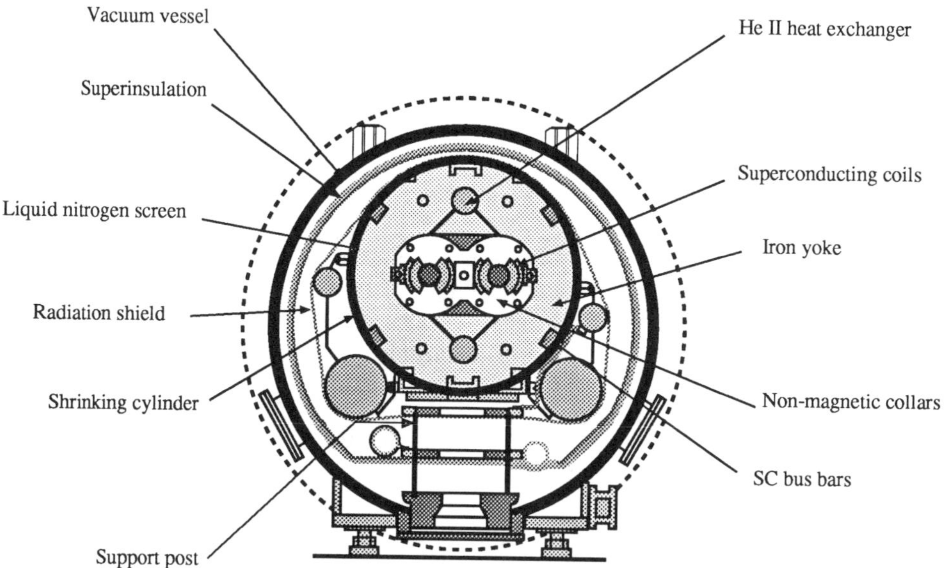

Figure 10. The LHC dipole magnet (cross-section).

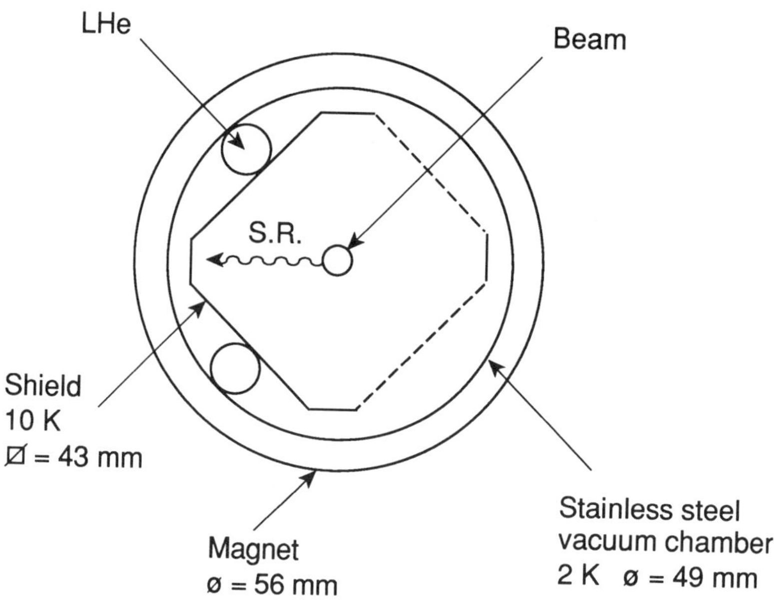

Figure 11. The LHC vacuum chamber.

Table 2. LHC main parameters.

Parameter	pp mode	Lead–lead mode	ep mode
$\sqrt{s}$ (TeV)	14	1150	1.3
$\mathscr{L}$ (cm^{-2} s^{-1})	10^{34}	2×10^{27}	2×10^{32}
Δt(ns)	25	135	165
Experiments	2	1	1

The LHC machine will make it possible in the first years of the next century to explore the mass range where new phenomena related to the mass generation mechanism are expected to occur. Some approximate discovery limits within its reach are listed in Table 3.

The ATLAS and CMS detectors are shown in Figure 12. Their expected performance in various physics topics is illustrated in Figure 13. The very large values of the luminosity and of the event multiplicity result in very high density counting rates (several particles per msr × ms). The challenge to cope with such conditions is enormous. It imposes the use of a very large number of independent detection channels (in the 10^6 range) and results in very large global dimensions for the detectors. Moreover, fast processing of the information which they carry is essential for a prompt selection of potentially interesting events, i.e. collisions in which two partons experience a hard interaction. Much of this processing has to be done in situ, implying the use of radiation-hard electronics at the limit of – or even beyond – present-day technology.

Table 3. The mass reach of LHC (TeV).

Process		pp	ep
H^0 (standard model) $m(H^0) > 200$ GeV		0.6–1.0	–
Heavy lepton	$L \rightarrow W + \nu$	$\lesssim 0.5$	–
Heavy quarks	$Q \rightarrow W + q$	0.8	0.1
New gauge bosons			
	charged W'	4.5	1.2
	charged W_R	–	1.5
	neutral Z'	4.0	0.5
Leptoquark	$D \rightarrow \ell + q$	2.0	1.6
SUSY			
	scalar quark	1	0.7
	gluino	1	–
	scalar electron	0.3	0.35
	Wino	0.4	–
Compositeness			
	$m(q^*)$	5	–
	$m(e^*)$	4	1.5
	$\Lambda(qq)$	12	–
	$\Lambda(eq)$	20	8–13

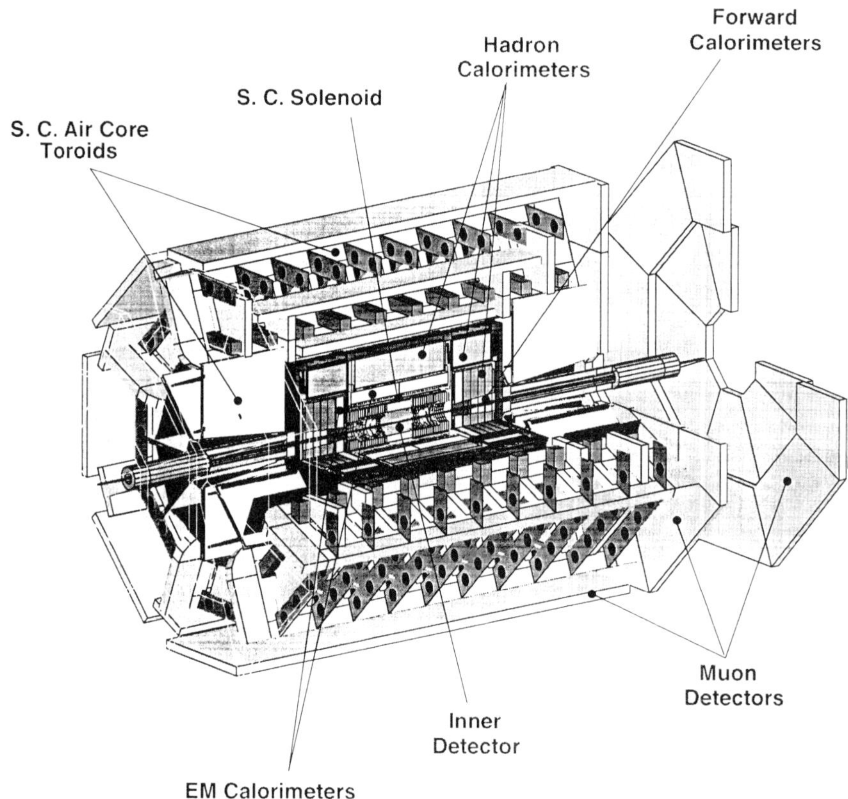

Figure 12. The ATLAS and CMS detectors.

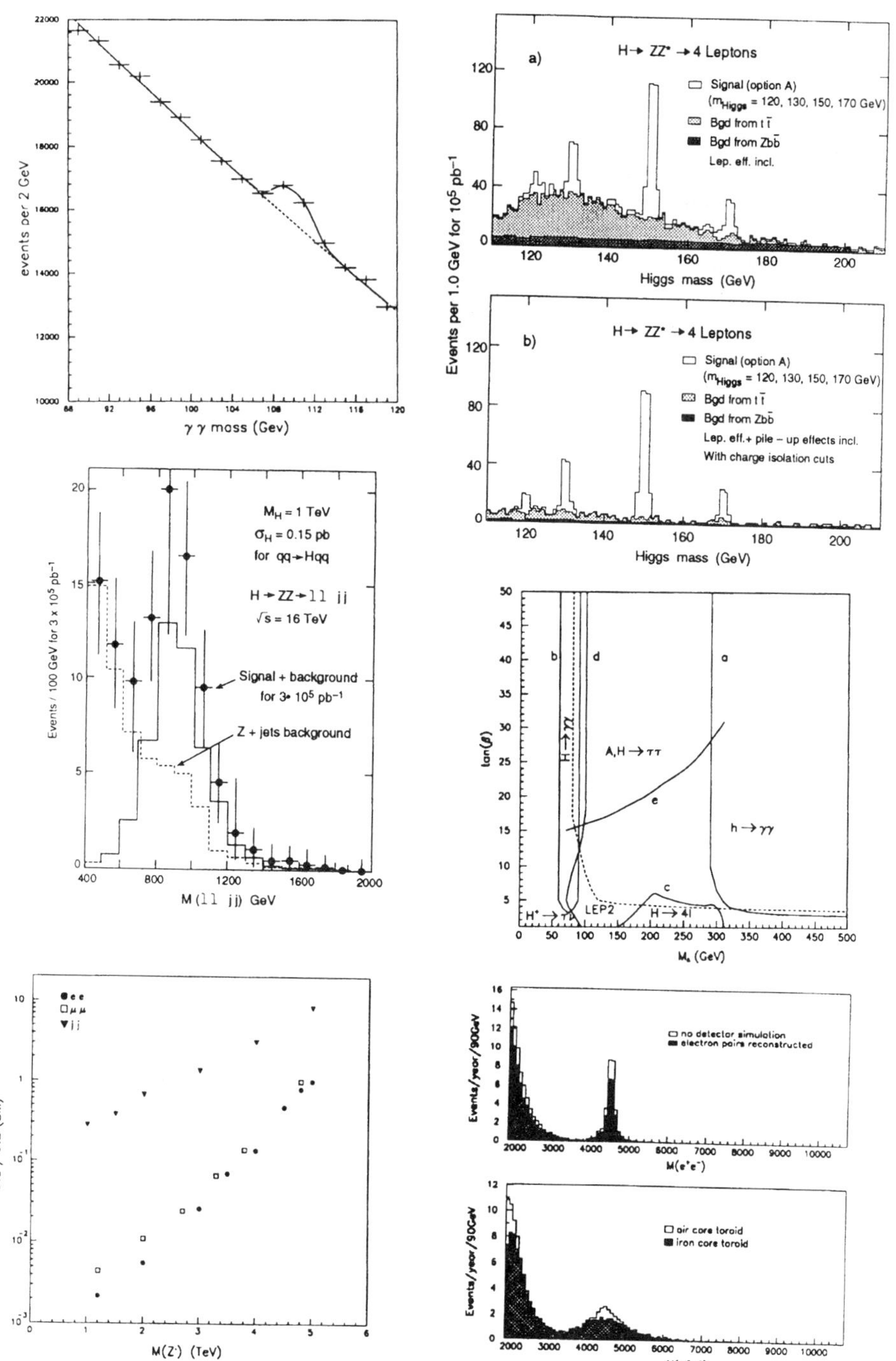

Figure 13. Top left: the $m\,(\gamma\gamma)$ spectrum for $H \to \gamma\gamma$ above irreducible background for $m_H = 110$ GeV and $\mathcal{L}_{\text{int}} = 10^5$ pb^{-1} (ATLAS). Top right: the four-lepton signal in $H \to ZZ^*$ before and after requiring track isolation (CMS). Middle left: the $H \to ZZ \to 2$ leptons 2-jet signal for $\mathcal{L}_{\text{int}} = 3 \times 10^5$ pb^{-1} (CMS). Middle right: 5σ discovery contours in the $(m_{A^0}, \tan\beta)$ plane for various Higgs signals (ATLAS). Bottom left: discovery mass limits for a heavy gauge boson $Z' \to ee, \mu\mu, jj$ (ATLAS). Bottom right: reconstructed dilepton mass for $Z' \to ee, \mu\mu$ with $m_{Z'} = 4.5$ TeV.

LINEAR e⁺e⁻ COLLIDERS

While we know today how to build a proton collider able to explore the domain at which we are aiming, such is not the case for a linear electron collider. Existing constraints on accelerator technology limit the average current to a few μA (compared with nearly 1 A for LHC). In order to make up for this missing factor, beam dimensions at the collision point have to be drastically reduced (a few nm rather than μm). A vigorous R&D programme is currently under way in order to develop large-gradient/high-current acceleration schemes, to master the problems associated with the design of the final focus, and to preserve as much as possible a small emittance all along the accelerating structure.

We noted in the section on limitations in accelerator performance that the luminosity is proportional to the product of the beam power by the beam–beam lens convergence, $\mathcal{L} \propto PC$. For obvious reasons the power taken from the line must remain well below 1 GW (which corresponds to a typical nuclear plant) and we must rely – in order to maximize P – on improving the transfer efficiency from the line to the RF system and from the RF system to the beam. This global efficiency is typically of the order of some per mil and means to increase it are being actively sought.

The R&D programmes under current study can be classified in three families. They differ by their scope, their ambition, and their time-scale (Table 4).

Table 4. Three approaches to the design of an e⁺e⁻ linear collider at $\sqrt{s} = 500$ GeV.

Parameter	TESLA	NLC	CLIC
Luminosity (10^{33} cm^{-2} s^{-1})	2.6	6	1–3
Length (km)	20	14	6.6
Power (from the line) (MW)	137	152	175
$N(10^{10})$	5.2	0.7	0.6
Δt (ns)	1000	1.4	0.33
Bunches per pulse	800	90	1–4
Beam size at collision (nm)^2	640×100	300×3	90×8
RF frequency (GHz)	1.3	11.4	30
RF gradient (MV/m)	25	50	80

The least ambitious approach is that of TESLA (DESY) which uses state-of-the-art technology as much as possible and aims at a machine which could be built soon to explore part of the high mass domain, say up to $\sqrt{s} \approx 500$ GeV. It uses superconducting cavities and its relatively high cost makes it an unlikely candidate for reaching higher energies.

The most ambitious approach is that of CLIC (CERN) using warm cavities of high gradient and frequency fed by a companion high-intensity, low-energy beam. Its design is well beyond the state of the art but it aims at the highest energies. It is premature to state a realistic time-scale for such a futuristic project (Fig. 14).

Between these two extremes, several projects are under study with the aim of reaching $\sqrt{s} = 500$ GeV in a first phase and $\sqrt{s} \gtrsim 1$ TeV in a second phase. One of those is the NLC (Fig. 15) which builds on existing SLC experience, with respect to which it represents an extrapolation by 1 to 2 orders of magnitude. Several laboratories around the world (SLAC, KEK, etc.) are collaborating on this R&D programme. They aim at designing a successor to LHC.

It is difficult today to tell which of these approaches best matches our needs. It will very much depend upon the energy range where new phenomena will occur. Meanwhile the pursuit of each of these R&D projects is an excellent investment in the future of particle physics.

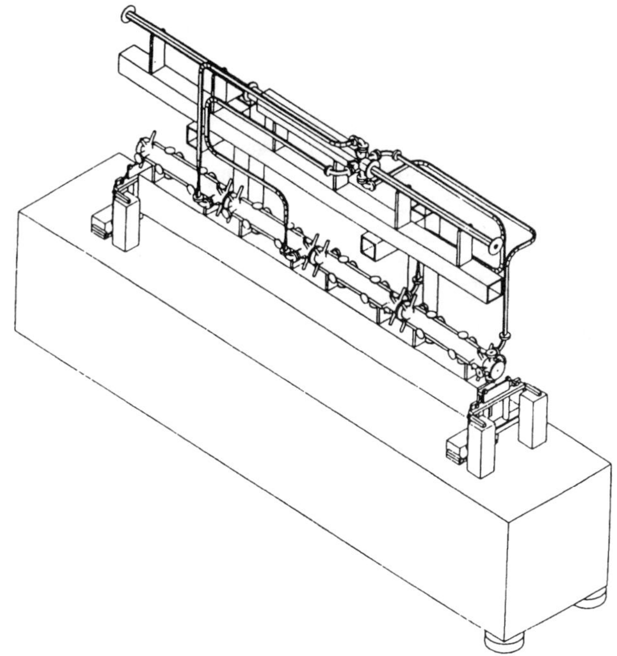

Figure 14. A 1.4-m module of CLIC showing the drive beam (above) feeding four accelerating sections (below).

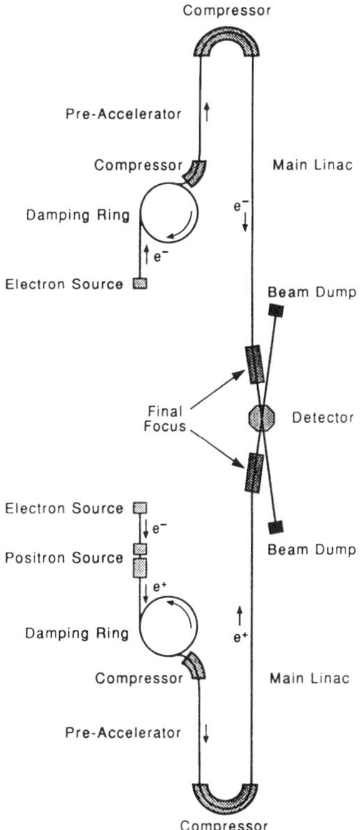

Figure 15. Schematic diagram of the NLC.

CONCLUSIONS

Our current knowledge of particle physics points clearly to the mass range of 100–1000 GeV as the domain on which to concentrate our efforts for future exploration. While the Tevatron and LEP2 will continue searching for new phenomena at the lower end of this range, new tools must be prepared to succeed them. Two complementary approaches are currently being considered, a circular proton collider and a linear e^+e^- collider. The former is that of the LHC, a $\sqrt{s} = 14$ TeV machine reaching a luminosity of 10^{34} cm^{-2} s^{-1}. Its construction and exploitation are actively being prepared with the aim of starting experimentation in 2003. The latter approach faces important technological issues currently addressed by a vigorous international R&D effort: the control of very dense nanometric beams and the development of high-gradient and high-current accelerating structures are among the most challenging.

INDEX

$150.00 (2)

PH 10